T0401991

ALLE ZEIT WACH
1842

C. Larsson I.M. Møller (Eds.)

The Plant Plasma Membrane

Structure, Function and Molecular Biology

With 79 Figures

Springer-Verlag
Berlin Heidelberg New York
London Paris Tokyo Hong Kong

Dr. CHRISTER LARSSON
Department of Plant Biochemistry
University of Lund
Box 7007
S-220 07 Lund
Sweden

Dr. IAN M. MØLLER
Department of Plant Physiology
University of Lund
Box 7007
S-220 07 Lund
Sweden

The schematic drawing on the cover illustrates a number of important functions of the plant plasma membrane. It was drawn by Dr. Michael G. Palmgren.

ISBN-13:978-3-642-74524-9 e-ISBN-13:978-3-642-74522-5
DOI: 10.1007/978-3-642-74522-5

Library of Congress Cataloging-in-Publication Data. The Plant plasma membrane : structure, function, and molecular biology / C. Larsson, I. M. Møller (eds.). p. cm. 1. Plant cell membranes. I. Larsson, C. (Christer), 1945– . II. Møller, I. M. (Ian Max), 1950– . QK725.P573 1989 581.87'5–dc20 89-28009

Softcover reprint of the hardcover 1st edition 1990

Typesetting: International Typesetters Inc., Makati, Philippines

2131/3145-543210 – Printed on acid-free paper

Preface

The plasma membrane forms the living barrier between the cell and its surroundings. For this reason it has a wide range of important functions related to the regulation of the composition of the cell interior and to communication with the cell exterior. The plasma membrane has therefore attracted a lot of research interest. Until the early 1970's it was only possible to study the plasma membrane in situ, its structure e.g. by electron microscopy and its function e.g. by uptake of radioactively labeled compounds into the intact cell or tissue. The first isolation of plant protoplasts by enzymatic digestion of the cell wall in the early 1970's was an important step forward in that it provided direct access to the outer surface of the plasma membrane. More importantly, T. K. Hodges and R. J. Leonard in 1972 published the description of a method by which a fraction enriched in plasma membranes could be isolated from plant tissues using sucrose gradient centrifugation. As a result, the 1970's saw a leap forward in our understanding of the structure and function of the plasma membrane. In 1981, S. Widell and C. Larsson published the first of a series of papers in which plasma membrane vesicles of high yield and purity were isolated from a wide range of plant tissues using aqueous polymer two-phase partitioning. The access to these highly purified fractions led to a second leap forward in our understanding of the plasma membrane. A limitation with these vesicles, however, was that they were predominantly apoplastic side-out, but later improvements in the separation technique (1988) made it possible to isolate also pure cytoplasmic side-out vesicles. This finally made it possible to study plasma membrane transport in vitro using well-defined vesicles.

When we, in 1987, were asked by an international publisher whether we could suggest a suitable subject area for a book, we thought that a book on the plant plasma membrane would be timely. We wrote a synopsis to Springer-Verlag in which all aspects of the plasma membrane were covered in 16 chapters and were very pleased to have it accepted. Most of the colleagues we invited to write the chapters accepted and we were fortunate enough to find excellent replacements in the few cases where the originally invited author could not, after all, write the chapter. We have learnt a great deal about the plasma membrane through editing this book and we hope the reader will also find the book informative and easy to read.

Finally, we should like to thank all chapter authors for a fruitful collaboration and for making it possible for us (almost) to meet the deadlines. We should also like to thank our colleagues at the Department of Plant Physiology and in particular Drs. Michael G. Palmgren, Marianne Sommarin, and Susanne Widell for many helpful suggestions, and the staff at Springer-Verlag for an efficient and professional handling of the production of the book.

Lund, November 1989

Christer Larsson
Ian M. Møller

Contents

Contributors

You will find the addresses at the beginning of the respective contribution

Chapter 1 Introduction to the Plant Plasma Membrane – Its Molecular Composition and Organization

C. LARSSON[1], I.M. MØLLER[2], and S. WIDELL[2]

1 Introduction

The plant cell consists of a number of organelles and membrane systems including chloroplasts, mitochondria, peroxisomes, Golgi, tonoplast(s), endoplasmic reticulum, a variety of more or less well-defined small vesicles, and a nucleus, all embedded in the cytoplasm and enclosed by the plasma membrane, which in turn is surrounded by a cell wall. While the relatively rigid cell wall confers stability and protection against mechanical damage to the cell, the plasma membrane provides a relatively constant milieu for the intracellular metabolism by performing a balanced exchange of metabolites with the rest of the organism and its surroundings.

An H^+-ATPase (Chap. 6) in the plasma membrane pumps protons from the cytoplasm to the apoplastic space, thus creating an electrochemical gradient across the plasma membrane (negative on the inside). This gradient is in turn used to drive uptake and extrusion of solutes. Thus, specific carriers or channels in the plasma membrane mediate the movement of small solutes (ions, photosynthetic products, etc.) into and out of the cell (Chaps. 7, 8, and 16). This

[1]Department of Plant Biochemistry and [2]Department of Plant Physiology, University of Lund, P.O. Box 7007, S-220 07 Lund, Sweden

Abbreviations: DGDG, diglycosyldiacylglycerol; MGDG, monoglycosyldiacylglycerol; PA, phosphatidic acid; PC, phosphatidylcholine; PE, phosphatidylethanolamine; PG, phosphatidylglycerol; PI, phosphatidylinositol; PS, phosphatidylserine

C. Larsson, I.M. Møller (Eds):
The Plant Plasma Membrane

movement is regulated by several factors such as the concentration of free Ca^{2+} in the cytoplasm and the membrane potential (Chap. 8). A redox chain in the plasma membrane may also be involved in the regulation of the solute transport (Chap. 5). Larger molecules, such as proteins and polysaccharides, are transported within vesicles which are either formed from the plasma membrane to enclose extracellular material for internalization, or fuse with the plasma membrane from the inside to excrete products (Chap. 10). Alternatively, polysaccharides are synthesized in the plasma membrane from monomers available in the cytoplasm and the polymer product excreted through the membrane for cell wall synthesis (Chap. 11) or repair (Chap. 14). The position as the outer, permeability barrier of the plant cell, and the first part of the cell to sense changes in the environment, suggests an important role for the plasma membrane also in interactions with symbionts (Chap. 15) and pathogens (Chap. 14), in frost hardiness (Chap. 13), and in signal sensing and transduction (Chaps. 9 and 16). Known and putative functions of the plant plasma membrane are indicated in Fig. 1.

To understand the complex and dynamic functions of the plasma membrane it is necessary to have at least a general picture of its composition and structure. In this chapter we will therefore attempt to summarize what is presently known about the molecular composition and organization of the plant plasma membrane to give a framework for the functions of the plasma membrane discussed in the following chapters.

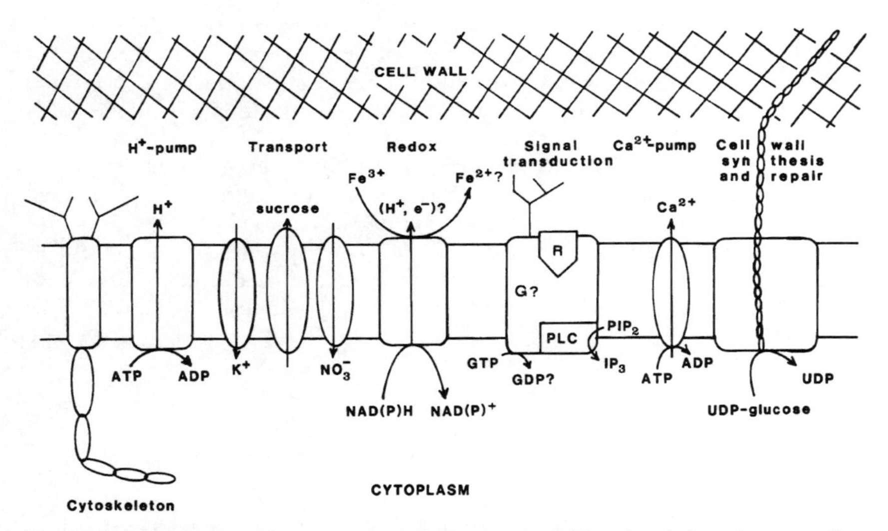

Fig. 1. Model of the plant plasma membrane with a number of functions indicated: some well established, others more uncertain as detailed in the following chapters. The carbohydrate residues (*branched structures*) on the anchor protein for the cytoskeleton and the putative G-protein (*G?*) are by analogy with mammalian plasma membranes. *R* receptor; *PLC* phospholipase C

2 General Properties

2.1 Overall Composition, Density, and Thickness

The main components of the plasma membrane are lipids, proteins, and carbohydrates. Based on elementary (C,H,N) analyses plasma membranes from spinach and barley leaves contain 40 and 30% (w/w) protein, respectively (Kjellbom and Larsson 1984). How the remaining 60–70% is divided between lipids and carbohydrates is uncertain. A figure of 20% carbohydrate has been reported (H.D. Grimes, unpublished results; cited from Grimes and Breidenbach 1987), which may be compared with the 8% carbohydrate of the human erythrocyte membrane (Guidotti 1972). These values give protein-to-lipid ratios that agree fairly well with the reported value of 0.53 (Lynch and Steponkus 1987).

The density of plasma membrane vesicles as determined by sucrose density gradient centrifugation lies in the middle of the range of 1.10–1.24 g cm^{-3} covered by plant membranes (Table 1: Quail 1979). Since the plasma membrane contains a relatively high percentage of carbohydrate (see above) and since the density of a membrane vesicle is dependent on its relative content of protein (1.35 g cm^{-3}), lipid (1.0 g cm^{-3}), carbohydrate (1.6 g cm^{-3}), and water (1.0 g cm^{-3}) (Hinton and Dobrota 1976), it follows that the plasma membrane would be expected to contain a relatively low protein-to-lipid ratio compared to plant membranes of similar densities.

Table 1. General physicochemical properties of the plasma membrane

Parameter	Method	Range	References
Density	Sucrose gradient centrifugation	1.13–1.18 g cm^{-3}	Quail 1979
	Percoll gradient centrifugation	1.05 g cm^{-3}	Widell and Larsson 1983
Thickness	Electron microscopy	9–11 cm	Morre and Bracker 1976; Chap. 4
Isoelectric point	Cross-partitioning	pH 3.4[a]	Westrin et al. 1983
		pH 3.6[a]	Körner et al. 1985
	Isoelectric focusing	pH 4.4[a]	Griffing and Quatrano 1984
		pH 4.2 and 4.4[b]	Steele and Stout 1986
Net negative surface charge density	9-Aminoacridine fluorescence	16–29 mC m^{-2}	Møller et al. 1984; Körner et al. 1985
	Particle electrophoresis	6–9 mC m^{-2} [c]	Gibrat et al. 1985; Grouzis et al. 1987

[a] The outer, apoplastic surface.
[b] Two peaks of plasma membrane marker activity.
[c] Microsomal membranes.

When plant plasma membranes are stained with lead or uranyl acetate they are appreciably thicker (9–11 nm) in electron micrographs than other plant membranes both in vivo and in vitro (Chap. 4). Factors contributing to this might be the high carbohydrate content (see above) and the high sterol content (Sect. 3) of the plasma membrane. The thickness of the plasma membrane is decreased by 1 μM indoleacetic acid (Morre and Bracker 1976) and increased by flower induction (Auderset et al. 1986). Neither the biochemical basis nor the physiological significance of these modifications of plasma membrane thickness are known.

2.2 *Surface Charges*

Biological membranes have fixed positive and negative charges on proteins, lipids, and carbohydrates. The observation that free-flow electrophoresis toward a cathode can be used to separate the membranes in a microsomal fraction (Chap. 3) indicates that most, if not all, membranes carry a net negative charge at neutral pH. In keeping with this, the isoelectric point (the pH at which the number of negative and positive charges is equal and the net charge zero) of mitochondrial and chloroplast membrane surfaces is at pH 4.1–5.4 as determined by cross-partitioning (Westrin et al. 1983; Petit et al. 1987). The plant plasma membrane has an even lower isoelectric point as determined by the same methods (Table 1). It is important to note that the pI in no way is related to the net negative surface charge density at neutral pH.

Inside-out plasma membrane vesicles travel more slowly toward the cathode than right side-out plasma membrane vesicles and other microsomal membranes during free-flow electrophoresis (Canut et al. 1988). This indicates that inside-out plasma membrane vesicles are the least negative microsomal particles. Consistent with these observations the surface charge densities determined for the outer surface of the plasma membrane by fluoresecence quenching (Table 1) are not markedly different from those of mitochondrial and thylakoid membranes (Barber 1982; Itoh and Nishimura 1986).

The surface charge densities determined by particle electrophoresis are often much lower than those determined by fluorescence quenching (e.g., Table 1). This is inherent in the methods (see, e.g., Barber 1982; Itoh and Nishimura 1986). However, using the same method, the surface charge density of the plasma membrane has been reported to vary between different species (Møller et al. 1984) and between organs of the same species (Körner et al. 1985).

It must be kept in mind that all surface charge densities are average values for the whole membrane and may hide very large lateral heterogeneities. It is not known to what extent proteins, lipids, and carbohydrates contribute to the charge of the plasma membrane. Barber (1982) concluded that most of the negative charges on the thylakoid membrane are contributed by glutamic and aspartic acid residues on the proteins. The outer surface of plasma membranes from barley leaves is significantly more negative than that of barley roots (–29 and –19 mC m^{-2}, respectively; Körner et al. 1985). Their polypeptide composition is very similar (Sect. 4) whereas the leaf plasma membrane contains about 50% more charged phospholipids than the root plasma membrane (Sect. 3). It is tempting to suggest that the difference in charge density is due to the difference in charged phospholipids.

The net negative charge on a membrane gives rise to a negative surface potential (note: different from the transmembrane potential). The size of the surface potential depends on the number of negative charges as well as on the concentration and type (mono-, di,- or trivalent) of cations in the medium with which the membrane is in contact. The nature of the anion is unimportant. By attracting cations and repelling anions the surface potential shifts the apparent pH optimum of all membrane-bound enzymes or carriers upwards and affects the apparent K_m of membrane-bound enzymes or carriers with charged substrates. Likewise the membrane potential affects the redox potential of membrane-bound enzymes, binding of proteins to membranes as well as the interaction between membrane surfaces (for reviews see Barber 1980, 1982; Itoh and Nishimura 1986). The relevance of such electrostatic effects to the plasma membrane was demonstrated by Gibrat et al. (1985), who showed that the apparent K_m(ATP) of the H^+-ATPase decreased by 50% when the size of the surface potential of plasma membrane vesicles was decreased by the addition of salt. The effect can be explained by assuming that the concentration of the negatively charged substrate MgATP at the membrane surface near the active site of the enzyme is decreased by the membrane potential.

Plasma membrane vesicles isolated from barley, wheat, and oat roots have both Ca^{2+} and Mg^{2+} bound to their outer apoplastic surface in roughly equal amounts (Körner et al. 1985; Møller and Lundborg 1985). On the other hand, a small difference was detected between the Ca^{2+}/Mg^{2+} ratio in plasma membrane vesicles from roots and leaves of barley (Körner et al. 1985). Since these plasma membranes were isolated without being exposed to any chelators, it is possible that the complement of bound divalent cations could be similar to that found in vivo. The role of these bound divalent cations is not known. They may stabilize the plasma membrane or be cofactors for enzymes and they certainly modulate the surface potential.

2.3 Overall Surface Properties

Aqueous polymer two-phase partitioning separates membranes with respect to both charge and hydrophobic/hydrophilic surface properties (Albertsson 1986). Right side-out plasma membrane vesicles have the highest affinity of all membrane vesicles for the upper, polyethylene glycol-rich phase, a property found for both plant and animal plasma membranes (review, Larsson 1983; Chap. 3) which has been suggested to be due to a similar lipid composition (cf.

Sect. 3). This similarity in surface properties between the outer surface of plant and animal plasma membranes may also be due to the carbohydrates of glycoproteins and glycolipids exposed on this surface (Rothman and Lenard 1977; Chap. 4).

3 Lipid Composition

All biological membranes consist of a lipid bilayer, which constitutes the structural framework in which the membrane proteins are anchored. Due to the fluidity of the lipid bilayer, lipids and proteins may move relatively freely in the plane of the bilayer unless their motion is restricted by interactions, e.g., with the cytoskeleton (Chap. 12). This is the so-called fluid mosaic model of biological membranes (Singer and Nicholson 1972). Knowledge of the lipid composition of the plasma membrane should be important since the activities of integral membrane proteins are a function of properties of the bilayer just as the activities of water-soluble enzymes depend on properties of the surrounding aqueous medium, such as pH and ionic composition (Carruthers and Melchior 1986).

Sterols, glycolipids, and phospholipids are the main lipid classes in both plant and mammalian plasma membranes (Table 2). Most of the sterols usually appear as free sterols but sterol esters, glycosides, and acylated glycosides are also found, and seem to be more abundant in plant than in mammalian plasma membranes. Sterol glycosylation is a reaction occurring in the plasma membrane, catalyzed by UDP-glucose:sterol glycosyltransferase, a useful marker for the plant plasma membrane (Chap. 2).

Glucocerebroside is the major glycolipid, but small amounts of monogalactosyldiacylglycerol (MGDG) and digalactosyldiacylglycerol (DGDG) are also found (Table 2). Also the latter two are true constituents of the plasma membrane and not due to contamination by chloroplast membranes, as shown by specific lectin agglutination of wheat protoplasts (Kogel et al. 1984). In mammalian plasma membranes the glycolipids are exclusively localized in the outer leaflet of the bilayer (Rothman and Lenard 1977), and this may be expected also for the plant plasma membrane in agreement with the agglutination experiments above.

Phosphatidylcholine (PC) and phosphatidylethanolamine (PE) are the major phospholipids in plant as well as mammalian plasma membranes (Table 2). However, PC is sometimes substituted by the equivalent sphingomyelin (both have choline head groups) in mammalian plasma membranes (review, Le and Doyle 1984), whereas sphingomyelin has not been reported in plants. Phosphatidic acid (PA) is a minor constituent of mammalian plasma membranes and it is usually regarded as a degradation product and an artifact of the preparation procedure. In plant plasma membranes, however, PA is often third to PC and PE in abundance. This could be due to higher phospholipase D activities in plants, but the relatively fruitless attempts to decrease the PA

Table 2. Lipid composition (mol%) of plasma membranes from various species and tissues

Component	Barley[a]		Spinach[a]	Cauliflower[a]	Mung bean[b]	Human[c]	Rat[c]
	Root	Leaf	Leaf	Bud	Hypocotyl	Erythrocyte	Liver
Sterols							
Free sterols	57	35	7	32	40		
Sterol esters	–	–	–	0.5	–		
Acylated sterol glycosides	–	–	13	3	2		
Sterol glycosides	7	–	–	7	2		
Subtotal	64	35	20	43	44	40	41
Glycolipids							
Glucocerebroside	9	16	14	3	7		
MGDG	1	4	–	0.3	0.2		
DGDG	0.3	1	3	1	1		
Subtotal	10	21	17	4	8	<5	7
Phospholipids							
PC	8	18	25	18	16	15	19
Sphingomyelin	–	–	–	–	–	16	9
PE	7	11	24	14	19	18	11
PA	2	9	6	13	8	–	2
PG	0.8	2	3	3	2	–	–
PI } PS	9	4	5	5			
PI					3	5	4
PS					2	5	5
Lyso-PC	–	–	–	–	–	–	2
Subtotal	26	44	64	53	49	43	43

[a] Rochester et al. (1987a).
[b] Yoshida and Uemura (1986).
[c] Le and Doyle (1984).

content with phospholipase inhibitors suggests that PA may be a true constituent of the plant plasma membrane (Yoshida and Uemura 1986). Phosphatidylinositol, which may play an important role in signal transduction via the phosphoinositide cycle (Chaps. 9 and 16), is present in minor amounts.

In mammalian plasma membranes PC and its equivalent sphingomyelin are mainly located in the outer leaflet of the bilayer, whereas PE and phosphatidylserine (PS) are mainly located in the inner, cytoplasmic leaflet. The total amount of phospholipid in the two leaflets is the same, however (Rothman and Lenard 1977). One would expect a similar lipid asymmetry to be found in the plant plasma membrane, although there is at present no evidence available. The very low isoelectric point of the apoplastic surface of plant plasma membranes (Table 1) has been suggested to be due to a distribution of the very acidic phospholipids, phosphatidylglycerol (PG), and PA, mainly to the outer leaflet of the bilayer (Rochester et al. 1987a).

Table 3. Composition (mol%) of free sterols in plasma membranes isolated from various species and tissues

Sterol	Barley[a]		Oat[b]	Rye[b]	Spinach[a]	Cauliflower[a]
	Root	Leaf	Leaf	Leaf	Leaf	Inflorescence
Campesterol	32	16	6	30	–	26
Stigmasterol	10	24	23	6	–	7
Sitosterol	54	60	10	62	–	67
Cholesterol	–	–	29	2	–	–
Spinasterol	–	–	–	–	79	–
Stigmastanol	–	–	–	–	17	–
7-Stigmastenol	–	–	–	–	4	–
Cycloartenol	–	–	33	–	–	–
Unidentified	4	–	–	–	–	–

[a] Rochester et al. (1987a).
[b] Cooke et al. (1989).

Free sterols make up a large proportion of total lipid in plasma membranes from most species (Table 2). Sitosterol, campesterol, and stigmasterol are most abundant in plants (Table 3), in contrast to mammals where cholesterol dominates (review, Le and Doyle 1984). More unusual sterols are found in some species: in oat shoot plasma membrane cycloartenol, a precursor of the other sterols, contitutes one-third of total sterols; and the spinach plasma membrane has a unique set of sterols (Table 3).

The composition of free sterols in the plasma membrane may differ between different organs, such as root and leaf (Table 3). However, there does not seem to be any difference in sterol composition between different membranes within the same cell, although free sterols are enriched in the plasma membrane (Rochester et al. 1987b). Thus, the sterol composition is not a characteristic of a certain membrane but rather of a certain organ (Rochester et al. 1987a). Environmental factors, such as low temperature, affect the sterol composition as well as the overall lipid composition (Chap. 13).

The major glycerolipid fatty acids in plant plasma membranes are palmitic (16:0), linoleic (18:2), and linolenic (18:3) acid, which contrasts with mammalian plasma membranes where palmitic, stearic (18:0), and arachidonic (20:4) acid dominate (Table 4). In spite of these differences, the degree of unsaturation is similar in plant and animal plasma membranes. Compared to the bulk intracellular membranes in plants the plasma membrane has more saturated fatty acids (Rochester et al. 1987a; Chap. 13). Furthermore, the hydrocarbon chains of the glucocerebroside are not included in Table 4. Thus, in the barley leaf plasma membrane with its relatively high content of glucocerebroside (Table 2) the hydroxy fatty acid and long-chain hydrocarbon (with one double bond only) moieties of the cerebroside together constitute ca. 30% of total hydrocarbon chains, considerably decreasing the proportion of

Table 4. Fatty acid composition (mol%) of the total phospholipids in plasma membrane from various species and tissues

Fatty acid	Barley[a]		Spinach[a]	Cauliflower[a]	Mung bean[b]	Orchard[c]	Liver[d]
	Root	Shoot	Leaf	Inflorescence	Shoot	Grass	
16:0	44	32	30	21	35	26	22
16:1	2	1	1	2	–	tr.	2
18:0	3	4	3	3	6	0.3	27
18:1	2	3	10	13	9	3	9
18:2	34	33	27	17	21	49	10
18:3	13	26	27	42	19	20	–
20:0	0.4	tr.	–	1	–	2	–
20:1	2	tr.	1	1	2	–	–
20:2	–	–	–	–	2	–	–
20:3	–	–	–	–	2	–	2
20:4	–	–	–	–	–	–	24
20:5	–	–	–	–	–	–	4
22:0	1	2	–	tr.	–	–	–
22:1	–	–	–	–	1	–	–
24:1	–	–	–	–	–	–	2

[a] Rochester et al. (1987a).
[b] Yoshida and Uemura (1986).
[c] Table 3 in Yoshida and Uemura (1984).
[d] Mahler et al. (1988).

polyunsaturated hydrocarbon chains in the bilayer (Rochester et al. 1987a; Chap. 13). Surprisingly, except for extremely cold-hardy species, the degree of fatty acid unsaturation is usually not increased by cold acclimation contrary to what has often been assumed (Chap. 13).

Traditionally, investigations on the lipid composition of plant membranes have focused on the fatty acid composition, and particularly on the degree of unsaturation of the fatty acids, which has been interpreted in terms of "fluidity" or "microviscosity" of the membranes. However, when attempting to understand how the lipids of a bilayer affect the activity of membrane enzymes at the molecular level, parameters such as "fluidity" are not particularly useful (Carruthers and Melchior 1986). Rather, recent results indicate that the order of importance for bilayer features is: lipid head group > lipid acyl chain length and saturation/unsaturation (affecting bilayer thickness) > lipid backbone ≫ bilayer "fluidity" (Carruthers and Melchior 1986 and references therein).

In view of the large differences in lipid composition between plant plasma membranes from different sources (Table 2), which contrasts with the similar functions expected from them, one may question the importance of the lipid bilayer composition for these functions. The same enzymes can clearly function in seemingly very different environments. However, since the features of the bilayer are determined by the interaction of all components in the bilayer (Carruthers and Melchior 1986), the final differences may be smaller than the bulk lipid analyses suggest. Thus, the sterol cholesterol has a moderating effect on both bilayer thickness and "fluidity", and may also create domains in the membrane enriched in certain lipids (review, Yeagle 1985), properties most probably shared by the common plant sterols (Table 3). Furthermore, the possible presence of so-called boundary lipids (review, Edidin 1981) may provide each protein with a more specific and constant lipid environment.

The only plant plasma membrane enzyme well studied with respect to its lipid requirements is the H^+-ATPase, which in reconstitution experiments does not show any strong dependence on the bulk lipid environment (Chap. 6). However, in the native membrane H^+ transport is very specifically stimulated by μM concentrations of palmitoyl-lysophosphatidylcholine (Palmgren and Sommarin 1989), similarly to Ca^{2+} transport exhibited by the Ca^{2+}-ATPase of the erythrocyte plasma membrane (Sarkadi et al. 1982). This points at the importance of the lipid environment for membrane enzyme activities.

4 Polypeptide Composition

Comparing the polypeptide pattern of plasma membranes from different species reveals large similarities (Fig. 2). Thus, the most prominent bands are found in the 50 and 30 kD regions, and some additional distinct bands between 60 and 100 kD (Kjellbom and Larsson 1984). Figure 2 shows the polypeptide pattern of plasma membranes from leaves of one monocotyledon and three dicotyledons. Since the leaf plasma membrane is expected to have similar functions in different species the overall similarities in polypeptide pattern are not unexpected. However, comparing the polypeptide pattern of plasma membranes from different organs, such as leaf and root, from the very same plants reveals even closer similarities (Fig. 3). Indeed, no striking qualitative differences are detected, rather the differences are quantitative, suggesting that the functions of the plasma membrane are similar in roots and leaves (Körner et al. 1985).

About 50 polypeptide bands are visible when plasma membranes are analyzed by one-dimensional SDS-polyacrylamide electrophoresis gel (Figs. 2–4). Some of these (ca. 20) are stained with concanavalin A indicating that they are glycosylated (Grimes and Breidenbach 1987). At least twice as many polypeptide bands can be distinguished by two-dimensional compared to one-dimensional electrophoresis gel (Chap. 13). Of these many polypeptides only a few have been identified to date: The H^+-ATPase with a molecular weight of ca. 100 kD (Chap. 6), a Ca^{2+}-ATPase of 140 kD (Briars et al. 1989), a putative sucrose transporter of 42 kD (Gallet et al. 1989), and the fusicoccin receptor consisting of two different polypeptides of 30 and 31 kD (de Boer et al. 1989).

Membrane polypeptides may be divided into two main groups: integral (intrinsic) polypeptides and peripheral (extrinsic) polypeptides. The integral polypeptides have relatively long hydrophobic sequences and span the membrane one or several times. For the H^+-ATPase a model of its transmembrane structure has been presented (Chap. 6). Integral polypeptides are thus firmly

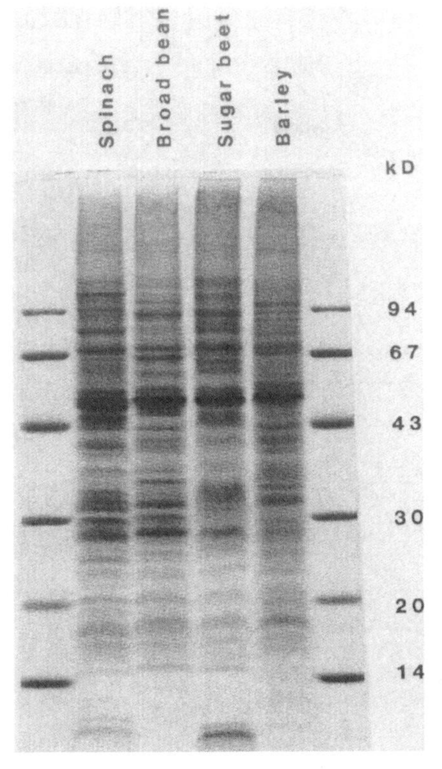

Fig. 2. Polypeptide patterns of leaf plasma membranes obtained from 4 different species. Note the overall similarities with the most prominent bands in the 50 and 30 kD regions, and some additional distinct bands between 60 and 100 kD. The gel was stained with Coomassie brilliant blue (Unpublished data of K. Fredrikson, R. Lemoine, and C. Larsson)

anchored in the membrane and can only be released by detergents or similar treatments. The peripheral polypeptides are hydrophilic and are bound to the surface of the membrane, often by electrostatic forces to integral proteins, and are usually removed by washing the membranes in high salt and pH. Analyses of the spinach leaf plasma membrane shows that ca. 80% of the protein is found in integral polypeptides, and the remaining 20% in peripheral polypeptides (Fig. 4). The latter are probably almost exclusively bound to the cytoplasmic surface of the plasma membrane (Kjellbom et al. 1989). Altogether, this suggests a protein organization very similar to that of mammalian plasma membranes (Lodish and Rothman 1979).

The integral polypeptides are of particular interest since they would constitute (or be subunits of) proteins involved in e.g., transport of solutes, cell wall synthesis, signal transduction, and electron transport, i.e., functions related to communication across the membrane as depicted in Fig. 1.

Many biological membranes are known to show lateral heterogeneity, i.e., certain proteins and lipids are found primarily in specific areas or domains. An

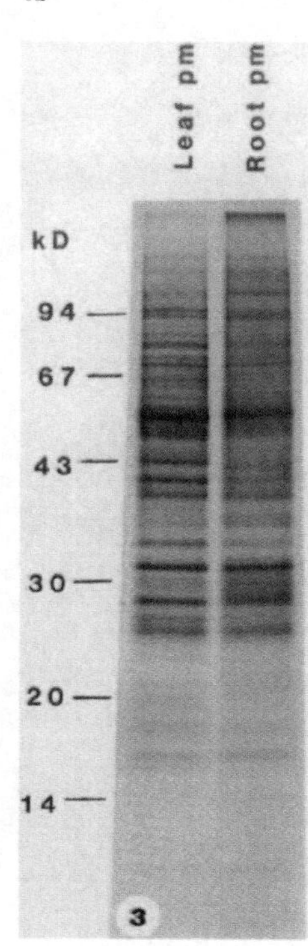

Fig. 3. Polypeptide patterns of barley leaf and root plasma membranes. Note the close similarity with essentially only quantitative differences. The gel was silver stained (Data from Körner et al. 1985)

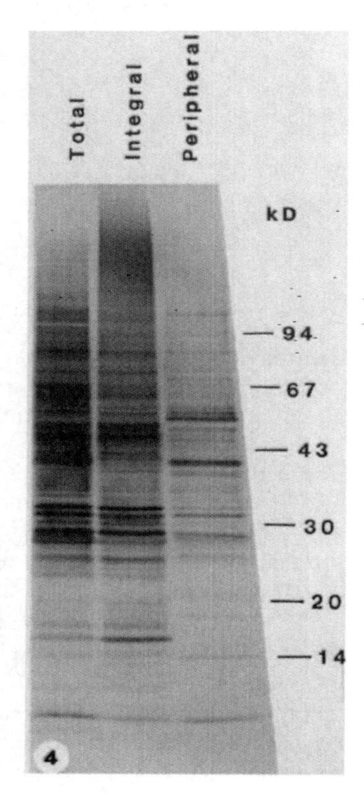

Fig. 4. Integral (*middle lane*) and peripheral (*right*) polypeptides of the spinach leaf plasma membrane as analyzed by Triton X-114 fractionation of total plasma membranes (*left*). Note that most of the polypeptides behaved either as integral or peripheral polypeptides resulting in a highly complementary pattern. Approx. 80% of total protein was recovered as integral polypeptides, the remaining 20% as peripheral. The gel was silver stained (Data from Kjellbom et al. 1989)

extreme case in point is the thylakoid membrane (review, Anderson and Andersson 1988). Considering the well-known polarity of many plant cells and the presence of coated pits (Chap. 10) and plasmodesmata one would expect lateral heterogeneity to be an important factor in the function of the plant plasma membrane.

5 Concluding Remarks

The plant plasma membrane shows both many similarities with and some differences from the mammalian plasma membrane. This is evident although much less is presently known about the plant plasma membrane. The development has been very rapid during the last decade, however, and has provided us with a firm basis for future research as reflected in the following chapters.

Acknowledgments. We are grateful to the Swedish Natural Science Research Council, the Swedish Council for Forestry and Agricultural Research, and the Carl Tesdorpf Foundation for economic support.

References

Albertsson P-Å (1986) Partition of cell particles and macromolecules, 3rd edn, Wiley-Interscience, New York

Anderson JM, Andersson B (1988) The dynamic photosynthetic membrane and regulation of solar energy conversion. Trends Biochem Sci 13:351–355

Auderset G, Sandelius AS, Penel C, Brightman A, Greppin H, Morré DJ (1986) Isolation of plasma membrane and tonoplast fractions from spinach leaves by preparative free-flow electrophoresis and effect of photoinduction. Physiol Plant 68:1–12

Barber J (1980) Membrane surface charges and potentials in relation to photosynthesis. Biochim Biophys Acta 594:253–308

Barber J (1982) Influence of surface charges on thylakoid structure and function. Annu Rev Plant Physiol 33:261–295

Briars SA, Kessler F, Evans DE (1989) The calmodulin-stimulated ATPase of maize coleoptiles is a 140,000 Mr polypeptide. Planta 176:283–285

Canut H, Brightman A, Boudet AM, Morré DJ (1988) Plasma membrane vesicles of opposite sidedness from soybean hypocotyls by preparative free-flow electrophoresis. Plant Physiol 86:631–637

Carruthers A, Melchior DL (1986) How bilayer lipids affect membrane protein activity. Trends Biochem Sci 11:331–335

Cooke DT, Burden RS, Clarkson DT, James CS (1989) Xenobiotic induced changes in membrane lipid composition: effects on plasma membrane ATPases. In: Mechanisms and regulation of transport processes. British plant growth regulator group, Monograph 18:41–53

de Boer AH, Watson BA, Cleland RE (1989) Purification and identification of the fusicoccin binding protein from oat root plasma membrane. Plant Physiol 89:250–259

Edidin M (1981) Molecular motions and membrane organization and function. In:Finean JB, Michell RH (eds) Membrane structure. Elsevier, Amsterdam, pp 37–82

Gallet O, Lemoine R, Larsson C, Delrot S (1989) The sucrose carrier of the plant plasma membrane. I. Differential affinity labeling. Biochim Biophys Acta 978:56–64

Gibrat R, Grouzis J-P, Rigaud J, Grignon C (1985) Electrostatic characteristics of corn root plasmalemma: effect on the Mg^{2+}-ATPase activity. Biochim Biophys Acta 816:349–357
Griffing LR, Quatrano RS (1984) Isoelectric focusing of plant cell membranes. Proc Natl Acad Sci USA 81:4804–4808
Grimes HD, Breidenbach RW (1987) Plant plasma membrane proteins. Immunological characterization of a major 75 kilodalton protein group. Plant Physiol 85:1048–1054
Grouzis J-P, Gibrat R, Rigaud J, Grignon C (1987) Study of sidedness and tightness to H^+ of corn root plasmalemma vesicles: preparation of a fraction enriched in inside-out vesicles. Biochim Biophys Acta 903:449–464
Guidotti G (1972) Membrane proteins. Annu Rev Biochem 41:731–752
Hinton R, Dobrota M (1976) Density gradient centrifugation. North-Holland, Amsterdam
Itoh S, Nishimura M (1986) Rate of redox reactions related to surface potential and other surface-related parameters in biological membranes. Methods Enzymol 125:58–86
Kjellbom P, Larsson C (1984) Preparation and polypeptide composition of chlorophyll-free plasma membranes from leaves of light-grown spinach and barley. Physiol Plant 62:501–509
Kjellbom P, Larsson C, Rochester CP, Andersson B (1989) Integral and peripheral proteins of the spinach leaf plasma membrane. Plant Physiol Biochem 27:1–6
Kogel H, Ehrlich-Rogozinski S, Reisener HJ, Sharon N (1984) Surface galactolipids of wheat protoplasts as receptors for soybean agglutinin and their possible relevance to host-parasite interaction. Plant Physiol 76:924–928
Körner LE, Kjellbom P, Larsson C, Møller IM (1985) Surface properties of right side-out plasma membrane vesicles isolated from barley roots and leaves. Plant Physiol 79:72–79
Larsson C (1983) Partition in aqueous polymer two-phase systems: a rapid method for separation of membrane particles according to their surface properties. In: Hall JL, Moore AL (eds) Isolation of membranes and organelles from plant cells. Academic Press, London, pp 277–309
Le AV, Doyle D (1984) General theory of membrane structure and function. In: Venter JC, Harrison L (eds) Membranes, detergents, and receptor solubilization. Alan R Liss, New York (Receptor biochemistry and methodology, vol 1, pp 1–25)
Lodish HF, Rothman JE (1979) The assembly of cell membranes. Sci Am 240:38–53
Lynch DV, Steponkus PL (1987) Plasma membrane lipid alterations associated with cold acclimation of winter rye seedlings (*Secale cereale* L. cv Puma). Plant Physiol 83:761–767
Mahler SM, Wilce PA, Shanley BC (1988) Studies on regenerating liver and hepatoma plasma membranes — I. Lipid and protein composition. Int J Biochem 20:605–611
Møller IM, Lundborg T (1985) Electrostatic surface properties of plasmalemma vesicles from oat and wheat roots. Ion binding and screening investigated by 9-aminoacridine fluorescence. Planta 164:354–361
Møller IM, Lundborg T, Bérczi A (1984) The negative surface charge density of plasmalemma vesicles from wheat and oat roots. FEBS Lett 167:181–185
Morré DJ, Bracker CE (1976) Ultrastructural alteration of plant plasma membranes induced by auxin and calcium ions. Plant Physiol 58:544–547
Palmgren MG, Sommarin M (1989) Lysophosphatidylcholine stimulates ATP dependent proton accumulation in isolated oat root plasma membrane vesicles. Plant Physiol 90:1009–1014
Petit PX, Edman KA, Gardeström P, Ericson I (1987) Some properties of mitochondria, mitoplasts and submitochondrial particles of different polarities from plant tissues. Biochim Biophys Acta 890:377–386
Quail PH (1979) Plant cell fractionation. Annu Rev Plant Physiol 30:425–484
Rochester CP, Kjellbom P, Larsson C (1987a) Lipid composition of plasma membranes from barley leaves and roots, spinach leaves and cauliflower inflorescences. Physiol Plant 71:257–263
Rochester CP, Kjellbom P, Andersson B, Larsson C (1987b) Lipid composition of plasma membranes isolated from light-grown barley (*Hordeum vulgare*) leaves: identification of cerebroside as a major component. Arch Biochem Biophys 255:385–391
Rothman JE, Lenard J (1977) Membrane asymmetry. Science 195:743–753
Sarkadi B, Enyedi A, Nyers A, Gardos G (1982) The function and regulation of the calcium pump in the erythrocyte membrane. Ann NY Acad Sci 402:329–348

Singer SJ, Nicholson GL (1972) The fluid mosaic model of the structure of cell membranes. Science 175:720–731

Steele ME, Stout RG (1986) Isoelectric focusing of oat root membranes. Plant Physiol 82:327–329

Westrin H, Shanbhag VP, Albertsson P-Å (1983) Isoelectric points of membrane surfaces of three spinach chloroplast classes determined by cross-partition. Biochim Biophys Acta 732:83–91

Widell S, Larsson C (1983) Distribution of cytochrome b photoreductions mediated by endogenous photosensitizer or methylene blue in fractions from corn and cauliflower. Physiol Plant 57:196–202

Yeagle PL (1985) Cholesterol and the cell membrane. Biochim Biophys Acta 822:267–287

Yoshida S, Uemura M (1984) Protein and lipid compositions of isolated plasma membranes from orchard grass (*Dactylis glomerata* L.) and changes during cold acclimation. Plant Physiol 75:31–37

Yoshida S, Uemura M (1986) Lipid composition of plasma membranes and tonoplasts isolated from etiolated seedlings of mung bean (*Vigna radiata* L.). Plant Physiol 82:807–812

Note Added in Proof. Using immunoblotting two additional polypeptides have recently been identified in isolated plasma membranes. Calmodulin, with a molecular weight of 18 kD, constitutes 0.5 to 1% of total protein in pea plasma membranes [Collinge M, Trewavas AJ (1989) J Biol Chem 264:8865–8872]. The major NADH-ferricyanide reductase in spinach leaf plasma membranes is identical to cytochrome b_5 reductase and has a molecular weight of 43 kD (P. Askerlund and P. Laurent, personal communication).

Chapter 2 A Critical Evaluation of Markers Used in Plasma Membrane Purification

S. WIDELL[1] and C. LARSSON[2]

[1] Department of Plant Physiology and [2] Department of Plant Biochemistry, University of Lund, P.O. Box 7007, S-220 07 Lund, Sweden

C. Larsson, I.M. Møller (Eds):
The Plant Plasma Membrane

1 Introduction

A membrane marker is any property that can be used to identify a specific membrane. In the ideal case, the marker used is not only confined to a single membrane, but is also uniformly distributed in the membrane, thus constituting an absolute marker for that membrane (Quail 1979). Meaning, that wherever the marker is found the membrane is present in direct proportion to the marker. Very few, if any, of the markers currently in use fulfill the above requirements. Rather, most markers have one primary location and one, or several, secondary locations. Furthermore, the marker may be non-uniformly distributed in the membrane due to membrane lateral heterogeneity, or may be absent in certain cell types. Two markers which are close to ideal are cytochrome c oxidase of the mitochondrial inner membrane and chlorophyll for chloroplast thylakoids, although lateral heterogeneity is a striking feature of the latter membrane (review Andersson and Anderson 1980). By contrast, antimycin A-insensitive NAD(P)H-cytochrome c reductase, the most commonly used marker for the endoplasmic reticulum, is in fact found in several membranes (review, Møller and Lin 1986) and is far from an ideal marker.

A constant, but often overlooked, problem is the structural latency of markers caused by the transverse asymmetric organization of membrane constituents together with the fact that isolated membranes appear as closed vesicles in aqueous solutions. Thus, depending on the orientation of the vesicle, a marker, e.g. the active site of an enzyme, may either be exposed to the bulk solution or hidden inside the vesicle. The latter situation may give rise to completely misleading results if proper measures are not taken.

Below, we will try to evaluate the markers used in plasma membrane purification, and penetrate the problems arising due to multiple location of markers, structural latency, suboptimal assay conditions, inactivation, etc. Rather than being comprehensive we will concentrate on the most commonly used markers and particularly on the most useful markers (see also Chaps. 3 and 4).

Abbreviations: BSA, bovine serum albumin; BTP, bis-Tris propane; CCD, countercurrent distribution; DCCD, N,N'-dicyclohexylcarbodiimide; DTT, dithiothreitol, GS I, glucan synthase I or 1,4-β-glucan synthase; GS II, glucan synthase II or 1,3-β-glucan synthase; HEPES, N-(2-hydroxyethyl)piperazine-N'-2-ethanesulfonic acid; MES, 2-(N-morpholino)ethanesulfonic acid; MOPS, 3-(N-morpholino)propanesulfonic acid; NAA, naphthylacetic acid; NPA, naphthylphthalamic acid; PMSF; phenylmethylsulfonyl fluoride; PTA, phosphotungstic acid; PVPP, polyvinylpolypyrrolidone; SDS, sodium dodecyl sulfate; STA, silicotungstic acid; Tris, tris(hydroxymethyl)aminomethane; UDPG, UDP-glucose.

2 Markers for the Plasma Membrane

2.1 H^+-ATPase

This is the classical enzyme marker for the plasma membrane and was originally characterized as a K^+-stimulated, Mg^{2+}-dependent ATPase (Hodges et al. 1972). Plant cells contain several different phosphohydrolases, both soluble and membrane-bound, which are dependent on divalent cations like Mg^{2+} and can use ATP as substrate. Therefore, properties relatively specific to the plasma membrane-bound H^+-ATPase (EC 3.6.1.35) have been exploited to obtain a specific assay for the enzyme. These include stimulation by K^+, inhibition by vanadate, a slightly acidic pH optimum, and insensitivity to azide, molybdate, and NO_3^-, which are inhibitors of mitochondrial and thylakoid ATPase (EC 3.6.1.34), soluble acid phosphatase (EC 3.1.3.2), and tonoplast ATPase, respectively (Gallagher and Leonard 1982).

2.1.1 K^+-Stimulated ATPase

At acidic pH, the plant plasma membrane ATPase is stimulated to a certain degree by K^+ (Fig. 1). This effect was initially thought to be due to a direct coupling between ATP hydrolysis and K^+ translocation across the plasma membrane by analogy with the Na^+/K^+-pump of animal plasma membranes (Hodges et al. 1972), an interpretation that is probably incorrect (Chap. 6). Nevertheless, the increase in Mg^{2+}-ATPase activity on addition of K^+ became the first established enzyme marker for the plant plasma membrane. However,

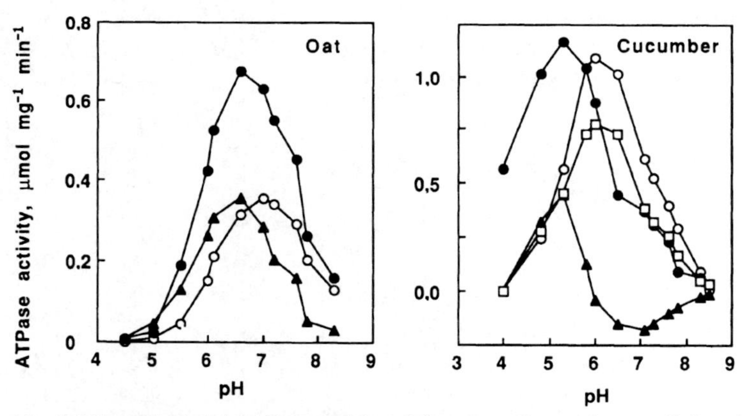

Fig. 1. Effect of pH on the ATPase activity of plasma membranes from oat (*left*) and cucumber (*right*) roots. *Symbols*: activity measured in the presence of Mg^{2+} (○), Mg^{2+} + K^+ (●), and Mg^{2+} + Ca^{2+} (□); and K^+ stimulation of the activity (▲). Note the low pH optimum of the K^+ stimulation and that it constitutes at most 100% of the basal Mg^{2+}-dependent activity. Note also that the presence of 50 μM Ca^{2+} in the assay is inhibitory (After Sommarin et al. 1985 and Memon et al. 1987)

K^+ also affects the activity of the mitochondrial ATPase (Hendriks 1977) and Golgi ATPase (Chanson et al. 1984). The mitochondrial activity has a higher pH optimum (pH 8–9) and can be suppressed by azide, but may still be a problem in crude homogenates containing a large proportion of mitochondrial membranes compared to plasma membranes. The Golgi ATPase has a somewhat higher pH optimum than the plasma membrane ATPase (Chanson et al. 1984) and it is not sensitive to vanadate (Hager and Biber 1984; Ali and Akazawa 1986). However, the ATPase activity of Golgi-derived secretory vesicles seems to be indistinguishable from the plasma membrane ATPase activity, and is possibly due to the same enzyme (Racusen 1988). The main problem with K^+-stimulated ATPase as a plasma membrane marker is that the K^+ stimulation is rather small with many species and organs even when using pure plasma membranes; it is often below 50%, and usually does not exceed 100% (Fig. 1). The pH optimum for the K^+ stimulation is lower than for the basal Mg^{2+}-dependent activity and varies between materials (Fig. 1; see also Chap. 6).

2.1.2 Vanadate-Inhibited ATPase

The plasma membrane H^+-ATPase is a member of the (E-P) class of membrane-bound ATPases and thus forms a phosphorylated intermediate during ATP hydrolysis. This is a property not shared by the (F_0F_1)-ATPases of tonoplasts, mitochondria, and plastids (Chap. 6). Since vanadate resembles the transition state form of phosphate, vanadate inhibits the ATPase activity of (E-P)-ATPases (Macara 1980). Vanadate-inhibited ATPase has therefore been widely used as a marker for the plasma membrane. However, all other phosphohydrolases of the plant cell which form phosphorylated intermediates are also vanadate-sensitive. Examples are the Ca^{2+}-ATPase of the endoplasmic reticulum (EC 3.6.1.38) (Bush and Sze 1986), and the plasma membrane (Gräf and Weiler 1989), and the Mg^{2+}-NTPase of the inner membrane of the chloroplast (McCarthy et al. 1984) and the amyloplast (Harinasut et al. 1988) envelope, which may all be confused with the plasma membrane H^+-ATPase.

The Mg^{2+}-NTPase of the plastid envelope may be a particular problem. Its role is unknown at present, but it has been suggested to be involved in ATP-dependent transport across the envelope (including H^+-transport; Maury et al. 1981), although there is no direct evidence for this (McCarthy and Selman 1986). The activity is inhibited by vanadate but only slightly by molybdate (McCarthy et al. 1984; Harinasut et al. 1988). It will thus be assayed as vanadate-inhibited ATPase, and can not be distinguished from the plasma membrane H^+-ATPase without control experiments. Differences between the two ATPases are the following: The plasma membrane ATPase usually shows a high latency, i.e., it is stimulated severalfold by the addition of detergent (Fig. 2), since the predominant fraction of plasma membrane vesicles exposes the apoplastic side, thereby making the substrate binding site inaccessible to MgATP (Larsson et al. 1984; Sect. 3). The envelope ATPase is much less specific towards ATP and it is not inhibited by N,N′-dicyclohexylcarbodiimide (DCCD)

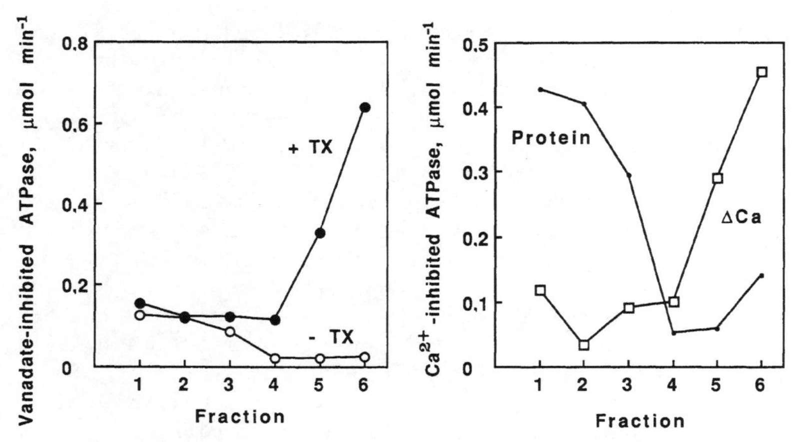

Fig. 2. Distribution of (*left*) vanadate-inhibited K^+, Mg^{2+}-ATPase activity, and (*right*) Ca^{2+}-inhibited K^+, Mg^{2+}-ATPase activity (□) and protein (•) after counter current distribution of a microsomal fraction from maize roots in an aqueous polymer two-phase system (cf. Figs. 4 and 6). The microsomal fraction was loaded in tube 1, and 5 transfers were made of the upper phase leaving the lower phase + interface as stationary phase. Thus membrane vesicles partitioning mainly to the interface or into the lower phase (intracellular membranes and inside-out plasma membrane vesicles) will end up in early fractions, whereas vesicles partitioning mainly into the upper phase (right side-out plasma membrane vesicles) will end up in late fractions. Vanadate-inhibited K^+, Mg^{2+}-ATPase activity was assayed either in the absence (○) or presence (●) of Triton X-100 (*TX*). Ca^{2+}-inhibited K^+, Mg^{2+}-ATPase activity was assayed in the presence of Triton. Note that the data obtained in the absence of detergent suggest that the plasma membranes are recovered in fractions *1–3*, whereas the true location is in fractions *5* and *6*, due to the fact that most of the plasma membranes are right side-out and the ATPase activity therefore highly latent (After Bérczi et al. 1989)

(McCarthy et al. 1984). It is furthermore not stimulated by K^+, but stimulated by low concentrations of Ca^{2+} (Nguyen et al. 1987). This Ca^{2+} stimulation is in contrast to the inhibition of the plasma membrane H^+-ATPase by Ca^{2+} (Fig. 1; Chap. 6). Indeed, Ca^{2+}-inhibited ATPase may turn out to be a more specific plasma membrane marker than vanadate-inhibited ATPase (Fig. 2). A complicating factor is the presence of also a Ca^{2+}-ATPase in the plasma membrane (Robinson et al. 1988; Gräf and Weiler 1989). The activity of this Ca^{2+}-ATPase, however, is often low compared to the H^+-ATPase (Robinson et al. 1988; Gräf and Weiler 1989; Chap. 6); although Kylin and Sommarin (1986) found relatively high ATPase activity in wheat root plasma membranes, even when Mg^{2+} was substituted with Ca^{2+}.

An additional problem with the vanadate-inhibited ATPase is that soluble acid phosphatases are also vanadate-sensitive (Gallagher and Leonard 1982). Membrane-bound phosphohydrolase activity (measured at pH 6.5–7 with ATP as substrate) constitutes only ca. 20% of the total phosphohydrolase activity in a homogenate (Kylin and Sommarin 1986). Therefore, if soluble acid phosphatases have some tendency to become pelleted (e.g., either enclosed within vesicles or adsorbed on the outside), even this activity can be a problem.

Finally, Golgi-derived secretory vesicles contain vanadate-inhibited ATPase activity, which probably is due to plasma membrane H^+-ATPase en route to its final location (Racusen 1988).

2.1.3 Design of Assays

The plasma membrane H^+-ATPase can be distinguished from the various vanadate-sensitive phosphohydrolyzing enzymes only by detailed analyses, e.g. of pH optimum, nucleotide specificity, K^+ stimulation, and sensitivity to inhibitors (Chap. 6). Although such a detailed approach in the final characterization of ATPase activity in plasma membrane preparations is possible, it is not practical for routine marker analyses. However, some properties relatively specific to the plasma membrane H^+-ATPase have been exploited to obtain a more specific assay in crude fractions. Thus, the following assay modified from Gallagher and Leonard (1982) is useful with many materials:

The vanadate-inhibited ATPase activity is determined using 2–25 μg protein in 120 μl of 3 mM ATP, 4 mM $MgSO_4$, 50 mM KNO_3, 1 mM sodium azide, 0.1 mM sodium molybdate, 50 mM MES-Tris (or MES-BTP), pH 6.5, 0–0.33 M sucrose, 0.1 mM EDTA, 1 mM DTT, 0.015–0.05% (w/v) Triton X-100, ± 0.1 mM orthovanadate, freshly prepared and boiled in buffer prior to addition. The reaction is initiated by adding ATP and it is run at 20–38 °C for 5–30 min. Blanks lacking $MgSO_4$ are run in parallel. With pure plasma membranes isolated by phase partitioning activities of 0.2–4 μmol Pi released (mg protein)$^{-1}$ min^{-1} (20 °C) are obtained depending on the source of membranes.

Azide, molybdate, and NO_3^- are included in the assay to inhibit mitochondrial and thylakoid ATPases, acid phosphatases, and tonoplast ATPase, respectively (Fig. 3). Na^+-molybdate is used rather than NH_4^+-molybdate, since NH_4^+ may inhibit the plasma membrane ATPase (Memon et al. 1987). A pH of 6–7 is chosen (Fig. 1), which is lower than the optimum for mitochondrial and thylakoid ATPases (pH 8–9) and higher than that of acid phosphatases (pH 4–6). Triton X-100 at 0.015–0.05% (w/v; the optimal concentration of detergent should be determined by titration) is used to permeabilize the vesicles and expose also the active sites on the inner surface of right side-out (apoplastic side-out) plasma membrane vesicles, which constitute the main proportion in microsomal fractions (Fig. 2). DTT and EDTA stabilize the activity by protecting essential SH-groups (Chap. 6). EDTA also lowers the background (activity minus Mg^{2+}) by complexing endogenous divalent cations. However, the EDTA concentration should be kept to a minimum, since EDTA forms a complex with vanadate, thus decreasing vanadate inhibition (Bowman and Slayman 1979). K^+ should be included even when the K^+-stimulation is low, since it enhances vanadate inhibition (Bowman and Slayman 1979; Gibrat et al. 1989).

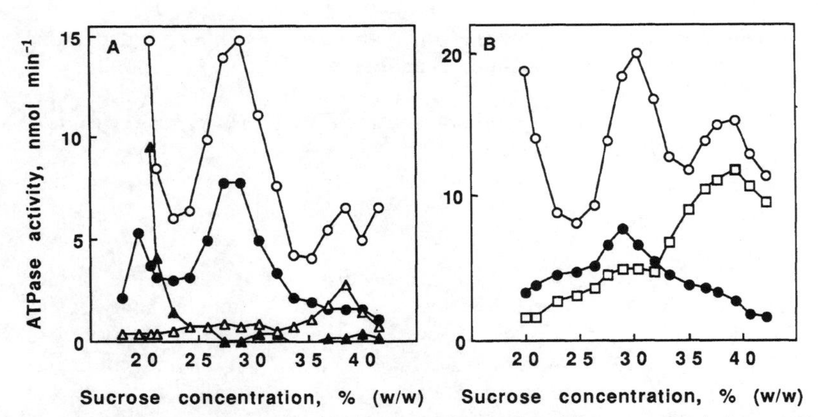

Fig. 3A,B. ATPase activities of microsomal membranes from *Zea mays* separated on a sucrose gradient. **A** ATPase measured at pH 7.5. Total ATPase (○), molybdate-sensitive (▲), azide-sensitive (△), and NO_3^--sensitive (●) activity. Inhibitor concentrations: 0.1 mM molybdate, 1 mM azide, 25 mM NO_3^-. Molybdate-sensitive activity is characteristic for soluble fractions (acid phosphatase), azide-sensitivity for heavy fractions (mitochondrial ATPase), and NO_3^--sensitivity for light fractions (tonoplast ATPase). **B** ATPase measured at pH 7.0. Total ATPase (○), NO_3^--sensitive (●), and vanadate-sensitive (□) ATPase. Inhibitor concentrations: 25 mM NO_3^- and 0.5 mM vanadate. Vanadate-sensitive ATPase (putative plasma membrane marker, but cf. Fig. 2) shows a broad distribution and peaks close to azide-sensitive ATPase (cf. **A**). ATPase activity not sensitive to any of the inhibitors (**A,B**) and located in a membrane with an intermediate density could reflect Golgi-ATPase (After Hager and Biber 1984)

DTT, EDTA (5 mM), PMSF (phenylmethylsulfonyl fluoride, protease inhibitor), and BSA and casein (competing substrates for proteases), as well as PVPP (binds phenolic compounds) should be included in the homogenizing medium to preserve ATPase activity (Palmgren and Sommarin 1989). DTT and EDTA should be used also in the subsequent purification steps.

Taking the above measures, vanadate-inhibited ATPase is a useful marker with many materials. However, with some materials, such as wheat root, the major part of the vanadate-inhibited ATPase activity in the microsomal fraction is probably not due to plasma membrane ATPase (Bérczi et al. 1989).

If the vanadate inhibition is not sufficiently specific, the K^+ stimulation may be measured using essentially the same assay. Instead of measuring ± vanadate, the assay is performed using ± 25 mM K_2SO_4. KNO_3 should not be used, since NO_3^- inhibits the tonoplast ATPase and this effect would counteract the stimulatory effect of K^+ in crude fractions; neither should KCl be used since Cl^- stimulates the tonoplast ATPase. Note also that the pH optimum for K^+ stimulation is lower than for the total K^+, Mg^{2+}-ATPase activity (Fig. 1).

The liberated phosphate is usually measured by complexing it with molybdate followed by reduction with ascorbate and spectrophotometric determination of the colored product. However, phosphate slowly liberated from the biological material and ATP in the assay mixture after addition of molybdate will also be complexed, thereby elevating the background. A modified method for detection of inorganic phosphate, the Baginski method, has turned out to be very useful. Here, excess molybdate is complexed with arsenite to form a colorless product (Baginski et al. 1967). Indeed, it is not possible to measure ATPase activity with some materials (e.g., spruce; S. Widell, unpublished results) without using this modified procedure.

The enzyme reaction is stopped by addition of 1 ml of ice-cold ascorbate/molybdate solution [3% (w/v) ascorbic acid, 1% (w/v) sodium dodecyl sulfate (SDS), 0.5 M HCl mixed 15 min before use with freshly made 8% (w/v) ammonium heptamolybdate, in a ratio of 10:1; the reagent should be light-yellow]. Keep samples on ice for at least 10 min. Excess molybdate is then complexed with 1.5 ml arsenite-citrate reagent [2% (w/v) sodium citrate dihydrate, 2% (w/v) sodium metaarsenite, and 2% (w/v) glacial acetic acid; the salts are dissolved in water before glacial acetic acid is added]. Move samples to a water bath at 37°C. The absorbance at 850 nm is measured after 25–60 min. A phosphate standard curve is made using 2 mM KH_2PO_4 diluted 4–40 times. SDS is included in the ascorbate/molybdate solution to form mixed micelles with Triton X-100, which otherwise gives cloudy solutions with high light scattering.

2.2 *1,3-β-Glucan Synthase*

1,3-β-Glucan synthase (EC 2.4.1.34) catalyzes the incorporation of glucose units from UDP-glucose (UDPG) into 1,3-β-polyglucan. It was described as a marker for the plasma membrane by Van Der Woude et al. (1974) and termed glucan synthase II (GS II, Ray 1977) [glucan synthase I (GS I) is treated in Sect. 4.5]. The assay has been considerably improved by Kauss and co-workers (Kauss et al. 1983; Kauss and Jeblick 1985) and 1,3-β-glucan synthase is at present probably the most reliable enzyme marker for the plasma membrane. The modification of the assay includes addition of detergent (digitonin), Ca^{2+}, DTT, and spermine to the assay, thereby increasing the activity at least 20-fold. 1,3-β-Glucan synthase activity is not exclusively located in the plasma membrane. Thus, some membrane structures closely related to the plasma membrane, such as coated vesicles, also appear to have 1,3-β-glucan synthase activity (Chap. 10). The activity is not easily distinguished from other glucan synthases, e.g., 1,4-β-glucan synthase (GS I; EC 2.4.1.12) (Sect. 4.5), but this is usually not a problem since the plasma membrane-bound 1,3-β-glucan synthase dominates completely. The following assay modified from Kauss et al. (1983) and Kauss and Jeblick (1985) is therefore usually satisfactory.

1,3-β-Glucan synthase is assayed in a final volume of 100 μl containing 1–25 μg protein, 50 mM HEPES-KOH, pH 7.25, 0–0.33 M sucrose, 0.8 mM spermine, 16 mM cellobiose, 4 mM EGTA/4 mM $CaCl_2$ (giving 80 μM free Ca^{2+}), DDT and 0.006–0.015% (w/v) digitonin (use disposable plastic test tubes). The reaction is initiated by addition of UDP-[^{3}H]-glucose (ca. 20 GBq mol^{-1}) to a final concentration of 2 mM. The reaction is run for 30 min at 25°C, and is terminated by immersing the test tubes in boiling water. The samples are transferred to paper filters (Whatman 3MM) which then are washed 2 × 1 h in 0.35 M ammonium acetate, pH 3.6, 30% (v/v) ethanol. The filter background is determined by running blanks without protein. A radioactivity standard is prepared by adding 10 μl [^{3}H]-UDPG (half the amount used in the assay) to a wetted glass fibre filter (Whatman GF/F). Paper filters cannot be used for the standard, since the low-molecular-weight UDPG will be absorbed into the paper filter and the radioactivity quenched. The filters are dried and radioactivity measured by liquid scintillation counting. With pure plasma membranes isolated by phase partitioning activities of 0.2–2 μmol (mg protein)$^{-1}$ min^{-1} are obtained depending on the source of membranes.

Digitonin is included in the assay to solubilize right side-out plasma membrane vesicles thereby exposing all active sites. Without digitonin, most active sites will be hidden (Fig. 4). The active site, as well as the binding sites for spermine, Ca^{2+}, and cellobiose are located on the cytoplasmic surface of the plasma membrane (Larsson et al. 1984; Fredrikson and Larsson 1989; Chap. 14). Spermine and Ca^{2+} activate the enzyme, whereas cellobiose possibly acts as a primer for the growing glucose chain. Cellobiose can be replaced by sucrose. The optimal concentration of digitonin depends on the amount of protein used as well as on the batch of digitonin and should be determined by titration. DTT stabilizes the activity probably by protecting essential SH-groups and should be included throughout the preparation procedure. The requirements of the preparation media are similar to those for the H^+-ATPase (Sect. 2.1.3).

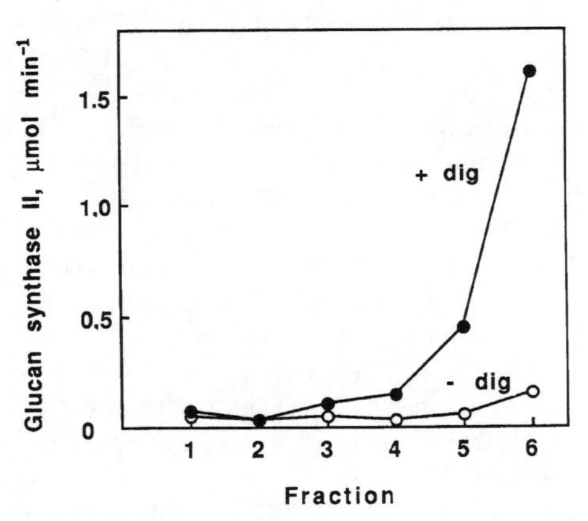

Fig. 4. The distribution of 1,3-β-glucan synthase after countercurrent distribution of a maize root microsomal fraction (see Fig. 2). The activity was assayed in the absence (*– dig*) or presence (*+ dig*) of digitonin. Note that the activity is very low in the absence of detergent. In the presence of digitonin a strong stimulation of activity is seen in fractions *5* and *6*, and by far most of the activity is recovered in these fractions, showing that most of the plasma membrane vesicles are right side-out (After Bérczi et al. 1989; data from the same experiment as shown in Fig. 2)

2.3 UDP-Glucose: Sterol Glucosyltransferase

This enzyme catalyzes the transfer of glucose units from UDPG to sterols, and is mainly located in the plasma membrane (Hartmann-Bouillon and Benveniste 1978). However, activity has also been reported in Golgi (Green 1983) and tonoplasts (Chap. 3). The assay is performed in the presence of detergent to obtain mixing of the hydrophobic substrate (cholesterol) and the membrane-bound enzyme, and therefore this activity does not show any latency. The assay (Hartmann-Bouillon and Benveniste 1978) is based on the incorporation of radioactive glucose into sterol glucosides. Total lipids are extracted into an organic phase and the radioactivity incorporated determined by liquid scintillation counting.

2.4 Phosphotungstic/Silicotungstic Acid Staining

The plasma membrane stains positively when exposed to phosphotungstic acid (PTA) or silicotungstic acid (STA) at low pH (Roland 1978). These stains are highly specific for the plant plasma membrane (Chap. 4), and have the additional great advantage that the actual percentage of plasma membrane in a fraction (the absolute purity) can be determined by calculating the number of stained versus unstained membranes. This is in contrast to enzyme markers which only give relative values of enrichment. The procedure was originally developed using PTA (Roland et al. 1972), but STA was later proposed to give an even more specific staining (Roland 1978). We have used STA, adjusted to pH 3 with HCl rather than with chromic acid, to avoid any unspecific staining by the latter heavy metal ion. Using this stain we have been able to extend the time of staining to 10 min at 37 °C, without loss of specificity (Widell et al. 1982; Widell and Larsson 1983; Kjellbom and Larsson 1984). Subsequent rinsing of the Ni-grids is in 1 mM HCl rather than distilled water (to maintain a low pH throughout). However, as pointed out by D.J. Morré (Chap. 4) there is no firm evidence for any particular advantage of one of the two closely related procedures over the other. The original procedure for PTA staining is found in Chapter 4, our modifications are found above. Regardless of the procedure used, this is probably the most reliable plasma membrane marker, and the only one which permits a direct assessment of the purity of an isolated fraction.

2.5 Naphthylphthalamic Acid (NPA) Binding

NPA is a potent inhibitor of auxin transport by binding to presumptive auxin transport sites in the plasma membrane. This binding was shown to correlate with PTA staining in a study by Lembi et al. (1971), and NPA binding has since been used as a plasma membrane marker. It also correlates well with 1,3-β-glucan synthase (Fig. 5). However, NPA binding is also associated with

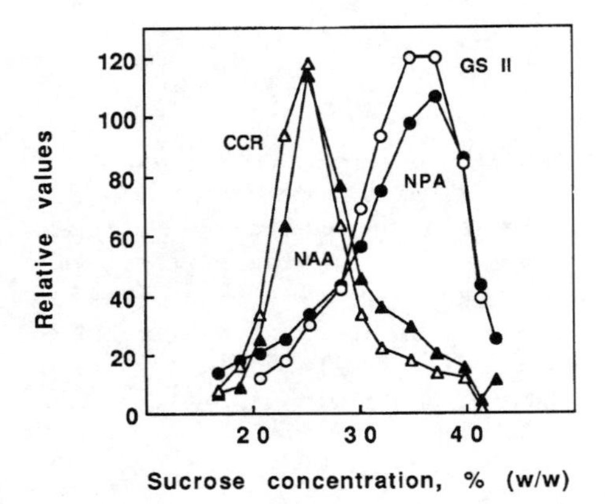

Fig. 5. Distribution of naphthylacetic acid (*NAA*, ▲) binding and naphthylphthalamic acid (*NPA*, ●) binding, 1,3-β-glucan synthase (*GS II*, ○) and NADH-cytochrome c reductase (*CCR*, △) after separation of a maize coleoptile microsomal fraction on a sucrose gradient. NPA binding correlates well with GS II activity, and NAA binding with CCR. Note, however, that a minor proportion of NAA binding as well as CCR activity is recovered with the plasma membrane markers and that NPA binding also overlaps with the main peak of CCR activity (After Jacobs and Hertel 1978)

membranes of lower density than the plasma membrane (Hartmann-Bouillon and Benveniste 1978; Chanson et al. 1984). Since binding sites for auxin are located also on the endoplasmic reticulum (Ray 1977), this membrane could harbor the other NPA binding site(s) (Fig. 5). The assay is based on the specific binding of radioactive NPA to the isolated membrane vesicles. A blank is run where the NPA binding to high affinity sites is competed out with non-labelled NPA. The pH optimum is relatively low (ca. pH 4; Uemura and Yoshida 1983). Michalke (1982) has suggested that the low pH optimum is due to the requirement for NPA to be uncharged and membrane permeable to reach its binding site on the inner surface of right side-out plasma membrane vesicles. The location of the NPA binding site(s) on the plasma membrane is, however, unknown.

2.6 Others

2.6.1 Blue Light-Induced Absorbance Change (LIAC)

A blue light-induced absorbance change (LIAC) reflecting the reduction of b-type cytochrome was used as a marker for the plasma membrane in the papers originally outlining the usefulness of aqueous two-phase partitioning for the isolation of plant plasma membranes (Lundborg et al. 1981; Widell and Larsson 1981; Widell et al. 1982). The usefulness of this marker relied largely on the fact that the plasma membranes could be separated from all intracellular membranes by counter current distribution (CCD) resulting in baseline-resolved

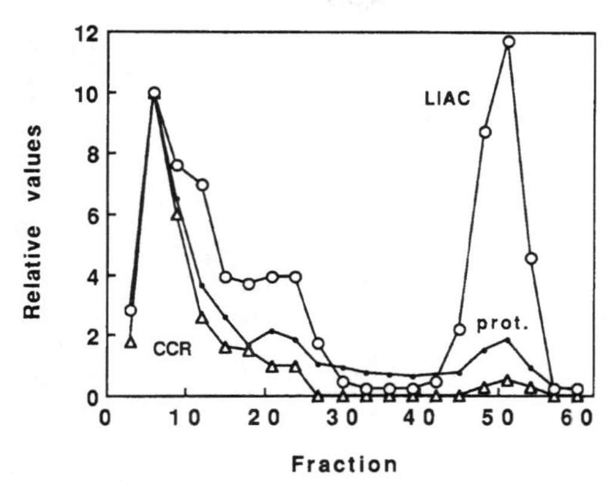

Fig. 6. Distribution of light-reducible cytochrome b (*LIAC*, ○), NADH-cytochrome c reductase (*CCR*, △), and protein (*prot.*, •) after countercurrent distribution of a wheat root microsomal fraction (cf. Figs. 2 and 4). The peak of LIAC around fraction *50* is due to plasma membrane. Note that a minor peak of CCR activity (well resolved from the main peak) is recovered with the plasma membranes (After Lundborg et al. 1981)

peaks of LIAC (Lundborg et al. 1981; Fig. 6). One of these (ca. 40% of total) coincided with material extremely distributed to the upper phase of the two-phase system. This was the expected behavior of right side-out plasma membrane vesicles based on earlier experience with plasma membrane-bounded multiorganelle complexes (Larsson et al. 1971; Larsson and Andersson 1978), protoplasts (Hallberg and Larsson 1981), and animal plasma membrane (López-Pérez et al. 1981). Thus, the high affinity of the plasma membrane vesicles for the upper phase was in fact used as an additional marker in these investigations. However, blue light-reducible cytochrome is also present in intracellular membranes, e.g., the endoplasmic reticulum, and the different types can only be distinguished by careful spectral analyses (Widell et al. 1983). LIAC is therefore not a useful plasma membrane marker in crude fractions.

2.6.2 5′-Nucleotidase

5′-Nucleotidase (EC 3.1.3.5) catalyzes the dephosphorylation of e.g. AMP, and is widely used as a marker for animal plasma membranes, although its biological function is unknown. This marker has only been used in a few plant systems. Flynn et al. (1987) could correlate 5′-nucleotidase activity to a membrane fraction distributed to the upper phase in a two-phase system, and which furthermore only contained membranes stained positively with PTA. These results were obtained with the diatom *Phaeodactylum tricornutum*, and it remains to be seen whether the activity is also present in the plasma membrane of higher plants. If so, this may become a very useful marker, since the assay (Avruch and Wallach 1971) is simple and very sensitive, and not least because this activity in animal plasma membrane is located on the outer surface of the membrane. It would therefore provide us with an enzyme marker for the apoplastic surface of the plant plasma membrane which at present is missing (Sect. 3).

2.6.3 Membrane Thickness

Membrane thickness may be used as a morphological marker for plasma membrane vesicles. In electron micrographs the plasma membrane appears as the thickest (ca. 10 nm) and best contrasted of all cellular membranes (Chap. 4).

3 Determination of Vesicle Sidedness

Isolated membranes appear for thermodynamic reasons as closed vesicles in aqueous solutions (Tanford 1973). This means that isolated plasma membranes appear as vesicles exposing either the apoplastic surface or the cytoplasmic surface to the bulk medium. The former may be termed apoplastic side-out, outside-out, right side-out or cytoplasmic side-in vesicles, and the latter cytoplasmic side-out or inside-out vesicles. Since most marker assays involve substrates (or ligands) that are water soluble and do not penetrate membranes, vesicle orientation will be of profound importance for the final outcome of an assay (Figs. 2 and 4). However, this problem can be used to our advantage, and the orientation of isolated vesicles can be determined by assaying for enzyme latency or ligand accessibility. Preparations highly enriched in vesicles of either sidedness are very important tools in studies on transport and membrane topology, and have just recently become available (Canut et al. 1988; Larsson et al. 1988; Chap. 3). Useful markers for either surface are listed in Table 1. These surface characteristics have been used to determine the orientation of isolated vesicles.

Table 1. Markers for plasma membrane surfaces

Marker	Representative references
Apoplastic surface	
Glycoproteins and glycolipids	Canut et al. 1988
	Monk et al. 1989
Cytoplasmic surface	
H^+-ATPase	Larsson et al. 1984
H^+ pumping	Larsson et al. 1988
1,3-β-Glucan synthase	Larsson et al. 1984
NAD(P)H-acceptor reductase	Askerlund et al. 1988; Buckhout and Hrubec 1986

3.1 Assays for Right Side-Out Vesicles

3.1.1 Concanavalin A Binding

So far only the carbohydrate moieties of glycoproteins and glycolipids have been used as markers for the apoplastic surface. Thus, right side-out vesicles are selectively stained by the lectin concanavalin A conjugated with peroxidase, and the proportion of such vesicles can thereby be determined by electron microscopy on thin sections (Canut et al. 1988; Chap. 4). Recently an assay has been designed where the amount of bound lectin is more directly determined from the peroxidase activity. The use of microtiter plates makes this a relatively rapid and handy assay. The values obtained correlate well with ATPase latency (see below) for yeast plasma membranes (Monk et al. 1989).

3.1.2 Enzyme Latency

More rapid assays for right side-out vesicles are based on the latencies of enzyme activities located on the cytoplasmic surface of the plasma membrane. These assays also give information on the amount of inside-out vesicles. The enzyme is assayed $\pm$ detergent and the percentage of latent activity calculated as:

$$\frac{\text{activity in the presence of detergent} - \text{activity in the absence of detergent}}{\text{activity in the presence of detergent}} \times 100$$

The percentage latent activity obtained will directly correspond to the percentage of sealed, right side-out vesicles, provided that the detergent used does not have any other effect than to make sealed vesicles permeable to the substrate(s).

3.1.2.1 H^+-ATPase

For the plasma membrane-bound H^+-ATPase, Triton X-100 at a concentration of 0.015–0.050% (w/v) is a suitable detergent for permeabilizing the membrane (Larsson et al. 1984). The optimal concentration of detergent depends on the amount of protein used in the assay; it also seems to vary with plant material, and should therefore be determined for each source of plasma membrane. In addition to exposing latent ATP binding sites, Triton also slightly inhibits the ATPase. The measured latency will therefore always be the sum of activation (through permeabilization of right side-out vesicles) and inactivation, and thus an underestimate of the proportion of right side-out vesicles. Lysophosphatidylcholine has therefore been proposed as an alternative to Triton, since it is not inhibitory (Grouzis et al. 1987). However, lysophosphatidylcholine has recently been shown to directly activate the ATPase by more than 100% at the concentrations employed in latency assays (Palmgren et al. 1988a,b; Palmgren and Sommarin 1989), and should therefore not be used in latency studies (for ATPase assays see Sects. 2.1.3 and 3.2.2).

3.1.2.2 1,3-β-Glucan Synthase
1,3-β-Glucan synthase is assayed ± 0.006–0.015% (w/v) digitonin (Sect. 2.2), the actual concentration being dependent not only on protein concentration and species, but also on the source of detergent. Excess digitonin is slightly inhibitory, whereas the concentrations optimal for activity probably stimulate also by some other mechanism than just permeabilizing the vesicles. Thus, 1,3-β-glucan synthase activity in preparations of inside-out vesicles from sugar beet leaves show a latency of ca. 50%, although other measures for sidedness suggest 20% contamination by right side-out vesicles (Palmgren et al. 1990a,b).

3.1.2.3 NADH-Acceptor Oxidoreductase
NADH-ferricyanide reductase measured ± ca. 0.025% (w/v) Triton X-100 is probably the most rapid and convenient assay for right side-out vesicles (Buckhout and Hrubec 1986; Askerlund et al. 1988). Triton is not inhibitory, at least not in moderate excess, but as for 1,3-β-glucan synthase the activity seems to be slightly stimulated (30–40%), leading to an overestimation of the amount of right side-out vesicles. Other electron acceptors besides ferricyanide can be used, e.g., cytochrome c and phenyl-p-benzoquinone (Askerlund et al. 1988). Electron transport to cytochrome c is inhibited by excess Triton, however, and the "unspecific" stimulation observed with ferricyanide is also found with cytochrome c and phenyl-p-benzoquinone at low Triton concentrations; thus these electron acceptors do not provide a better alternative.

Assay: The reduction of ferricyanide is measured as $\Delta (A_{420}-A_{500})$ using a dual-wavelength spectrophotometer operated in the dual-beam mode. The reaction mixture contains 20–50 μg protein, 25 mM HEPES-KOH, pH 7.3, 0.33 M sucrose, 1 mM $K_3[Fe(CN)_6]$, 0.4 μM antimycin A, 1 mM KCN, ± 0.025% (w/v) Triton X-100 in a total volume of 1 ml. The reaction is initiated by the addition of 0.25 mM NADH (final concentration). Correction should be made for non-enzymatic reduction of ferricyanide.

3.2 Assays for Inside-Out Vesicles

3.2.1 Non-Latent Enzyme Activities

Non-latent activities of H^+-ATPase, 1,3-β-glucan synthase, and NADH-acceptor oxidoreductase are useful markers for inside-out vesicles. A problem is that particularly the two latter activities seem to be stimulated by detergent also by some mechanism unrelated to membrane sidedness, leading to an underestimation of the proportion of inside-out vesicles.

3.2.2 ATP-Dependent H^+ Pumping

At least theoretically, non-latent activities of H^+-ATPase, 1,3-β-glucan synthase, and NADH-acceptor oxidoreductase may also be due to leaky right side-out vesicles or membrane sheets. Thus, a positive marker for sealed, inside-out vesicles is needed. ATP-dependent H^+-accumulation is such a marker (Sze 1985).

Recently an assay has been designed which permits the simultaneous measurement of non-latent ATPase activity and H^+ pumping (Palmgren and Sommarin 1989) exploiting the fact that ATPase activity cannot only be assayed as phosphate release (Sect. 2.1.3) but also as ADP release in a coupled enzyme assay (Nørby 1988). In the latter assay, hydrolyzed ATP is regenerated by the consumption of phosphoenolpyruvate mediated by pyruvate kinase (EC 2.7.1.40). The pyruvate formed is reduced to lactate by NADH-dependent lactate dehydrogenase (EC 1.1.1.27). Thus, the phosphorylation of ADP to ATP can be measured as NADH oxidation (absorbance decrease at 340 nm). This assay permits continuous ATPase determinations even on fractions in media containing, e.g., phosphate and Percoll. A further advantage is that H^+ pumping (assayed according to Vianello et al. 1982) and ATPase activity can be monitored simultaneously, by alternating between measuring the absorbance at 495 and 340 nm (Palmgren and Sommarin 1989). For this combined ATPase and H^+-pump assay, a spectrophotometer is needed, which can automatically move between the two wavelengths in several cycles (e.g., a Shimadzu UV-160).

The H^+-ATPase assay is performed in the following way (Palmgren and Sommarin 1989): 50–100 μg protein (in a total volume of 80 μl) is added to 850 μl of medium resulting in the following concentrations in the final assay system (1 ml): 10 mM MOPS-BTP, pH 7.0, 2 mM ATP-BTP, pH 7.0, 140 mM KCl, 1 mM EDTA-BTP, pH 7.0, 1 mM DTT, 20 μM acridine orange, 1 mg BSA (fatty acid free). Then, 5 μl valinomycin (1 mg ml^{-1} in ethanol) is added under vigorous mixing, followed by 42.5 μl substrate mixture (6 mM NADH, 24 mM phosphoenolpyruvate, 10 mM MOPS-BTP, pH 7.0) and 8.5 μl enzyme mixture (3 mg ml^{-1} puruvate kinase, 1.5 mg ml^{-1} lactate dehydrogenase, 10 mM MOPS-BTP, pH 7.0; enzymes in ammonium sulfate cannot be used since NH_4^+ acts as an uncoupler). The reaction mixture is incubated at 20°C for 5 min and then transferred to a cuvette. The spectrophotometer is set to measure the absorbance at 495 and 340 nm, in cycles (10-s interval, 18 cycles). After 1 min of registration, the reaction is started by the addition of 15 μl 272 mM $MgCl_2$ (final concentration 4 mM). The slopes (absorbance decreases at 495 and 340 nm) are then measures of H^+ pumping and ATPase activity, respectively. The H^+-pump activity, i.e., decrease in absorbance at 495 nm, of ca. 80% pure inside-out plasma membrane vesicles from sugar beet leaves in this assay system is ca. 1 absorbance unit (mg protein)$^{-1}$ min^{-1}, and the corresponding ATPase activity ca. 1 μmol NADH oxidized (ATP consumed) (mg protein)$^{-1}$ min^{-1}.

Acridine orange serves as Δ pH indicator, which accumulates in the vesicles where its absorbance is quenched when H^+ is pumped across the membrane. Valinomycin and the high concentration of KCl collapse the electrical gradient formed by H^+ pumping. BSA binds free fatty acids which can act as uncouplers in the reaction (Palmgren et al. 1990b) and DTT and EDTA keep SH groups intact (for composition of preparation media see Sect. 2.1.3). The coupled assay cannot be used in the presence of vanadate, since vanadate interferes with NADH absorbance. Furthermore, K^+ and Mg^{2+} are essential for the activity of the added pyruvate kinase, which means that these ions cannot be omitted. When NADH, phosphoenolpyruvate, pyruvate kinase, and lactate dehydrogenase are omitted, the same assay can be used to measure proton pumping only. The absorbance at 495 nm is then monitored continuously (cf. Chap. 7). Note that the ATPase activity (and not only H^+ pumping) with inside-out vesicles is inhibited already at very low Triton concentrations in the coupled assay (Palmgren et al. 1990a,b), in contrast to results with the phosphate release assay (Sect. 2.1.3) where a slight stimulation is observed using identical vesicle preparations.

3.2.3 Trypsin Inhibition

Trypsin inhibition of ATPase activity has recently been used as an alternative assay for inside-out vesicles (Gräf and Weiler 1989). The rationale is that trypsin will inhibit the ATPase activity of inside-out vesicles, whereas the activity associated with right side-out vesicles can be measured after the sequential addition of trypsin inhibitor and detergent. This procedure also permits the determination of minor amounts of right side-out vesicles in preparations of inside-out vesicles, and thus a more accurate determination of the proportion of inside-out vesicles. Similarly to the ATPase and in contrast to NADH-ferricyanide reductase, NADH-cytochrome c reductase is also sensitive to trypsin treatment. Since NADH-cytochrome c reductase is measured by a simple spectrophotometric assay, this is a convenient alternative (Fig. 7). The assay is the same as for NADH-ferricyanide reductase (Sect. 3.1.2.3) except that 40 μM cytochrome c (final concentration) is substituted for ferricyanide and the activity is recorded at 550 minus 600 nm.

3.2.4 Others

As mentioned above (Sect. 2.6.2) 5′-nucleotidase is a potentially useful marker for the apoplastic surface. If this activity is present also in the plasma membrane of higher plants and associated with the outer surface as in animals, latent activity would be a marker for sealed, inside-out vesicles and non-latent activity a marker for right side-out vesicles.

4 Markers for Contaminants

On homogenization of plant material all membranes to some degree form small vesicles. This is true also for membranes of large organelles, such as chloroplasts and mitochondria, even when all measures are taken to keep them intact. Thus, the type of microsomal fraction normally used as starting material for plasma membrane purification will be extremely diverse. Indeed, it will be a mixture of

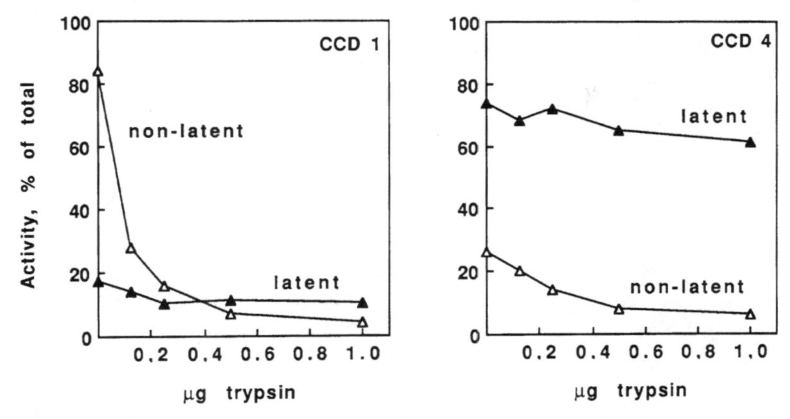

Fig. 7. The effect of trypsin on NADH-cytochrome c reductase activity of a fraction enriched in inside-out plasma membrane vesicles (*CCD 1*), and another fraction enriched in right side-out plasma membrane vesicles (*CCD 4*). In both fractions, the non-latent activity associated with inside-out vesicles is inhibited by the trypsin treatment, whereas the latent activity associated with right-side out vesicles is largely uneffected. This procedure permits a more exact determination of the polarity of the vesicles than enzyme latency alone (After Palmgren et al. 1990a)

vesicles derived from all the different membranes of the cell. Considering the relatively few parameters exploited for purification of membranes compared to those usually needed to obtain a pure protein, even the most pure plasma membranes are probably contaminated by vesicles of a number of other membranes. The detection and quantification of contamination are crucial for the interpretation of data, since a property ascribed to the plasma membrane may in fact be due to a minor but highly active contaminant. Unfortunately, reliable markers for possible contaminants are often missing as will be evident below.

4.1 Mitochondria

Cytochrome c oxidase (EC 1.9.3.1) is one of the few markers with only one location, i.e., on the inner mitochondrial membrane. It is often claimed that practically all mitochondrial membranes are pelleted (e.g., at 10,000 g for 10 min) before the microsomal membranes are collected. However, mitochondrial membranes still constitute the same proportion in microsomal fractions as the plasma membrane, and therefore cannot be neglected. Cytochrome c oxidase is assayed as the oxidation of reduced cytochrome c in a very reliable assay (Appelmans et al. 1955). To measure the reaction under optimal conditions, Triton X-100 (ca. 0.02%) should be included to overcome permeability barriers, and 50–100 mM KCl to screen membrane charges (Wigge and Gardeström 1987). Reducing agents such as DTT interfere strongly with the assay by re-reducing the substrate. Another useful marker for the mitochondrial inner membrane is succinate dehydrogenase (EC 1.3.99.1).

There is no good marker for the outer mitochondrial membrane available, but as many other membranes it contains an antimycin A-insensitive NADH-cytochrome c reductase (review, Møller and Lin 1986).

4.2 Plastids

Thylakoids from green plants contain chlorophyll and can thus be easily detected, even by eye. This is probably one reason why dark-grown material has been so popular for plasma membrane purification – the heavy contamination by other membranes is simply not visible! Prothylakoids from dark-grown plants can be detected by their carotenoid content (Goodwin 1955), although this is a property shared by the plastid envelope. The latter is a membrane not often mentioned in papers dealing with the separation and characterization of components in microsomal fractions. The plastid envelope is a double membrane of low density (similar to the smooth endoplasmic reticulum and the tonoplast), which can be distinguished from the plasma membrane by its content of carotenoids (Douce et al. 1973). More specific markers are: for the inner membrane, acyl-CoA thioesterase (Andrews and Keegstra 1983), and for the outer membrane, UDP-galactose: diacylglycerol galactosyltransferase (EC 2.4.1.46) (Joyard and Douce 1976; Cline and Andrews 1983); the latter is an enzyme involved in the synthesis of monogalactosyldiacylglycerol.

4.3 Endoplasmic Reticulum

The most commonly used marker for the endoplasmic reticulum is antimycin A-insensitive NAD(P)H-cytochrome c reductase. However, both NADH and NADPH activities are also found in other membranes (Fig. 8), e.g., the plasma membrane (Lundborg et al. 1981), the tonoplast (Barr et al. 1986), the mitochondrial outer membrane (Day and Wiskich 1975), and the glyoxysomal membrane (Donaldson et al. 1981). The specific activity of the plasma membrane dehydrogenase is five to ten times lower than that of the endoplasmic reticulum (S. Widell and M. Sommarin, unpublished results). It is even lower in isolated plasma membranes in the absence of detergent, since the binding site for both donor and acceptor is on the cytoplasmic side of the membrane (Askerlund et al. 1988) and most plasma membrane vesicles are oriented apoplastic side-out on isolation (Larsson et al. 1984). Thus, NAD(P)H-cytochrome c reductase may serve as a marker for the endoplasmic reticulum during preparation of endoplasmic reticulum, since the bulk activity is localized in this membrane, particularly when assayed without a detergent. However, it cannot be used to detect contamination of endoplasmic reticulum in plasma membrane fractions, since the plasma membrane also contains this activity (see Chap. 5).

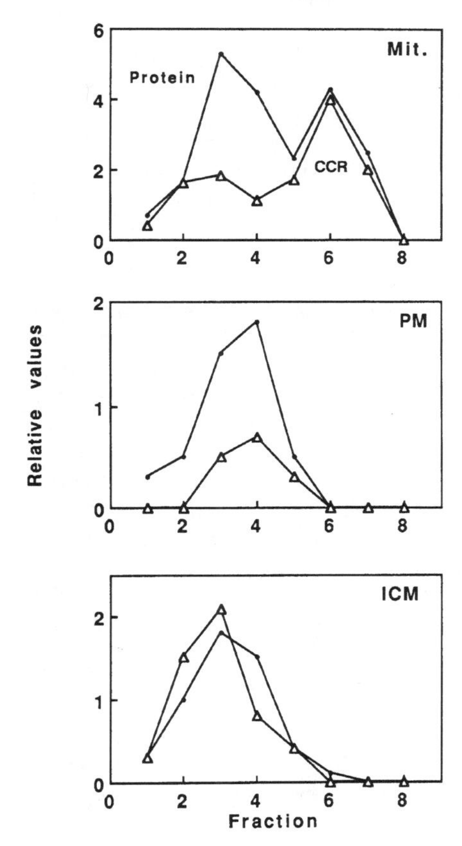

Fig. 8. The distribution of antimycin A-insensitive NADH cytochrome c reductase (*CCR*, △) and *protein* (•) after separation on a Percoll gradient (higher density to the *right*). Fractions from cauliflower inflorescences enriched in mitochondria (*Mit.*), plasma membranes (*PM*), and intracellular membranes (*ICM*), respectively, were used and the tubes centrifuged in parallel. Protein peaks at about the same fraction with all three materials (the high-density peak in the *top panel* corresponds to intact mitochondria), whereas CCR differs more in its distribution. This reflects the fact that CCR is a constituent of most membranes of the cell (After Widell and Larsson 1983)

Several enzymes engaged in the biosynthesis of phospholipids have their primary location in the endoplasmic reticulum, e.g., choline phosphotransferase (EC 2.7.8.2) (Lord et al. 1973). This enzyme catalyzes phosphatidylcholine synthesis from CDP-choline and diacylglycerol. However, this activity is also found in other membranes, such as the mitochondrial inner and outer membranes (Sparace and Moore 1981), the Golgi apparatus (Montague and Ray 1977), and possibly also the plasma membrane (P. Askerlund, unpublished results).

Binding of the auxin analogue naphthylacetic acid (NAA) is a relatively useful marker for the endoplasmic reticulum, and correlates well with NAD(P)H-cytochrome c reductase on sucrose gradients (Fig. 5). However, there are also binding sites for auxin on the plasma membrane, and auxin binding therefore cannot be used to determine the contamination of plasma membrane preparations by endoplasmic reticulum.

4.4 Tonoplasts

An H^+-translocating, Cl^--stimulated, NO_3^--inhibited, Mg^{2+}-dependent ATPase resides in the tonoplast (Sze 1985), a membrane of much lower density than the plasma membrane (tonoplast, 1.10–1.12 g cm^{-3}; plasma membrane, 1.17–1.18 g cm^{-3}; Fig. 3). NO_3^--inhibited Mg^{2+}-ATPase activity is therefore commonly used as a marker for the tonoplast. The tonoplast ATPase is of the (F_0F_1)-type, and thus insensitive to vanadate, unlike the plasma membrane ATPase. Rather it shares some properties with the mitochondrial ATPase, which is also NO_3^- sensitive (Wang and Sze 1985). However, if azide is included in the assay the mitochondrial activity is largely suppressed while the tonoplast activity is unaffected. In crude preparations, NO_3^- inhibition is usually small due to the large amount of other phosphohydrolases, and therefore difficult to measure accurately. Similarly, NO_3^- inhibition in a plasma membrane-enriched fraction will be small compared to the plasma membrane ATPase activity, and therefore difficult to estimate. Inclusion of 0.1–0.2 mM vanadate in the assay should abolish most of the plasma membrane ATPase activity and increase the sensitivity of the assay. The same assay as for the vanadate-inhibited ATPase (Sect. 2.1.3) can be used, except that a pH of 7.5 is more optimal and the reaction is run ± 50 mM BTP-NO_3^- rather than ± vanadate (Sze 1985).

An H^+-translocating pyrophosphatase is also present in the tonoplast (Chanson et al. 1985; Rea and Poole 1985) and has been used as a marker, although pyrophosphatase activity is also associated e.g., with the plastid envelope inner membrane (McCarthy et al. 1984), a membrane of about the same density as the tonoplast.

Other possible markers are α-mannosidase (EC 3.2.1.24) and acid phosphatase, which are soluble vacuolar enzymes (Wagner 1985). They are useful markers for intact vacuoles, but since some activity should be trapped in cytoplasmic side-out vesicles of the tonoplast they might be useful markers also for the latter.

4.5 Golgi

The Golgi apparatus is a heterogeneous structure with membrane vesicles similar to both the endoplasmic reticulum and the plasma membrane. It is therefore not surprising that the enzyme profile is also heterogeneous. Golgi cisternae are characterized by fucosyl transferase (Camirand et al. 1987) and cisternae as well as secretory vesicles by 1,4-β-glucan synthase (Chanson et al. 1984).

1,4-β-Glucan synthase (glucan synthase I, GS I; Ray 1977) is a widely used marker for Golgi membranes. It is assayed similarly to the 1,3-β-glucan synthase of the plasma membrane (glucan synthase II, GS II; Sect. 2.2), the differences being a much lower UDPG concentration (μM vs mM) and the presence of mM $MgCl_2$ in the 1,4-β-glucan synthase assay (Van Der Woude et al. 1974).

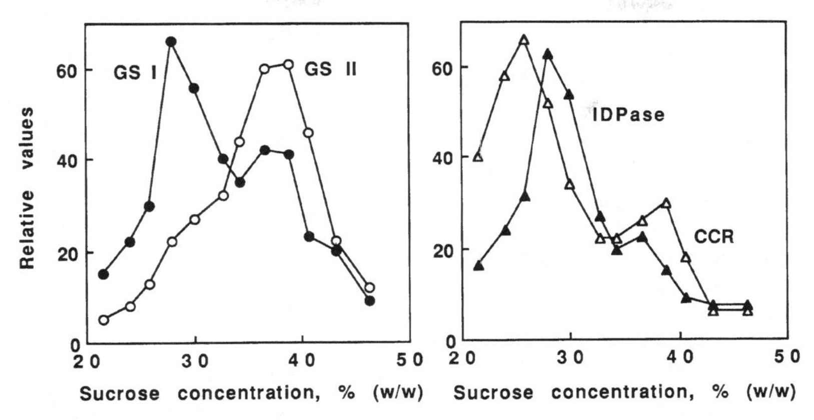

Fig. 9. The distribution of 1,4-β-glucan synthase (*GS I*, ●), 1,3-β-glucan synthase (*GS II*, ○), NADH-cytochrome c reductase (*CCR*, △), and *IDPase* (▲) after separation of a 1000 g supernatant from maize coleoptiles on a sucrose gradient. Note that GS I and II overlap due to the overlap of the assays. Note also that a minor peak of both IDPase and CCR activity coincides with the plasma membrane (GS II) peak (After Ray 1979)

1,4-β-Glucan synthase is furthermore not stimulated by Ca^{2+} or spermine, therefore these are omitted. A major problem is this similarity between the two assays, which makes it impossible to measure 1,4-β-glucan synthase activity without simultaneously measuring 1,3-β-glucan synthase activity and vice versa (Fig. 9). This overlap in the assays makes it particularly difficult to measure Golgi contamination in plasma membrane preparations, since the 1,3-β-glucan synthase activity is usually a 1000-fold higher than the 1,4-β-glucan synthase activity already in the microsomal fraction (e.g., Widell et al. 1982) and the proportions will be even more unfavorable in an enriched plasma membrane fraction. This large difference in activity is usually not evident in published tables and figures, due to the unfortunate habit of showing cpm values rather than true activities, and the fact that the labelled UDPG is diluted 1000-fold with cold UDPG in the 1,3-β-glucan synthase assay but used undiluted in the 1,4-β-glucan synthase assay.

Fucosyltransferase is a potentially useful marker for Golgi cisternae (Camirand et al. 1987). However, the activity is extremely low even in Golgi-enriched fractions, and the assay often relies on endogenous receptor only. In view of the low activity and the poor stimulation by added receptors, it may be questioned whether the enzyme is really measured under optimal conditions. Thus, the incorporation of labelled fucose may reflect concentration of receptor at the active site rather than the amount of enzyme in the present assay. This, in turn, may be due to the fact that detergent is not included in the assay. This reaction should show a high latency in analogy with the glycosylation steps in animal Golgi, where these reactions all occur on the lumenal side of the membrane (review, Doyle et al. 1988).

Another suggested Golgi marker is latent IDPase (or UDPase; Ray et al. 1969). The reaction has either been localized solely to secretory vesicles (Chanson et al. 1984) or to Golgi as a whole (Camirand et al. 1987). There is, however, convincing cytochemical evidence for NDPase activity associated also with the cytoplasmic surface of the plasma membrane (Chap. 4). This activity would be latent with right side-out plasma membrane vesicles, and thus indistinguishable from the Golgi activity. Latent NDPase is therefore not a useful Golgi marker in plasma membrane-enriched fractions. Indeed, latent UDPase has even been used as a plasma membrane marker and found to correlate with 1,3-β-glucan synthase, NPA binding, and UDPG:sterol glucosyltransferase on sucrose gradients (Hartmann-Bouillon and Benveniste 1978; M'Voula-Tsieri et al. 1981). This plasma membrane-bound NDPase activity could be due to the H^+-ATPase of the plasma membrane, since its nucleotide specificity is not absolute (Chap. 6). Another possible candidate is the Ca^{2+}-ATPase, which has a relatively low nucleotide specificity (Gräf and Weiler 1989).

4.6 Microbody and Nuclear Membranes

To date, no useful markers for either microbody or nuclear membranes have been presented. These membranes seem to share some properties with the endoplasmic reticulum, e.g., NAD(P)H-cyt c reductase (Donaldson et al. 1981). Possibly, soluble microbody enzymes, such as catalase or glycolate oxidase of leaf peroxisomes, which may become trapped within right side-out vesicles, can be used.

4.7 Soluble Enzymes

Soluble enzymes may become adsorbed to membrane surfaces, or trapped within vesicles. This is particularly evident with plasma membranes isolated from green leaves, which always contain significant amounts of the very abundant, chloroplast stromal enzyme, ribulose-1,5-bisphosphate carboxylase (EC 4.1.1.39), easily identified after SDS-polyacrylamide gel electrophoresis by its typical small and large subunit (molecular weights of ca. 14 and 54 kD, respectively). Any well-characterized soluble enzyme should be suitable as a marker for this type of contamination. Phosphoenolpyruvate carboxylase (EC 4.1.1.31) is often used as a marker for contaminating cytoplasm in, e.g., chloroplast preparations (e.g., Hallberg and Larsson 1983), and ribulose-1,5-bisphosphate carboxylase similarly as a marker for chloroplast stroma. NADH- and NADPH-glyceraldehyde-3-phosphate dehydrogenase, activities found both in the cytoplasm and the stroma, were recently used as markers for contaminating soluble enzymes in plasma membranes isolated from leaves of *Commelina communis* (Gräf and Weiler 1989).

5 Concluding Remarks

When the plasma membrane markers above are critically evaluated, 1,3-β-glucan synthase and PTA/STA staining stand out as the markers to use. It is also evident that vanadate-inhibited ATPase should be used with great care and never alone as a marker for the plasma membrane. Furthermore, a minor contamination by intracellular membranes in plasma membrane fractions is not easily determined. The problem being that most markers for intracellular membranes are not absolute markers but are also present in the plasma membrane, or are assayed in a way which overlaps with assays of plasma membrane activities. This makes PTA/STA staining an indispensible tool, since it gives an absolute value for the proportion of plasma membranes in a preparation. Since PTA/STA staining is a rather slow procedure, it is usually only used to evaluate the quality of the final plasma membrane fraction, whereas 1,3-β-glucan synthase is more useful for monitoring the enrichment of plasma membranes through the different steps of a preparation procedure.

During the preparation of plasma membranes, great care should be taken to preserve enzyme activities by inclusion of various protective agents in the media (Sect. 2.1.3; Chap. 3). It should be noted, however, that the requirements are different for different activities and sometimes contradictory. For example, the H^+-ATPase is stabilized by DTT, whereas DTT interferes with measurement of cytochrome c oxidase and NAD(P)H cytochrome c reductase. Such problems can be solved by resuspending the membranes in a minimal medium, dividing it into several aliquots, and then immediately supplementing as required. It may be wise to freeze some aliquots in liquid N_2, so that assays can always be performed on relatively fresh samples. Note, however, that assays for vesicle sidedness should always be performed on unfrozen material, since freezing and thawing changes the polarity (Larsson et al. 1988), particularly repeated freeze-thaw cycles (Palmgren et al. 1990a).

Finally, during the development of a preparation procedure, a balance sheet must be prepared presenting the distribution of total activities, protein and chlorophyll between all fractions as well as specific activities, so that both recovery and enrichment of different markers can be properly evaluated.

Acknowledgments. We would like to thank Drs. P. Askerlund, K. Fredrikson, I.M. Møller, M.G. Palmgren, and M. Sommarin (Departments of Plant Physiology and Plant Biochemistry, University of Lund, Sweden) for valuable discussions concerning this review. We are also indebted to a large number of colleagues elsewhere for providing us with relevant reprints and preprints. Work in the authors' laboratories was supported by grants from the Swedish Natural Science Research Council, the Swedish Council for Forestry and Agricultural Research, and the Carl Tesdorpf Foundation.

References

Ali MD, Akazawa T (1986) Association of H^+-translocating ATPase in the Golgi membrane system from suspension-cultured cells of sycamore (*Acer pseudoplatanus* L.). Plant Physiol 81:222–227

Andersson B, Anderson JM (1980) Lateral heterogeneity in the distribution of chlorophyll-protein complexes of the thylakoid membranes of spinach chloroplasts. Biochim Biophys Acta 593:427–440

Andrews J, Keegstra K (1983) Acyl-CoA synthetase is located in the outer membrane and acyl-CoA thioesterase in the inner membrane of pea chloroplast envelopes. Plant Physiol 72:735–740

Appelmans F, Wattiaux R, de Duve C (1955) Tissue fractionation studies. 5. The association of acid phosphatase with a special class of cytoplasmic granules in rat liver. Biochem J 59:438–455

Askerlund P, Larsson C, Widell S (1988) Localization of donor and acceptor sites of NADH dehydrogenase activities using inside-out and right-side-out plasma membrane vesicles from plants. FEBS Lett 239:23–28

Avruch J, Wallach DFH (1971) Preparation and properties of plasma membrane and endoplasmic reticulum fragments from isolated rat fat cells. Biochim Biophys Acta 233:334–347

Baginski ES, Foa PP, Zak B (1967) Determination of phosphate: study of labile organic phosphate interference. Clin Chim Acta 15:155–158

Barr R, Sandelius AS, Crane FL, Morré DJ (1986) Redox reactions of tonoplast and plasma membranes isolated from soybean hypocotyls by free-flow electrophoresis. Biochim Biophys Acta 852:254–261

Bérczi A, Larsson C, Widell S, Møller IM (1989) On the presence of inside-out plasma membrane vesicles and vanadate-inhibited K^+ Mg^{2+}-ATPase in microsomal fractions from wheat and maize roots. Physiol Plant 77:12–19

Bowman BJ, Slayman CW (1979) The effects of vanadate on the plasma membrane ATPase of *Neurospora crassa*. J Biol Chem 254:2928–2934

Buckhout TJ, Hrubec TC (1986) Pyridine nucleotide-dependent ferricyanide reduction associated with isolated plasma membranes of maize (*Zea mays* L.) roots. Protoplasma 135:144–154

Bush DR, Sze H (1986) Calcium transport in tonoplast and endoplasmic reticulum vesicles isolated from cultured carrot cells. Plant Physiol 80:549–555

Camirand A, Brummell D, MacLachlan G (1987) Fucosylation of xyloglucan: localization of the transferase in dictyosomes of pea stem cells. Plant Physiol 84:753–756

Canut H, Brightman A, Boudet AM, Morré DJ (1988) Plasma membrane vesicles of opposite sidedness from soybean hypocotyls by preparative free-flow electrophoresis. Plant Physiol 86:631–637

Chanson A, McNaughton E, Taiz L (1984) Evidence for a KCl-stimulated, Mg^{2+}-ATPase on the Golgi of corn coleoptiles. Plant Physiol 76:498–507

Chanson A, Fichmann J, Spear D, Taiz L (1985) Pyrophospate-driven proton transport by microsomal membranes of corn coleoptiles. Plant Physiol 79:159–164

Cline K, Andrews J (1983) Galactosyltransferases involved in galactolipid biosynthesis are located in the outer membrane of pea chloroplast envelopes. Plant Physiol 71:366–372

Day DA, Wiskich JT (1975) Isolation and properties of the outer membrane of plant mitochondria. Arch Biochem Biophys 171:117–123

Donaldson RP, Tully RE, Young OA, Beevers H (1981) Organelle membranes from germinating castor bean endosperm. II. Enzymes, cytochromes, and permeability of the glyoxysome membrane. Plant Physiol 67:21–25

Douce R, Holz RB, Benson AA (1973) Isolation and properties of the envelope of spinach chloroplasts. J Biol Chem 248:7215–7222

Doyle D, Bujanover Y, Petell JK (1988) Plasma membrane: biogenesis and turnover. In: Arias IM, Jacoby WB, Popper H, Schachter D, Shafritz DA (eds) The liver and pathobiology, 2nd Edn. Raven, New York, pp 141–163

Flynn KJ, Öpik H, Syrett PJ (1987) The isolation of plasma membrane from the diatom *Phaeodactylum tricornutum* using an aqueous two-polymer phase system. J Gen Microbiol 133:93–101

Fredrikson K, Larsson C (1989) Activation of 1,3-β-glucan synthase by Ca^{2+}, spermine and cellobiose. – Localization of activator sites using inside-out plasma membrane vesicles. Physiol Plant 77:196–201
Gallagher SR, Leonard RT (1982) Effect of vanadate, molybdate, and azide on membrane-associated ATPase and soluble phosphatase activities of corn roots. Plant Physiol 70:1335–1340
Gibrat R, Grouzis J-P, Rigaud J, Galthier N, Grignon C (1989) Electrostatic analysis of effects of ions on the inhibition of corn root plasma membrane Mg^{2+}-ATPase by the bivalent orthovanadate. Biochim Biophys Acta 979:46–52
Goodwin TW (1955) Carotenoids. In: Paech K, Tracey MV (eds) Modern methods of plant analysis, vol 3, Springer, Berlin Göttingen Heidelberg, pp 272–311
Gräf P, Weiler EW (1989) ATP-driven Ca^{2+}-transport in sealed plasma membrane vesicles prepared by aqueous two-phase partitioning from leaves of *Commelina communis* L. Physiol Plant 75:469–478
Green JR (1983) The Golgi apparatus. In: Hall JL, Moore AL (eds) Isolation of membranes and organelles from plant cells. Academic Press, London, pp 135–152
Grouzis J-P, Gibrat R, Rigaud J, Grignon C (1987) Study of the sidedness and tightness to H^+ of corn root plasmalemma vesicles: preparation of a fraction enriched in inside-out vesicles. Biochim Biophys Acta 903:449–464
Hager A, Biber W (1984) Functional and regulatory properties of H^+-pumps at the tonoplast and plasma membranes of *Zea mays* coleoptiles. Z Naturforsch 39c:927–937
Hallberg M, Larsson C (1981) Compartmentation and export of $^{14}CO_2$ fixation products in mesophyll protoplasts from the C_4 plant *Digitaria sanguinalis*. Arch Biochem Biophys 208:121–130
Hallberg M, Larsson C (1983) Highly purified intact chloroplasts from mesophyll protoplasts of the C_4 plant *Digitaria sanguinalis*. Inhibition of the phosphoglycerate reduction by orthophosphate and by phosphoenolpuruvate. Physiol Plant 57:330–338
Harinasut P, Takabe T, Akazawa T, Tagaya M, Fukui T (1988) Characterization of an ATPase associated with the inner envelope membrane of amyloplasts from suspension-cultured cells of sycamore (*Acer pseudoplatanus* L.). Plant Physiol 88:119–124
Hartmann-Bouillon M-A, Benveniste P (1978) Sterol biosynthetic capability of purified membrane fractions from maize coleoptiles. Phytochemistry 17:1037–1042
Hendriks T (1977) Multiple location of K^+-ATPase in maize coleoptiles. Plant Sci Lett 9:351–363
Hodges TK, Leonard RT, Bracker CE, Keenan TW (1972) Purification of an ion-stimulated adenosine triphosphatase from plant roots: association with plasma membranes. Proc Natl Acad Sci USA 69:3307–3311
Jacobs M, Hertel R (1978) In vitro auxin binding to subcellular fractions from *Cucurbita* hypocotyls: in vitro evidence for an auxin transport carrier. Planta 142:1–10
Joyard J, Douce R (1976) Préparation et activités enzymatiques de l'enveloppe des chloroplastes d'Épinard. Physiol Vég 14:31–48
Kauss H, Jeblick W (1985) Activation by polyamines, polycations and ruthenium red of the Ca^{2+}-dependent glucan synthase from soybean cells. FEBS Lett 185:226–230
Kauss M, Köhle M, Jeblick W (1983) Proteolytic activation and stimulation by Ca^{2+} of glucan synthase II from soybean cells. FEBS Lett 158:84–88
Kjellbom P, Larsson C (1984) Preparation and polypeptide composition of chlorophyll-free plasma membranes from leaves of light-grown spinach and barley. Physiol Plant 62:501–509
Kylin A, Sommarin M (1986) ATPases and membrane properties in relation to ecological differences. In: Trewavas AJ (ed) Molecular and cellular aspects of calcium in plant development. Plenum, New York, pp 261–268
Larsson C, Andersson B (1978) Two-phase methods for chloroplasts, chloroplast elements and mitochondria. In: Reid E (ed) Plant organelles. Ellis Horwood, Chichester, pp 35–46
Larsson C, Collin C, Albertsson P-Å (1971) Characterization of three classes of chloroplasts obtained by counter-current distribution. Biochim Biophys Acta 245:425–438
Larsson C, Kjellbom P, Widell S, Lundborg T (1984) Sidedness of plant plasma membrane vesicles purified by partitioning in aqueous two-phase systems. FEBS Lett 171:271–276

Larsson C, Widell S, Sommarin M (1988) Inside-out plasma membrane vesicles of high purity obtained by aqueous polymer two-phase partitioning. FEBS Lett 229:289–292
Lembi CA, Morré DJ, St-Thomson D, Hertel R (1971) N-1-naphthylphthalamic-acid-binding activity of a plasma membrane-rich fraction from maize coleoptiles. Planta 99:37–45
López-Pérez MJ, París G, Larsson C (1981) Highly purified mitochondria from rat brain prepared by phase partition. Biochim Biophys Acta 635:359–368
Lord JM, Kagawa T, Moore TS, Beevers H (1973) Endoplasmic reticulum as the site of lecithin formation in castor bean endosperm. J Cell Biol 57:659–667
Lundborg T, Widell S, Larsson C (1981) Distribution of ATPases in wheat root membranes separated by phase partition. Physiol Plant 52:89–95
Macara IG (1980) Vanadium, an element in search of a role. Trends Biochem Sci 5:92–94
Maury WJ, Huber SC, Moreland DE (1981) Effects of magnesium on intact chloroplasts. II Cation specificity and involvement of the envelope ATPase in (sodium)potassium/proton exchange across the envelope. Plant Physiol 68:1257–1263
McCarthy DR, Selman BR (1986) Properties of a partially purified nucleoside triphosphatase (NTPase) from the chloroplast envelope of pea. Plant Physiol 80:908–912
McCarthy DR, Keegstra K, Selman BR (1984) Characterization and localization of the ATPase associated with pea chloroplast envelope membranes. Plant Physiol 76:584–588
Memon AR, Sommarin M, Kylin A (1987) Plasmalemma from the roots of cucumber: isolation by two-phase partitioning and characterization. Physiol Plant 69:237–243
Michalke W (1982) pH-shift dependent kinetics of NPA-binding in two particulate fractions from corn coleoptile homogenates. In: Marmé D, Marrè E, Hertel R (eds) Plasmalemma and tonoplast: their function in the plant cell. Elsevier Biomedical, Amsterdam, pp 129–135
Møller IM, Lin W (1986) Membrane-bound NAD(P)H dehydrogenases in higher plant cells. Annu Rev Plant Physiol 37:309–334
Monk BC, Montesinos C, Leonard K, Serrano R (1989) Sidedness of yeast plasma membrane vesicles and mechanisms of activation of the ATPase by detergents. Biochim Biophys Acta 981:226–234
Montague MJ, Ray PM (1977) Phospholipid-synthesizing enzymes associated with Golgi dictyosomes from pea tissue. Plant Physiol 59:225–230
M'Voula-Tsieri M, Hartmann-Bouillon MA, Benveniste P (1981) Properties of nucleosides diphosphatases in purified membrane fractions from maize coleoptiles. I. Study of latency. Plant Sci Lett 20:379–386
Nguyen TD, Miguel M, Dubacq J-P, Siegenthaler P-A (1987) Localization and some properties of a Mg^{2+}-dependent ATPase in the inner membrane of pea chloroplast envelopes. Plant Sci 50:57–63
Nørby JG (1988) Coupled assay of Na^{+}, K^{+}-ATPase activity. Methods Enzymol 156:116–119
Palmgren MG, Sommarin M (1989) Lysophosphatidylcholine stimulates ATP dependent proton accumulation in isolated oat root plasma membrane vesicles. Plant Physiol 90:1009–1014
Palmgren MG, Sommarin M, Ulvskov P, Jørgensen PL (1988a) Modulation of plasma membrane H^{+}-ATPase from oat roots by lysophosphatidylcholine, free fatty acids and phospholipase A_2. Physiol Plant 74:11–19
Palmgren MG, Sommarin M, Jørgensen PL (1988b) Substrate stabilization of lysophosphatidylcholine-solubilized plasma membrane H^{+}-ATPase from oat roots. Physiol Plant 74:20–25
Palmgren MG, Askerlund P, Fredrikson K, Widell S, Sommarin M, Larsson C (1990a) Sealed inside-out and right-side-out plasma membrane vesicles: – Optimal conditions for formation and separation. Plant Physiol, in press
Palmgren MG, Sommarin M, Ulvskov P, Larsson C (1990b) Effect of detergents on the H^{+}-ATPase activity of inside-out and right-side-out plasma membrane vesicles. Biochim Biophys Acta, in press
Quail PH (1979) Plant cell fractionation. Annu Rev Plant Physiol 30:425–484
Racusen RH (1988) Separation of dense, polysaccharide-containing vesicles from secreting, cultured oat cells. Characterization of a putative secretory vesicle fraction. Physiol Plant 74:752–762

Ray PM (1977) Auxin-binding sites of maize coleoptiles are localized on membranes of the endoplasmic reticulum. Plant Physiol 59:594–599

Ray PM (1979) Maize coleoptile cellular membranes bearing different types of glucan synthetase activity. In: Reid E (ed) Plant organelles. Ellis Horwood, Chichester, pp 135–146

Ray PM, Shininger TL, Ray MM (1969) Isolation of β-Glucan synthetase particles from plant cells and identification with Golgi membranes. Proc Natl Acad Sci USA 64:605–612

Rea PA, Poole RJ (1985) Proton-translocating inorganic pyrophosphatase in red beet (*Beta vulgaris* L) tonoplast vesicles. Plant Physiol 77:46–52

Robinson C, Larsson C, Buckhout TJ (1988) Identification of a calmodulin-stimulated (Ca^{2+} + Mg^{2+})-ATPase in a plasma membrane fraction isolated from maize (*Zea mays*) leaves. Physiol Plant 72:177–184

Roland J-C (1978) General preparation and staining of thin sections. In: Hall JL (ed) Electron microscopy and cytochemistry of plant cells. Elsevier, Amsterdam, pp 1–62

Roland J-C, Lembi CA, Morré DJ (1972) Phosphotungstic acid-chromic acid as a selective electron-dense stain for plasma membranes of plant cells. Stain Technol 47:195–200

Sommarin M, Lundborg T, Kylin A (1985) Comparison of K, MgATPases in purified plasmalemma from wheat and oat. – Substrate specificities and effects of pH, temperature and inhibitors. Physiol Plant 65:27–32

Sparace SA, Moore Jr TS (1981) Phospholipid metabolism in plant mitochondria. II Submitochondrial sites of synthesis of phosphatidylcholine and phosphatidylethanolamine. Plant Physiol 67:261–265

Sze H (1985) H^+-translocating ATPases: advances using membrane vesicles. Annu Rev Plant Physiol 36:175–208

Tanford CH (1973) The hydrophobic effect. Wiley, New York

Uemura M, Yoshida S (1983) Isolation and identification of plasma membrane from light-grown winter rye seedlings (*Secale cereale* L. cv. Puma). Plant Physiol 73:586–597

Van Der Woude WJ, Lembi CA, Morré DJ, Kindinger JI, Ordin L (1974) β-Glucan synthetases of plasma membrane and Golgi apparatus from onion stem. Plant Physiol 54:333–340

Vianello A, Dell'Antone P, Macrì F (1982) ATP-dependent and ionophore-induced proton translocation in pea stem microsomal vesicles. Biochim Biophys Acta 689:89–96

Wagner GJ (1985) Vacuoles. In: Linskens HF, Jackson JF (eds) Modern methods of plant analysis, new series, vol 1, Cell components. Springer Berlin Heidelberg New York, pp 105–133

Wang Y, Sze H (1985) Similarities and differences between the tonoplast-type and the mitochondrial H^+-ATPases of oat roots. J Biol Chem 260:10434–10443

Widell S, Larsson C (1981) Separation of presumptive plasma membranes from mitochondria by partition in an aqueous polymer two-phase system. Physiol Plant 51:368–374

Widell S, Larsson C (1983) Distribution of cytochrome b photoreductions mediated by endogenous photosensitizer or methylene blue in fractions from corn and cauliflower. Physiol Plant 57:196–202

Widell S, Lundborg T, Larsson C (1982) Plasma membranes from oats prepared by partition in an aqueous polymer two-phase system. On the use of light-induced cytochrome b reduction as a marker for the plasma membrane. Plant Physiol 70:1429–1435

Widell S, Caubergs RJ, Larsson C (1983) Spectral characterization of light-reducible cytochrome in a plasma membrane-enriched fraction and in other membranes from cauliflower inflorescences. Photochem Photobiol 38:95–98

Wigge B, Gardeström P (1987) The effects of different ionic-conditions on the activity of cytochrome c oxidase in purified plant mitochondria. In: Moore AL, Beechey RB (eds) Plant mitochondria. Structural, functional and physiological aspects. Plenum, New York, pp 127–130

Chapter 3 Plasma Membrane Isolation

A.S. SANDELIUS[1] and D.J. MORRÉ[2]

[1]Department of Plant Physiology, University of Göteborg, Carl Skottsbergs Gata 22, S-413 19 Göteborg, Sweden
[2]Department of Medicinal Chemistry, Purdue University, West Lafayette, IN 47907, USA

C. Larsson, I.M. Møller (Eds):
The Plant Plasma Membrane

1 Introduction

The plasma membrane exists as the interface between the cytosol and the environment in all living cells and is probably the most complex and differentiated membrane. The membrane differs between cells of different organs or tissues, and in several animal cell types, the plasma membrane consists of distinct domains of different structural, compositional, and functional properties. Microdomains of coated membranes and various categories of intercellular junctions also are present. Additionally, the cytosolic surface of the membrane differs from the environmental surface. Consequently, methods to isolate plasma membrane fractions from animal cells or organs are extremely varied; domain-specific isolation methods have also been developed (Evans 1987).

The diversity of plant plasma membranes is less well recorded, but differences between different cell types are expected. For example, on plant tissue sections, immunological studies have demonstrated that a protein binding the auxin transport inhibitor N-1-naphthylphthalamic acid (NPA) was localized predominantly in the plasma membrane of parenchyma cells sheathing the vascular bundle (pea stems; Jacobs and Gilbert 1983), while an auxin-binding protein of corn coleoptiles was localized mainly at theplasma membrane of the outer epidermal cells (Löbler and Klämbt 1985). Concerning microdomains, the NPA-binding protein of pea stems was restricted to the basal ends of the cells (Jacobs and Gilbert 1983) and coats have been demonstrated on regions of the cytoplasmic leaflet of plasma membranes of protoplasts isolated from suspension cultured tobacco cells (Robinson and Depta 1988). A lateral asymmetry of the plant plasma membrane has been observed by electron microscopy of photoinduced spinach leaves (Auderset et al. 1986).

Most protocols for plasma membrane isolation from plants utilize whole plant organs; usually leaves, roots, or stem segments are used. The presence of cell walls requires either mechanical disruption (razor-blade chopping, mortar and pestle, blender) or enzymatic digestion of cell walls and most procedures fragment the plasma membrane into numerous small vesicles. As a result, the vesicles may be derived not only from different cell types but also from different plasma membrane domains of the same cell. In addition, both cytoplasmic side-in and cytoplasmic side-out plasma membrane vesicles may be formed. The fact that techniques based on membrane surface properties in some cases can be used with only minor modifications on different plant materials [e.g., Kjellbom and Larsson 1984; Møller et al. 1984; Sommarin et al. 1985; Rochester

Abbreviations: BSA, bovine serum albumine; DTE, dithioerythritol; DTT, dithiothreitol; EDTA, ethylenediaminetetraacetic acid; EGTA, ethylene-bis(oxy-ethylenenitrilo)tetraacetic acid; HEPES, N-(2-hydroxyethyl)piperazine-N′-(2-ethanesulfonic acid); INT, *p*-iodonitrotetrazolium violet; MES, 2-(N-morpholino)ethanesulfonic acid; MOPS, 3-(N-morpholino)propanesulfonic acid; NPA, N-1-naphthylphthalamic acid; PMSF, phenylmethylsulfonyl fluoride; Tris, tris(hydroxymethyl)-aminomethane; TES, N-tris (hydroxymethyl)methyl-2-aminoethanesulfonic acid.

et al. 1987 (two-phase partition); Auderset et al. 1986; Sandelius et al. 1986 (free-flow electrophoresis)] reflects either a basic similarity of the plasma membrane from different sources, or that the homogenization technique and the subsequent isolation procedures favor isolation of a particular population of plasma membranes.

The two basic approaches to isolate plasma membranes from plants rely on intrinsic properties of the membrane for isolation and identification or on surface-labeling of isolated protoplasts prior to disruption and purification. This chapter will emphasize isolation of plasma membrane fractions after mechanical disruption of the tissue, with assessments of fraction purity relying on intrinsic markers (Chaps. 2 and 4). A brief description of the advantages and disadvantages of protoplast isolation prior to plasma membrane isolation also is included.

2 Isolation of Plasma Membrane Fractions After Mechanical Disruption of the Plant Tissue

To obtain a plasma membrane fraction from mechanically disrupted tissues, a plasma membrane-containing microsomal membrane fraction is isolated and used for isolation of plasma membrane fractions, utilizing methods based either on membrane density or on surface properties of the membranes.

2.1 Isolation of a Microsomal Membrane Fraction

2.1.1 Homogenization Medium

The minimum requirements of a homogenization medium is that it should contain an osmoticum to minimize swelling and rupture of organelles, a buffer at pH 7–8 to minimize activities of hydrolytic enzymes and to counteract the low pH of the vacuole, and means to control the level of free divalent metal cations. Two examples of commonly used homogenization media are: (1) 0.25 M sucrose, 3 mM EDTA, and 25 mM Tris-MES, pH 7.5 (Hodges and Leonard 1974) and (2) 0.30 M sucrose, 1 mM EDTA, and 10 mM Tris-HCl, pH 7.5 (Widell et al. 1982). Sucrose can be substituted with sorbitol or mannitol and other buffers, e.g., HEPES or MOPS (25–50 mM), may be preferred. Further additions necessary to maintain the integrity of the plasma membrane during the isolation procedure depend on the intended use for the isolated plasma membranes and can also differ among different plant tissues.

Agents which protect membrane proteins against degradation during the isolation procedure include sulfhydryl group protectants [mercaptoethanol, dithioerythritol (DTE), dithiothreitol (DTT), cysteine], protease inhibitors [phenylmethylsulfonyl fluoride (PMSF), leupeptin, chymostatin], and protease substrates [bovine serum albumin (BSA), casein]. Usually 1–5 mM DTE or DTT and up to ten times higher concentrations of mercaptoethanol are used. PMSF

(1–5 mM) is added at the onset of tissue disruption as a methanol or isopropanol solution with a final concentration of alcohol preferably less than 0.1% (v/v) and BSA normally makes up 0.1–0.5% (w/v) of the homogenization medium. None of the protease inhibitors are uniformly effective against proteases of all plants studied (Gallagher et al. 1986). PMSF alkylates a reactive group of serine-type proteases and sulfhydryl proteases (Ryan and Walker-Simmons 1981), and as with the sulfhydryl group protectants, it may also affect a desired enzyme activity of the membrane.

Inhibitors of different lipid-degrading activities are not often included in homogenization buffers, despite the fact that the lipid environment probably is of great importance for many of the plasma membrane-localized enzymes studied. Phospholipase D inhibitors include general enzyme inhibitors as KF, NaF (around 10 mM), and *p*-chloromercuribenzoate (0.1–1 mM), but these are not always effective on their own and could also disturb subsequent enzyme assays of the isolated membrane (Wilkinson et al. 1987). Phospholipase D activity requires Ca^{2+} (Heller 1978; Galliard 1980), as do the plant phospholipase C enzymes investigated (Irvine et al. 1980; Connett and Hanke 1986; Helsper et al. 1986; Melin et al. 1987; Pfaffmann et al. 1987), but 3 mM EDTA (Wilkinson et al. 1987) or 20 mM EDTA or EGTA (Scherer and Morré 1978a) was insufficient to protect against phospholipid degradation. However, both EDTA and EGTA (5 mM each) together with 10 mM NaF at pH 7.5 reduced phospholipase activities (Whitman and Travis 1985). It should in this context be noted that reagent grade sucrose contains traces of $CaCl_2$, as does reagent grade $MgCl_2$ and KCl. Phospholipid degradation could also be inhibited by 4% (w/v) each of choline and ethanolamine to the homogenization medium (Scherer and Morré 1978a) or by 25 mM phosphorylcholine (Wilkinson et al. 1987). Maintaining a high pH of the homogenization buffer (e.g., pH 7.5) also reduces phospholipase D activity (Whitman and Travis 1985). Defatted BSA adsorbs free fatty acids which would otherwise stimulate acyl hydrolase activities (Galliard 1971), and formation of lysophospholipids during membrane isolation can be markedly reduced by between 0.15 (Scherer 1984) and 10 mM nupercaine (Scherer and Morré 1978b). To inhibit phosphatidic acid phosphatase, glycerol-1-phosphate is added (10 mM, Scherer and Morré 1978a; 0.4 M, Scherer 1984).

Other additions include polyvinylpolypyrrolidone (usually 0.1–1%, w/v) and polyvinylpyrrolidone (up to 5%, w/v), which are added to adsorb phenolic compounds; antioxidants, e.g., ascorbate, potassium metabisulfite, and/or butylated hydroxytoluene (2,[6]-di-tert-butyl-*p*-cresol), which are added at the onset of tissue disruption; and KCl, salicylhydroxamic acid, glycerol, and dextran. It can be beneficial to include a substrate to protect a particular enzyme under investigation, and for this reason, divalent cations may need to be present in the isolation medium.

2.1.2 Tissue Collection

The manner in which the tissue is collected may influence the membrane isolation. However, this aspect has been little studied and is seldom reported in detail as part of published methods. We harvest, e.g., soybean hypocotyls, by cutting the segments desired into ice-cold distilled water. Common variations are to harvest into ice-cold homogenization medium or to put the harvested segments on moist paper towels on ice. Ideally, the harvest should be fast and avoid damaging temperatures and anaerobic conditions. The yield of plasma membrane-derived vesicles may be increased by a short vacuum infiltration of the harvested tissue prior to disruption (Kjellbom and Larsson 1984).

2.1.3 Mechanic Disruption of the Tissue

The tissue should be broken open by cutting rather than shearing. Hand chopping with a razor-blade is too time-consuming for large-scale isolations where a mechanized razor-blade chopper (Morré 1971), a Polytron, Ultraturrax, or a blender are preferred. The knife-holder of a blender can be rebuilt to use replaceable razor-blades (Kannangara et al. 1977). Knives or razor-blades should be sharpened or changed regularly.

The medium to tissue ratio should be between 2:1 and 4:1 (ml medium:g fresh weight). Higher ratios may prove beneficial in some cases, but the ratio of medium to tissue should be about 2:1 for best morphological preservation. When razor-blade chopping is used, the homogenizing medium is added gradually but when a Polytron, Ultraturrax, or blender is used, the entire amount of homogenization medium is added at the outset. For some purposes, the choice is to mash the tissue with a mortar and pestle after it has been chopped with a razor-blade. To maintain organelle integrity and thus facilitate their removal from the homogenate, the tissue should be mashed and not ground and no abrasive should be added.

The choice of homogenization method will affect not only the purity of the plasma membrane fraction isolated but also the vesicle size, intactness, and orientation, i.e., whether the cytoplasmic surface of the plasma membrane will face inwards or outwards. For example, compared with razor-blade chopping, use of mortar and pestle favors cytoplasmic side-out plasma membrane vesicles (Canut et al. 1988).

2.1.4 Isolation of a Microsomal Membrane Fraction from the Homogenate

When the tissue has been homogenized, the first step is to remove incompletely broken tissue fragments, cell wall fragments, and large debris by filtration. One or several layers of cotton gauze (cheesecloth), Miracloth (a porous, nonwoven cellulosic fabric, Chickopee Mills, New York, NY), or nylon cloth may be used. Too large pore size permits cell wall debris to pass through, while too small pore size facilitates breaking of plastids if the material is squeezed through the cloth.

The rationale for keeping plastids and mitochondria as intact as possible during the isolation procedure is that intact organelles are more readily separated from plasma membrane vesicles, e.g., by differential centrifugation, than are organelle fragments.

The next step is to remove as much as possible of the starch grains, nuclei, and unbroken plastids and mitochondria. Usually a centrifugation at 6000–10 000 g for 10–20 min is used. The greater of these centrifugation forces needed to pellet most of the unbroken mitochondria also pellets a large proportion of the plasma membrane-derived vesicles. If possible, a swinging bucket rotor should be used to minimize dissociation of the plastid outer membranes along the centrifuge tube wall. The speed and time chosen depends on the size and density of the majority of the plasma membrane vesicles obtained and may need to be optimized.

The pellet from the first differential centrifugation step is discarded and the membranes in the supernatant are concentrated and separated from soluble proteins, usually by centrifugation. It is often sufficient to use a centrifugation force of approximately 6000–7000 g hours (e.g., 6000–7000 g for 1 h or 20 000 g for 20 min), but higher forces (e.g., 50 000 g for 30 min) are often used. Too high speeds result in an overly compact pellet that is difficult to resuspend with the membrane morphology maintained intact. If subsequent separation steps follow, it is advisable to collect the membranes on a sucrose cushion to reduce the possibility that aggregations of different membranes are formed (Morré et al. 1965; Ray 1979).

The supernatant from the first centrifugation can also be subjected to gel filtration, e.g., through Sepharose 4B-CL (Jones 1980) or Sepharose 2B-CL (Scherer 1982) to separate the membranes, which will elute with the void volume, from soluble proteins. The gel-filtered membranes can be concentrated by centrifugation as above. The pellet or gel-filtered membranes will be referred to as the microsomal membrane fraction. The composition of the buffer in which the microsomal membrane fraction is suspended depends on the method chosen for plasma membrane isolation.

To isolate the plasma membrane vesicles from the microsomal membrane fraction, the principle of isolation can be based on membrane vesicle size and density or on membrane vesicle surface properties, or on a combination of both.

2.2 Plasma Membrane Isolation, Based on Vesicle Size and Density

Isolation of a plasma membrane fraction based on differences in size and density between membrane vesicles of plasma membrane origin and vesicles of different origin, involves centrifugation of the suspended microsomal membrane fraction through a continuous or discontinuous density gradient. Continuous gradients need to be centrifuged for 12–24 h to approach equilibrium for every membrane vesicle category. However, long centrifugation times are not always compatible with the preservation of membrane structure and function, and therefore, centrifugation times leading to separation, but not necessarily equilibrium conditions, more often are used. The results of continuous density gradient centrifugations are used to create step gradients, which are easier to prepare and require shorter centrifugation times to reach a reproducible separation.

Continuous density gradient separations are used not only to yield information needed to construct step gradients, but are of great value in assignment studies. Discontinuous density gradients are also utilized as part of isolation procedures, either to obtain a crude plasma membrane fraction prior to a second gradient centrifugation (Scherer and Fischer 1985) or as an additional cleaning step after a two-phase partition (Yoshida et al. 1986). Continuous or discontinuous density gradients alone have not yielded plasma membrane fractions exceeding 75–80% fraction purity. The most commonly used gradient material for plasma membrane isolation is sucrose. Other materials used include glycerol, renograffin, metrizamide, and Percoll.

2.2.1 Sucrose Gradients

Isolation of plasma membrane fractions from plants by sucrose gradient centrifugation to a large extent still relies on the protocol (with minor modifications) of Hodges and co-workers (Hodges et al. 1972; Hodges and Leonard 1974; Hodges and Mills 1986), which was developed to isolate a plasma membrane fraction from oat roots. Based on results from continuous gradient centrifugations, a step gradient was developed, which consisted of 4 ml of 45% (w/w) sucrose, followed by 6.4 ml each of 38, 34, 30, 25 and 20% (w/w) sucrose in 1 mM $MgSO_4$, and 1 mM Tris-MES, pH 7.2, with or without 2.5 mM DTT. A 13 000 g × 15 min – 80 000 g × 60 min microsomal membrane fraction was applied to the gradient, which was centrifuged in a swinging bucket rotor for 2 h at 80 000 g. Plasma membrane-enriched fractions were obtained at the 34/38 and 38/45% interfaces and in both fractions, staining of thin sections for electron microscopy with phosphotungstic acid/chromic acid showed that more than 75% of the membrane vesicles were of plasma membrane origin (Hodges et al. 1972). Similar fraction purities were obtained also with gradients of fewer steps, where the plasma membrane fraction was recovered at a 34/45% interface (Hodges et al. 1972; Hodges and Mills 1986).

In parallel, similar methods employing sucrose step gradients were used to isolate plasma membrane fractions from stem tissues of soybean (Hardin et al. 1972) and onion (Morré and VanDerWoude 1974). Again, fraction purity was assessed by specific staining of plasma membranes with phosphotungstic acid at low pH (Roland et al. 1972).

Concerning the applicability of their method to different plant materials, Hodges and Leonard (1974) reported that corn and barley root membranes were distributed on a discontinuous gradient in the same manner as the oat root membranes, but $MgSO_4$ had to be eliminated from the gradient when corn or barley membranes were separated as it caused aggregation also of other membranes at the interfaces where the plasma membrane vesicles collected. Often modifications of the composition of the homogenization medium and/or of the density gradient are necessary also when the same plant tissue is used, but in another laboratory (e.g., plasma membranes from barley roots; Hodges and Leonard 1974; Nagahashi et al. 1978; DuPont and Hurkman 1985).

2.2.2 Fraction Purity and Orientation of Vesicles Obtained from Sucrose Density Gradient Centrifugations

The purity of the obtained plasma membrane fraction is evaluated using biochemical and morphological markers (Chaps. 2 and 4). For descriptions of methods to evaluate intactness and sidedness of the plasma membrane vesicles, see also Sections 2.4.1 and 2.4.2.

Sucrose gradient centrifugation has been reported to yield fractions consisting of more than 75% plasma membrane-derived vesicles (Hodges et al. 1972). Few fractionation schemes include data on all possible contaminants, but mitochondria and Golgi-derived vesicles have been reported to make up a large proportion of the nonplasma membrane-derived vesicles (Hodges and Mills 1986). The presence of tonoplast-derived vesicles was not determined. In plasma membrane fractions isolated from soybean (Hardin et al. 1972) or onion (Morré and VanDerWoude 1974), morphological analyses showed that for both, the major contaminants were smooth membranes, approximately 70 nm thick, which did not stain with phosphotungstic acid at low pH and most likely were of tonoplast origin (Twohig et al. 1974).

Both cytoplasmic side-in and cytoplasmic side-out vesicles are present in plasma membrane fractions isolated by sucrose gradient centrifugation, as suggested from ATPase latency studies (Bérczi and Møller 1986; Hodges and Mills 1986; Grouzis et al. 1987). However, with plasma membrane fractions not exceeding 75–80% fraction purity, a nonlatent ATPase activity of a cytoplasmic side-out plasma membrane vesicle might easily be confused with an ATPase activity of the cytoplasmic surface of a contaminating membrane. Neither does lack of structural latency, e.g., of an ATPase, distinguish between tightly sealed cytoplasmic side-out and leaky cytoplasmic side-in plasma membrane vesicles.

2.2.3 Other Gradient Materials

As an alternative to sucrose, glycerol has been employed as a gradient material but applications to plasma membrane isolation are limited. Scherer (Scherer 1984; Scherer and Fisher 1985) reported a preparative procedure involving two consecutive centrifugation steps, a sucrose step gradient and an isopycnic glycerol gradient, to isolate a tonoplast vesicle fraction reduced in plasma membrane contamination.

On the premise that the high osmotic pressures of a sucrose gradient may have deleterious effects on various membrane properties, gradient materials which are osmotically inactive are often employed for membrane isolation from animal tissues and cells (Evans 1987). Examples of such materials are renograffin (=Urograffin), an iodinated derivative of benzoic acid; Metrizamide and Nycodenz (Nygaard), materials similar to renograffin but uncharged at physiological pH; and Percoll (Pharmacia, Sweden), a colloidal silica sol with a polyvinylpyrrolidone coating. Of these, only Percoll and renograffin have been used with plant membranes. Percoll has been used extensively for isolation of plant organelles but not for plasma membrane isolation (see Morré et al. 1987 for a review). A procedure based on 10–38% continuous or discontinuous renograffin gradients was developed by Boss and Ruesink (1979) for fractionation of a 1000 g supernatant of ruptured carrot protoplasts to yield a plasma membrane-enriched fraction and Ray (1979) used renograffin as gradient material to separate mitochondria and plasma membrane vesicles of maize coleoptiles.

2.3 Plasma Membrane Isolation, Based on Membrane Vesicle Surface Properties

Plasma membranes can be isolated from other membranes of a microsomal membrane fraction by utilizing their different surface properties. Several methods have been employed. Of these aqueous polymer two-phase partition has been most commonly used. Other techniques involve preparative free-flow electrophoresis and isoelectric focusing. For surface-labeling of isolated protoplasts, see Section 3.2.

2.3.1 Two-Phase Partition

The underlying principle of aqueous two-phase partition is that numerous water-soluble, high-molecular-weight polymers do not mix above a certain concentration, but will form separate phases, each composed of more than 85% water. This makes the polymer-containing phases suitable for biological material. Added membrane vesicles partition between the aqueous polymer phases according to differences in surface properties between membrane vesicles of different origins. The procedure was first developed by Albertsson (1958, 1986) and he and co-workers have adapted the procedure to the isolation of various plant membrane fractions (Albertsson et al. 1982; Albertsson 1986).

The procedure of how to isolate a plant plasma membrane fraction by aqueous polymer two-phase partition has been reviewed several times, as have the principles involved in adapting the technique to new plant materials (Larsson 1983, 1985; Hodges and Mills 1986; Larsson et al. 1987). The reader is referred to these papers for detailed descriptions of the procedures outlined here.

The two polymers used for plasma membrane isolation are Dextran T 500 (Pharmacia) and Polyethylene Glycol 3350 (Union Carbide, USA). They are usually added at the same concentrations, between 5.5 and 6.5% (w/w) of each. The system also contains a buffer, an osmoticum, and often monovalent salts and other additions. Usually 5 mM potassium phosphate, pH 7.8, is used to buffer the system but 10–15 mM of organic buffers have also been used, e.g., Tris-maleic acid (Yoshida et al. 1983; Flynn et al. 1987), Tris-MES (Clément et al. 1986), and TES-NaOH (Fink et al. 1987). Sucrose or sugar alcohols can be included in the system without altering the separatory properties. Inclusion of NaCl or KCl (up to 50 mM) will modify the properties of the phases. BSA or DTE/DTT do not seem to affect the system and may be included.

It is practical to prepare stock solutions of the phase system components and store in aliquots in a freezer. The stock solutions of dextran and polyethylene glycol are made on a weight basis, while sucrose, buffer, and salt stock solutions are made on a molar basis. Dextran T-500 is hygroscopic, and to make a precise stock solution, access to a polarimeter for determination of the optical rotation is necessary (for description of how to make the Dextran T-500 stock solution, see Larsson 1985). To ensure uniformity of the Dextran T-500 used in a series of experiments, a large batch should made up at one time and this becomes especially necessary if a polarimeter is not used to correct for water content of the dextran. Here, the stock solution is made assuming a Dextran T-500 water content of 5%. The molecular weight distribution differs between different lots of Dextran T-500. This means that the phase system composition that worked well to isolate a pure plasma membrane fraction using one lot of Dextran T-500 might not function if a new lot of Dextran T-500 is used. With every new lot of Dextran T-500, membrane separation in the phase system must be evaluated and necessary adjustments of the composition of the phase system made (see below).

A two-phase system is constituted on a weight basis. The size depends on how much membrane material is to be loaded on the system and how many washes will be required. As an example, Table 1 gives the amount of stock solutions to weigh in to create a 16 g sample system with two washes to isolate a plasma membrane fraction from soybean hypocotyl microsomes (Sandelius et al. 1987). In this example, the membrane fraction to be added to the sample

Table 1. Sample system and wash systems used to isolate a plasma membrane fraction from dark-grown soybean hypocotyls (Sandelius et al. 1987)[a]

Stock solution	Sample system	Wash system[b]
20% (w/w) Dextran T-500	5.12 g	5.12 g
40% (w/w) Polyethylene Glycol 3350	2.56 g	2.56 g
1.00 M Sucrose	4.00 ml	4.00 ml
200 mM Potassium phosphate, pH 6.8	375 μl	400 μl
Add distilled water to weight	15.00 g	16.00 g
Membranes, suspended in 5.00 mM potassium phosphate, pH 6.8	1.00 g	---

[a] The final composition of the systems are: 6.4% (w/w) Dextran T-500, 6.4% (w/w) Polyethylene Glycol 3350, 0.25 M sucrose, and 5.0 mM potassium phosphate, pH 6.8. For characteristics of the obtained plasma membrane fraction, see Table 3.
[b] Prepare two such systems.

system was suspended in 5 mM potassium phosphate, pH 6.8, only. The rationale behind leaving the osmoticum out at this step was based on the fact that two-phase partition is also used to separate intact organelles from plasma membrane-enclosed organelles (Larsson and Albertsson 1974). Therefore, the microsomal membrane fraction was suspended without osmoticum to aid swelling and rupture of the membranes to minimize the presence of plasma membrane-enclosed organelles in this fraction. If the suspension buffer had included sucrose, the weighing-in of the sample system should have been adjusted concerning the amount of included sucrose, e.g., if the membrane suspension buffer had contained 0.25 M sucrose, 3.75 instead of 4.00 ml 1.00 M sucrose should have been included (cf. Table 1).

Two-phase partition is temperature-dependent, so it is best to work consistently either in a cold room or from an ice-bucket. When the sample system with added microsomal membranes has reached the desired temperature, its contents are mixed vigorously by shaking the tube 30–40 times without allowing the temperature of the system to increase. It is essential that the contents of the tube do mix and that a cushion of dextran does not slide along the wall of the tube without getting mixed into the system. The system may be allowed to separate out into two phases at unit gravity, which can take several hours, depending on system size, amount of membranes, etc., or the separation can be accelerated by centrifugation, e.g., 250–1500 g for 3–10 min.

Ideally, plasma membrane-derived vesicles will partition preferentially to the polyethylene glycol-rich upper phase, while membrane vesicles originating from other membranes will partition to the lower phase or the interface. To purify the plasma membrane fraction of the upper phase, this phase is repartitioned against a fresh lower phase. This washing step can be repeated as many times as needed. If more than three washes are needed to obtain a $>$ 90% pure plasma membrane fraction, the system is either overloaded or not optimized. To increase recovery, a fresh upper phase is partitioned against the original lower phase. This second upper phase is then partitioned against the same lower washing phases as was used against the original upper phase. Figure 1 shows the basic principle of a two-phase separation. Before the procedure has become routine, the fresh upper and lower phases are derived from wash systems of the same size as the sample system (Table 1). When the composition of the two-phase system has been optimized, a scaled up two-phase wash system can be prepared in a separatory funnel, the system allowed to separate overnight at the appropriate temperature, and the two phases collected separately and used in the procedure (Albertsson et al. 1982).

The upper phase should be diluted three- to fivefold before pelleting the plasma membrane fraction by centrifugation. For assessments of membrane recoveries, the other phases obtained should also be centrifuged. To pellet membranes out of a dextran-containing solution, an eight- to tenfold dilution is advisable. If needed, the membranes can be further washed of polymers by a second centrifugation.

Polyethylene glycol has been reported to interfere with certain assays of enzyme activity, e.g., with determinations of cytochrome c oxidase and NADH-cytochrome c reductase (Wheeler and Boss 1987) and with the standard Lowry procedure (Albertsson et al. 1982). The latter can be overcome by using a modified Lowry including sodium dodecyl sulfate (Markwell et al. 1978) or by using a procedure based on Coomassie Brilliant Blue (e.g., Read and Northcote 1981).

When two-phase partition is used for a new plant material, a system from the literature that has been successfully employed on a similar plant material should be chosen as the point of reference. Table 2 lists examples of plant tissues and cells where two-phase partition has been used to isolate plasma membrane fractions. If the tested two-phase system yields a contaminated plasma membrane fraction, the system can be modified to move more of the contaminating membranes to the lower phase while maintaining the plasma membrane vesicles in the upper phase. Modifications include increasing the polymer concentrations and adding or increasing the concentration of NaCl/KCl.

The first step in determining optimal conditions is to partition the microsomal membrane fraction in a series of phase systems of increasing polymer concentrations. Usually a system size of 3–5 g is enough. After phase separation, the upper and lower phases are analyzed separately to determine the distribution of plasma membranes and other membranes between the phases. It is not necessary to obtain a pure plasma membrane fraction at this step, but the plasma membranes should preferentially partition to the upper phase while all contaminating membranes should partition preferentially to the lower phase. If this is achieved with a single partition, consecutive washes will result in a pure plasma membrane fraction. If increasing the polymer composition does not result in a purer plasma membrane fraction, a series of increasing KCl or NaCl concentrations should be tried. If, on the other hand, the plasma membrane fraction initially obtained was pure, but the yield too low, the polymer or the salt concentrations may have been too high and could be reduced.

Either of two approaches may be used: The "high polymers – low salt" systems which contain 6.2–6.5% (w/w) of each polymer and 0–5 mM KCl (e.g., Widell et al. 1982; Kjellbom and Larsson 1984; Sommarin et al. 1985; Sandstrom et al. 1987), or the "low polymers – high salt" systems which contain 5.6–6.0% (w/w) of each polymer and 30–50 mM KCl or NaCl (e.g., Yoshida et al. 1983; Uemura and Yoshida 1983; Brüggemann and Janiesch 1987; Flynn et al. 1987; Galtier et al. 1988). The same system often works for different organs from the same plant (Körner et al. 1985; Sandelius and Sommarin 1986) but to isolate plasma membranes from *Vicia faba* leaves, 6.3% (w/w) of each polymer and 4 mM KCl was used (together with buffer and osmoticum) to isolate the mesophyll plasma membranes, while isolation of plasma membranes from the leaf epidermal cells required lower polymer concentrations (5.6%, w/w) and higher salt concentrations (17.5 mM KCl) (Blum et al. 1988). To isolate a plasma membrane fraction from dark-grown soybean hypocotyls (Tables 1 and 3), it

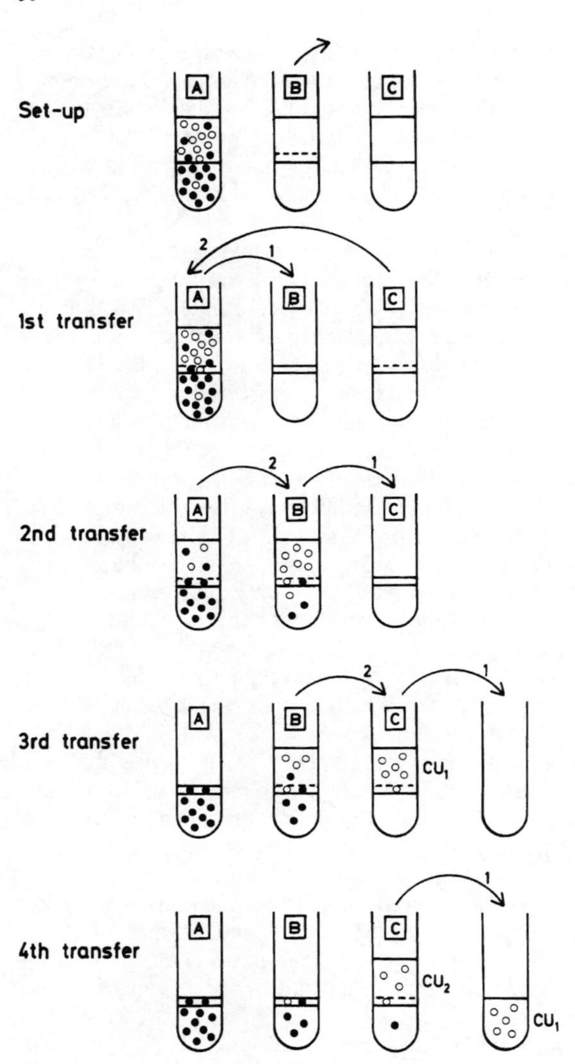
Set-up
A
B
C
1st transfer
2
1
2nd transfer
3rd transfer
4th transfer
CU$_1$
CU$_2$

Fig. 1

Table 2. Examples of plant tissues which have been used to isolate plasma membrane fractions by two-phase partition. If stated, fraction purity is given, together with method for quantitation

Plant source	Fraction purity	Reference
ANGIOSPERMS		
Monocotyledons		
Barley, light-grown		
Leaves	[a]	Kjellbom and Larsson 1984
	–	Körner et al. 1985
Roots	–	Körner et al. 1985
Barley, dark-grown		
Roots	–	Caldwell and Whitman 1987
Leek, dark-grown		
Seedlings	–	Moreau et al. 1988
Maize, light-grown		
Leaves	–	Robinson et al. 1988
Maize, dark-grown		
Coleoptiles	–	Widell and Larsson 1981
	[a]	Widell and Larsson 1983
Shoots	–	Clément et al. 1986
Roots	≥92%[b]	Buckhout and Hrubec 1986
	–	Galtier et al. 1988
Oat, light-grown		
Leaves	–	Larsson et al. 1987
Oat, dark-grown		
Roots	–	Widell et al. 1982
	–	Sandstrom et al. 1987

Fig. 1. Purification of plasma membranes by aqueous two-phase separation. *Set-up*: The microsomal membrane fraction, consisting of plasma membranes (○) and other membranes (●), is added to a preweighed sample system (labeled *A*) to give a phase system of the desired composition (cf. Table 1). The sample system (*A*) and the wash systems (*B* and *C*) are thoroughly mixed by 30–40 inversions and the phases allowed to separate at unit gravity or by a low speed centrifugation. Then 90% of the upper phase of tube *B* (*dashed line*) is removed and discarded. *First transfer*: 90% of the plasma membrane-enriched upper phase of tube *A* (*dashed line*) is transferred to the fresh lower phase in tube *B* and the system is thoroughly mixed with the purpose of increasing fraction purity. Ninety percent of the upper phase of the wash system *C* is transferred to tube *A* and the system is thoroughly mixed. The purpose here is to increase the yield of plasma membrane. The phases are allowed to separate. *Second transfer*: 90% of the upper phase of tube *B* is transferred to the fresh lower phase in tube *C* to further wash the plasma membrane fraction. Ninety percent of the plasma membrane-enriched second upper phase of tube *A* is transferred to tube *B* to be washed of contaminating membranes. Tubes *B* and *C* are thoroughly mixed and the phases are allowed to separate. *Third transfer*: 90% of the upper phase of tube *C*, containing washed plasma membranes, is collected (CU_1). Ninety percent of the plasma membrane-containing upper phase of tube *B* is transferred to tube *C* to be further washed of contaminating membranes. Tube *C* is mixed and the phases are allowed to separate. *Fourth transfer*: 90% of the upper phase of tube *C*, containing the second batch of washed plasma membranes, is collected (CU_2). Before pooling CU_1 and CU_2, it needs to be established that both fractions are of similar fraction purity. In all transfers, care should be taken not to disturb the interphase of the phase system. When the composition of a two-phase system has been worked out to yield the desired plasma membrane fraction, the fresh upper and lower phases of the wash systems are obtained from a bulk phase system prepared separately (After Larsson 1985)

Table 2. *Continued*

Plant source	Fraction purity	Reference
Orchard grass		
Leaves	[a]	Yoshida et al. 1983
	–	Yoshida and Uemura 1984
Rye, light-grown		
Seedlings	[a]	Uemura and Yoshida 1983
	–	Lynch and Steponkus 1987
Wheat, dark-grown		
Roots	–	Lundborg et al. 1981
	–	Sommarin et al. 1985
Dicotyledons		
Broadbean leaves		
Mesophyll	–	Blum et al. 1988
Epidermis	–	Blum et al. 1988
Carrot		
Suspension-cultured cells	–	Wheeler and Boss 1987
Cauliflower		
Inflorescences	–	Rochester et al. 1987
Cucumber, light-grown		
Roots	–	Memon et al. 1987
Jerusalem artichoke		
Tubers	–	Uemura and Yoshida 1986
Mullberry tree		
Bark	[a]	Yoshida 1984
Mungbean, dark-grown		
Hypocotyls	[a]	Yoshida et al. 1986
	–	Yoshida and Uemura 1986
Plantago sp.	–	Brüggemann and Janiesch 1987
	–	Staal et al. 1987
Soybean, dark-grown		
Hypocotyls	95%[b]	Sandelius et al. 1987 (see Table 3)
Suspension-cultured cells	–	Fink et al. 1987
Spinach, light-grown		
Leaves	–	Kjellbom and Larsson 1984
Sugar beet		
Suspension-cultured cells	95%[b]	Gaivoronskaya et al. 1987
Light-grown leaves	–	Gallet et al. 1989
Tomato		
Suspension-cultured cells	–	Grimes and Breidenbach 1987
Green fruits	–	Grimes and Breidenbach 1987
GYMNOSPERMS		
Spruce		
needles	[a]	Hellergren et al. 1983
ALGAE		
Chlamydomonas reinhardtii	–	Dolle 1988
Dunaliella salina	–	Einspahr et al. 1988
Phaeodactylum tricornutum	[a]	Flynn et al. 1987

– denotes only biochemical markers (positive and negative) assayed.

[a] Phosphotungstic acid/chromic acid or silicotungstic acid staining of sections for transmission electron microscopy. Qualitative data are presented.

[b] Quantitative data from phosphotungstic acid/chromic acid or silicotungstic acid staining.

Table 3. Specific activities of marker enzymes and morphological characteristics of a plant plasma membrane fraction, isolated from dark-grown soybean hypocotyls as described in Fig. 1 and Table 1[a]

Marker	Microsomal membrane fraction	Plasma membrane fraction	Recovery %
NADH-cytochrome c reductase[b]	5–10	0.4–2.8	0.2–1.0
NADPH-cytochrome c reductase[b]	1.1–1.8	0.3–0.5	2–4
Succinate-INT reductase[b]	1.1–3.3	0.00–0.02	0–0.1
Latent IDPase[b]	7–29	2–6	1–3
Class I membranes (11–12 nm)[c]	–	≥ 95	–
Reactivity with phosphotungstic acid at low pH[c]	–	≥ 95	–
Protein	–	–	3–7

[a] The results are compared to the corresponding activities of the parent microsomal membrane fraction and are also given as percentage recovery. The values presented are maximum and minimum values obtained from 3–5 independent experiments. The assays were performed as described in Sandelius et al. (1986).
[b] μmol (mg protein)$^{-1}$ h^{-1}.
[c] Profiles/100 profiles.

was necessary to decrease the pH of the two-phase system to below pH 7, as changing the polymer or salt concentrations did not result in a pure enough fraction above pH 7.

2.3.2 Fraction Purity and Vesicle Sidedness of Plasma Membrane Fractions Obtained by Two-Phase Partition

Fraction purities of around 95% can be obtained with carefully adapted two-phase systems, but quantitative determinations of the purity of the plasma membrane fractions are scarce (Table 2). Usually, only the presence of positive markers and the relative absence, compared to the microsomal fraction, of negative markers are reported.

The plasma membrane vesicles that are collected from the upper phase of a two-phase system are mainly or exclusively of a cytoplasmic side-in orientation as judged from assays of the plasma membrane ATPase with or without the addition of detergents to determine latency (Larsson et al. 1984; Bérczi and Møller 1986; Buckhout and Hrubec 1986; Hodges and Mills 1986; Galtier et al. 1988). Treating the membrane population with trypsin in the presence or absence of detergents also showed that most vesicles were of a cytoplasmic side-in orientation (Sandstrom et al. 1987).

However, Canut et al. (1987, 1988) observed that when microsomes of soybean hypocotyls were subjected to a two-phase separation, the upper phase, or rather the upper phase + interphase, contained plasma membrane vesicles of cytoplasmic side-out as well as of cytoplasmic side-in orientation. It was also observed that when a fraction of cytoplasmic side-out plasma membrane vesicles, isolated by preparative free-flow electrophoresis (Sect. 2.3.4), was applied to a two-phase system, the vesicles collected at the interphase rather than in the upper phase. The possibility to recover cytoplasmic side-out vesicles at the interphase should depend on the tendency of the plasma membrane of the tissue used to form vesicles of both orientations during tissue homogenization and on the descrimination of the two-phase system. Normally, only approximately 90% of the upper phase of a two-phase system is collected (see Fig. 1), and only cytoplasmic side-in plasma membrane vesicles are obtained. The tendency of cytoplasmic side-out vesicles to partition to the interphase of the two-phase system (cf. Canut et al. 1987, 1988) has been utilized to develop a method to isolate both membrane vesicle orientations separately by two-phase partition (Larsson et al. 1988).

2.3.3 Isolation of a Cytoplasmic Side-Out Plasma Membrane Fraction by Two-Phase Partition

The procedure of Larsson et al. (1988) initially involved the isolation of a plasma membrane fraction from sugar beet leaves using a two-phase system composed of 6.5% (w/w) each of Dextran T-500 and Polyethylene Glycol 3350, 0.33 M sucrose, 5 mM KCl, 1 mM DTT, and 5 mM potassium phosphate, pH 7.8 (Gallet et al. 1989). The plasma membranes obtained were suspended in 0.33 M sucrose, 1 mM DTT, 5 mM potassium phosphate, pH 7.8, and stored in liquid N_2. The low-temperature freezing and subsequent thawing of the plasma membrane vesicles were essential as this treatment increased the proportion of cytoplasmic side-out vesicles.

The thawed membranes were subjected to a modified two-phase procedure called countercurrent distribution (for a general description, see Albertsson et al. 1982). The final two-phase composition was as above, except that 6.2% (w/w) of each polymer was used. After the first mixing and phase separation, 90% of the upper phase was transferred to a fresh lower phase, and a fresh upper phase was added to the first tube. After the second mixing and separation, the upper phase of the second tube was transferred to a fresh lower phase in a third tube, the upper phase of tube 1 was transferred to tube 2 and a fresh upper phase added to the first tube. After mixing and separating, the procedure was repeated once again (the upper phase of tube 3 was transferred to a fresh lower phase in tube 4, etc.) to produce four tubes containing complete phase systems and the separated plasma membrane fraction. In tube 1, membranes with a low affinity for the polyethylene glycol-rich upper phase collected, while membrane vesicles with high affinity for the upper phase collected preferentially in tube 4. The contents of each tube were diluted ten-fold and the membranes collected by centrifugation. When proton pumping (decrease in acridine orange absorbance) and latency of K^+,Mg^{2+}-ATPase was monitored, the results showed that approximately 90% of the membrane vesicles in tube 1 were cytoplasmic side-out and that approximately 90% of the membrane vesicles in tube 4 were cytoplasmic side-in (Larsson et al. 1988; Fig. 2).

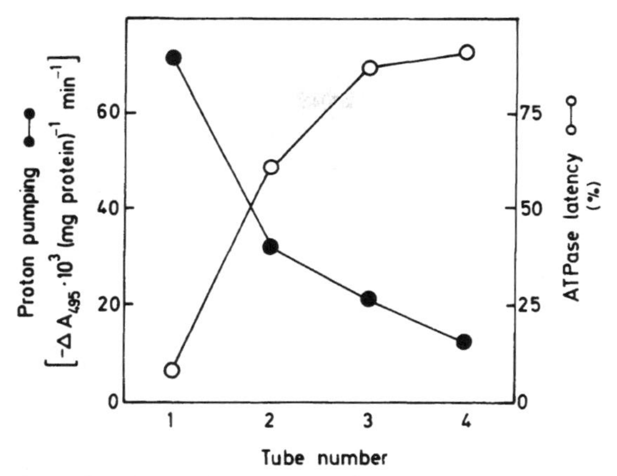

Fig. 2. Distribution of proton pumping (●) and latency of the K^+,Mg^{2+}-ATPase activity (○) after countercurrent distribution of a plasma membrane fraction obtained from leaves of sugar beet. Proton pumping was measured as the decrease in acridine orange absorbance and the latency of the ATPase was calculated as 100 × [(activity with 0.025% Triton X-100) – (activity without Triton X-100)]/(activity with 0.025% Triton X-100). The plasma membranes used as starting material showed an ATPase latency of 75% (Redrawn from Larsson et al. 1988; see Sect. 2.3.3 for a description of the method used)

2.3.4 Preparative Free-Flow Electrophoresis

In this technique, described by Hannig and Heidrich (1974, 1977), a mixture of components to be separated is continuously injected into a separation buffer moving through an electrical field (Fig. 3). The separation chamber is approximately 60 cm high, 10 cm wide, and 1 mm deep and there is no solid support. The chamber is lined by glass on front and back, and by the electrode gaskets on the vertical sides. The buffer inlet is at the top of the chamber and the rate of buffer flow is controlled by a peristaltic pump at the bottom of the chamber, where approximately 90 tubes collect the chamber outflow. The electrodes run the entire vertical length of the chamber and are continuously washed by an electrode buffer. The chamber, as well as the holder of the injection syringe (operated by a pump) and the fraction collector (test tube rack), are temperature controlled. Presently, there is one manufacturer of the equipment for preparative use (Bender and Hobein, Munich, FRG).

The first application of preparative free-flow electrophoresis to isolate plant plasma membranes utilized a microsomal membrane fraction (6000 g × 10 min – 60 000 g × 30 min) of soybean hypocotyls as starting material (Sandelius et al. 1986). The membranes were washed in free-flow chamber buffer (see below) as the presence of ions other than those of the chamber buffer may interfere with the separation (Hannig and Heidrich 1974). The washed membranes were suspended in chamber buffer, at approximately 1 ml per 5–10 mg of microsomal membrane protein, and were injected into the separation chamber. Injection of

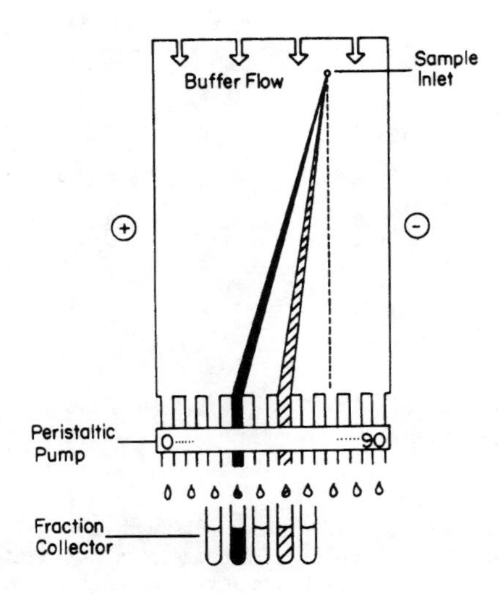

Fig. 3. Principles of preparative free-flow electrophoresis (Redrawn and reprinted with permission from Bender and Hobein, Munich, FRG; from Morré et al. 1984)

a sample in a buffer of density differing from that of the chamber buffer will cause the injected membranes to float up (too low density of injected sample) or to drop straight down (too high density of injected sample; see dashed line in Fig. 3). The free-flow electrophoretic conditions for isolating plasma membrane vesicles from soybean hypocotyls are as follows: the electrophoresis chamber buffer contains 0.25 M sucrose, 2.0 mM KCl, 10 μM $CaCl_2$, 10 mM triethanolamine, and 10 mM acetic acid, adjusted to pH 7.5 with NaOH. The $CaCl_2$ was added at a time when Ca^{2+} was shown to enhance survival of certain enzyme activities but can probably be omitted. However, exchanging $CaCl_2$ with $MgCl_2$ resulted in a somewhat narrower separation profile (A.S. Sandelius, K. Safranski, and D.J. Morré, unpublished results). The electrode buffer contains 100 mM triethanolamine and 100 mM acetic acid adjusted to pH 7.5 with NaOH. The separation is carried out with a preparative free-flow electrophoresis unit (VaP-22, Bender and Hobein, Munich, FRG, or equivalent) under conditions of constant voltage (800 V/9.2 cm field), 165 $\pm$ 10 mA, buffer flow 1.7 ml fraction^{-1} h^{-1}, sample injection 2.7 ml h^{-1} and a constant temperature of 6°C.

The distribution of membranes is monitored from the absorbance of the fractions at 280 nm (Figs. 4 and 5). The membranes are collected from individual or pooled fractions (A-E of Fig. 5a) by centrifugation for determination of protein content and other constituents or activities. The least electronegative fractions contained mainly plasma membrane-derived vesicles (Fig. 4). Slower injection times and lower density of the injected sample can increase the recovery of a plasma membrane fraction of high purity. For instance, fewer fractions can be combined to constitute the pooled E fraction (Fig. 5a) if a concentrated sample has been injected than if the separation has been slow with a diluted sample.

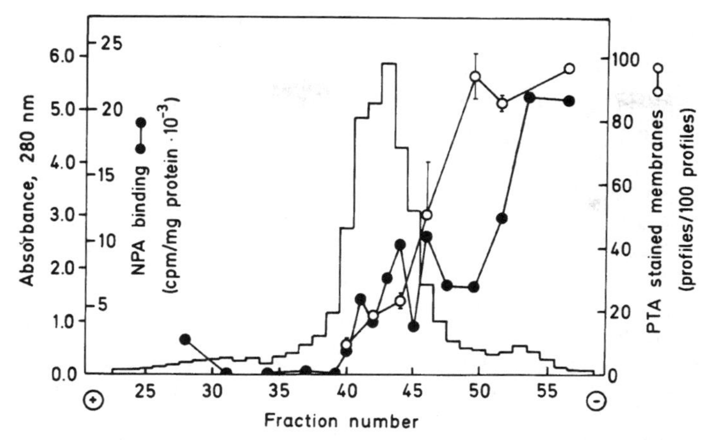

Fig. 4. The distribution of total membranes (measured as the absorbance at 280 nm; –) and the plasma membrane markers, binding of N-1-naphthylphthalamic acid (NPA binding; ●—●) and reactivity with phosphotungstic acid at low H (PTA staining; ○—○) of fractions from a free-flow electrophoretic separation of microsomal membranes of dark-grown soybean hypocotyls (Redrawn from Sandelius et al. 1986)

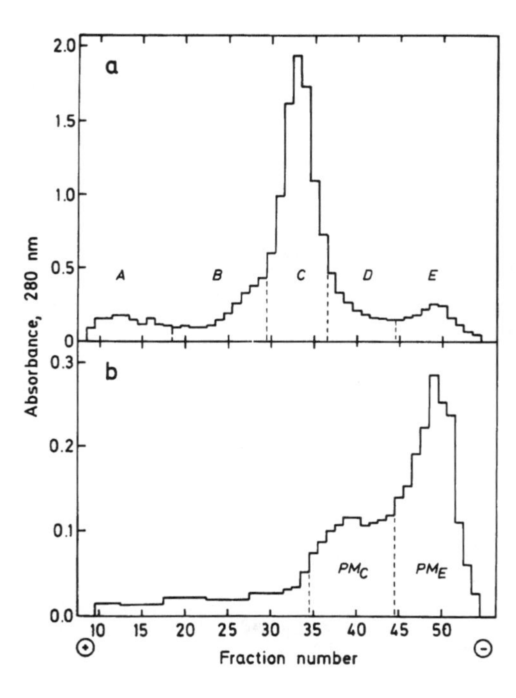

Fig. 5. The absorbance at 280 nm of fractions from free-flow electrophoretic separations of *a* a microsomal membrane fraction, and *b* a plasma membrane fraction obtained from the upper phase + interphase from a two-phase partition of a microsomal membrane fraction isolated from dark-grown soybean hypocotyls. In *a* the A_{280} profile was used as a basis for pooling the fractions into fractions *A*-*E* (cf. Table 4). In *b* the A_{280} was used as a basis to separate PM_C (cytoplasmic side-in plasma membrane vesicles) from PM_E (cytoplasmic side-out plasma membrane vesicles). Redrawn from Canut et al. (1988)

2.3.5 Fraction Purity and Orientation of the Vesicles Obtained from a Free-Flow Electrophoretic Separation

So far, the technique has been used to isolate plasma membranes from a handful of different plant tissues only, and the reported plasma membrane fraction purities of the E fraction (Fig. 5a) are 92/97% (morphometry/staining with phosphotungstic acid at low pH; Sandelius et al. 1986) or at least 89% (filipin-label; Scherer and vom Dorp 1988) for dark-grown soybean hypocotyls, 87/89% for photoinduced spinach leaves (morphometry/staining with phosphotungstic acid at low pH; Auderset et al. 1986), and at least 84% for cress roots (filipin-label; Scherer and vom Dorp 1988).

As evident from Table 4, plasma membrane-derived vesicles separate as two populations during the free-flow electrophoretic separation, one moving with various endomembranes in fraction C and one obtained as the more than 90% pure plasma membrane fraction E (Figs. 4 and 5; Sandelius et al. 1986). When both fractions were separately subjected to two-phase partition, a very pure plasma membrane fraction was recovered also from the C fraction (approximately 95% fraction purity; Canut et al. 1987, 1988). Structure-linked latency of the ATPase activities and concanavalin A-peroxidase labeling of the two plasma membrane fractions showed that the plasma membrane vesicles in the free-flow C fraction were of cytoplasmic side-in orientation, while those in the E fraction were of cytoplasmic side-out orientation (Canut et al. 1987, 1988). The observation that fraction C plasma membranes subjected to two-phase partition partitioned into the upper phase, while fraction E plasma membrane vesicles tended to stay close to the interphase during a two-phase partition

Table 4. Distribution in percent of recovered protein and plasma membrane markers between fractions A-E (see Fig. 5a) obtained from free-flow electrophoretic separations of microsomal membrane fractions, isolated from dark-grown soybean hypocotyls. (All data from Sandelius et al. 1986)[a]

Plasma membrane marker	Fraction				
	A	B	C	D	E
Sterol glycoside synthesis	9	14	21	7	50
Glucan synthetase II	<1	3	39	25	32
NPA binding	2	1	46	9	40
Class I membranes (11-12 nm)	<1	2	57	7	33
Phosphotungstic acid staining at low pH	<1	3	57	8	32
Protein	7	8	71	6	8

[a] The distribution of the morphological markers (reactivity of phosphotungstic acid at low pH and Class I membranes) was calculated from the percentage membrane in one fraction multiplied by the percentage morphological character in the same fraction and divided by the total (membrane percentage times morphological character percentage) for fractions A-E.

(Canut et al. 1987, 1988), provided a basis for the above described procedure for isolating a cytoplasmic side-out plasma membrane fraction with two-phase partition (Sect. 2.3.3).

2.3.6 Isoelectric Focusing

Isoelectric focusing, used widely to purify proteins on the basis of surface charge, has also been used to purify particles of very high molecular weight when employed together with a density gradient as an anticonvection medium. Griffing and Quatrano (1984) have applied this method to plasma membrane isolation from etiolated pea epicotyls.

2.4 The Isolated Plasma Membrane Fraction

To characterize the isolated plasma membrane fraction, biochemical and morphological markers are utilized to estimate fraction purity, and the intactness and sidedness of the majority of the vesicles can be determined as described below (see also Chaps. 2 and 4).

2.4.1 Separation of Tightly Sealed and Leaky Membrane Vesicles

Tightly sealed vesicles can be separated from leaky ones by their different buoyancy in high polymer density gradients (Steck 1974; Steck and Kant 1974). High-molecular-weight polymers do not penetrate membranes and they exert a low osmotic pressure relative to their density contribution (Steck 1974). Since the density of a tightly sealed vesicle is determined mainly by its intravesicular content, whereas the density of a leaky vesicle depends on its membrane density, sealed and leaky vesicles can be separated using a dextran density cushion of higher density than the medium in which the vesicles were formed. The sealed vesicles will not penetrate the cushion, while vesicles leaky to dextran will pellet. The vesicles which do not penetrate the dextran cushion appear to have higher transport activities than vesicles prepared by gradient centrifugation only (Sze 1985).

2.4.2 Determination of Membrane Vesicle Sidedness

The proportion of plasma membrane-derived vesicles with the cytoplasmic side facing inwards or outwards, respectively, is usually determined by the structural latency of the K^+, Mg^{2+}-ATPase of the cytoplasmic leaflet of the membrane. This has been assayed as the activity remaining after trypsin digestion of the vesicle population with or without detergent present (Grouzis et al. 1987; Sandstrom et al. 1987), but most often, the $(K^+)Mg^{2+}$-ATPase activity with or without detergent addition has been used for assignments of sidedness. Triton X-100 has been the detergent of choice and the data are usually interpreted as the result of the detergent opening up the membrane to give the enzyme free access to its substrate. However, Grouzis et al. (1987) showed that Triton X-100 inhibited the Mg^{2+}-ATPase of corn root plasma membrane vesicles compared with lysophosphatidylcholine. The latter detergent affected substrate-enzyme availability without altering the properties of the enzyme. Already the concentration of Triton X-100 which caused maximal stimulation (250 μM = 0.015% w/v) was shown to be inhibitory. Also De Michelis et al. (1988) reported effects of Triton X-100 on plasma membrane ATPase activities that could not be due to active site accessibility only. In this report, Triton X-100 showed maximal stimulation at 0.02% (v/v) and 0.04% was inhibitory compared to the control (De Michelis et al. 1988). The results probably reflect binding of Triton X-100 to proteins (McIntosh and Davidson 1984; Nicholson and McMurray 1986), whereas lysophosphatidylcholine interacts with lipid domains of the membranes (Tokumura et al. 1985). When Grouzis et al. (1987) incubated fractions of mainly cytoplasmic side-in plasma membrane vesicles with 0.1% (w/v) Triton X-100 and pelleted the membranes by centrifugation, a proportion of the vesicles now appeared to be cytoplasmic side-out, as determined by their ability to quench quinacrine fluorescence in the presence of ATP (De Michelis and Spanswick 1986; Grouzis et al. 1987). Galtier et al. (1988) suggested that the Triton X-100-washed membranes had lost their asymmetrical structure. Palmgren and Sommarin (1989) have shown that the H^+-ATPase activity of isolated plasma membrane vesicles is stimulated by added lysophosphatidylcholine. The various effects of detergents on plasma membrane ATP hydrolysis make interpretation of vesicle sidedness difficult and demonstrate the need for more than one assay to monitor vesicle sidedness. In addition to trypsin treatment (Sandstrom et al. 1987) and structure-linked latency based on detergent treatment (e.g., Larsson et al. 1984; Bérczi and Møller 1986; Hodges and Mills 1986), concanavalin A-peroxidase labeling has been used (Canut et al. 1988; Chap. 4). Plasma membrane asymmetry, as revealed by electron microscopy (Auderset et al. 1986), has also been applied to isolated membranes (Canut et al. 1987; Chap. 4).

2.4.3 Comparison Between Sucrose Gradient Centrifugation and Two-Phase Partition to Isolate a Plasma Membrane Fraction

Two reports deal with the comparison of these two methods: Bérczi and Møller (1986) and Hodges and Mills (1986). Both reports come to the same conclusion: the plasma membrane fractions obtained by two-phase partition are much purer than the plasma membrane fractions which can be obtained from a sucrose gradient centrifugation. Both reports concluded that the two-phase plasma membrane fraction consisted primarily of cytoplasmic side-in vesicles, while the fraction isolated by sucrose gradient centrifugations contained vesicles of both orientations. Again, it should be noted that for the less pure plasma membrane fractions obtained by sucrose gradient centrifugation, the non latent ATPase activities measured may reflect not only cytoplasmic side-out plasma membrane vesicles but also cytoplasmic side-out vesicles of other origin. In view of the recent findings concerning the effects of Triton X-100 (Sect. 2.4.2) and that sucrose has been reported to exert a slightly inhibitory effect on plasma membrane ATPase activities (Bérczi and Møller 1987; Grouzis et al. 1987), it is probably not possible to make a detailed comparison between the methods concerning percentage latency or leakiness versus tightness of the vesicles obtained. By comparing the identity of contaminating membranes, Hodges and Mills (1986) found that membranes of mitochondrial and endoplasmic reticulum origin were recovered to a higher degree in the sucrose gradient plasma membrane fraction than in the two-phase plasma membrane fraction. The fate of tonoplast-derived membrane vesicles was not determined for either method of isolation.

2.4.4 Comparison Between Two-Phase Partition and Preparative Free-Flow Electrophoresis to Isolate a Plasma Membrane Fraction

As evident from Sections 2.3.2 and 2.3.5, both two-phase partition and preparative free-flow electrophoresis can result in isolation of plasma membrane fractions of at least 90%, often 95%, plasma membrane-derived vesicles. The fraction obtained by stringent two-phase partition (CU_1+CU_2; Fig. 1) is composed mainly of cytoplasmic side-in vesicles, while the fraction E obtained by free-flow electrophoresis (Fig. 5a) is composed mainly of cytoplasmic side-out vesicles. When the methods are combined, plasma membrane fractions of both vesicle sidedness can be obtained separately (Fig. 5b; Canut et al. 1987, 1988). No direct comparison between the methods has been published, but one-dimensional sodium dodecylsulfate-polyacrylamide gel electrophoresis showed identical polypeptide patterns for plasma membrane fractions prepared from soybean hypocotyls by either method (Sandelius et al. 1986). The identity of the 5–10% contaminating membranes probably differs depending on the method. While a virtually chlorophyll-free plasma membrane fraction can be obtained from green spinach leaves by two-phase partition (Kjellbom and Larsson 1984), the plasma membrane fraction E from a free-flow electrophoretic separation of spinach microsomes was faintly green (Auderset et al. 1986).

3 Isolation of Plasma Membranes After Enzymatic Digestion of the Tissue

3.1 Isolation of Plant Protoplasts

Plant protoplasts may offer advantages for plasma membrane isolation when suspension-cultured cells are used or when a certain cell type is desired, e.g., mesophyll but not epidermal cells of a leaf. Protoplasts have been isolated from a wide range of plant tissues and cells (e.g., see Ruesink 1980; Vasil and Vasil 1980; Eriksson 1985; Nishimura et al. 1987) and only a general description of the method will be given here.

The plant material (e.g., cells, leaves with the lower epidermis removed, 1–2 mm thick tissue slices, epidermal strips) is incubated with microbial hydrolases to digest the cell walls. Usually, both a cellulase and a pectinase are necessary. The incubation medium also contains an osmoticum (usually 0.4–0.7 M sorbitol or mannitol) to prevent protoplast lysis and a buffer, usually at pH 5–6. Further additions may prove necessary to counteract the activities of contaminating enzymes (e.g., proteases, phosphatases) present in microbial hydrolases.

3.2 Isolation of Plasma Membrane Fractions from Isolated Protoplasts

Following lysis of the protoplasts, plasma membrane fractions may be isolated from the protoplast lysates by any of the techniques described for plasma membrane isolation from mechanically ruptured tissues. For example, aqueous polymer two-phase partition has been utilized to isolate a plasma membrane fraction from protoplasts of suspension-cultured carrot cells (Blowers et al. 1988). However, the initiation of cell wall digestion resulted in changes in plasma membrane protein phosphorylation (Blowers et al. 1988). Proteases present in the cell wall digesting hydrolase may also cause alterations in the plasma membrane surface, which would require phase systems of other compositions than those used for mechanically ruptured cells.

One advantage that protoplasts offer to plasma membrane isolation is that different labels can be applied to the protoplast surface (for a recent review, see Hall 1987). The labels can be used to identify the isolated plasma membrane fractions or they can be used to alter the properties of the plasma membrane to aid in the isolation procedure. Examples of labels that have been used to facilitate identification of plasma membrane fractions, isolated from protoplasts, are: ^{125}I-iodosulfanilic acid (soybean, Gailbraith and Northcote 1977; corn, Perlin and Spanswick 1980), ^{14}C-acetyl concanavalin A (carrot, Boss and Ruesink 1979), lanthanum ions (tobacco, Taylor and Hall 1979; soybean, Canut et al. 1987), and ^{125}I-labeled myeloma protein J539 (ryegrass, Schibeci et al. 1982; Schibeci 1985). The separation of the plasma membrane from the endomembranes may be facilitated by binding positively charged silica microbeads to the surfaces of intact protoplasts and collecting a plasma membrane-bead pellet after lysis of the protoplasts. This approach has been used successfully to isolate plasma membrane fractions from animal cells (Jacobson 1980; Chaney and Jacobson 1983) and yeast protoplasts (Schmidt et al. 1983). A magnetic isolation of plasma membrane vesicles from *Dictyostelium discoideum* was achieved after coating the cells with polyvinylamine-magnetite colloids (Patton et al. 1985). The silica-microbead technique has been applied to protoplasts isolated from etiolated *Pisum sativum* stems (Polonenko and Maclachlan 1984) and tomato cells (Grimes et al. 1986). None of these reports give information on the quantitative purity of the plasma membrane fractions obtained.

4 Concluding Remarks

Plasma membrane fractions of high yield and fraction purity are readily isolated by aqueous two-phase partition. The use of positive and negative enzyme markers are necessary to verify the qualitative purity of the isolated plasma membrane fraction, but it should be remembered that not all enzymatic markers for plasma membranes are absolute nor do they necessarily apply to all species and tissue types (Chap. 2). Quantitative assessment, however, still remains problematic. Morphological criteria for membrane identification, such as the staining of plasma membranes with phosphotungstic acid at low pH in sections for transmission electron microscopy, provide one basis for plasma membrane quantification and have a general applicability to a wide range of species, tissues, and isolation conditions (Chap. 4). The continuing practice of publishing results from putative plasma membrane fractions in the absence of either morphological and biochemical verification of fraction composition and purity (including negative markers), or both, is difficult to reconcile with the vast array of approaches now available for making these assessments.

The general applicability of aqueous polymer two-phase partition and preparative free-flow electrophoresis permits the isolation of plasma membrane fractions of high yield and fraction purity from virtually any plant source. As these separation techniques are based on surface properties of the membranes, they also permit the isolation of plasma membrane vesicles of defined absolute orientations, i.e., cytoplasmic side-in plasma membrane vesicles separately from cytoplasmic side-out plasma membrane vesicles. As several laboratories now have access to preparative free-flow electrophoresis equipment and as two-phase partition has no unusual equipment requirements, it can be expected that future plasma membrane research will benefit greatly from a general use of plasma membrane fractions of 90–95% fraction purity and of defined vesicle orientation.

Acknowledgments. The authors wish to acknowledge the Swedish Natural Science Research Council (ASS) and the National Institutes of Health (DJM) for financial support.

References

Albertsson P-Å (1958) Particle fractionation in liquid two-phase systems. The composition of some phase systems and the behaviour of some model particles in them. Application to the isolation of cell walls from microorganisms. Biochim Biophys Acta 27:378–395

Albertsson P-Å (1986) Partition of cell particles and macromolecules, 3rd edn. Wiley, New York

Albertsson P-Å, Andersson B, Larsson C, Åkerlund H-E (1982) Phase partition – a method for purification and analysis of cell organelles and membrane vesicles. In: Glick D (ed) Methods of biochemical analysis, vol 28, Wiley-Interscience, New York, pp 115–150

Auderset G, Sandelius AS, Penel C, Brightman A, Greppin H, Morré DJ (1986) Isolation of plasma membrane and tonoplast fractions from spinach leaves by preparative free-flow electrophoresis and effect of photoinduction. Physiol Plant 68:1–12

Bérczi A, Møller IM (1986) Comparison of the properties of plasmalemma vesicles purified from wheat roots by phase partitioning and by discontinuous sucrose gradient centrifugation. Physiol Plant 68:59–66

Bérczi A, Møller IM (1987) Mg^{2+}-ATPase activity in wheat root plasma membrane vesicles: time-dependence and effect of sucrose and detergents. Physiol Plant 70:583–589

Blowers DP, Boss WF, Trewavas AJ (1988) Rapid change in plasma membrane protein phosphorylation during initiation of cell wall digestion. Plant Physiol 86:505–509

Blum W, Key G, Weiler EW (1988) ATPase activity of plasmalemma-rich vesicles obtained by aqueous two-phase partitioning from *Vicia faba* mesophyll and epidermis: characterization and influence of abscisic acid and fusicoccin. Physiol Plant 72:279–287

Boss WF, Ruesink AW (1979) Isolation and characterization of concanavalin A-labeled plasma membranes of carrot protoplasts. Plant Physiol 64:1005–1011

Brüggemann W, Janiesch P (1987) Characterization of plasma membrane H^+-ATPase from salt-tolerant and salt-sensitive *Plantago* species. J Plant Physiol 130:395–411

Buckhout TJ, Hrubec TC (1986) Pyridine nucleotide-dependent ferricyanide reduction associated with isolated plasma membranes of maize (*Zea mays* L.) roots. Protoplasma 135:144–154

Caldwell CR, Whitman CE (1987) Temperature-induced protein conformational changes in barley root plasma membrane-enriched fraction. Plant Physiol 84:918–923

Canut H, Brightman AO, Boudet AM, Morré DJ (1987) Determination of sidedness of plasma membrane and tonoplast vesicles isolated from plant stems. In: Leaver C, Sze H (eds) Plant membranes: structure, function, biogenesis. Alan R Liss, New York, pp 141–159

Canut H, Brightman A, Boudet AM, Morré DJ (1988) Plasma membrane vesicles of opposite sidedness from soybean hypocotyls by preparative free-flow electrophoresis. Plant Physiol 86:631–637

Chaney LK, Jacobson BS (1983) Coating cells with colloidal silica for high yield isolation of plasma membrane sheets and identification of transmembrane proteins. J Biol Chem 258: 10062–10072

Clément J-D, Blein J-P, Rigaud J, Scalla R (1986) Characterization of ATPase from maize shoot plasma membrane prepared by partition in an aqueous polymer two phase system. Physiol Vég 24:25–35

Connett RJA, Hanke DE (1986) Breakdown of phosphatidylinositol in soybean callus. Planta 169:216–221

De Michelis MI, Spanswick RM (1986) H^+-pumping driven by the vanadate-sensitive ATPase in membrane vesicles from corn roots. Plant Physiol 81:542–547

De Michelis MI, Olivari C, Pugliarello MC, Rasi-Caldogno F (1988) Effect of Mg^{2+}, Triton X-100 and temperature on basal and FC-stimulated plasma membrane ATPase activity. Plant Sci 54:117–124

Dolle R (1988) Isolation of plasma membrane and binding of the Ca^{2+} antagonist nimodipine in *Chlamydomonas reinhardtii*. Physiol Plant 73:7–14

DuPont FM, Hurkman WJ (1985) Separation of the Mg^{2+}-ATPase from the Ca^{2+}-phophatase activity of microsomal membranes prepared from barley roots. Plant Physiol 77:857–862

Einspahr KJ, Peeler TC, Thompson GA Jr (1988) Rapid changes in polyphosphoinositide metabolism associated with the response of *Dunaliella salina* to hypoosmotic shock. J Biol Chem 263:5775–5779

Eriksson TR (1985) Protoplast isolation and culture. In: Fowke LC, Constabel F (eds) Plant protoplasts. CRC Press, Boca Raton, FL, pp 1–20

Evans WH (1987) Organelles and membranes of animal cells. In: Findlay JBC, Evans WH (eds) Biological membranes: a practical approach. IRL, Oxford, pp 1–35

Fink J, Jeblick W, Blaschek W, Kauss H (1987) Calcium ions and polyamines activate the plasma membrane-located 1,3-β-glucan synthase. Planta 171:130–135

Flynn KJ, Öpik H, Syrett PJ (1987) The isolation of plasma membrane from the diatom *Phaeodactylum tricornutum* using an aqueous two-polymer phase system. J Gen Microbiol 133:93–101

Gailbraith DW, Northcote DH (1977) The isolation of plasma membrane from protoplasts of soybean suspension cultures. J Cell Sci 24:295–310

Gaivoronskaya LM, Timonida VN, Trofimova MS, Dzyubenko VS (1987) Isolation and characterization of sealed vesicles of the plasma membrane from sugar-beet cell suspension culture. Fiziol Rast 34:13–27
Gallagher SR, Carroll EJ Jr, Leonard RT (1986) A sensitive diffusion plate assay for screening inhibitors of protease activity in plant cell fractions. Plant Physiol 81:869–874
Gallet O, Lemoine R, Larsson C, Delrot S (1989) The sucrose carrier of the plant plasma membrane. I. Differential affinity labeling. Biochim Biophys Acta 978:56–64
Galliard T (1971) Enzymic deacylation of lipids in plants. The effects of free fatty acids on the hydrolysis of phospholipids by the lipolytic acyl hydrolase of potato tubers. Eur J Biochem 21:90–98
Galliard T (1980) Degradation of acyl lipids: hydrolytic and oxidative enzymes. In: Stumpf PK (ed) The biochemistry of plants, Vol 4, Lipids: Structure and function, Academic Press, New York, pp 85–116
Galtier N, Belver A, Gibrat R, Grouzis J-P, Rigaud J, Grignon C (1988) Preparation of corn root plasmalemma with low Mg-ATPase latency and high electrogenic H^+ pumping activity after phase partitioning. Plant Physiol 87:491–497
Griffing LR, Quatrano RS (1984) Isoelectric focusing of plant cell membranes. Proc Natl Acad Sci USA 81:4804–4808
Grimes HD, Breidenbach RW (1987) Plant plasma membrane proteins. Immunological characterization of a major 75 kilodalton protein group. Plant Physiol 85:1048–1054
Grimes HD, Watanabe NM, Breidenbach RW (1986) Plasma membrane isolated with a defined orientation used to investigate protein topography. Biochim Biophys Acta 862:165–177
Grouzis J-P, Gibrat R, Rigaud J, Grignon C (1987) Study of the sidedness and tightness to H^+ of corn root plasmalemma vesicles: preparation of a fraction enriched in inside-out vesicles. Biochim Biophys Acta 903:449–464
Hall JL (1987) Possible approaches to surface labeling of the plasma membrane. Methods Enzymol 148:568–575
Hannig K, Heidrich H-G (1974) The use of continuous preparative free-flow electrophoresis for dissociating cell fractions and isolation of membranous components. Methods Enzymol 21:746–761
Hannig K, Heidrich H-G (1977) Continuous free-flow electrophoresis and its applications in biology. In: Bloemendal H (ed) Cell separation methods, Part IV. Elsevier/North Holland Biomedical, Amsterdam, pp 93–116
Hardin JW, Cherry JH, Morré DJ, Lembi CA (1972) Enhancement of RNA polymerase activity by a factor released by auxin from plasma membrane. Proc Natl Acad Sci USA 63:3146–3150
Heller M (1978) Phospholipase D. Adv Lipid Res 16:267–326
Hellergren J, Widell S, Lundborg T, Kylin A (1983) Frosthardiness development in *Pinus sylvestris*: the involvement of a K^+-stimulated Mg^{2+}-dependent ATPase from purified plasma membranes of pine. Physiol Plant 58:7–12
Helsper JPFG, de Groot PFM, Linskens F, Jackson JF (1986) Phosphatidylinositol phospholipase C activity in pollen of *Lilium longiflorum*. Phytochemistry 25:2053–2055
Hodges TK, Leonard RT (1974) Purification of a plasma membrane-bound adenosine triphosphatase from plant roots. Methods Enzymol 32:392–406
Hodges TK, Mills D (1986) Isolation of the plasma membrane. Methods Enzymol 118:41–54
Hodges TK, Leonard RT, Bracker CE, Keenan TW (1972) Purification of an ion-stimulated adenosine triphosphatase from plant roots: association with plasma membranes. Proc Natl Acad Sci USA 69:3307–3311
Irvine RF, Letcher AJ, Dawson RMC (1980) Phosphatidylinositol phosphodiesterase in higher plants. Biochem J 192:279–283
Jacobs M, Gilbert SF (1983) Basal localization of the presumptive auxin transport carrier in pea stem cells. Science 220:1297–1300
Jacobson BS (1980) Improved methods for isolation of plasma membrane on cationic beads. Biochim Biophys Acta 600:769–780
Jones RL (1980) The isolation of endoplasmic reticulum from barley aleurone layers. Planta 150:58–69

Kannangara CG, Gough SP, Hansen B, Rasmussen JN, Simpson DJ (1977) A homogenizer with replaceable razor blades for bulk isolation of active barley plastids. Carlsberg Res Commun 42:431–439

Kjellbom P, Larsson C (1984) Preparation and polypeptide composition of chlorophyll-free plasma membranes from leaves of light-grown spinach and barley. Physiol Plant 62:501–509

Körner LE, Kjellbom P, Larsson C, Møller IM (1985) Surface properties of right side-out plasma membrane vesicles isolated from barley roots and leaves. Plant Physiol 79:72–79

Larsson C (1983) Partition in aqueous polymer two-phase systems: a rapid method for separation of membrane particles according to their surface properties. In: Hall JL, Moore AL (eds) Isolation of membranes and organelles from plant cells. Academic Press, London, pp 277–309

Larsson C (1985) Plasma membranes. In: Linskens HF, Jackson JF (eds) Modern methods of plant analysis, New Series, Vol 1. Springer, Berlin Heidelberg New York Tokyo, pp 85–104

Larsson C, Albertsson P-Å (1974) Photosynthetic $^{14}CO_2$ fixation by chloroplast populations isolated by a polymer two-phase technique. Biochim Biophys Acta 357:412–419

Larsson C, Kjellbom P, Widell S, Lundborg T (1984) Sidedness of plasma membrane vesicles purified by partition in aqueous two-phase systems. FEBS Lett 171:271–276

Larsson C, Widell S, Kjellbom P (1987) Preparation of high-purity plasma membranes. Methods Enzymol 148:558–568

Larsson C, Widell S, Sommarin M (1988) Inside-out plant plasma membrane vesicles of high purity obtained by aqueous two-phase partitioning. FEBS Lett 229:289–292

Löbler M, Klämbt D (1985) Auxin-binding protein from coleoptile membranes of corn (*Zea mays* L.). II. Localization of a putative auxin receptor. J Biol Chem 260:9854–9859

Lundborg T, Widell S, Larsson C (1981) Distribution of ATPases in wheat root membranes separated by phase partition. Physiol Plant 52:89–95

Lynch DV, Steponkus PL (1987) Plasma membrane lipid alterations associated with cold acclimation of winter rye seedlings (*Secale cereale* L. cv. Puma). Plant Physiol 83:761–767

Markwell MAK, Haas JM, Beiber LL, Tolbert NE (1978) A modification of the Lowry procedure to simplify protein determination in membrane and liposome samples. Anal Biochem 87:206–210

McIntosh DB, Davidson GA (1984) Effects of nonsolubilizing and solubilizing concentrations of Triton X-100 on Ca^{2+} binding and Ca^{2+}-ATPase activity of sarcoplasmic reticulum. Biochemistry 23:1959–1965

Melin P-M, Sommarin M, Sandelius AS, Jergil B (1987) Identification of Ca^{2+}-stimulated polyphosphoinositide phospholipase C in isolated plant plasma membranes. FEBS Lett 232:87–91

Memon AR, Sommarin M, Kylin A (1987) Plasmalemma from the roots of cucumber: isolation by two-phase partitioning and characterization. Physiol Plant 69:237–243

Møller IM, Lundborg T, Bérczi A (1984) The negative surface charge density of plasmalemma vesicles from wheat and oat roots. FEBS Lett 167:181–185

Moreau P, Juguelin H, Lessire R, Cassagne C (1988) Plasma membrane biogenesis in higher plants: in vivo transfer of lipids to the plasma membrane. Phytochemistry 27:1631–1638

Morré DJ (1971) Isolation of Golgi apparatus. Methods Enzymol 22:130–148

Morré DJ, VanDerWoude WJ (1974) Origin and growth of cell surface components. In: Hay ED, King TJ, Papaconstantinou J (eds) Macromolecules regulating growth and development, Academic Press, New York, pp 81–111

Morré DJ, Mollenhauer HH, Chambers JE (1965) Glutaraldehyde stabilization as an aid to Golgi apparatus isolation. Exp Cell Res 38:672–675

Morré DJ, Creek KE, Matyas GR, Minnifield N, Sun I, Baudoin P, Morré DM, Crane FL (1984) Free-flow electrophoresis for subfractionation of rat liver Golgi apparatus. BioTechniques 2:224–233

Morré DJ, Brightman AO, Sandelius AS (1987) Membrane fractions from plant cells. In: Findlay JBC, Evans WH (eds) Biological membranes: a practical approach. Alan R Liss, Oxford, pp 37–72

Nagahashi G, Leonard RT, Thomson WW (1978) Purification of plasma membranes from roots of barley. Specificity of the phosphotungstic acid – chromic acid stain. Plant Physiol 61:993–999

Nicholson DW, McMurray WC (1986) Triton solubilization of proteins from pig liver mitochondrial membranes. Biochim Biophys Acta 856:515–525

Nishimura M, Hara-Nishimura I, Akazawa T (1987) Preparation of protoplasts from plant tissues for organelle isolation. Methods Enzymol 148:27–34

Palmgren MG, Sommarin M (1989) Lysophosphatidylcholine stimulates ATP dependent proton accumulation in isolated oat root plasma membrane vesicles. Plant Physiol 90:1009–1014

Patton WF, Kim J, Jacobson BS (1985) Rapid, high-yield purification of cell surface membrane using colloidal magnetite coated with polyvinylamine: sedimentation versus magnetic isolation. Biochim Biophys Acta 816:83–92

Perlin DS, Spanswick RM (1980) Proton transport in plasma membrane vesicles. In: Spanswick RM, Lucas WJ, Dainty J (eds) Plant membrane transport: current conceptual issues. Elsevier/North-Holland Biomedical, Amsterdam, pp 529–530

Pfaffmann H, Hartmann E, Brightman AO, Morré DJ (1987) Phosphatidylinositol specific phosphalipase C of plant stems. Membrane associated activity concentrated in plasma membranes. Plant Physiol 85:1151–1155

Polonenko DR, Maclachlan GA (1984) Plasma-membrane sheets from pea protoplasts. J Exp Bot 35:1342–1349

Ray PM (1979) Maize coleoptile cellular membranes bearing different types of glucan synthetase activity. In: Reid E (ed) Plant organelles. Ellis Horwood, Chichester, pp 135–146

Read SM, Northcote DH (1981) Minimization of variation in the response to different proteins of the Coomassie Brilliant Blue G dye-binding assay for protein. Anal Biochem 116:53–64

Robinson C, Larsson C, Buckhout TJ (1988) Identificationof a calmodulin-stimulated (Ca^{2+} + Mg^{2+})-ATPase in a plasma membrane fraction isolated from maize (*Zea mays*) leaves. Physiol Plant 72:177–184

Robinson DG, Depta H (1988) Coated vesicles. Annu Rev Plant Physiol Plant Mol Biol 39:53–99

Rochester CP, Kjellbom P, Larsson C (1987) Lipid composition of plasma membranes from barley leaves and roots, spinach leaves and cauliflower inflorescences. Physiol Plant 71:257–263

Roland JC, Lembi CA, Morré DJ (1972) Phosphotungstic acid-chromic acid as a selective electron-dense stain for plasma membranes of plant cells. Stain Technol 47:195–200

Ruesink A (1980) Protoplasts of plant cells. Methods Enzymol 69:69–84

Ryan CA, Walker-Simmons M (1981) Plant proteinases. In: Stumpf PK, Conn EE (eds) The biochemistry of plants, vol 6. Academic Press, New York, pp 321–350

Sandelius AS, Sommarin M (1986) Phosphorylation of phosphatidylinositols in isolated plant membranes. FEBS Lett 201:282–286

Sandelius AS, Penel C, Auderset G, Brightman A, Millard M, Morré DJ (1986) Isolation of highly purified fractions of plasma membrane and tonoplast from the same homogenate of soybean hypocotyls by free-flow electrophoresis. Plant Physiol 81:177–185

Sandelius AS, Barr R, Crane FL, Morré DJ (1987) Redox reactions of plasma membranes isolated from soybean hypocotyls by phase partition. Plant Sci 48:1–10

Sandstrom RP, de Boer AH, Lomax TL, Cleland RE (1987) Latency of plasma membrane H^+-ATPase in vesicles isolated by aqueous phase partitioning. Plant Physiol 85:693–698

Scherer GFE (1982) A new method to prepare membrane fractions containing ionophore-stimulated ATPase from pumpkin hypocotyls (*Cucurbita maxima* L.). Z Naturforsch 37c:550–552

Scherer GFE (1984) H^+-ATPase and auxin-stimulated ATPase in membrane fractions from zucchini (*Cucurbita pepo* L.) and pumpkin (*Cucurbita maxima* L.) hypocotyls. Z Pflanzenphysiol 114:233–237

Scherer GFE, Fischer G (1985) Separation of tonoplast and plasma membrane H^+-ATPase from zucchini hypocotyls by consecutive sucrose and glycerol gradient centrifugation. Protoplasma 129:109–119

Scherer GFE, Morré DJ (1978a) Action and inhibition of endogenous phospholipases during isolation of plant membranes. Plant Physiol 62:933–937

Scherer GFE, Morré DJ (1978b) In vitro stimulation by 2,4-dichlorophenoxyacetic acid of an ATPase and inhibition of phosphatidate phosphatase of plant membranes. Biochem Biophys Res Commun 84:238–247

Scherer GFE, vom Dorp B (1988) Isolation by free-flow electrophoresis and identification of tonoplast and plasma membrane for studies of plant membrane traffic. In: Morré DJ, Howell KE, Cook GMW, Evans WH (eds) Cell-free analysis of membrane traffic. Alan R Liss, New York, pp 269–279

Schibeci A (1985) Isolation of plasma membrane from ryegrass (*Lolium multiflorum*) endosperm protoplasts. In: Pilet PE (ed) The physiological properties of plant protoplasts. Springer, Berlin Heidelberg New York Tokyo, pp 37–44

Schibeci A, Fincher GB, Stone BA, Wardrop AB (1982) Isolation of plasma membrane from protoplasts of *Lolium multiflorum* (ryegrass) endosperm cells. Biochem J 205:511–519

Schmidt R, Ackermann R, Kratky Z, Wasserman B, Jacobson B (1983) Fast and efficient purification of yeast plasma membranes using cationic silica microbeads. Biochim Biophys Acta 732:421–427

Sommarin M, Lundborg T, Kylin A (1985) Comparison of K, Mg ATPases in purified plasmalemma from wheat and oat. – Substrate specificities and effects of pH, temperature and inhibitors. Physiol Plant 65:27–32

Staal M, Hommels C, Kuiper D (1987) Characterization of the plasmalemma ATPase activity from roots of *Plantago major* ssp. *pleiosperma*, purified by the two-phase partition method. Physiol Plant 70:461–466

Steck TL (1974) Preparation of impermeable inside-out and right-side-out vesicles from erythrocyte membranes. Methods Membr Biol 2:245–281

Steck TL, Kant JA (1974) Preparation of impermeable ghosts and inside-out vesicles from human erythrocyte membranes. Methods Enzymol 31:172–180

Sze H (1985) H^+-translocating ATPases: advances using membrane vesicles. Annu Rev Plant Physiol 36:175–208

Taylor ARD, Hall JL (1979) An ultrastructural comparison of lanthanum and silicotungstic acid/chromic acid as plasma membrane stains of isolated protoplasts. Plant Sci Lett 14:139–144

Tokumura A, Mostafa MH, Nelson DR, Hanahan DJ (1985) Stimulation of (Ca^{2+} + Mg^{2+})-ATPase activity in human erythrocyte membranes by synthetic lysophosphatidic acids and lysophosphatidylcholines. Effects of chain length and degree of unsaturation of the fatty acid group. Biochim Biophys Acta 812:568–574

Twohig F, Morré DJ, Vigil EL (1974) Properties and subcellular distribution of peroxidases of onion stem. Proc Indiana Acad Sci for 1973 83:86–94

Uemura M, Yoshida S (1983) Isolation and identification of plasma membrane from light-grown winter rye seedlings (*Secale cereale* L. cv. Puma). Plant Physiol 73:586–597

Uemura M, Yoshida S (1986) Studies on freezing injury in plant cells. II. Protein and lipid changes in the plasma membranes of Jerusalem artichoke tubers during a lethal freezing in vivo. Plant Physiol 80:187–195

Vasil IK, Vasil V (1980) Isolation and culture of protoplasts. In: Vasil IK (ed) Perspectives in plant cell and tissue culture. Int Rev Cyt 11B. Academic Press, New York, pp 1–19

Wheeler JJ, Boss WF (1987) Polyphosphoinositides are present in plasma membranes isolated from fusogenic carrot cells. Plant Physiol 85:389–392

Whitman CE, Travis RL (1985) Phospholipid composition of a plasma membrane-enriched fraction from developing soybean roots. Plant Physiol 79:494–498

Widell S, Larsson C (1981) Separation of presumptive plasma membranes from mitochondria by partition in an aqueous polymer two-phase system. Physiol Plant 57:196–202

Widell S, Larsson C (1983) Distribution of cytochrome *b* photoreductions mediated by endogenous photosensitizer or methylene blue in fractions from corn and cauliflower. Physiol Plant 57:198–202

Widell S, Lundborg T, Larsson C (1982) Plasma membranes from oats prepared by partition in an aqueous polymer two-phase system. Plant Physiol 70:1429–1435

Wilkinson MC, Brearley CA, Galliard T, Laidman DL (1987) Inhibition of membrane damage due to phospholipase activity during fractionation of wheat aleurone tissue. Phytochemistry 26:1903–1908

Yoshida S (1984) Chemical and biophysical changes in the plasma membrane during cold acclimation of mulberry bark cells (*Morus bombycis* Koidz. cv. Goroji). Plant Physiol 76:257–265

Yoshida S, Uemura M (1984) Protein and lipid compositions of isolated plasma membranes from orchard grass (*Dactylis glomerata* L.) and changes during cold acclimation. Plant Physiol 75:31–37

Yoshida S, Uemura M (1986) Lipid composition of plasma membranes and tonoplasts isolated from etiolated seedlings of mungbean (*Vigna radiata* L.). Plant Physiol 82:807–812

Yoshida S, Uemura M, Niki T, Sakai A, Gusta LV (1983) Partition of membrane particles in aqueous two-polymer phase system and its practical use for purification of plasma membranes from plants. Plant Physiol 72:105–114

Yoshida S, Kawata T, Uemura M, Niki T (1986) Properties of plasma membrane isolated from chilling-sensitive seedlings of *Vigna radiata* L. Plant Physiol 80:152–160

Chapter 4 Plasma Membrane Cytochemistry

D.J. MORRÉ[1]

1 Introduction

Cytochemical staining reactions specific for the plasma membrane of plants have served important roles especially as a means to provide positive identification of plasma membranes to guide isolation procedures. The introduction, in 1972, of phosphotungstic acid at low pH to selectively stain the plasma membrane (Roland et al. 1972), has received widespread use for this purpose. It allowed identification of plasma membrane vesicles in the absence of the positional relationships of the intact cell. Thus, the plasma membrane-derived vesicles could be distinguished from tonoplast vesicles, from endoplasmic reticulum vesicles devoid of ribosomes as well as from vesicles of the outer mitochondrial membrane and other smooth membranes. The procedure facilitated identification of the first vesicle fractions of plasma membrane origin isolated from plant sources (Hardin et al. 1972; Hodges et al. 1972). It remains in use to augment plasma membrane markers such as vanadate-inhibited ATPase (EC 3.6.1.35) and glucan synthase II (EC 2.1.3.34), and the absence of

[1]Department of Medicinal Chemistry, Purdue University, West Lafayette, IN 47907, USA

Abbreviation: NPA, N-1-naphthylphthalamic acid

C. Larsson, I.M. Møller (Eds):
The Plant Plasma Membrane

specific markers for other membranes (1) to identify isolated vesicles of plant plasma membrane, and (2) as a reliable means to determine the absolute plasma membrane content of fractions by simple morphometric procedures.

Other labeling methods, such as binding of lectins and heavy metal cations, have been used to guide fractionation procedures to some advantage especially with plant protoplasts. Information based on enzyme cytochemistry of plant plasma membranes, other than imposed labels (e.g., ligand-linked peroxidase), however, is relatively scant.

2 Phosphotungstic Acid at Low pH

A cytochemical stain based on phosphotungstic acid at low pH originally was used to detect complex carbohydrates and glycoproteins of the Golgi apparatus and cell coats of animal cells (Marinozzi and Gautier 1961; Marinozzi 1967; Rambourg 1967; Rambourg et al. 1969). This same procedure was found to stain the cell wall and plasma membrane of plant cells (Roland 1969; Roland and Vian 1971). The principal difference between use of the stain for the animal and plant material was the very short staining time of 5 min needed to differentially stain the plasma membrane of plants (Roland et al. 1972). With mammalian cells, staining times of 30 min or longer were often employed (Marinozzi and Gautier 1961). With staining times of comparable length for plants, the differential staining reaction for the plasma membrane would be largely lost.

Briefly, thin sections of glutaraldehyde-osmium tetroxide fixed, Epon-embedded material are collected in a droplet held by a plastic loop with a 2.25 mm diameter hole, a wire loop, or on a nickel grid. The sections are destained by floating on 1% (w/v) periodic acid for 20–30 min. The sections are rinsed in five washes of 5–10 min each with distilled water and stained by floating the sections on 1% (w/v) phosphotungstic acid in 10% (w/v) chromic acid for 5 min. The staining step must be followed immediately by five washes of 5 to 10 min each on distilled water. The latter washes are especially critical as any stain remaining on the sections will continue to react. Inadequate rinsing may have led, in some instances, to claims of lack of specificity, i.e., tonoplast staining, attributed to the procedure. It is important when working with membrane fractions that specificity of the procedure be checked by including a tissue section of the starting material along with the fractions each time the procedure is applied (Nagahashi et al. 1978). Conditions should be adjusted so that the plasma membrane is the only membrane heavily stained. When properly conducted, the plasma membrane stains uniformly over its surface and is the darkly contrasted membrane (Fig. 1). Cell walls and ribosomes attract the stain as sometimes do osmiophilic regions in internal membranes where the lipid phase has separated from the membrane, i.e., so-called myelin figures and related forms. Vesicles in isolated fractions that stain with phosphotungstic acid at low pH are thus identified as plasma membrane (Fig. 2). Simple morphometric procedures may be used to quantify the proportion of stained

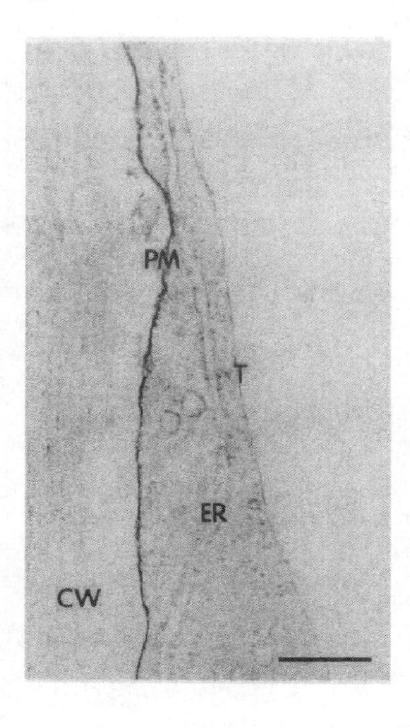

Fig. 1. Portion of a cortical cell of etiolated soybean hypocotyl segment fixed for electron microscopy using simultaneous osmium tetroxide plus glutaraldehyde (Franke et al. 1969) and then destained and restained with phosphotungstic acid at low pH according to Roland et al. (1972). The specificity of the staining procedure is shown by plasma membrane (*PM*) being the only membrane structure that attracted stain. Tonoplast (*T*), *ER*, and other internal membranes were unstained. Ribosomes and cell walls (*CW*) normally do stain but cannot be confused as membranes. *Bar* = 0.5 μm

vesicles. A transparent overlay bearing parallel lines 1 cm apart is placed over electron micrographs at a final magnification of about 35 000 ×. Intercepts of membranes with the lines are counted and data are expressed as intercepts of staining vesicles per 100 total intercepts witlh all vesicles (Loud 1962).

The constituents staining with phosphotungstic acid at low pH in plants have been shown to be lipid soluble (Yunghans et al. 1978). Therefore, best results are obtained with plant material or fractions dehydrated for very short times in acetone rather than ethanol. A suggested dehydration protocol for fractions would consist of 50, 70, 90, and 100% (v/v) acetone 5 min each followed by 100% acetone for 10 min followed by 50% (w/v) plastic in acetone for 30 min, 75% plastic in acetone for 30 min, and finally 100% plastic for several hours to overnight. Use of Spurr (1969) plastic correlates with nonuniform plasma membrane staining and is not recommended.

Plasma membranes of all higher plant species and organs examined specifically attract the stain as do those of algae and some fungi. A few fungal species have plasma membranes that stain unreliably, e.g., *Gilbertella persicaria* (Powell et al. 1982), such that it has been necessary to use other approaches to identify the plasma membrane with these organisms. Except for plasma membranes of mature sperm (Morré et al. 1974) and of an organelle resembling the Golgi apparatus unique to male germ cells, the dictyosome-like structure (Mollenhauer and Morré 1981), animal membranes do not stain with phosphotungstic acid at low pH when a staining time of 5 min is used as described above.

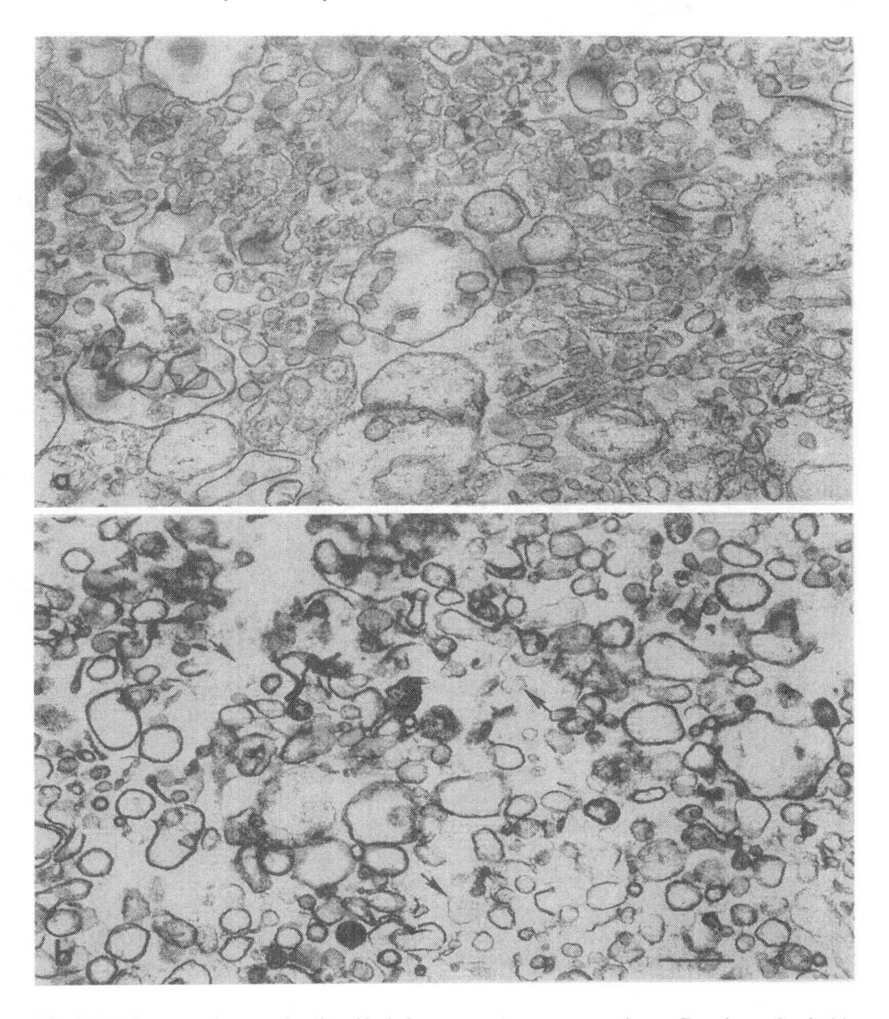

Fig. 2a,b. Electron micrographs of purified plasma membrane preparations. **a** Fraction stained with lead citrate to show all vesicles. **b** A section from the same preparation destained and restained with phosphotungstic acid at low pH (Roland et al. 1972) to accentuate vesicles of plasma membrane origin. More than 95% of the membranes attracted the stain. *Arrows* in **b** denote unstained vesicles. *Bar* = 0.5 μm

Some authors (Kjellbom and Larsson 1984) favor silicotungstic acid (Roland 1978) over phosphotungstic acid for plasma membrane staining. Comparisons of the two tungstic acids used to stain plasma membrane in the author's laboratory have not indicated any particular advantage of one over the other.

3 Measurements of Membrane Thickness and Membrane Morphology

In well-fixed preparations of isolated membranes poststained on thin sections with lead or uranyl acetate, plasma membrane vesicles can often be distinguished from all other membranes and vesicles on the basis of morphology alone. This is because the plasma membrane, with a thickness of about 10 nm, is the thickest of all cellular membranes and illustrates the dark-light-dark pattern of the membrane most clearly of all the cellular membranes (Morré and Ovtracht 1977; Table 1). Except for membranes of the trans-most secretory vesicles of the Golgi apparatus destined to fuse with plasma membranes, overlap occurs only with tonoplast (7–9 nm diameter and a moderate dark-light-dark pattern) that would tend to complicate a morphological analysis. All other membranes are thinner (5–7 nm) and show the dark-light-dark pattern much less clearly.

Measurements of membrane thickness in combination with phosphotungstic acid staining have been used as a method of membrane identification with purified fractions from preparative free-flow electrophoresis (Auderset et al. 1986; Sandelius et al. 1986). The vesicles with 7–10 nm thick membranes that also stain with phosphotungstic acid at low pH may be unequivocally identified as plasma membrane (Fig. 2). In contrast, vesicles with 7–10 nm thick membranes and not stained with phosphotungstic acid at low pH correlate with the tonoplast marker NO_3^--inhibited ATPase. Thin membranes, 5–6 nm in thickness, also unstained by phosphotungstic acid at low pH, are then those derived from endoplasmic reticulum, nuclear envelope, mitochondria, or plastid envelopes.

To facilitate these types of measurements, electron microscope negatives photographed at a magnification of about 60 000 × are projected at a final magnification of about 10^6 × with thickness established by direct measurement with a ruler or from the averaged densitometer trace from an image analysis system (Fig. 3). Another means for morphological identification and quantitation of plasma membrane vesicles in isolated membrane fractions utilized by vom Dorp et al. (1986) employs the use of filipin, a polyene antibiotic that combines with membrane sterols to form characteristic lesions. In combination with freeze-fracturing of the membranes, sterol-rich plasma membrane vesicles are identified from the distributions of membrane particles and the presence of lesions after filipin treatment.

Fig. 3. Densitometry traces of fraction E (plasma membrane) from vesicle preparations of free-flow electrophoresis comparing nonphotoinduced (**A**) and photoinduced spinach leaves (**B**). The microdensitometric measurements were averaged from negatives using a LeMont Model DV 4400 Oasys image analysis system equipped with an IMP program. Examples of the images generated for analysis are illustrated in the *upper panels.* The peak to peak distance of the membranes was little affected by photoinduction. Rather the membrane broadening was due to an extra layer at the interior of the vesicles. To generate the values of Table 1, for example, membrane thicknesses were estimated from the midpoints of the densities of each of the membrane leaflets as indicated by the positions of the *small arrows* (Penel et al. 1988)

Table 1. Dimensions of different membrane plant tissues as a criterion for identification of plasma membranes in isolated fractions (Morré et al. 1987a)

	Membrane thickness, nm ± SD				
Cell component	Soybean hypocotyl[a] (in situ)	Spinach leaf[b] (in situ)	Onion stem[c] (in situ)	*Pythium aphanidermatum*[d] (in situ)	(isolated)
Plasma membrane	10.1 ± 0.7	10.5 ± 0.3	9.3	9.2 ± 0.4	9.6 ± 0.45
Tonoplast membrane	7.2 ± 0.8	8.1 ± 0.9	8.0	9.3 ± 0.5	9.7 ± 0.45
Endoplasmic reticulum	5.7 ± 0.7	6.3 ± 0.8	5.3	6.0 ± 0.1	6.0 ± 0.1
Nuclear envelope					
Outer membrane	5.8 ± 0.8	6.8 ± 0.6	–	–	–
Inner membrane	6.5 ± 0.6	5.8 ± 0.7	–	–	–
Average of both membranes	–	–	5.6	6.1 ± 0.5	6.1 ± 0.45
Mitochondria					
Outer membrane	5.0 ± 0.6	5.4 ± 0.8	5.5	6.0 ± 0.2	6.4 ± 0.3
Inner membrane	6.2 ± 0.7	6.0 ± 0.9	6.0	6.2 ± 0.2	6.2 ± 0.4
Etioplast					
Outer envelope membrane	5.0 ± 0.6	–	–	–	–
Inner envelope membrane	6.1 ± 0.8	–	–	–	–
Thylakoid	8.0 ± 1.0	–	–	–	–
Chloroplast					
Outer envelope membrane	–	5.1 ± 0.8	–	–	–
Inner envelope membrane	–	6.3 ± 0.8	–	–	–
Grana thylakoid[e]	–	13.0 ± 1.1	–	–	–
Peroxisome	5.5 ± 0.6	7.0 ± 0.4	–	–	–
Golgi apparatus	6–9	6.6 ± 1.8	5–9	7–9[f]	7–9[f]

[a] From Sandelius et al. (1986). Simultaneous glutaraldehyde-osmium tetroxide fixation.
[b] From Auderset et al. (1986). Simultaneous glutaraldehyde-osmium tetroxide fixation.
[c] From a study with F. Twohig, Purdue University. Sequential glutaraldehyde-osmium tetroxide fixation.
[d] From Powell et al. (1982). Sequential glutaraldehyde-osmium tetroxide fixation.
[e] Both appressed membranes together.
[f] Golgi apparatus equivalent. Includes secretory vesicles.

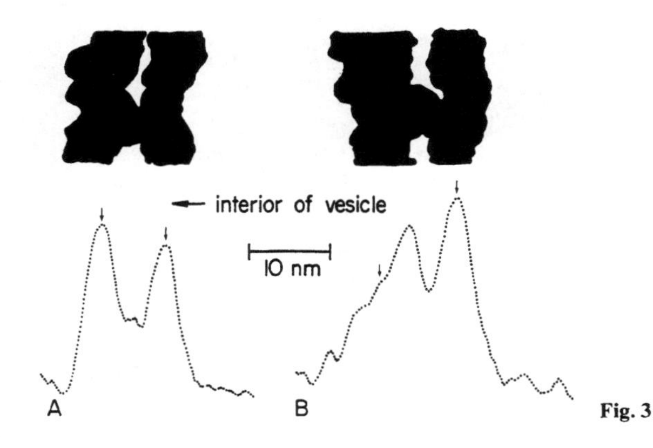

Fig. 3

4 Imposed Labels

Using imposed labels, some property of the plasma membrane is employed to permit selective tagging of the cell surface prior to cell breakage (see Hall 1987 for a recent review of approaches to surface labeling of plasma membranes, including chemical methods). The imposed labels then permit the recognition of the membrane fragments belonging to the plasma membrane after the cells are broken Labels selected should not only specifically label the plasma membrane but, preferably, should be nondestructive to permit both labeling under physiological conditions and the later evaluation of membrane activities. In practice, this has been difficult to achieve.

Because of the cell wall, imposed labels have thus far been applied successfully with plants only to protoplasts or broken cell preparations. Since surface labels may also react with internal membranes, care must be taken to avoid broken or leaky protoplasts in the preparations before or during labeling. Membranes from these sources are potential sources of serious artifacts.

4.1 Lanthanum Ions

Protoplasts exposed to concentrated (2.5%, w/v) solutions of $LaCl_3$ or $La(NO_3)_3$ bind electron dense deposits on their external surfaces (Taylor and Hall 1978, 1979; Fig. 4). The La^{3+} does not penetrate unbroken cells and appears to be irreversibly bound with little or no redistribution after the protoplasts are broken. Both electron microscopy (Fig. 4) and neutron activation analysis (Taylor and Hall 1980) have been used to analyze the distribution of La^{3+} within membrane separations from sucrose gradients. A major problem with La^{3+} is that it alters both the density and surface properties of the membrane vesicles such that the La^{3+}-labeled vesicles may fractionate differently from vesicles not labeled with La^{3+}.

4.2 Lectin Binding

Lectins are now available with a spectrum of well-defined sugar specificities (Alroy et al. 1984). Lectins offer advantages for plasma membrane studies since carbohydrate chains of various glycoconjugates (glycoprotein and glycolipids) of integral membrane constituents are exposed at the external cell surface. Applications to plants again have been limited because of the cell wall as a barrier to lectin penetration. Despite the potential loss or alteration of carbohydrate moieties of plasma membrane glycoproteins as a result of cell wall digestion during protoplast preparation (Randall and Ruesink 1983), plant protoplasts still bind concanavalin A (Burgess and Linstead 1976; Williamson et al. 1976a). Lectins are readily conjugated with peroxidase, ferritin, or other markers such as hemocyanin (Williamson et al. 1976b) to permit localization in the electron microscope.

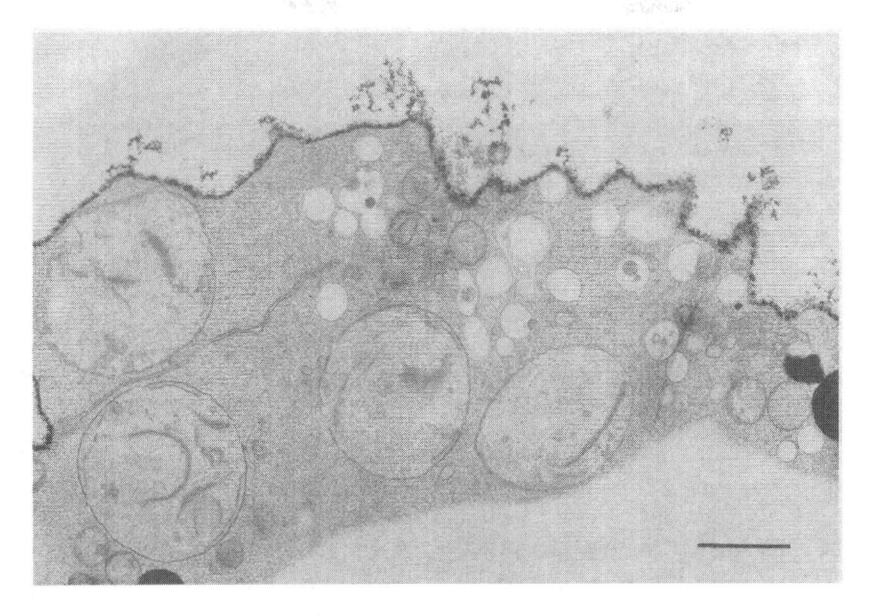

Fig. 4. Portion of a soybean protoplast in which the plasma membrane has been labeled with 2.5% lanthanum nitrate. The lanthanum is electron dense and imparts to the cell surface an imposed electron density. *Bar* = 0.5 μm (Canut et al. 1987)

Concanavalin A binding to carrot (Boss and Ruesink 1979) and tobacco (Taylor and Hall 1980) protoplasts was employed to identify putative plasma membrane fractions. In these studies the concanavalin A was bound to the cell surface prior to protoplast breakage.

Specific binding of concanavalin A as an aid in the identification of plasma membrane vesicles once isolated from the cell has also been attempted. Travis and Berkowitz (1980) used ferritin-labeled concanavalin A with fractions from soybean roots. However, internal membranes also bound varying amounts of the ferritin-labeled concanavalin A (Berkowitz and Travis 1981; Frederick et al. 1981) to preclude the use of the method to identify plasma membrane vesicles in crude mixtures from broken cells. Because of the putative orientation of carbohydrate moieties of glycoconjugates of the cell surface toward the outside, lectin binding or binding of lectin-marker conjugates would be expected, however, to be useful in determining the absolute orientation (cytoplasmic side-in or cytoplasmic side-out) of vesicles known to be plasma membrane-derived. For example, plasma membrane fractions obtained by aqueous two-phase partition and containing 90% vesicles that stain with phosphotungstic acid at low pH were resolved by Canut et al. (1987) into two fractions of differing electronegativity using preparative free-flow electrophoresis. Each fraction retained reactivity with phosphotungstic acid at low pH but only one (putative cytoplasmic side-in = right side-out vesicles) reacted strongly with concanavalin A conjugated with peroxidase (Canut et al. 1987; Sect. 5).

5 Determination of Vesicle Orientation

Various transport and assignment studies would be aided by the availability of plasma membrane vesicles of homogeneous absolute orientation (cytoplasmic side-in = right side-out vs. cytoplasmic side-out = inside-out). Both free-flow electrophoretic (Canut et al. 1987, 1988) and aqueous two-phase partition methods (Larsson et al. 1988) have been described for this purpose. However, progress has been hampered by the lack of methods to validate orientation and to detect and quantitate low levels of contaminating vesicles of the orientation opposite to that desired. For example, strong ATPase (or NADH-ferricyanide oxidoreductase) latency is indicative of a vesicle with a cytoplasmic side-in orientation. Cytoplasmic side-out vesicles would lack latency as would leaky cytoplasmic side-in vesicles that were permeable to ATP (or NADH) (Chap. 2). Also, latency measurements (activity determinations before and after solubilization of the vesicles with detergent) normally are not absolute. Even so-called pure fractions of sealed cytoplasmic side-in vesicles catalyze significant ATP hydrolysis prior to detergent treatment. This may be due to low levels of externally oriented hydrolase, to low rates of ATP entry into the vesicles, or to contaminating cytoplasmic side-out vesicles. Conversely, so-called pure fractions of cytoplasmic side-out vesicles may also exhibit some latency. This may be due to an external hydrolase (now present on the inside of the vesicle), to contaminating cytoplasmic side-in vesicles, or to a small detergent activation of the hydrolase by a mechanism unrelated to classic structure-linked latency. Also for the enzymatic analyses to be meaningful requires the virtual absence of vesicles of nonplasma membrane origin in the starting preparations. Therefore, independent morphological methods based on some characteristic of one or the other of the two membrane leaflets of the bilayer would have wide utility in establishing vesicle orientation and in assessing fraction composition.

In some plant tissues and cells and under certain conditions of fixation, the plant plasma membrane is naturally asymmetric (Auderset et al. 1986; Canut et al. 1987). In spinach leaves, an asymmetry enhanced by simultaneous glutaraldehyde-osmium tetroxide fixation (Franke et al. 1969) increases markedly with a photoinductive light period (Auderset et al. 1986). The latter is expressed as an enhancement or broadening of the electron density of the apoplastic leaflet of the plasma membrane bilayer (Figs. 5 and 6; Auderset et al. 1986). Asymmetric plasma membranes have also been encountered in suspension cultures of soybean (Canut et al. 1987) and the asymmetric vesicles have been employed as an aid in identifying electrophoretic separations of differently oriented plasma membrane vesicles. Since plasma membranes are the only membranes in these cells showing marked asymmetry, the parameter not only defines polarity but identifies the vesicles as plasma membrane-derived. However, not all cells show the marked asymmetry. As a result, many vesicles of plasma membrane origin are present that show insufficient membrane asymmetry to permit either their identification as plasma membrane or an assignment of absolute polarity.

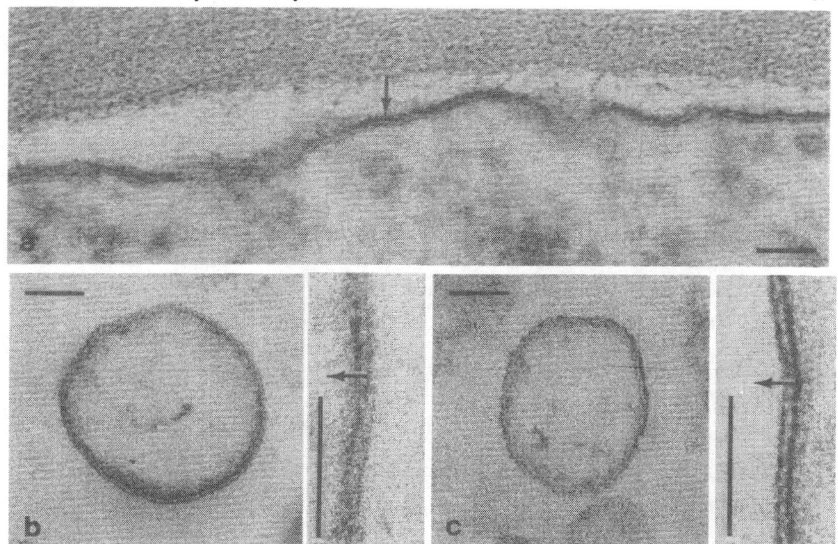

Fig. 5a-c. Electron micrographs (standard staining procedure) of plasma membrane vesicles of cultured soybean cells showing preparations of vesicles of different absolute orientation. **a** Intact cell with the dark-staining layer of plasma membrane adjacent to the cell wall (*arrow*). **b** Vesicle from free-flow electrophoresis fraction E. Membrane asymmetry shows the vesicle to be cytoplasmic side-out. **c** Plasma membrane vesicles from free-flow electrophoresis fraction C purified by aqueous two-phase partitioning. Membrane asymmetry shows the opposite (cytoplasmic side-in) orientation. *Arrows* indicate direction of vesicle interiors. *Bars* = 0.05 μm (Canut et al. 1987 and an unpublished study with G. Auderset, University of Geneva)

As an alternative procedure to determine sidedness, the isolated plasma membrane vesicles have been reacted with concanavalin A linked to peroxidase. With soybean plasma membranes prepared by aqueous two-phase partition, two populations of vesicles were observed (Fig. 7). One reacted strongly with the lectin and the other reacted less strongly. When separated by preparative free-flow electrophoresis, the most electronegative fractions contained the strongly reactive vesicles with high ATPase latency and were assigned a cytoplasmic side-in orientation. The fractions of lower electronegativity contained the less reactive vesicles of low ATPase latency and were assigned a cytoplasmic side-out orientation. Clearly, the problem of vesicle orientation is one in which a cytochemical approach at the level of electron microscopy using either natural or imposed markers has considerable utility.

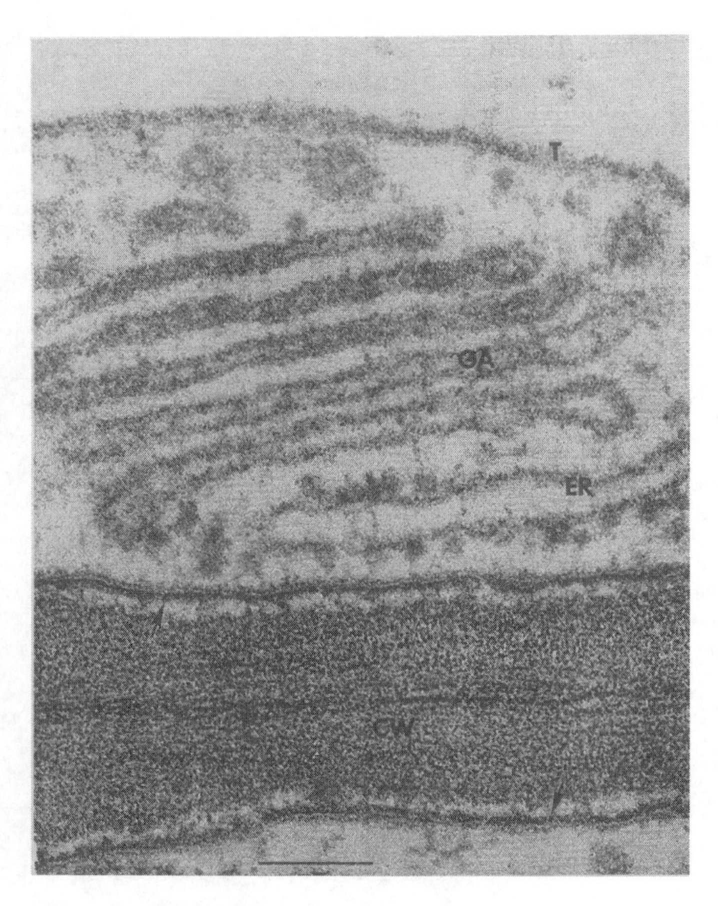

Fig. 6. High magnification electron microscope view of a portion of two adjacent mesophyll cells of a photoinduced spinach leaf after simultaneous glutaraldehyde and osmium tetroxide fixation (Franke et al. 1969). The additional density of the dark-light-dark layer (*arrows*) of the plasma membrane is seen to be directed toward the cell wall (*CW*) in the intact cell. *T* Tonoplast; *GA* Golgi apparatus; *ER* endoplasmic reticulum. *Bar* = 0.1 μm (Penel et al. 1988)

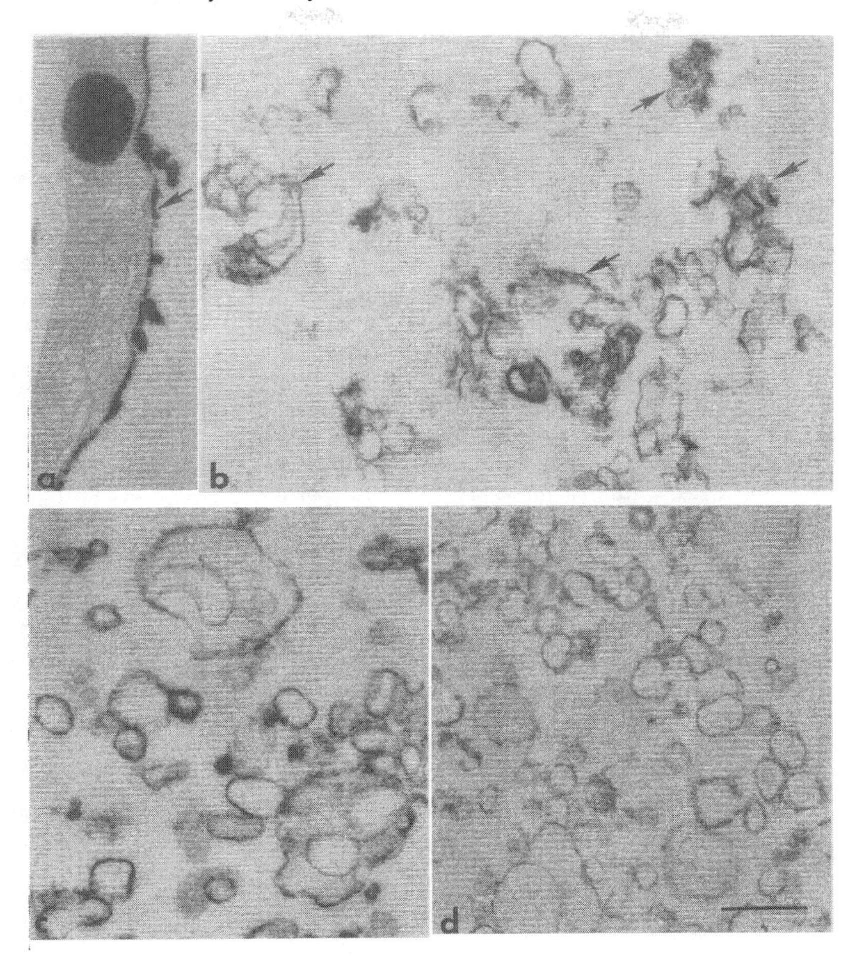

Fig. 7a-d. Concanavalin A-peroxidase labeling to show sidedness of isolated plasma membrane vesicles. **a** Soybean protoplast prepared by enzymatic digestion from a soybean cell suspension culture. The external membrane surface exposed to the reagent yielded a strong reaction (*arrow*). **b** Fraction obtained by aqueous two-phase partition was a mixture of strongly (*arrows*) and weakly reactive vesicles; **c** free-flow electrophoresis fraction C of an aqueous two-phase separation contained only strongly reactive vesicles indicating a right side-out (cytoplasmic side-in) orientation; **d** free-flow electrophoresis fraction E from an aqueous two-phase separation contained only weakly reactive vesicles indicative of the opposite, cytoplasmic side-out orientation. *Bar* = 0.5 μm

6 Enzyme Cytochemistry

A nucleoside diphosphatase (EC 3.6.1.6) activity at neutral pH was first found associated with the plant plasma membrane by electron microscope cytochemistry in a now classical study of Poux (1967). These findings were later confirmed and extended by work in several laboratories (Dauwalder et al. 1969; Goff 1973). The activity was not specific to the plasma membrane but was shown also by the endoplasmic reticulum, Golgi apparatus, and developing vacuoles. While IDP was a favored substrate, other nucleoside diphosphates as well as ATP and other triphosphates together with thiamine pyrophosphate will serve as substrates for plasma membrane-located hydrolases. An inosine diphosphatase (IDPase) exhibiting structure-linked latency (activated in response to detergent solubilization) has often been assumed to represent a Golgi apparatus-specific marker (Chap. 2). However, based on the nucleoside diphosphate phosphatase cytochemistry, plasma membrane vesicles oriented cytoplasmic side-in should also exhibit a similar latency as vesiculated Golgi apparatus membranes. A latency of nucleoside diphosphate phosphatase overcome by storage of fractions at low temperatures for several days (Ray et al. 1969) may be more specific for the Golgi apparatus and is not shown by plasma membrane-enriched cell fractions (Morré et al. 1971).

NADH-ferricyanide oxidoreductase is an enzymatic activity that has been localized cytochemically to the inner surface of the plasma membrane of etiolated hypocotyls of soybean by electron microscopy. Ferricyanide reduction is coupled to the formation of deposits of reduced copper ferrocyanide (Hatchett's brown) (Morré et al. 1987b). The latter have been used widely as a general method for the electron microscope cytochemical localization of NADH-linked oxidoreductases (Karnovsky and Roots 1964). The activity of the plasma membrane is unaffected by fixation with 0.1% glutaraldehyde under conditions in which those of the endoplasmic reticulum and mitochondria are inactivated (Morré et al. 1978). Cytochemically demonstrated NADH-ferricyanide oxidoreductase activity is also associated with the cytoplasmic surface of the tonoplast.

7 Immunolocalization

Electron microscope immunolocalization using gold-, ferritin-, or peroxidase-linked conjugates has not yet been exploited to investigate plasma membrane constituents in plants. Again, the presence of the cell wall would restrict the use of surface labeling prior to fixation, dehydration, and embedment to protoplasts. However, even the standard analyses in which gold-conjugated antibodies are presented to Lowicryl-embedded sections have been little used presumably due to the lack of appropriate antibodies. Antibodies have been generated to the receptor for N-1-naphthylphthalamic acid (NPA) (Jacobs and Gilbert 1983), a plasma membrane-located marker (Lembi et al. 1971; Hertel et al. 1972). The use of those antibodies in subcellular localization, however, has been restricted to light microscope (immunofluorescence) applications. Based on these studies, the NPA receptor would be expected to be most concentrated in the basal ends of parenchymal cells sheathing the vascular bundles. Similarly, antisera to an auxin-binding protein of maize coleoptiles has been used together with immunofluorescence light microscopy to localize the protein to the plasma membranes of the outer epidermal cells (Löbler and Klämbt 1985).

8 Summary

Cytochemistry at the level of electron microscopy has served an important role in the study of plasma membranes from plants. Unique properties of plasma membranes, as compared to tonoplasts or other internal membrane systems, were established first. These unique properties, as well as the relatedness of plasma membranes and membranes of mature secretory vesicles of the Golgi apparatus, were indicated by measurements of membrane thickness from electron microscope preparations. Confirmation was provided subsequently from the specific reaction of the plasma membrane with phosphotungstic acid at low pH introduced by Roland et al. (1972). This procedure guided development of methods for plasma membrane isolation and, most recently, for plasma membrane subfractionation. Surface labeling with antibody conjugates, lectin conjugates, or with other electron-dense organic or inorganic ions, so useful for animal cells, has with plants been restricted to protoplasts and isolated vesicles due to the presence of the cell wall. The specific labeling of the plant cell surface by imposed labels as a means to identify plasma membrane vesicles in mixtures has been hampered by the paucity of knowledge concerning plasma membrane-specific determinants to which ligands might be uniquely directed. The domain-specific receptor for NPA would be a possible example of such a determinant.

The resolution of cytoplasmic side-in and cytoplasmic side-out plasma membrane vesicles for use in transport and assignment studies requiring vesicles of known absolute orientation offers opportunities for development of new cytochemical procedures to surface label, mark, or identify one or both of the two bilayer leaflets of the plasma membrane. Progress has already been made using both natural and imposed markers but methods of greater specificity and/or generality would facilitate their applications as cytochemical methods to confirm latency measurements.

Acknowledgments. I thank Dorothy Werderitsh for technical assistance and Dr. Charles Bracker for use of electron microscope facilities.

References

Alroy J, Ucci AA, Pereira MEA (1984) Lectins: histochemical probes for specific carbohydrate residues. In: DeLellis RA (ed) Advances in immunohistochemistry. Masson, New York, pp 67–88

Auderset G, Sandelius AS, Penel C, Brightman A, Greppin H, Morré DJ (1986) Isolation of plasma membrane and tonoplast fractions from spinach leaves by preparative free-flow electrophoresis and effect of photoinduction. Physiol Plant 68:1–12

Berkowitz RL, Travis RL (1981) Characterization and quantitation of concanavalin A binding by plasma membrane enriched fractions from soybean root. Plant Physiol 68:1014–1019

Boss WF, Ruesink AW (1979) Isolation and characterization of concanavalin A-labeled plasma membranes of carrot protoplasts. Plant Physiol 64:1005–1011

Burgess J, Linstead PJ (1976) Ultrastructural studies of the binding of concanavalin A to the plasmalemma of higher plant protoplasts. Planta 130:73–79

Canut H, Brightman AO, Boudet AM, Morré DJ (1987) Determination of sidedness of plasma membrane and tonoplast vesicles isolated from plant stems. In: Leaver C, Sze H (eds) Plant membranes: structure, function, biogenesis. Alan R. Liss, New York, pp 141–159

Canut H, Brightman A, Boudet AM, Morré DJ (1988) Plasma membrane vesicles of opposite sidedness from soybean hypocotyls by preparative free-flow electrophoresis. Plant Physiol 86:631–637

Dauwalder M, Whaley WG, Kephart JE (1969) Phosphatases and differentiation of the Golgi apparatus. J Cell Sci 4:455–498

Franke WW, Krien S, Brown RM Jr (1969) Simultaneous glutaraldehyde-osmium tetroxide fixation with postosmication. An improved fixation procedure for electron microscopy of plant and animal cells. Histochemie 19:162–164

Frederick SE, Nies B, Gruber PJ (1981) An ultrastructural search for lectin-binding sites on surfaces of spinach leaf organelles. Planta 152:145–152

Goff CW (1973) Localization of nucleoside diphosphatase in the onion root tip. Protoplasma 78:397–416

Hall JL (1987) Possible approaches to surface labeling of the plasma membrane. Methods Enzymol 148:568–575

Hardin JW, Cherry JH, Morré DJ (1972) Enhancement of RNA polymerase activity by a factor released by auxin from plasma membrane. Proc Natl Acad Sci USA 63:3146–3150

Hertel R, Thomson K-St, Russo VEA (1972) In vitro auxin binding to particulate cell fractions from corn coleoptiles. Planta 107:325–340

Hodges TK, Leonard RT, Bracker CE, Keenan TW (1972) Purification of an ion-stimulated adenosine triphosphatase from plant roots: association with plasma membranes. Proc Natl Acad Sci USA 69:3307–3311

Jacobs M, Gilbert SF (1983) Basal localisation of the presumptive auxin transport carrier in pea stem cells. Science 220:1297–1300

Karnovsky MJ, Roots L (1964) A "direct coloring" thiocholine method for cholinesterase. J Histochem Cytochem 12:219–221

Kjellbom P, Larsson C (1984) Preparation and polypeptide composition of chlorophyll-free plasma membranes from leaves of light grown spinach and barley. Physiol Plant 62:501–509

Larsson C, Widell S, Sommarin M (1988) Inside-out plant plasma membrane vesicles of high purity obtained by aqueous two-phase partitioning. FEBS Lett 229:289–292

Lembi CA, Morré DJ, Thomson KS, Hertel R (1971) 1-N-Naphthylphthalamic acid (NPA) binding of a plasma membrane-rich fraction from maize coleoptiles. Planta 99:37–45

Löbler M, Klämbt D (1985) Auxin-binding protein from coleoptile membranes of corn (*Zea mays* L.) J Biol Chem 260:9854–9859

Loud AV (1962) A method for the quantitative estimation of cytoplasmic structures. J Cell Biol 15:481–487

Marinozzi V (1967) Reaction de l'acide phosphotungstique avec la mucine et les glycoproteines des plasmamembranes. J Microsc 6:68a

Marinozzi V, Gautier A (1961) Essais de cytochemie ultrastructurale. Du rôle de l'osmium reduir dans les "coloration" electronique. CR Acad Sci Paris 253:1180–1182

Mollenhauer HH, Morré DJ (1981) Selective staining of dictyosome-like structures (DLS) from spermatocytes of the guinea pig using phosphotungstic acid at low pH. Eur J Cell Biol 25:340–345

Morré DJ, Ovtracht L (1977) The dynamics of Golgi apparatus: membrane differentiation and membrane flow. Int Rev Cytol Suppl 5:61–188

Morré DJ, Clegg ED, Lunstra DD, Mollenhauer HH (1974) An electron-dense stain for isolated fragments of plasma and acrosome membranes from porcine sperm. Proc Soc Exp Biol Med 145:1–6

Morré DJ, Lembi CA, Van Der Woude WJ (1977) A latent inosine-5′-diphosphatase associated with Golgi apparatus-rich fractions from onion stem. Eur J Cell Biol (Cytobiologie) 16:72–81

Morré DJ, Vigil EL, Frantz C, Goldenberg H, Crane FL (1978) Cytochemical demonstration of glutaraldehyde-resistant NADH-ferricyanide oxido-reductase activities in rat liver plasma membranes and Golgi apparatus. Eur J Cell Biol (Cytobiologie) 18:213–230

Morré DJ, Brightman AO, Sandelius AS (1987a) Membrane fractions from plant cells. In: Findlay JBC, Evans WH (eds) Biological membranes: a practical approach. Alan R. Liss, Oxford, pp 37–72

Morré DJ, Auderset G, Penel C, Canut H (1987b) Cytochemical localization of NADH-ferricyanide oxido-reductase in hypocotyl segments and isolated membrane vesicles of soybean. Protoplasma 140:133–140

Nagahashi G, Leonard RT, Thomson WW (1978) Purification of plasma membranes from roots of barley. Specificity of the phosphotungstic acid – chromic acid stain. Plant Physiol 61:993–999

Penel C, Auderset G, Bernardine N, Castillo FJ, Greppin H, Morré DJ (1988) Compositional changes associated with plasma membrane thickening during floral induction of spinach. Physiol Plant 73:134–146

Poux N (1967) Localisation d'activités enzymatiques dans les cellules du méristeme radiculaire de *Cucumis sativus* L. I. Activités phosphatasiques neutres dans les cellules du protoderme. J Microsc 6:1043–1058

Powell MJ, Bracker CE, Morré DJ (1982) Isolation and ultrastructural identification of membranes from the fungus *Gilbertella persicaria*. Protoplasma 111:87–106

Rambourg A (1967) Detection des glycoprotéines et microscopie électronique coloration de la surface cellulaire et de l'appareil de Golgi par un mélange acide chromique-phosphotungstique. CR Acad Sci Paris 265:1426–1428

Rambourg A, Hernandez W, Leblond CP (1969) Detection of complex carbohydrates in the Golgi apparatus of rat cells. J Cell Biol 40:395–414

Randall SK, Ruesink AR (1983) Orientation and integrity of plasma membrane vesicles obtained from carrot protoplasts. Plant Physiol 73:385–391

Ray PM, Shininger TL, Ray MM (1969) Isolation of β-glucan synthetase particles from plant cells and identification with Golgi membranes. Proc Natl Acad Sci USA 64:605–612

Roland J-C (1969) Mise en évidence sur coupes ultrafines de formations polysaccharidiques directement associées au plasmalemme. CR Acad Sci Paris 269:939–942

Roland J-C (1978) General preparation and staining of thin sections. In: Hall JL (ed) Electron microscopy and cytochemistry of plant cells. Elsevier, Amsterdam, pp 1–62

Roland J-C, Vian B (1971) Réactivité du plasmalemme végétal. Etude cytochimique. Protoplasma 73:121–137

Roland J-C, Lembi CA, Morré DJ (1972) Phosphotungstic acid-chromic acid as a selective electron-dense stain for plasma membranes of plant cells. Stain Technol 47:195–200

Sandelius AS, Penel C, Auderset G, Brightman A, Millard M, Morré DJ (1986) Isolation of highly purified fractions of plasma membrane and tonoplast from the same homogenate of soybean hypocotyls by free-flow electrophoresis. Plant Physiol 81:177–185

Spurr AR (1969) A low-viscosity epoxy resin embedding medium for electron microscopy. J Ultrastruct Res 26:31–43

Taylor ARD, Hall JL (1978) Fine structure and cytochemical properties of tobacco leaf protoplasts and comparison with the source tissue. Protoplasma 96:113–126

Taylor ARD, Hall JL (1979) An ultrastructural comparison of lanthanum and silicotungstic acid/chromic acid as plasma membrane stains of isolated protoplasts. Plant Sci Lett 14:139–144
Taylor ARD, Hall JL (1980) Labeling and isolation of plasma membrane from higher plant protoplasts. In: Ferenczy L, Farkas GL (eds) Advances in protoplast research. Pergamon, Oxford, pp 463–468
Travis RL, Berkowitz RL (1980) Characterization of soybean plasma membrane during development. Plant Physiol 65:871–879
vom Dorp B, Volkmann D, Scherer GFE (1986) Identification of tonoplast and plasma membrane in membrane fractions from garden cress (*Lepidium sativum* L.) with and without filipin treatment. Planta 168:151–160
Williamson FA, Fowke LC, Constabel FC, Gamborg OL (1976a) Labelling of concanavalin A sites on the plasma membrane of soybean protoplasts. Protoplasma 89:305–316
Williamson FA, Morré DJ, Shen-Miller J (1976b) Inhibition of 5′-nucleotidase by concanavalin A: evidence for localization in the outer plasma membrane surface. Cell Tissue Res 170:477–484
Yunghans WN, Clark JE, Morré DJ, Clegg ED (1978) Nature of the phosphotungstic acid-chromic acid (PACP) stain for plasma membranes of plants and mammalian sperm. Eur J Cell Biol 17:165–172

Chapter 5 Redox Processes in the Plasma Membrane

I.M. Møller[1] and F.L. Crane[2]

[1]Department of Plant Physiology, University of Lund, Box 7007, S-220 07 Lund, Sweden
[2]Department of Biological Sciences, Purdue University, West Lafayette, IN 47907, USA

Abbreviations: DCPIP, 2,6-dichlorophenolindophenol; Fe^{3+}CN, ferricyanide [= hexacyanoferrat (III)]; Fe^{2+}CN, ferrocyanide [= hexacyanoferrat (II)]; LIAC, light-induced absorbance change; SHAM, salicylhydroxamic acid.

C. Larsson, I.M. Møller (Eds):
The Plant Plasma Membrane

1 Introduction

As early as 1945, Lundegårdh proposed that redox processes took place in the plasma membrane of plants. He further envisaged that ion uptake, specifically anion uptake, across the plasma membrane of root cells was directly coupled to the flow of electrons to oxygen. This anion respiration was thought to be catalyzed by the respiratory redox components known at that time, mainly the cytochromes. When, around 1950, these were discovered to reside in the mitochondrion, and not in the plasma membrane, the theory of Lundegårdh became less tenable (see Lundegårdh 1955) and it gradually lost prominence. However, in hindsight we can now see that his main idea was correct; there is plenty of evidence that the plasma membrane of both animals and plants contains redox components which can participate in a number of redox processes (see Crane et al. 1985a,b; Lüttge and Clarkson 1985; Møller and Lin 1986; Møller et al. 1988b for reviews). This area of plant physiology/ biochemistry is attracting a great deal of interest as evidenced by three recent workshops with published proceedings (Ramirez 1987; Crane and Møller 1988; Crane et al. 1988a). The aim of the present review is to discuss critically the properties, location, and possible physiological roles of these redox processes in the plant plasma membrane.

2 Choice of Experimental Material

An intact plant tissue or organ, such as a root, consists of a number of different cell types, e.g., epidermis, cortex, phloem, etc., at different stages of development and with different functions. The plasma membrane would therefore be expected to differ between different cell types. The transfer cell is an important case in point (see Sect. 5.1). However, even in a given cell the plasma membrane is not necessarily homogeneous, rather it may have a lateral heterogeneity (secondary level of organization, Møller 1988). This may be caused, e.g., by coated pits (see Chap. 10), plasmodesmata between cortical cells, or by polarity as in the phloem cells. There may also be problems of penetration in experiments with intact tissues (see, e.g., Jensén et al. 1987). Single cell cultures have often been used to circumvent these problems although only a limited number of cell cultures exist. In addition, cells in cell culture will not resemble any given cell type in the original tissue since they represent dedifferentiated cells (Steward et al. 1964). Finally protoplasts, which are obtained by removal of the cell wall with various enzyme mixtures, may have modified plasma membranes since the enzymes may also change the outer surface of the plasma membrane, e.g., by removing carbohydrate moieties from glycoproteins. One must always bear these experimental limitations in mind when interpreting the results obtained with whole cells and intact tissues.

To identify, quantify, and characterize redox components in the plasma membrane, it is necessary to isolate the membrane without significant contamination. This is particularly important as similar redox components and activities are found in virtually all plant membranes (review, Møller and Lin 1986). Many reports on plasma membrane redox components in the literature are based on experiments with microsomal fractions or plasma membrane-enriched fractions obtained by (sucrose) gradient centrifugation. The former contain ca. 10% and the latter 50–60% plasma membrane on a protein basis (Bérczi and Møller 1986; Hodges and Mills 1986; Larsson et al. 1987). Neither of these preparations are sufficiently pure to allow unequivocal localization and identification of components. In the following we will therefore cite only work in which a high purity of the plasma membrane preparation is documented. The reader is referred to Chapters 2 and 3 for more details on marker enzymes and methods for the isolation of pure plasma membrane vesicles.

Thus we are faced with a classic dilemma: An intact tissue is very complex and it can be difficult to interpret the results. This will only be too clear in the following. It is, therefore, desirable to obtain simpler experimental systems. On the other hand, how can we be certain that purified plasma membrane vesicles accurately reflect the properties of the plasma membrane in vivo? Fortunately, evidence is available from experiments at different levels of biological complexity, viz. intact plants and plant organs, whole cells and protoplasts, and isolated plasma membrane vesicles. In the following we will use evidence at all these levels to get as complete a picture as possible of the redox components, the redox activities they participate in, and their physiological significance.

3 Redox Components in Purified Plasma Membranes

3.1 b-Type Cytochromes

The first cytochrome reported in plant plasma membranes was a blue light-reducible b-type cytochrome (Brain et al. 1977; Jesaitis et al. 1977). Their observations were made on plasma membrane-enriched fractions from corn coleoptiles, and light-reducible b-type cytochrome was later reported in plasma membrane from cauliflower inflorescences (Widell and Larsson 1981; 1983), spinach leaf (Kjellbom and Larsson 1984), and several other species, and seems to be a general constituent of plant plasma membranes (for reviews, see Widell 1987; Chap. 2). This blue light-reducible b-type cytochrome is measured as a light-induced absorbance change (LIAC) under low oxygen tension in the presence of a reductant such as EDTA. LIAC is thought to be caused by the transfer of electrons from the reductant via a flavin to a b-type cytochrome (see Widell 1987; and references therein). About half the total amount of LIAC in the homogenate is located in the plasma membrane (Widell and Larsson 1983). Low-temperature spectra of the light-reducible cytochrome show only one component in corn coleoptile plasma membrane (Leong et al. 1981), cauliflower plasma membrane (Widell et al. 1983), and spinach leaf plasma membrane (Kjellbom and Larsson 1984), with the peak of the α-band at 555–558 nm, which coincides with that of the dithionite-reducible b-type cytochrome, suggesting that the light-reducible b-type cytochrome might be identical to the major b-type cytochrome of the plasma membrane.

The amount of dithionite-reducible cytochrome b in purified plasma membrane as measured by difference spectrophotometry is 0.1–0.5 nmol (mg protein)$^{-1}$ (Table 1). Figure 1 shows the results of a more detailed study of the b-type cytochrome in plasma membrane from spinach leaves (Askerlund et al. 1989). A small part of the cytochrome was oxidized in the sample when ferricyanide was added to the reference cuvette (Fig. 1B). Apparently this cytochrome was partly reduced even in the absence of added reductants. The vesicles could contain a small amount of endogenous reductant possibly deriving from the reductants present in the homogenization medium (e.g., dithiothreitol). NADH reduced about 30% of the cytochrome b, but only a single

Table 1. Redox components identified with certainty in the plant plasma membrane

Component	Source	Concentration nmol (mg protein)$^{-1}$	Reference
Cytochrome b	Corn coleoptiles	0.12[a]	Jesaitis et al. (1977)
	Oat roots	0.23[b]	Ramirez et al. (1984)
	Barley roots	0.35	Askerlund et al. (1989)
	Spinach leaves	0.42	Askerlund et al. (1989)
	Sugar beet leaves	0.29	Askerlund et al. (1989)
	Soybean hypocotyls	0.5	Barr et al. (1986), Sandelius et al. (1987)
	Cauliflower inflorescences	1.9[c]	Caubergs et al. (1986)
		0.28	Askerlund et al. (1989)
Cytochrome b (LIAC)	Corn coleoptiles	0.03[d]	Leong et al. (1981)
	Cauliflower inflorescences	0.08[d]	Widell et al. (1983)
Cytochrome P-450	Cauliflower inflorescences	≤ 0.03	Askerlund et al. (1989)
Cytochrome P-420	Cauliflower inflorescences	≤ 0.06	Askerlund et al. (1989)
Flavins	Oat roots	0.21	Ramirez et al. (1984)
	Soybean hypocotyls	0.5	Barr et al. (1986)
		2.4	Sandelius et al. (1987)

[a] Calculated from Table III (Jesaitis et al. 1977) using an extinction coefficient for the Soret band of 171 mM^{-1} cm^{-1},
[b] Calculated from Fig. 2 using an extinction coefficient for the α-band of 20 mM^{-1} cm^{-1},
[c] Calculated from Fig. 8 (Caubergs et al. 1986) using an extinction coefficient for the α-band of 20 mM^{-1} cm^{-1},
[d] Calculated from data given in results.

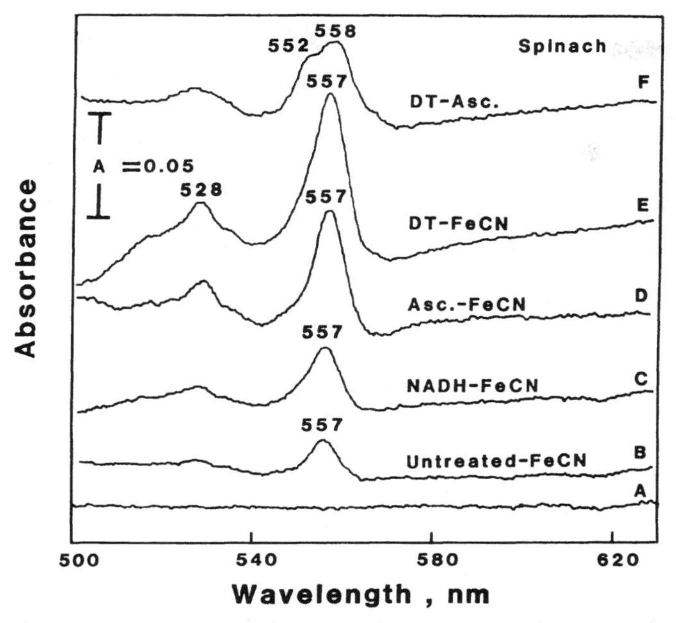

Fig. 1. Difference spectra (77 K) of plasma membrane vesicles from spinach leaves purified by phase partitioning as described by Kjellbom and Larsson (1984). The medium contained 0.1 M potassium phosphate, pH 7.3, ca. 30% (w/v) glycerol, and 2.9 mg protein ml^{-1}. *A* Baseline, air-oxidized minus air-oxidized; *B* air-oxidized minus ferricyanide-oxidized; *C* NADH (1 mM)-reduced minus ferricyanide-oxidized; *D* ascorbate (4 mM)-reduced minus ferricyanide-oxidized; *E* dithionite-reduced minus ferricyanide-oxidized; *F* dithionite-reduced minus ascorbate-reduced. Note the complete absence of cyt. c oxidase, which has a peak at 600 nm. (Reproduced from Askerlund et al. 1989 by permission of Physiologia Plantarum)

peak at 557 nm was observed (Fig. 1C). Ramirez et al. (1984) also found little NADH-reducible b-type cytochrome. Cytochrome b_5 has a split α-band and, as long as NADH-cytochrome b_5 reductase is present, it would be reduced by NADH (Bruder et al. 1978; Madyastha and Krishnamachary 1986). Ascorbate caused a greater reduction than NADH (Fig. 1D), but considerably less than dithionite (Fig. 1E). Similar results were obtained on cauliflower plasma membrane (Caubergs et al. 1986; Askerlund et al. 1989). The difference spectrum between dithionite- and ascorbate-reduced showed a split α-band (Fig. 1F); this was also observed in plasma membrane from cauliflower, sugar beet, barley leaves, and barley roots (Askerlund et al. 1989). This split α-band is probably due to cyt b_5 since ascorbate does not necessarily reduce cytochrome b_5 (Madyastha and Krishnamachary 1986). Approximately 20% of the cytochrome with a split α-band was reducible by NADH in the presence of 0.015% Triton X-100 in plasma membrane vesicles from cauliflower inflorescences (Askerlund et al. 1989).

Some of the b-type cytochrome in the plant plasma membrane may be cytochrome P-450/420 (Kjellbom et al. 1985; Askerlund et al. 1989). Although the ascorbate-reducible cytochrome in spinach and cauliflower plasma membrane is spectrally similar to cytochrome P-420 (Lemberg and Barret 1973; Bruder et al. 1978), it has been concluded that they are not identical (Caubergs et al. 1986; Askerlund et al. 1989).

Heme staining of gels shows one band with plasma membranes from cauliflower inflorescences (Kjellbom et al. 1985); however, this band disappears together with most of the peroxidase activity upon washing of the plasma membrane at high pH, indicating that the band is due to a contaminating peroxidase (Askerlund et al. 1989). The noncovalently bound heme in the b-type cytochrome thus appears to be lost upon solubilization of the plasma membrane for gel electrophoresis. Only with barley leaf plasma membrane is a heme band found (at 94 kD), which can not be washed off the membrane (Askerlund et al. 1989).

Redox titrations of the b-type cytochrome in purified plasma membrane vesicles from plants have been published only in a few cases. Leong et al. (1981) reported the presence of one component in the plasma membrane from corn coleoptiles with a midpoint potential of -65 mV. Caubergs et al. (1986), Asard et al. (1989), and Askerlund et al. (1989) all reported the presence of two or three b-type cytochromes with different redox potentials in plasma membrane from a total of five species including one monocotyledon. In plasma membrane from cauliflower inflorescence (Caubergs et al. 1986; Asard et al. 1989; Askerlund et al. 1989), spinach leaves, and barley roots (Askerlund et al. 1989) the major cytochrome (ca. 70% of the total) had a midpoint potential around 150 mV, whereas one or two minor components had a midpoint potential below 0 mV. In plasma membrane from zucchini and bean leaves, on the other hand, the high-potential cytochrome b was less dominant (39 and 46% of the total, respectively) and a second major b-type cytochrome with a midpoint potential of -77 and 99 mV, respectively, was observed (Asard et al. 1989). Even in the latter cases, three components may have been present.

3.2 Flavins

Extraction and quantification of noncovalently bound flavin from purified plasma membrane vesicles have yielded values of 0.2–2.4 nmol (mg protein)$^{-1}$ (Table 1). We assume that this flavin derives from one or several flavoproteins. It is possible that there are also covalently bound flavins in the plasma membrane, but there is no experimental evidence available.

3.3 Other Redox Components

Ubiquinones have been reported to be present in certain animal plasma membranes (Zambiano et al. 1975; Crane and Morré 1977; Crane et al. 1985a,b; Kalin et al. 1987), but very little information is available from plants. Lin (1984) reported an NADH minus oxidized difference spectrum of a pentane extract of trypsin-released components from corn root protoplasts. It had minima at 280 and 345 nm and a maximum at 315 nm. This was interpreted to be due to ubiquinone (Lin 1984) in spite of the fact that the spectrum of ubiquinone looks quite different (Crane and Barr 1971). Furthermore, Lin (1984) did not exclude contamination by mitochondria or chloroplasts released by the trypsin treatment.

By using methanol/light petroleum extraction, a method which extracts all the ubiquinone from mitochondria (Kröger 1978), substances, which show a difference spectrum similar to that of ubiquinone-10, are extracted from pure plasma membrane vesicles (P. Askerlund, personal communication). If this is indeed ubiquinone, the concentration in the plasma membrane can be calculated to be around 1.5 nmol (mg protein)$^{-1}$ (Møller et al. 1988b) or about five times higher than for the cytochrome content in the same tissue (Table 1). This is about the same ratio as in the respiratory chain of mitochondria where ubiquinone has a central role. Thus, it is possible that a quinone also has a function in the plant plasma membrane. Other quinones that should be considered as possible plasma membrane redox carriers are phylloquinone and tocopherolquinones since they have been reported to be present in mammalian plasma membranes (e.g., the erythrocyte plasma membrane, Yamamato et al. 1985). Their location in plant cells is not well known.

It has been suggested that one of the functions of the plasma membrane redox chain is to keep SH-groups reduced (Crane et al. 1985a; Bienfait and Lüttge 1988). The human erythrocyte plasma membrane contains about 80 nmol SH-groups (mg protein)$^{-1}$ (Crane et al. 1985a). So far there are no studies available on plant plasma membranes, but in Section 4.3.2 a model is presented which includes the reversible oxidation-reduction of SH-groups.

Iron-sulfur proteins are important redox carriers in all the energy-transducing membranes, the inner mitochondrial membrane, the thylakoid membrane, and the bacterial plasma membrane. Plasma membrane from mammals contains 4–8 nmol iron (mg protein)$^{-1}$ (Crane et al. 1985a), but there are no reports on iron-sulfur proteins in plant plasma membrane. It appears worthwhile to analyze both sulfur-bound and nonsulfur-bound iron (Tangerås et al. 1980) in purified plasma membrane vesicles. Early ESR studies on mammalian plasma membrane showed no iron-sulfur signals (H. Beinert, personal communication), but Dreyer and Treichler (1988) reported that ESR signals were detectable only after detergent treatment of synaptic plasma membranes. Special techniques may therefore be necessary for the study of nonheme iron in the plasma membrane.

4 Redox Activities

4.1 Activities with Various Donors and Acceptors

4.1.1 Basic Properties

The plasma membrane has been reported to catalyze a number of redox processes in vivo and in vitro. Most prominent among the electron donors used have been NADH, NADPH, and ferrocyanide and among the electron acceptors ferricyanide, Fe^{3+}EDTA, O_2, DCPIP, and cytochrome c. In many cases the observed activity lacks physiological relevance; e.g., ferricyanide and ferrocyanide are not normally found in biological systems, and cytochrome c, although present in all plant cells, is only found in the intermembrane space of the mitochondria and would therefore not have access to the plasma membrane in the intact cell. Nonetheless, experiments in which these nonphysiological donors and acceptors are used provide information about the reaction mechanism and the components involved. The identity of the natural electron donors and acceptors will be addressed in Section 4.3.

In a number of studies the efficiency of several donors/acceptors have been compared and Table 2 shows a compilation of the results. NADH is often a more efficient electron donor than NADPH, but in general activity is observed with both donors irrespective of the acceptor. Among the acceptors ferricyanide almost always gives the highest rates of electron transfer often followed by DCPIP. In general, the higher the redox potential of the acceptor, the higher is the activity (Sandelius et al. 1987).

A compilation of the pH optima and substrate affinities of these redox processes are found in Tables 3 and 4. A general feature of the pH optima is that there is often no clear optimum or little or no pH dependence in the physiological range pH 6–9. When a pH optimum is observed, it is often at acidic pH (Table 3) which means that the activity will be near maximal under the pH conditions in the cell wall. No trend can be found for substrate affinity except perhaps that the K_m for NADH is often in the mM range in the absence of added electron acceptor (Table 4). These activities could be due to peroxidase(s) (see Sect. 4.1.2).

A word of caution should be given here concerning the determination of kinetic constants and pH optima for membrane-bound enzymes with charged substrates [NAD(P)H and ferricyanide are negatively charged and cytochrome c is positively charged at neutral pH]. All plant membranes have an isoelectric point at pH 3.5–5.0 and they therefore carry a net negative charge at neutral pH. The plasma membrane is no exception (Møller et al. 1984; Körner et al. 1985). This charge will affect the concentration of any charged substrate near the active site and thus the apparent K_m, as has been shown for the NADH dehydrogenase on the outer surface of the inner membrane of plant mitochondria (Edman et al. 1985). The apparent pH optimum is also affected by these electrostatic interactions. To minimize such interactions and to ensure that kinetic results (and

Table 2. The efficiency of various redox mediators in plasma membrane redox processes. Papers only included if electron mediators other than ferricyanide were tried (DQ, duroquinone; PMS, phenazine methosulfate)

Process	Experimental material	Redox mediators ranked (range of activity)	Reference
Part A. Electron donor			
Whole cells and intact tissues			
Donor oxidation	*Lemna gibba*	NADH and $Fe^{2+}CN$ had no effect on the membrane potential	Lass et al. (1986)
	Sugarcane protoplasts	NADH = NADPH (7.2)[a]	Thom and Maretzki (1985)
		α-NADH = β-NADH (5.9 – 6.4)[a]	Komor et al. (1987)
	Corn root protoplasts	Menadione[b] stimulated NADH oxidation (oxygen consumption) by 50%	Lin (1984)
	Guard cell protoplasts[c]	NADH = NADPH (44)[d]	Pantoja and Willmer (1988)
	Carrot cell culture	NADH > glutathione = NADPH > ascorbate (1.73 – 0.37)[e]	Misra et al. (1985)
	Carrot cell culture	$Fe^{2+}CN$ was oxidized (rate not given)	Chalmers et al. (1984)
	Mesophyll cells[f]	$Fe^{2+}CN$ was not oxidized	Neufeld and Bown (1987)
	Barley roots	NADH > $Fe^{2+}CN$ (3 – 2)[g]	Ivankina and Novak (1988)
	Scenedesmus acuminatus	$Fe^{2+}CN$ was oxidized (52 in the presence of 4 mM $CaCl_2$)[g]	Novak and Miklashevich (1986)
	Elodea and *Valisneria* leaves	External NADH not oxidized	Ivankina and Novak (1988)
Isolated plasma membrane vesicles			
Donor oxidation	Wheat roots	NADH = NADPH (20)[h]	Møller and Bérczi (1986)
	Soybean hypocotyls	NADH > NADPH (2 – 1)[i]	Barr et al. (1985b)
NAD(P)H-$Fe^{3+}CN$	Maize roots	NADH > NADPH (1000 – 300)[i]	Buckhout and Hrubec (1986)
	Oat roots	NADH > NADPH (800 – 250)[j]	Ramirez et al. (1984)
	Soybean hypocotyls	NADH > NADPH (333 – 238)[k]	Sandelius et al. (1987)
		NADH > NADPH (367 – 339)[k]	Barr et al. (1986)
NAD(P)H-cytochrome c	Soybean hypocotyls	NADH = NADPH (21 – 18)[k]	Sandelius et al. (1987)
		NADH = NADPH (6 – 10)[k]	Barr et al. (1986)

Table 2. *Continued*

Process	Experimental material	Redox mediators ranked (range of activity)	Reference
	Sugar beet leaves	NADH ≫ NADPH (340 – 28)[i]	Askerlund et al. (1989)
	Cauliflower inflorescences	NADH ≫ NADPH (170 – 22)[i]	Askerlund et al. (1989)
	Spinach leaves	NADH ≫ NADPH (220 – 56)[i]	Askerlund et al. (1989)
	Barley roots	NADH ≫ NADPH (110 – 15)[i]	Askerlund et al. (1989)
NADH-Fe^{3+}EDTA	Soybean hypocotyls	NADH > NADPH (56 – 40)[k]	Barr et al. (1986)
Part B. Electron acceptor			
Whole cells and intact tissues			
NADH-acceptor	Intact oat roots	Fe^{3+}CN > cyt. c > dehydroascorbate > DQ (331 – 98)[l]	Rubinstein et al. (1984)
	Detached pea roots[m]	NADH + Fe^{3+}CN > Fe^{3+}CN only (4 – 1)[g]	Ivankina and Novak (1988)
Acceptor reduction	*Lemna gibba* Fe-sufficient	Fe^{3+}CN = DCPIP = methylene blue ≫ Fe^{3+}EDTA (100 – 10 mV depolarization of MP)	Lass et al. (1986)
	Lemna gibba Fe-deficient	Fe^{3+}CN = DCPIP = methylene blue = Fe^{3+}EDTA (70 – 100 mV depolarization for all)	Lass et al. (1986)
	Bean roots Fe-deficient[n]	Fe^{3+}CN = Fe^{3+}EDTA = DCPIP (30 – 40 mV depolarization of MP)[n]	Sijmons et al. (1984b)
		Fe^{3+}CN = DCPIP = Fe^{3+}EDTA = PMS ≫ FMN = FAD = riboflavin[o]	Sijmons and Bienfait (1983)
		A range of Fe^{3+}-chelates – most of them reduced (20 – 140)[p]	Bienfait et al. (1983)
	Maize roots	Hexachloroiridate IV is reduced (2.5)[g]	Lüthen and Böttger (1988)
	Phytoplankton	Various copper complexes (0.02)[a]	Jones et al. (1987)
	Carrot cells	Range of acceptors; only those with a redox potential over 0 mV are reduced	Crane et al. (1988b)

Table 2. *Continued*

Process	Experimental material	Redox mediators ranked (range of activity)	Reference
Isolated plasma membrane vesicles			
NADH-acceptor	Sugar beet leaves	$Fe^{3+}CN$ > phenyl-p-benzoquinone > cyt. c (1560 – 200)[i]	Askerlund et al. (1988, 1989)
	Maize roots	$Fe^{3+}CN$ > cyt. c > INT > DCPIP (-Triton X-100) (160 – 20)[k]	Buckhout and Hrubec (1986)
		$Fe^{3+}CN$ > DCPIP > cyt. c > INT (+Triton X-100) (1000 – 80)[k]	Buckhout and Hrubec (1986)
	Oat roots	$Fe^{3+}CN$ > cyt. c > O_2 (800 – 7)[k]	Ramirez et al. (1984)
	Barley roots	$Fe^{3+}CN \gg$ cyt. c (1331 – 111)[i]	Askerlund et al. (1989)
	Cauliflower inflorescence	$Fe^{3+}CN \gg$ cyt. c (860 – 170)[i]	Askerlund et al. (1989)
	Spinach leaves	$Fe^{3+}CN \gg$ cyt. c (1120 – 220)[i]	Askerlund et al. (1989)
	Soybean hypocotyl	$Fe^{3+}CN$ > cyt. c = DCPIP > hexammine ruthenium = gallocyanine = ferric oxalate > methylene blue (333 – 2)[k,q]	Sandelius et al. (1987)
	Soybean hypocotyls	$Fe^{3+}CN$ > DCPIP > cyt. c = $Fe^{3+}EDTA$ (160 – 6)[k]	Barr et al. (1986)

[a] μmol NAD(P)H oxidized (10^6 protoplasts/cells)$^{-1}$ h^{-1}.
[b] A membrane-permeant and autoxidizable dye.
[c] From *Commelina communis*.
[d] Increase in oxygen consumption in nmol (mg protein)$^{-1}$ min^{-1} upon NAD(P)H addition.
[e] Increase in respiration in nmol O_2 (mg dry wt)$^{-1}$ min^{-1}.
[f] From *Asparagus sprengeri*.
[g] μmol oxidized (g fresh wt)$^{-1}$ h^{-1}.
[h] nmol O_2 (mg protein)$^{-1}$ min^{-1}.
[i] nmol acceptor reduced (mg protein)$^{-1}$ min^{-1} in the presence of Triton X-100.
[j] nmol donor oxidized (mg protein)$^{-1}$ min^{-1}.
[k] nmol acceptor reduced (mg protein)$^{-1}$ min^{-1}.
[l] Percent of NADH oxidase rate.
[m] *Nitella* gave similar results, whereas NADH did not increase $Fe^{3+}CN$ reduction in *Elodea*.
[n] $Fe^{3+}CN$ had a much smaller effect upon the membrane potential of Fe-sufficient roots.
[o] Only plus or minus reaction was reported.
[p] Rates given in percent of rate with $Fe^{3+}EDTA$.
[q] The rates with indigo tetrasulfonate, indigo trisulfonic acid, indigo carmine, and ferric pyrophosphate were $\leq$ 1 nmol reduced acceptor (mg protein)$^{-1}$ min^{-1}.

Table 3. pH optima for plasma membrane redox processes

Process	Species	Experimental material	pH optimum	Reference
		A. Intact tissues and protoplasts		
NADH oxidation[a]	Maize	Root protoplasts	pH 5, Sharp decrease above pH 6	Lin (1982b)
[b]	Oat	Intact roots	Constant pH 5.5–7.5	Rubinstein et al. (1984)
[b]	Sugarcane	Protoplasts from cell culture	Same rate at pH 5.5 and 6.9	Komor et al. (1987)
NADH-Fe^{3+}CN[a]	Oat	Intact roots	Increasing pH 5.5–8.3	Rubinstein et al. (1984)
Fe^{3+}CN reduction	Oat	Intact roots	Increasing pH 5.5–8.3	Rubinstein et al. (1984)
	Maize	Apical root segments	Slight increase 5.3–8.3	Qui et al. (1985)
	Bean	Intact roots	Constant pH 4–8	Sijmons et al. (1984b)
	Oat	Peeled leaf segments	Broad pH optimum at pH 6.0; an increase above pH 8.0	Dharmawardhane et al. (1987)
	Carrot	Suspension culture	pH 7.0	Barr et al. (1985a)
	Scenedesmus acuminatus	Cells	Optimum at pH 5, slow decrease pH 5–7.5	Novak and Miklashevich (1986)
	Yeast	Cells	Constant pH 6.0–8.7 decreases below pH 6.0	Crane et al. (1982)
Fe^{2+}CN oxidation	*Scenedesmus acuminatus*	Cells	Optimum at pH 4, sharp decrease pH 4–6.5	Novak and Miklashevich (1986)
Fe^{3+}-oxalate reduction	Bean	Intact Fe-def. roots	Optimum at pH 2, decrease to pH 5	Bienfait et al. (1983)
Fe^{3+}-EDTA reduction	Bean	Intact Fe-def. roots	Constant pH 2–4.5, decrease pH 4.5–6	Bienfait et al. (1983)
	Bean	Intact Fe-def. roots	Optimum at pH 5.5	Cakmak et al. (1987)
	Peanut	Intact Fe-def. roots	Optimum at pH 5, 30% left at pH 7	Römheld and Marschner (1983)

Table 3. *Continued*

Process	Species	Experimental material	pH optimum	Reference
$FeNH_4(SO_4)_2$ reduction	Bean	Intact Fe-def. roots	Optimum at pH 2, decrease to 0 at pH 5	Bienfait et al. (1983)
$FeCl_3$	Peanut	Intact Fe-def. roots	Steady decrease pH 3.5–8.5	Römheld and Marschner (1983)
		B. Isolated plasma membrane vesicles		
NADH oxidation[a]	Wheat	Roots	Steady rise with decreasing pH	Møller and Bérczi (1986)
+ SHAM			Optimum at pH 6.5	Møller and Bérczi (1986)
NADH oxidation[a] + ferulic acid	Cauliflower	Inflorescence	Optimum at pH 4.5	Askerlund et al. (1987)
NADH oxidation[b]	Soybean	Hypocotyl	Broad optimum pH 6.5–8.5	Barr et al. (1985b)
NADH-Fe^{3+}CN[c]	Soybean	Hypocotyl	Broad optimum pH 6.5–8.0	Barr et al. (1986)[a]
	Wheat	Roots	pH 6.8	Brüggemann and Moog (1989)

[a] Measured as oxygen consumption.
[b] Measured as NADH consumption.
[c] Measured as Fe^{3+}CN reduction.

Table 4. Kinetics of plasma membrane redox processes

Process	Species	Experimental material	K_m, μM (substrate)	Reference
		A. Intact tissues and protoplasts		
NADH oxidation	Maize	Root protoplasts	800 (NADH)	Lin (1982b)
	Oat	Intact roots	600–1000 (NADH)	Rubinstein et al. (1984)
	Barley	Roots	200 (NADH)	Ivankina and Novak (1988)
	Sugarcane	Protoplasts from a cell culture	> 1000 (NADH)	Komor et al. (1987)
Fe^{2+}CN oxidation	Barley	Roots	230 (Fe^{2+}CN)	Ivankina and Novak (1988)
	Scenedesmus	Cells	ca 200 (Fe^{2+}CN)	Novak and Miklashevich (1986)

Table 4. *Continued*

Process	Species	Experimental material	K_m, μM (substrate)	Reference
NADH-$Fe^{3+}CN$	Oat	Intact roots	100 ($Fe^{3+}CN$)	Rubinstein et al. (1984)
$Fe^{3+}CN$ reduction	Maize	Intact roots	1400 ($Fe^{3+}CN$)	Federico and Giartosio (1983)
	Maize	Apical root segments	170–200 ($Fe^{3+}CN$)	Qui et al. (1985)
	Oat	Intact roots	1000–1400 ($Fe^{3+}CN$)	Rubinstein et al. (1984)
	Oat	Peeled leaf segments	ca. 200 ($Fe^{3+}CN$)	Dharmawardhane et al. (1987)
	Sycamore	Suspension cells	170 ($Fe^{3+}CN$)	Blein et al. (1986)
	Carrot	Suspension cells	50 ($Fe^{3+}CN$)	Barr et al. (1985a)
	Lemna gibba	Intact plants	115 ($Fe^{3+}CN$)	Lass et al. (1986)
	Scenedesmus	Cells	ca. 500 ($Fe^{3+}CN$)	Novak and Miklashevich (1986)
	Green alga	Vegetative internodes	200 ($Fe^{3+}CN$)	Thiel and Kirst (1988)
Fe^{3+}-EDTA reduction	*Lemna gibba*	Intact plants	107 (Fe^{3+}-EDTA)	Lass et al. (1986)
Fe^{3+}-chelate reduction	Peanut	Intact iron-def. roots	50–100	Röheld and Marschner (1983)
$Fe^{3+}CN$-induced depolarization	*Lemna gibba*	Intact plants	100 ($Fe^{3+}CN$)	Lass et al. (1986)
$Fe^{3+}CN$-induced H^+-pumping	*Asparagus*	Cell culture	ca 300 ($Fe^{3+}CN$)[a] ca 50 ($Fe^{3+}CN$)[b]	Neufeld and Bown (1987)
H^+-pumping	Maize	Intact roots	40 (O_2)	Böttger and Lüthen (1986)
Inhibition of K^+ uptake by $Fe^{3+}CN$	Maize	Apical root segments	80 ($Fe^{3+}CN$)	Rubinstein and Stern (1986)
Reduction of Cu^{2+} complexes	Phytoplankton	Cell culture	2–15 (Different complexes)	Jones et al. (1987)
		B. Isolated plasma membrane vesicles		
NADH oxidation	Wheat	Roots	1500 (NADH)[c] 40 (SHAM)	Møller and Bérczi (1986)
	Cauliflower	Inflorescences	3000 (NADH)[d] 75 (ferulic acid)[e] 0.13–8.7 (DQ)[f]	Askerlund et al. (1987) Asard et al. (1987)
	Soybean	Hypocotyls	100 (NADH)	Barr et al. (1985b)
NAD(P)H-$Fe^{3+}CN$	Sugar beet	Leaves	25–33 (NADH)[g] 8 ($Fe^{3+}CN$)[h]	Askerlund et al. (1988)
	Maize	Roots	24 (NADH)[i] 20 (NADPH)[i]	Buckhout and Hrubec (1986)
	Soybean	Hypocotyls	77 (NADH) 29 ($Fe^{3+}CN$)	Barr et al. (1986)

Table 4. *Continued*

Process	Species	Experimental material	K_m, μM (substrate)	Reference
	Soybean	Hypocotyls	50 (NADH) 20 (NADPH) 50 ($Fe^{3+}CN$)[j] 100 ($Fe^{3+}CN$)[k]	Sandelius et al. (1987) Sandelius et al. (1987)
NADH-Fe^{3+}EDTA	Barley	Roots	125 (NADH) 120 (Fe^{3+}EDTA)	Brüggerman and Moog (1989)
NADH-cytochrome c	Sugar beet	Leaves	6–8 (NADH)[g] 4 (cyt. c)[h]	Askerlund et al. (1988)

[a] Illuminated cells.
[b] Nonilluminated cells.
[c] With SHAM (sigmoidal curve).
[d] With ferulic acid (sigmoidal curve).
[e] Similar K_m for coniferyl alcohol and propyl gallate, all in the presence of 2 mM NADH.
[f] Two K_m in the presence of duroquinone (DQ) and KCN.
[g] The same results obtained with inside-out vesicles plus or minus Triton X-100 and with right-side-out vesicles plus Triton X-100.
[h] With inside-out vesicles.
[i] In the presence of Triton X-100.
[j] With NADH.
[k] With NADPH.

pH optima) are comparable, experiments should be performed in the presence of a high concentration of cations (100 mM KCl or perhaps better 5–10 mM $MgCl_2$) which screen the negative charges of membrane surfaces and reduce the size of the surface potential. A number of studies have reported relatively unspecific stimulations of redox activities in the plasma membrane by cations (Craig and Crane 1981; Federico and Giartosio 1983; Rubinstein et al. 1984; Marigo and Belkoura 1985; Belkoura et al. 1986; Dharmawardhane et al. 1987; Askerlund et al. 1988; Böttger and Hilgendorf 1988). These could have been caused by electrostatic interactions.

In the following sections we will review the properties of the main redox activities in the plasma membrane.

4.1.2 Peroxidase and NAD(P)H Oxidase

When NAD(P)H is added to an intact tissue, such as a root (eg., Rubinstein et al. 1984), to cell cultures (e.g., Misra et al. 1985) or to protoplasts (e.g., Lin 1982a,b; Komor et al. 1987) NAD(P)H is oxidized and an increase in O_2 consumption is often observed. The subsequent addition of salicylhydroxamic acid (SHAM), an inhibitor of the alternative oxidase in plant mitochondria (Schonbaum et al. 1971) or similar phenol-like compounds such as ferulic acid causes a further five- to tenfold increase. This SHAM-stimulated O_2 consumption has a high K_m for NADH, it is cyanide- and azide-sensitive but antimycin-insensitive (Spreen Brouwer et al. 1985; Bingham and Farrar 1987; van der Plas et al. 1987). The vast majority of the ferulic acid- or SHAM-stimulated activity is found in the soluble fraction of tissue homogenates (Askerlund et al. 1987; van der Plas et al. 1987), but a small fraction is recovered in purified plasma membranes to which it appears to be strongly bound (Askerlund et al. 1987, 1989). There are a number of observations suggesting that both the soluble (Spreen Brouwer et al. 1985; Komor et al. 1987; van der Plas et al. 1987; Pantoja and Willmer 1988) and at least the major part of the plasma membrane-bound activity (Møller and Bérczi 1985, 1986; Askerlund et al. 1987) are caused by peroxidase(s) (EC 1.11.1). The soluble enzyme(s) probably derives from the cell wall (see, e.g., Elstner and Heupel 1976), whereas the plasma membrane-bound activity has been suggested to be due to peroxidase(s) in transit from its site of synthesis in the cytoplasm to its work station in the cell wall (Askerlund et al. 1987). The peroxidase-mediated O_2 consumption proceeds via a radical chain reaction involving both O_2^- and H_2O_2 (see Askerlund et al. 1987 and references therein). The presence of such SHAM-stimulated peroxidase(s) makes it difficult to use SHAM to assess the activity and capacity of the alternative oxidase in the mitochondria in vivo as discussed by Møller et al. (1988a). We do not know the source of extracellular reducing equivalents for this peroxidase(s) (see Sect. 5.4).

The peroxidase(s), however, is not the only redox enzyme in the plasma membrane which can use O_2 as the electron acceptor. Purified plasma membrane vesicles from soybean hypocotyl also contain an auxin-stimulated NADH oxidase (Barr et al. 1985b; Morré et al. 1988b). This NADH oxidase activity is not due to a peroxidase since KCN and superoxide dismutase (EC 1.15.1.1) have no inhibitory effect. It has been purified 80-fold from the plasma membrane vesicles and the purified enzyme preparation contains only three polypeptides. It does not catalyze NADH-ferricyanide activity and is completely inhibited by 0.01 μM actinomycin (Brightman et al. 1988), an inhibitor of redox processes in mammalian plasma membranes, which inhibits NADH-ferricyanide activity in plasma membrane vesicles from the same tissue by $< 39\%$ (Morré et al. 1988a). The purified enzyme retains the ability to be specifically stimulated by auxin and active auxin analogs (Brightman et al. 1988). The redox components involved, the location of the active sites for NADH and oxygen, as well as the regulatory site for auxin are all unknown.

Pupillo and co-workers (see Guerrini et al. 1987 and references therein) have characterized and purified (240-fold) an NAD(P)H oxidase from microsomal membranes of zucchini squash. The enzyme catalyzes duroquinone-dependent O_2 uptake in the presence of cyanide. Purified plasma membranes of cauliflower inflorescences and mung bean hypocotyls contain a similar enzyme activity, but the activities are ten times lower (Asard et al. 1987) than those reported by Pupillo and co-workers for their microsomal fraction. This suggests that the enzyme purified by Pupillo and co-workers does not originate in the plasma membrane. Quinacrine, which inhibits the purified NADH oxidase of Brightman et al. (1988) completely at 100 μM, did not have any effect on this duroquinone-dependent NADH oxidase at 400 μM (Asard et al. 1987), indicating either that duroquinone accepts electrons at a point before the quinacrine block or that the NADH oxidase and the duroquinone-dependent NADH oxidase are due to different enzymes.

4.1.3 Reduction of Ferricyanide and Other Fe^{3+} Chelates

Ferricyanide is reduced by intact tissues of both monocotyledons and dicotyledons such as leaves (e.g., Dharmawardhane et al. 1987; Elzenga and Prins 1987) and roots (see, e.g., Tables 3 and 4), by cell cultures (Craig and Crane 1981; Chalmers et al. 1984; Barr et al. 1985a; Blein et al. 1986), and by protoplasts (Thom and Maretzki 1985) in the absence of an added reductant. Other Fe^{3+} chelates like Fe^{3+}EDTA are normally reduced at a much lower rate except in iron-deficient plants (Table 2; Bienfait 1985 and references therein; see also Sect. 5.1). The activity does not usually show a sharp pH optimum (Table 3) and the K_m reported for ferricyanide varies from 50 to 1400 μM (Table 4). This large variation in K_m might partly reflect tissue and species differences, but it probably also reflects different assay conditions (see Sect. 4.1.1).

Since ferricyanide (e.g., Craig and Crane 1981), and probably other Fe^{3+} chelates, are membrane-impermeant this reduction must take place on the outside of the living cells. In *Elodea* leaves, which only have transfer cells (see Sect. 5.1) on the lower surface, ferricyanide is reduced only by the transfer cells on the lower leaf surface in the light. In the dark, on the other hand, ferricyanide reduction takes place on both the upper and the lower leaf surface (Elzenga and Prins 1987). One possible explanation for these results is that the supply of substrate, NAD(P)H, is restricted in the upper cell layer of the leaves in the light.

Investigation of trans-plasma membrane ferricyanide reduction in isolated plasma membrane vesicles will be treated in Section 4.1.4 since it is necessary to add an electron donor like NAD(P)H to measure the activity in this in vitro system. The question of the possible coupling of ferricyanide reduction to H^+ pumping will be treated in Section 4.2.

4.1.4 NAD(P)H-Acceptor (Other than Oxygen) Reductase

The addition of NAD(P)H to intact tissues, cell cultures, or protoplasts will increase the rate of ferricyanide reduction (Rubinstein et al. 1984; Thom and Maretzki 1985) and cause/induce the reduction of other electron acceptors like DCPIP, duroquinone, and cytochrome c (Rubinstein et al. 1984). However, due to the formation of oxygen radicals particularly in the presence of a peroxidase it is difficult to know whether these other acceptors are interacting directly with an NAD(P)H dehydrogenase on the membrane surface or whether the oxygen radicals act as intermediates in the electron transfer. For instance, superoxide dismutase inhibits ferricyanide reduction by NADH with intact Fe-deficient bean roots (Cakmak et al. 1987). In addition, use of intact cells only gives us access to the outer, apoplastic surface of the plasma membrane. For these reasons, we will in the following mainly discuss data obtained with purified plasma membrane vesicles.

Although antimycin A-insensitive NAD(P)H-cytochrome c reductase activity is commonly referred to as a marker for the endoplasmic reticulum, it is also present in most other plant cell membranes including the plasma membrane (review, Møller and Lin 1986; see also Chap. 2). Purified plasma membrane vesicles often contain a relatively low activity compared to the microsomal fraction (e.g., Lundborg et al. 1981; Widell and Larsson 1983; Buckhout and Hrubec 1986; Hodges and Mills 1986), however, in some tissues the plasma membrane contains up to five times higher specific activities of NAD(P)H-cytochrome c reductase than the microsomal fraction (Caubergs et al. 1986; Larsson et al. 1987). This variation probably depends on the relative amount of endoplasmic reticulum and plasma membrane in the microsomal fraction. In the endoplasmic reticulum NADH-cytochrome c reductase activity is based on electron transfer from the enzyme to cytochrome b_5 which in turn reduces cytochrome c nonenzymatically. NADPH-cytochrome c reductase activity is attributed to a different specific NADPH dehydrogenase which normally acts to reduce cytochrome P-450, but can also reduce added cytochrome c. Thus, oxidation of NADH and NADPH in the endoplasmic reticulum depends on two separate enzymes (De Pierre and Ernster 1977).

NAD(P)H-ferricyanide reductase activity is catalyzed by most flavoprotein dehydrogenases and the plasma membrane contains high activities (Barr et al. 1986; Buckhout and Hrubec 1986; Sandelius et al. 1987; Askerlund et al. 1988; Luster and Buckhout 1988; Bérczi et al. 1989). Both right side-out (apoplastic side-out) and inside-out (cytoplasmic side-out) plasma membrane vesicles can now be isolated (Canut et al. 1988; Larsson et al. 1988; Chap. 3). Right side-out plasma membrane vesicles from maize and wheat roots and sugar beet leaves show low NAD(P)H-ferricyanide, NAD(P)H-quinone, and NAD(P)H-cyt. c reductase activities which are greatly enhanced (by a factor of 5–10) upon addition of an optimal concentration of detergent (Buckhout and Hrubec 1986; Askerlund et al. 1988; Bérczi et al. 1989). Since neither NAD(P)H, ferricyanide, nor cytochrome c can cross the plasma membrane this indicates that the binding site for at least one of these substrates is located on the cytoplasmic surface of the plasma membrane. Likewise Morré et al. (1987) found stain deposits indicating NADH-ferricyanide reductase activity on the outer surface of inside-out plasma membrane vesicles after incubation in the staining solution, whereas no deposits were found on right side-out vesicles. Consistent with this they only found stain deposits on the inner, cytoplasmic surface of the plasma membrane of broken cells in soybean hypocotyl segments. After a thorough kinetic analysis on both right side-out and inside-out plasma membrane vesicles from sugar beet leaves, Askerlund et al. (1988) concluded that "...both donor and acceptor sites for [NADH-ferricyanide and NADH-cyt. c reductase] are located on the cytoplasmic surface of the plasma membrane, and that a possible trans-plasma membrane electron transport would constitute only a minor proportion of the total activity". Furthermore, Giannini and Briskin (1988), working with a crude plasma membrane fraction from red beet storage tissue, found no evidence for transmembrane electron transport from NADH to ferricyanide. It should be noted, however, that none of these studies were conducted on material from Fe-deficient roots where a relatively stronger trans-plasma membrane electron transport would be expected (see Sect. 5.1.1).

NADH-ferricyanide reductase activity reaches a maximum at 0.015% (w/v) Triton X-100 (this value will vary with the protein concentration) and then stays constant at higher detergent concentrations (Buckhout and Hrubec 1986; Askerlund et al. 1988). In contrast, NADH-cyt. c reductase activity is strongly inhibited at Triton X-100 concentrations above 0.015–0.020% and probably never reaches the true maximum (Askerlund et al. 1988). This indicates that several components (viz. a short electron transport chain containing a flavoprotein and a b-type cytochrome) are involved in NADH-cyt. c reductase activity and that these compounds are separated upon solubilization of the membrane. Fewer components (viz. a flavoprotein) are involved in NADH-ferricyanide reductase activity (Crane et al. 1985a; Askerlund et al. 1988; Møller et al. 1988b).

Luster and Buckhout (1988) and Buckhout and Luster (1988) solubilized purified plasma membrane vesicles from maize roots and attempted to separate and purify the different redox activities. NAD(P)H-ascorbate free radical reductase activity could be separated from NAD(P)H-ferricyanide and NAD(P)H-duroquinone reductase activities. A component catalyzing only NAD(P)H-ferricyanide reductase activity could then be separated from one catalyzing both NAD(P)H-ferricyanide and NAD(P)H-duroquinone activities. Thus, there are probably three distinct redox enzymes in this material. The component catalyzing both NAD(P)H-ferricyanide and -duroquinone reductase activities was purified to homogeneity showing one silver-stained polypeptide of 28 kD (Buckhout and Luster 1988).

Brüggemann and Moog (1989) solubilized plasma membrane from barley roots and found one band on a native gel that stained for NADH-Fe^{3+}EDTA reductase activity. This band was shown to consist of at least four polypeptides with molecular masses of 94–205 kD. Whether this redox system is at all related to that purified by Buckhout and Luster (1988) remains to be established.

4.1.5 Nitrate Reductase

The plasma membrane of barley seedlings contains significant NAD(P)H nitrate reductase (EC 1.6.6.2) activities. The activity is highly latent suggesting that the active site for nitrate is on the inner, cytoplasmic surface (Ward et al. 1988). Based on results with nonpermeant external electron acceptors and antibodies against nitrate reductase, Jones and Morel (1988) concluded that nitrate reductase spans the plasma membrane in the diatom *Thalassiosira.* They proposed that a membrane-spanning diaphorase is responsible for trans-plasma membrane electron transport and that cyt. b-557 and the molybdo-protein in nitrate reductase loop back to the cytoplasmic surface. Future research will show to what extent this nitrate reductase, its redox components and its activity are identical to previously observed redox components and activities.

4.2 Coupling to Proton Transport?

The possibility that the plasma membrane redox system(s) could be proton pumping and therefore involved in energization of the plasma membrane and transport of solutes across the plasma membrane has attracted a lot of attention since Lin (1982a,b; 1984) reported that addition of exogenous NADH increased (1) the rate of O_2 uptake, (2) the size of the membrane potential, and (3) efflux of H^+ and uptake of K^+ and P_i into protoplasts and segments of corn roots. The stimulation of K^+ uptake by the addition of NADH was confirmed by Misra et al. (1985) using carrot cells. However, the observations by Lin could not be repeated by Kochian and Lucas (1985) who also used corn root segments.

Before we discuss other evidence it will be helpful if we, as suggested by E. Marrè et al. (1988), consider various models and the predictions we can derive from them concerning the expected initial change in pH in the cytoplasm and the medium, the initial $\Delta H^+/e^-$ ratio in the medium and the *initial* effect on the membrane potential *without assuming a direct coupling to H^+ pumping*. We stress that these are initial effects because any net transmembrane electron transport must soon be coupled to another charge-compensating process to avoid extreme membrane potentials. Eight alternatives are presented in Fig. 2. In models I-IV the electron donor is on the inner surface of the plasma membrane (cytoplasmic side) and in models V-VIII on the outer surface (apoplastic side). Models IX and X consider the possible direct coupling of electron and proton transport. A key fact in most of the models is that NAD(P)H turnover always involves the appearance/disappearance of a H^+ depending on the electron acceptor. Thus, models I and II where cytoplasmic NAD(P)H is oxidized differ only by the location of the site of O_2 binding, i.e., on the inner surface in model I and on the outer surface in model II, but the predictions are completely different (Table 5). Models V and VI differ in the same way, only here the oxidation of NAD(P)H is on the outer surface. The results by Lin (see above) are consistent with the predictions of model VI, but the acidification of the external medium may only be due to NADH oxidation and not to H^+ pumping across the plasma membrane (this will depend on the H^+/e^- ratio). However, to our knowledge there is no other evidence to suggest that such a trans-plasma membrane electron transport from the outer to the inner surface takes place.

The H^+/e^- ratio for the appearance of H^+ in the external medium during ferricyanide reduction in a variety of tissues falls within the range 0.3–1.1 and it is often close to 1.0 (Table 6). This could indicate that the trans-plasma

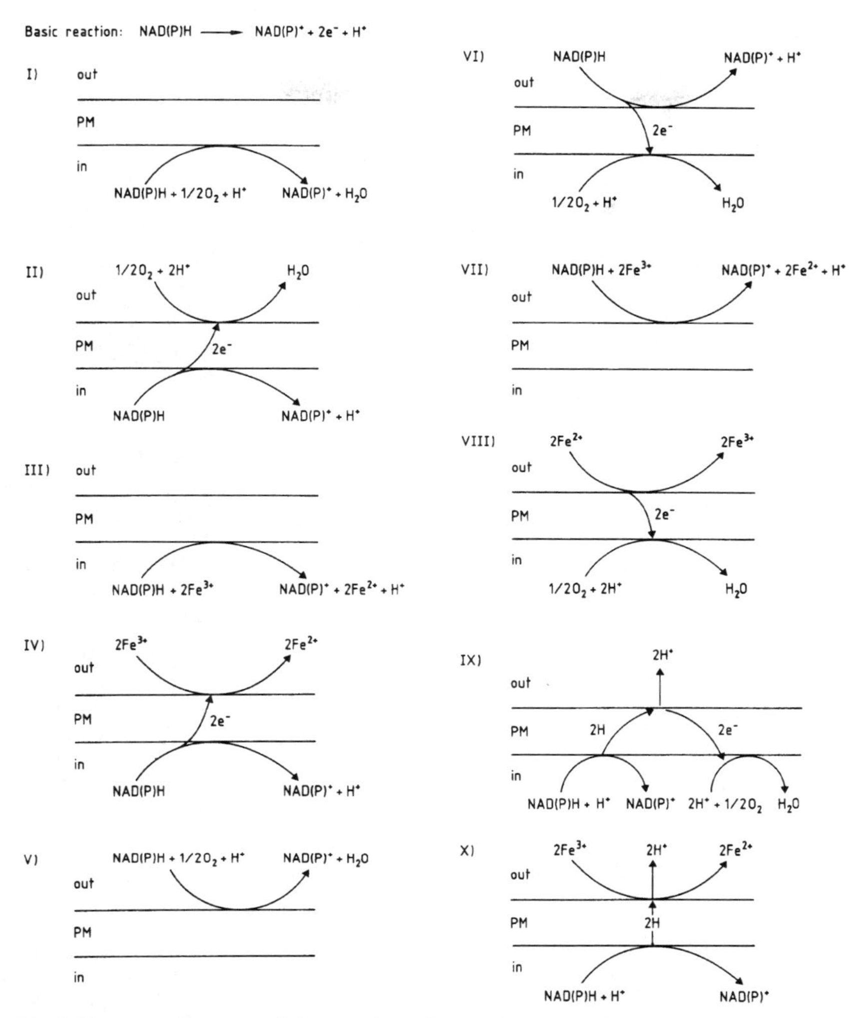

Fig. 2. Turnover of protons and electrons depending on the nature of the electron acceptor and on which side it interacts with the plasma membrane (*PM*). It is assumed that NAD(P)H is the electron donor in the cases in which a cytoplasmic electron donor is involved and that either O_2 or Fe^{3+} is the electron acceptor. The plasma membrane has a negative membrane potential (negative on the inside, positive on the outside). Fe^{2+} and Fe^{3+} represent any iron chelate. For explanation of models *I-X*, see text and Table 5

Table 5. Predictions from the models presented in Fig. 2

Model	Net change in pH		Medium $\Delta H^+/e^-$	Effect on membrane potential	Comments
	Medium	Cytoplasm			
I	Zero	Increase	Zero	None	Same principle for monooxygenase activity
II	Increase	Decrease	-1.0	Depolarization	This mechanism in effect collapses the proton gradient
III	None	Decrease	Zero	None	Major dehydrogenase activity in isolated plasma membrane vesicles
IV	None	Decrease	Zero	Depolarization	Iron reduction
V	Increase	Zero	-0.5	None	External NAD(P)H oxidase or peroxidase
VI	Decrease	Increase	+0.5	Hyperpolarization	The pH change in the medium independent of the nature of the acceptor
VII	Decrease	Zero	+0.5	None	Not likely mechanisms in vivo, but can occur in in vitro experiments (see Tables 2-4)
VIII	None	Increase	Zero	Hyperpolarization	
IX	Decrease	Increase	+ 1.0	Hyperpolarization	These mechanisms create a proton gradient
X	Decrease	Increase	+ 1.0	None per se but compared to IX it leads to a depolarization	

membrane redox system is directly coupled to electron transport. However, the addition of ferricyanide to intact cells and tissues normally causes a rapid depolarization of the trans-plasma membrane potential of 40–100 mV at high concentrations of ferricyanide (e.g., 1.0 mM) (Sijmons et al. 1984b; Lass et al. 1986; Novak and Miklashevich 1986; Elzenga and Prins 1987; Guern and Ullrich-Eberius 1988; Ivankina and Novak 1988; Thiel and Kirst 1988). This depolarization means that more electrons than cations (e.g., protons) pass across the plasma membrane, as already pointed out by Sijmons et al. (1984b), and it is, thus, consistent with model IV. The depolarization has also been explained on the basis of a transmembrane electron loop in which a protonated electron carrier transfers H^+ to the outside while the electron is returned to the inside by an electron carrier (e.g., a cytochrome) (Model IX; Novak and Ivankina 1983;

Table 6. Proton/electron ratios for plasma membrane redox processes. Unless otherwise indicated all data are measured as ferricyanide-induced increased acidification of the medium

Species	Experimental material	H^+/e^- Ratio	Reference
Maize	Intact roots	0.3	Federico and Giartosio (1983)
	Apical segments	0.5–0.8 (After 10 min lag)	Rubinstein and Stern (1986)
		0.01–0.05 (With inhibitors)	Rubinstein and Stern (1986)
Bean	Intact roots	0.5	Sijmons et al. (1984b)
Elodea	Leaves	< 1	M.T. Marrè et al. (1988)
Lemna gibba	Intact plants	0.58–0.64	Lass et al. (1986)
		0.51–0.86	Guern and Ullrich-Eberius (1988)
Carrot	Cell culture	1.1	Craig and Crane (1985)
Sycamore	Cell culture	0.93 0.94[a]	Blein et al. (1986)
Asparagus	Cell culture	1.2 (Light) 0.99 (Dark)	Neufeld and Bown (1987)
Scenedesmus	Cells	0.99 0.92[a]	Novak and Miklashevich (1986)
Lamprothamnium papulosum (green alga)	Cells	0.8–1.1	Thiel and Kirst (1988)

[a] H^+/e^- ratio for the decrease in medium acidification or increase in medium alkalization during $Fe^{2+}CN$ oxidation.

Ivankina et al. 1984; Böttger and Lüthen 1986). As shown in model X, when ferricyanide is present, both electrons and protons would be released to the outside so that the membrane potential maintained by the loop shown in model IX is lost. This type of electron loop is similar to the scheme of Mitchell (1966) for proton transfer in mitochondria.

In the above experiments both the H^+-ATPase and the redox system were active and H^+ appearing in the external medium could have been pumped by any one of the two systems. For this reason several investigators have used inhibitors to determine whether H^+ pumping has one or two components. Neufeld and Bown (1987) and Bown and Crawford (1988) showed that, in the absence of ferricyanide, addition of ATPase inhibitors such as dicyclohexylcarbodiimide could completely inhibit acidification of the medium. The subsequent addition of ferricyanide was then accompanied by a renewed acidification. Assuming that the ATPase was still inhibited in the presence of ferricyanide, it was then concluded that electron transport to ferricyanide was coupled to H^+ transport. However, it is still possible that the addition of ferricyanide caused a depolarization of the membrane potential across the plasma membrane and that this depolarization in some way was responsible for the increased H^+ efflux. It would therefore be interesting to monitor the membrane potential during such an experiment. In contrast to Bown and co-workers, Rubinstein and Stern (1986) found a complete inhibition of H^+ pumping by ATPase inhibitors in the presence of ferricyanide. However, this could have been due to the buffering capacity of the added inhibitors which would lead to an overestimation of the degree of inhibition as pointed out by Neufeld and Bown (1987).

If the initial flux of electrons to external ferricyanide is not accompanied by H^+, as strongly indicated by the depolarization observed, the question arises of where the protons, which eventually appear in the medium often in near stoichiometric amounts (Table 5), come from. Since a lag is sometimes observed in the appearance of the H^+ (e.g., Rubinstein and Stern 1986; Guern and Ullrich-Eberius 1988) and since K^+ together with H^+ appear to act as charge compensators rather than H^+ alone (Guern and Ullrich-Eberius 1988), it has been concluded that model IV is correct and that the acidification of the cytoplasm and the depolarization of the membrane potential activate the H^+-ATPase as well as K^+ channels (Rubinstein and Stern 1986; Guern and Ullrich-Eberius 1988; E. Marrè et al. 1988; M.T. Marrè et al. 1988). In favor of this interpretation is also the common observation that ferricyanide addition causes an inhibition of K^+ uptake or a stimulation of K^+ efflux (Sijmons et al. 1984b; Kochian and Lucas 1985; Blein et al. 1986; Lass et al. 1986; Novak and Miklashevich 1986; Rubinstein and Stern 1986; Guern and Ullrich-Eberius 1988; Thiel and Kirst 1988).

4.3 Natural Electron Donors and Acceptors

4.3.1 Electron Donors

When external ferricyanide is reduced by intact roots, a parallel lowering of the total amount of NADPH in the whole tissue is observed. Simultaneously, the level of $NADP^+$ increases, whereas the levels of NADH and NAD^+ are relatively constant (Sijmons et al. 1984a; Qui et al. 1985). This has been interpreted to mean that NADPH is the donor to the enzyme responsible for transmembrane electron transport. This approach has two major weaknesses: (1) The data obtained do not actually give any information about the cytoplasmic pools of free pyridine nucleotides which are directly in contact with the plasma membrane; (2) Even if the size of the cytoplasmic pool of NADPH were found to decrease, it would not be proof that NADPH acted as the direct donor. A change in pool size can be due *either* to a change in the rate of synthesis *or* to a change in the rate of consumption *or* to both. What is observed is a new steady-state level and only direct measurements of the actual rates of synthesis and consumption will allow further conclusions.

In another approach Craig and Crane (1981) and Böttger and Lüthen (1986) added compounds to the cells/tissues which would affect the rate of glycolysis and observed a correlation with the rate of ferricyanide reduction by the experimental material. Since NADH is one of the products of glycolysis, these results were interpreted to mean that NADH can be an electron donor. Again, this is indirect evidence.

Buckhout et al. (1989) isolated pure plasma membrane vesicles from Fe-sufficient and Fe-deficient tomato roots and found that both NADH-ferricyanide and NADH-Fe^{3+}citrate reductase activities were approximately twice as high in plasma membrane vesicles from Fe-deficient plants as in vesicles from Fe-sufficient plants. In contrast, the NADPH-receptor reductase activities were almost the same in the two types of vesicles. Since the Fe-deficient roots showed a strongly (sevenfold) increased rate of Fe^{3+} reduction these results indicate that an NADH-dependent enzyme is induced by iron deficiency (see also Sect. 5.1.1). The reason that the relative stimulation of NADH-receptor reductase activities was much lower than the increased rate of Fe^{3+} chelate reduction by the intact roots is probably that the latter primarily takes place in the transfer cells, the plasma membrane of which probably only constitutes a small proportion of the total plasma membrane extracted and purified from the tissue.

There are a number of other potential natural electron donors, e.g., ascorbate, semidehydroascorbate, reduced glutathione, but there is little or no information about their occurrence and concentration in the cytoplasm of plant cells or about their possible interaction with the plasma membrane redox system(s).

4.3.2 Electron Acceptors

Fe^{3+}-chelates are the natural substrates for the transmembrane redox system called the Turbo system (Sect. 5.1.2). When it comes to the natural acceptor for the Standard system we are still very much in the dark. Bienfait and Lüttge (1988) suggested that the function of the Standard system was to keep essential SH-groups reduced, that transport proteins were regulated by the reduction state of certain SH-groups, and that this reduction state in turn was controlled by the Standard redox system. Since the activity of the K^+,Mg^{2+}-ATPase is regulated by redox mediators (Elzenga et al. 1989), we have here an attractive model for the direct regulation of the ATPase by the redox system in the plasma membrane (Fig. 3). The final electron acceptor is O_2 in the model. This mode of action need not be restricted to transport proteins. The plasma membrane from mammalian cells contains a phospholipase c which is specific for hydrolysis of phosphatidylinositol phosphate. This enzyme has extensive sequence homology with thioredoxin including clustered SH-groups which can easily be reduced, e.g., by thioredoxin reductase. The oxidation-reduction of the paired

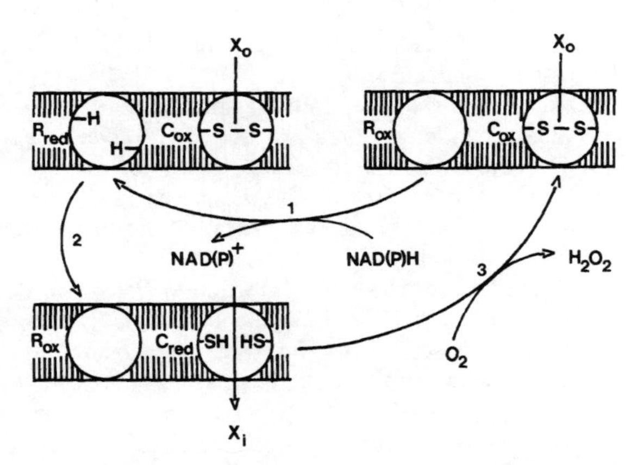

Fig. 3. Model of the putative operation of the 'Standard' reductase. *R*, plasma membrane oxido-reductase; *C*, a transporter, carrier, or permease (e.g., sucrose carrier or, in the reverse direction, H^+ extrusion); *X*, a transported solute; *o*, outside; *ox*, oxidized; *red*, reduced. *1* NAD(P)H reduces a group(s) on the reductase, probably a flavin. *2* The reductase reduces an -S-S- bond in an adjacent carrier or channel leading to activation. *3* The SH-groups in the carrier are (nonenzymatically?) reoxidized by oxygen and the carrier inactivated (After Bienfait and Lüttge 1988)

SH-groups could change the phospholipase from an active to an inactive form (Bennett et al. 1988).

Morré et al. (1986) reported that in purified plasma membrane vesicles from soybean hypocotyl semidehydroascorbate could act as acceptor for trans-plasma membrane electron transport from external NADH. However, they presented no direct evidence that the semidehydroascorbate was involved nor did they exclude O_2 as the actual acceptor.

5 Physiological Relevance

5.1 Iron and Ion Uptake

5.1.1 Reduction and Uptake of Iron

When dicotyledons, or nongraminaceous monocotyledons, begin to suffer from Fe-deficiency their roots develop several new characteristics (see Bienfait 1985 and Marschner 1986 for more details and references): (1) Transfer cells develop in the epidermis. These transfer cells have convoluted cell walls and the plasma membrane therefore has a greatly increased surface area. (2) The efflux of H^+ is greatly increased. (3) The capacity to reduce Fe^{3+} is greatly increased. At the same time the specificity changes such that a range of iron chelates can now be reduced instead of only ferricyanide. The result of the lowered pH in the immediate vicinity of the root is that Fe^{3+} is made more soluble; this Fe^{3+}, in the form of various chelates, can then be reduced at the surface of the roots; and finally the iron is taken up as Fe^{2+} without the chelator (Römheld and Marschner 1983). The observation that a tomato mutant that is unable to develop transfer cells is also unable to develop pH-lowering and Fe^{3+}-reduction capacity (Bienfait 1985) indicates that these characteristics are all linked.

Bienfait (1985) has suggested that all plant cells contain the so-called Standard redox system, whereas Fe-deficient plants also contain another, inducible system the so-called Turbo system. The former can only reduce ferricyanide or other oxidants with a redox potential greater than 0 mV (Crane et al. 1988b), whereas the latter may have a much broader specificity (Bienfait et al. 1983; Bienfait 1985). Both systems are thought to reside in the plasma membrane, but so far their separation, purification, and characterization have not been published and we therefore know nothing about the redox components involved.

5.1.2 Ion Uptake

The possibility that the plasma membrane redox system(s) could be H^+ pumping and thus participate in the creation of the membrane potential and the H^+ gradient used to drive solute transport across the plasma membrane has naturally attracted a lot of attention as described in Section 4.2. The conclusion above was that all of the results could be explained by a model in which trans-plasma membrane electron transport from NAD(P)H in the cytoplasm to an electron acceptor on the apoplastic side would depolarize the membrane potential and cause an acidification of the cytoplasm. This, in turn, would activate the H^+-ATPase resulting in an increased H^+ pumping. Such trans-plasma membrane electron transport also normally results in an efflux of K^+ as part of the charge neutralization. Another way in which the redox system could affect membrane transport is by the mechanism described in Section 4.3.2 and Fig. 3.

In conclusion, we do not think that the redox system(s) is directly coupled to H^+ pumping, membrane energization, and membrane transport. However, the redox system(s) may help to regulate these processes through effects on the H^+-ATPase.

5.2 Blue Light Response

The action spectrum of LIAC (see Sect. 3.1 for a definition) in purified plasma membrane is very similar to that of blue-light photomorphogenesis (Widell et al. 1983). Thus, the flavoprotein-cyt. b complex could be the receptor for the light signal (Kjellbom et al. 1985; see Widell 1987 for references). How the light signal is converted to an intracellular message is not known; however, it has been suggested that reduction of a specific compound with a high biological activity, such as a plant hormone, on the cytoplasmic surface of the plasma membrane could be involved (Kjellbom et al. 1985).

5.3 Growth Control

The growth of animal cell cultures is stimulated by ferricyanide, indicating that plasma membrane electron transport is somehow important to growth. In contrast, growth of carrot cells is inhibited by ferricyanide (Crane et al. 1985b and references therein). Adriamycin, *cis*-platin and actinomycin D are inhibitors of mammalian cell growth, so-called antiproliferative agents, and p-nitrophenylacetate is an inhibitor of redox processes in mammalian plasma membrane (Morré et al. 1988a,b). When these compounds were added to plasma membrane vesicles (a mixture of right-side out and inside-out orientation) from soybean hypocotyls at 10–100 μM they all inhibited NADH-ferricyanide reductase activity to some extent (30–55%), whereas only actinomycin D inhibited NADH oxidase significantly (by 100%) (Morré et al. 1988b). The compounds inhibiting NADH-ferricyanide activity also inhibited auxin-stimulated elongation growth of excised soybean hypocotyl segments (Morré et al. 1988a,b), but actinomycin D was the best inhibitor of growth. Since the NADH oxidase, but not the NADH-ferricyanide reductase, is stimulated by auxin the results with actinomycin are consistent with the view that there is a direct correlation between growth and the NADH oxidase (Morré et al. 1988b). However, the results with the other inhibitors are not consistent with this interpretation.

5.4 Cell Wall Synthesis

Peroxidases are involved in the polymerization of phenols excreted from the cell to form lignin (Mäder and Füssl 1982). The source of the electrons is not clear, but it has been suggested that malate is released from the cell and that malate dehydrogenase (EC 1.1.1.37), reported to be present in the cell wall, converts NAD^+ to NADH (Gross 1977). This NADH could then react with O_2 in a nonenzymatic process to produce H_2O_2 used in lignin synthesis.

An alternative model is that trans-plasma membrane electron transport reduces either O_2 to O_2^- or Fe^{3+} to Fe^{2+} on the outer surface of the plasma membrane (Cakmak et al. 1987). This O_2^- and Fe^{2+} could then form H_2O_2 either directly by dismutation of O_2^- or by reaction of Fe^{2+} with O_2:

$$Fe^{2+} + O_2 \rightarrow Fe^{3+} + O_2^-$$

followed by dismutation of the O_2^- to form H_2O_2.

5.5 Defence Against Pathogen Infection

The production of extracellular O_2^- and/or H_2O_2 by trans-plasma membrane electron transport could also be part of the plant cell's defence mechanism against invading pathogens as outlined in Chapter 14.

6 Summary

The plasma membrane of plant cells contain two or more b-type cytochromes, the major one of which has a midpoint potential of + 150 mV. The presence of cytochrome P-420, cytochrome P-450, and cytochrome b_5 has also been indicated. In addition, noncovalently bound flavin(s) and peroxidase(s) are found in the plasma membrane. These redox components, and possibly others like sulfhydryl groups, are involved in the following redox activities: (1)trans-plasma membrane electron transport from the cytoplasmic surface to Fe^{3+} chelates on the apoplastic surface. (2) NAD(P)H-ferricyanide reductase activity on the cytoplasmic surface. (3) NAD(P)H-quinone reductase activity on the cytoplasmic surface. (4) NAD(P)H-ascorbate free radical reductase activity on the cytoplasmic surface. (5) SHAM-stimulated NADH-dependent oxygen consumption by a peroxidase probably on the apoplastic surface. (6) Auxin-stimulated NADH oxidase, the location of which is not known. The trans-plasma membrane activity is probably not directly coupled to H^+ pumping. Instead it may activate the H^+-ATPase. The natural electron donor to all of these redox activities is probably either NADH or NADPH but thiols have not been excluded as possible electron donors. The only established physiological function of the redox systems is the reduction of Fe^{3+} chelates as a step in iron uptake by Fe-deficient roots of nongraminaceous plants, but the redox systems are probably involved in several other important functions.

Acknowledgments. We would like to thank Drs. Per Askerlund, Rita Barr, Ingo Dahse, Martin G. Klotz, Christer Larsson, D. James Morré, and Susanne Widell for many stimulating discussions. We are also grateful to a number of other colleagues who provided us with manuscripts of papers in press. The work in the authors' laboratories was supported by The National Institutes of Health (F.L.C.), The Swedish Natural Science Research Council, and the Carl Tesdorpf Foundation (I.M.M.).

References

Asard H, Caubergs R, Renders D, De Greef JA (1987) Duroquinone-stimulated NADH oxidase and b type cytochromes in the plasma membrane of cauliflower and mung beans. Plant Sci 53:109–119

Asard H, Venken M, Caubergs R, Reijnders W, Oltmann FL, De Greef JA (1989) b-Type cytochromes in higher plant plasma membranes. Plant Physiol 90:1077–1083

Askerlund P, Larsson C, Widell S, Møller IM (1987) NAD(P)H oxidase and peroxidase activities in purified plasma membranes from cauliflower inflorescences. Physiol Plant 71:9–19

Askerlund P, Larsson C, Widell S (1988) Localization of donor and acceptor sites of NADH dehydrogenase activities using inside-out and right-side-out plasma membrane vesicles from plants. FEBS Lett 239:23–28

Askerlund P, Larsson C, Widell S (1989) Cytochromes of plant plasma membranes. Characterization by absorbance difference spectrophotometry and redox titration. Physiol Plant 76:123–134

Barr R, Craig TA, Crane FL (1985a) Transmembrane ferricyanide reduction in carrot cells. Biochim Biophys Acta 812:49–54

Barr R, Sandelius AS, Crane FL, Morré DJ (1985b) Oxidation of reduced nucleotides by plasma membranes of soybean hypocotyl. Biochem Biophys Res Commun 131:943–948

Barr R, Sandelius AS, Crane FL, Morré DJ (1986) Redox reactions of tonoplast and plasma membranes isolated from soybean hypocotyls by free-flow electrophoresis. Biochim Biophys Acta 852:254–261

Belkoura M, Ranjeva R, Marigo G (1986) Cations stimulate proton pumping in *Catharanthus roseus* cells: implications of a redox system? Plant Cell Environ 9:653–656

Bennett CF, Balcarek JM, Varichio A, Crooke ST (1988) Molecular cloning and complete amino-acid sequence of form-I phosphoinositide-specific phospholipase C. Nature 334:268–270

Bérczi A, Møller IM (1986) Comparison of the properties of plasmalemma vesicles purified from wheat roots by phase partitioning and by discontinuous sucrose gradient centrifugation. Physiol Plant 68:59–66

Bérczi A, Larsson C, Widell S, Møller IM (1989) Separation of wheat root microsomal membranes by countercurrent distribution. An evaluation of plasma membrane markers. In: Loughman BC (ed) Structural and functional aspects of transport in roots. Kluwer Academic Publishers, pp 69–72

Bienfait HF (1985) Regulated redox processes at the plasmalemma of plant root cells and their function in iron uptake. J Bioenerg Biomembr 17:73–83

Bienfait HF, Lüttge U (1988) On the function of two systems that can transfer electrons across the plasma membrane. Plant Physiol Biochem 26:665–671

Bienfeit HF, Bino RJ, van der Bliek AM, Duivenvoorden JF, Fontaine JM (1983) Characterization of ferric reducing activity in roots of Fe-deficient *Phaseolus vulgaris*. Physiol Plant 59:196–202

Bingham IJ, Farrar JF (1987) Respiration of barley roots: assessment of activity of the alternative path using SHAM. Physiol Plant 70:491–498

Blein J-P, Canivenc M-C, De Cherade X, Bergon M, Calmon J-P, Scalla R (1986) Transplasma-membrane ferricyanide reduction in sycamore cells. Characterization of the system and inhibition by some phenyl biscarbamates. Plant Sci 46:77–85

Böttger M, Hilgendorf F (1988) Hormone action on transmembrane electron and H^+ transport. Plant Physiol 86:1038–1043

Böttger M, Lüthen H (1986) Possible linkage between NADH-oxidation and proton secretion in *Zea mays* L. roots. J Exp Bot 37:666–675

Bown AW, Crawford LA (1988) Evidence that H^+ efflux stimulated by redox activity is independent of plasma membrane ATPase activity. Physiol Plant 73:170–174

Brain RD, Freeberg JF, Weiss CV, Briggs WR (1977) Blue light-induced absorbance changes in membrane fractions from corn and *Neurospora*. Plant Physiol 57:948–952

Brightman AO, Barr R, Crane FL, Morré DJ (1988) Auxin-stimulated NADH oxidase purified from plasma membrane of soybean. Plant Physiol 86:1264–1269

Bruder G, Fink A, Jarasch E-D (1978) The b-type cytochrome in endoplasmic reticulum of mammary gland epithelium and milk fat globule membranes consists of two components, cytochrome b5 and cytochrome P-420. Exp Cell Res 117:207–217

Brüggemann W, Moog PR (1989) NADH-dependent Fe^{3+} EDTA and oxygen reduction by plasma membrane vesicles from barley roots. Physiol Plant 75:245–254

Buckhout TJ, Hrubec TC (1986) Pyridine nucleotide-dependent ferricyanide reduction associated with isolated plasma membranes of maize (*Zea mays* L.) roots. Protoplasma 135:144–154

Buckhout TJ, Luster DG (1988) Purification of NADH-ferricyanide and NADH-duroquinone reductase from maize (*Zea mays* L.) root plasma membranes. In: Crane FL, Morré DJ, Löw H (eds) Plasma membrane oxidoreductases in control of animal and plant growth. Plenum, New York, pp 81–88

Buckhout TJ, Bell BF, Luster DG, Chaney RL (1989) Iron-stress induced redox activity in tomato (*Lycopersicon esculentum* Mill) is localized on the plasma membrane. Plant Physiol 90:151–156

Cakmak I, van de Wetering DAM, Marschner H, Bienfait HF (1987) Involvement of superoxide radical in extracellular ferric reduction by iron-deficient bean roots. Plant Physiol 85:310–314

Canut H, Brightman A, Boudet AM, Morré DJ (1988) Plasma membrane vesicles of opposite sidedness from soybean hypocotyls by preparative free-flow electrophoresis. Plant Physiol 86:631–637

Caubergs RJ, Asard HH, De Greef JA, Leeuwerik FJ, Oltmann FL (1986) Light-inducible absorbance changes and vanadate-sensitive ATPase activity associated with the presumptive plasma membrane fraction from cauliflower inflorescences. Photochem Photobiol 44:641–649

Chalmers JDC, Coleman JOD, Walton NJ (1984) Use of an electrochemical technique to study plasma membrane redox reactions in cultured cells of *Daucus carota* L. Plant Cell Rep 3:243–246

Craig TA, Crane FL (1981) Evidence for a trans-plasma membrane electron transport system in plant cells. Proc Indiana Acad Sci 90:150–155

Craig TA, Crane FL (1985) The transplasma membrane redox system in carrot cells: inhibited by membrane-impermeable DABS and involved in H^+ release from cells. In: Randall DD, Blevins DG, Larsson RL (eds) Current topics in plant biochemistry and physiology, Vol 4, p 247

Crane FL, Barr R (1971) Determination of ubiquinones. Methods Enzymol 18C:137–165

Crane FL, Møller IM (eds) (1988) Plasmalemma redox functions in plants. Physiol Plant 73:161–200

Crane FL, Morré DJ (1977) Evidence for coenzyme Q function in Golgi membranes. In: Folkers K, Yamamato Y (eds) Biomedical and biochemical aspects of coenzyme Q. Elsevier, Amsterdam, pp 3–14

Crane FL, Roberts H, Linnane AW, Löw H (1982) Transmembrane ferricyanide reduction by cells of the yeast, *Saccharomyces cerevisiae*. J Bioenerg Biomembr 14:191–205

Crane FL, Löw H, Clark MG (1985a) Plasma membrane redox enzymes. In: Martonosi AN (ed) The enzymes of biological membranes, vol 4, Plenum, New York, pp 465–510

Crane FL, Sun IL, Clark MG, Grebing C, Löw H (1985b) Transplasma-membrane redox systems in growth and development. Biochim Biophys Acta 811:233–264

Crane FL, Morré DJ, Löw H (eds) (1988a) Plasma membrane oxidoreductases in control of animal and plant growth. Plenum, New York

Crane FL, Barr R, Craig TA, Morré DJ (1988b) Transplasma membrane electron transport in relation to cell growth and iron uptake. J Plant Nutr 11:1117–1126

De Pierre JW, Ernster L (1977) Enzyme topology of intracellular membranes. Annu Rev Biochem 46:201–262

Dharmawardhane S, Stern AI, Rubinstein B (1987) Light-stimulated transplasmalemma electron transport in oat mesophyll cells. Plant Sci 51:193–201

Dreyer JL, Treichler T (1988) NADH-dehydrogenase in synaptic plasma membranes. In: Crane FL, Morré DJ, Löw H (eds) Plasma membrane oxidoreductases in control of animal and plant growth. Plenum, New York, pp 406–407

Edman K, Ericson I, Møller IM (1985) The regulation of exogenous NAD(P)H oxidation in spinach leaf mitochondria by pH and cations. Biochem J 232:471–477

Elstner EF, Heupel A (1976) Formation of hydrogen peroxide by isolated cell walls from horseradish (*Armoracia lapathifolia* Gilib.). Planta 130:175–180

Elzenga JTM, Prins HBA (1987) Light induced polarity of redox reactions in leaves of *Elodea canadensis* Michx. Plant Physiol 85:239–242

Elzenga JTM, Staal M, Prins HBA (1989) ATPase activity of isolated plasmalemma vesicles of leaves of *Elodea* as affected by thiol reagents and NADH/NAD^+ ratio. Physiol Plant 76:379–385

Federico R, Giartosio CE (1983) A transplasma membrane electron transport system in maize roots. Plant Physiol 73:182–184

Giannini JL, Briskin DP (1988) Pyridine nucleotide oxidation by a plasma membrane fraction from red beet (*Beta vulgaris*) storage tissue. Arch Biochem Biophys 260:653–660

Gross GG (1977) Cell wall-bound malate dehydrogenase from horseradish. Phytochemistry 16:319–321

Guern J, Ullrich-Eberius CI (1988) The ferricyanide-driven redox system at the plasmalemma of plant cells: origin of the proton production and reappraisal of the stoichiometry e^-/H^+. In: Crane FL, Morré DJ, Löw H (eds) Plasma membrane oxidoreductases in control of animal and plant growth. Plenum, New York, pp 253–262

Guerrini F, Valenti V, Pupillo P (1987) Solubilization and purification of NAD(P)H dehydrogenase of *Cucurbita* microsomes. Plant Physiol 85:828–834

Hodges TK, Mills D (1986) Isolation of the plasma membrane. Methods Enzymol 118:41–54

Ivankina NG, Novak VA (1988) Transplasmalemma redox reactions and ion transport in photosynthetic and heterotrophic plant cells. Physiol Plant 73:161–164

Ivankina NG, Novak VA, Miklashevich AI (1984) Redox reactions with active H^+ transport in the plasmalemma of *Elodea* leaf cells. In: Cram WJ, Janacek K, Rybova R, Sigler K (eds) Membrane transport in plants. Wiley, Chichester, pp 404–405
Jensén P, Erdei L, Møller IM (1987) K^+ uptake in plant roots: experimental approach and influx models. Physiol Plant 70:743–748
Jesaitis AJ, Heners PR, Hertel R, Briggs WR (1977) Characterization of a membrane fraction containing a b-type cytochrome. Plant Physiol 59:941–947
Jones GJ, Morel FMM (1988) Plasmalemma redox activity in the diatom *Thalassiosira*. A possible role for nitrate reductase. Plant Physiol 87:143–147
Jones GJ, Palenik BP, Morel FMM (1987) Trace metal reduction by phytoplankton: the role of plasmalemma redox enzymes. J Phycol 23:237–244
Kalin A, Norling B, Appelkvist EL, Dallner G (1987) Ubiquinone synthesis in the microsome fraction of rat liver. Biochim Biophys Acta 926:70–78
Kjellbom P, Larsson C (1984) Preparation and polypeptide composition of chlorophyll-free plasma membranes from leaves of light-grown spinach and barley. Physiol Plant 62:501–509
Kjellbom P, Larsson C, Askerlund P, Schelin C, Widell S (1985) Cytochrome P-450/420 in plant plasma membranes: a possible component of the blue-light-reducible flavoprotein-cytochrome complex. Photochem Photobiol 42:779–783
Kochian L, Lucas WJ (1985) Potassium transport in corn roots. III. Perturbation by exogenous NADH and ferricyanide. Plant Physiol 77:429–436
Komor E, Thom M, Maretzki A (1987) The oxidation of extracellular NADH by sugarcane cells; coupling to ferricyanide reduction, oxygen uptake and pH change. Planta 170:34–43
Körner LE, Kjellbom P, Larsson C, Møller IM (1985) Surface properties of right side-out plasma membrane vesicles isolated from barley roots and leaves. Plant Physiol 79:72–79
Kröger A (1978) Determination of contents and redox states of ubiquinone and menaquinone. Methods Enzymol 53:579–591
Larsson C, Widell S, Kjellbom P (1987) Preparation of high-purity plasma membranes. Methods Enzymol 148:558–568
Larsson C, Widell S, Sommarin M (1988) Inside-out plant plasma membrane vesicles of high purity obtained by aqueous two-phase partitioning. FEBS Lett 229:289–292
Lass B, Thiel G, Ullrich-Eberius CI (1986) Electron transport across the plasmalemma of *Lemna gibba* G1. Planta 169:251–259
Lemberg R, Barret J (1973) Cytochromes. Academic Press, London
Leong T-Y, Vierstra RD, Briggs WR (1981) A blue light-sensitive cytochrome-flavin complex from corn coleoptiles. Further characterization. Photochem Photobiol 34:697–703
Lin W (1982a) Isolation of NADH oxidation system from the plasmalemma of corn root protoplasts. Plant Physiol 70:326–328
Lin W (1982b) Responses of corn root protoplasts to exogenous reduced nicotinamide adenine dinucleotide: oxygen consumption, ion uptake, and membrane potential. Proc Natl Acad Sci USA 79:3773–3776
Lin W (1984) Further characterization on the transport property of plasmalemma NADH oxidation system in isolated corn root protoplasts. Plant Physiol 74:219–222
Lundborg T, Widell S, Larsson C (1981) Distribution of ATPase in wheat root membranes separated by phase partition. Physiol Plant 52:89–95
Lundegårdh H (1945) Absorption, transport and exudation of inorganic ions by the roots. Ark Bot 32A(12):1–125
Lundegårdh H (1955) Mechanisms of absorption, transport, accumulation, and secretion of ions. Annu Rev Plant Physiol 6:1–24
Luster DG, Buckhout TJ (1988) Characterization and partial purification of multiple electron transport activities in plasma membranes from maize (*Zea mays*) roots. Physiol Plant 73:339–347
Lüthen H, Böttger M (1988) Hexachloroiridate IV as an electron acceptor for a plasmalemma redox system in maize roots. Plant Physiol 86:1044–1047
Lüttge U, Clarkson DT (1985) Mineral nutrition: plasmalemma and tonoplast redox activities. Prog Bot 47:73–86

Mäder M, Füssl R (1982) Role of peroxidase in lignification of tobacco cells II. Regulation by phenolic compounds. Plant Physiol 70:1132–1141
Madyastha KM, Krishnamachary N (1986) Purification and partial characterization of microsomal cytochrome b555 from the higher plant *Catharanthus roseus*. Biochem Biophys Res Commun 136:570–576
Marigo G, Belkoura M (1985) Cation stimulation of the proton-translocating redox activity at the plasmalemma of *Catharanthus roseus* cells. Plant Cell Rep 4:311–314
Marrè E, Marrè MT, Albergoni FG, Trockner V, Moroni A (1988) Electron transport at the plasma membrane and ATP-driven H^+ extrusion in *Elodea densa* leaves. In: Crane FL, Mooré DJ, Löw H (eds) Plasma membrane oxidoreductases in control of animal and plant growth. Plenum, New York, pp 233–242
Marrè MT, Moroni A, Albergoni FG, Marrè E (1988) Plasmalemma redox activity and H^+ extrusion. I. Activation of the H^+-pump by ferricyanide-induced potential depolarization and cytoplasm acidification. Plant Physiol 87:25–29
Marschner H (1986) Mineral nutrition in higher plants. Academic Press, London
Misra PC, Craig TA, Crane FL (1985) A link between transport and plasma membrane redox system(s) in carrot cells. J Bioenerg Biomembr 16:143–152
Mitchell PM (1966) Chemiosmotic coupling in oxidative and photosynthetic phosphorylation. Glynn Laboratories, Bodmin, England, p 192
Møller IM (1988) The organization of biological membranes. Physiol Plant 73:153–157
Møller IM, Bérczi A (1985) Oxygen consumption by purified plasmalemma vesicles from wheat roots. Stimulation by NADH and salicylhydroxamic acid. FEBS Lett 193:180–184
Møller IM, Bérczi A (1986) Salicylhydroxamic acid-stimulated NADH oxidation by purified plasmalemma vesicles from wheat roots. Physiol Plant 68:67–74
Møller IM, Lin W (1986) Membrane-bound NAD(P)H dehydrogenases in higher plant cells. Annu Rev Plant Physiol 37:309–334
Møller IM, Lundborg T, Bérczi A (1984) The negative surface charge density of plasmalemma vesicles from wheat and oat roots. FEBS Lett 167:181–185
Møller IM, Bérczi A, van der Plas LHW, Lambers H (1988a) Measurement of the activity and capacity of the alternative pathway in intact plant tissues: identification of problems and possible solutions. Physiol Plant 72:642–649
Møller IM, Askerlund P, Larsson C, Bérczi A, Widell S (1988b) Redox components in the plant plasma membrane. In: Crane FL, Morré DJ, Löw H (eds) Plasma membrane oxidoreductases in control of animal and plant growth. Plenum, New York, NY, pp 57–69
Morré DJ, Navas P, Penel C, Castillo FJ (1986) Auxin-stimulated NADH oxidase (semidehydroascorbate reductase) of soybean plasma membrane: role in acidification of cytoplasm? Protoplasma 133:195–197
Morré DJ, Auderset G, Penel C, Canut H (1987) Cytochemical localization of NADH-ferricyanide oxido-reductase in hypocotyl segments and isolated membrane vesicles of soybean. Protoplasma 140:133–140
Morré DJ, Crane FL, Barr R, Penel C, Wu L-Y (1988a) Inhibition of plasma membrane redox activities and elongation growth of soybean. Physiol Plant 72:236–240
Morré DJ, Brightman AO, Wu L-Y, Barr R, Leak B, Crane FL (1988b) Role of plasma membrane redox activities in elongation growth in plants. Physiol Plant 73:187–193
Neufeld E, Bown AW (1987) A plasmamembrane redox system and proton transport in isolated mesophyll cells. Plant Physiol 83:895–899
Novak VA, Ivankina NG (1983) Influence of nitroblue tetrazolium on membrane potential and ion transport in *Elodea*. Sov Plant Physiol 30:845–853
Novak VA, Miklashevich AI (1986) Ferricyanide reductase and ferrocyanide oxidase activities of the microalga *Scenesdesmus acuminatus*. Sov Plant Physiol 32:694–702
Pantoja O, Willmer CM (1988) Redox activity and peroxidase activity associated with the plasma membrane of guard-cell protoplasts. Planta 174:44–50
Qui Z-S, Rubinstein B, Stern AI (1985) Evidence for electron transport across the plasma membrane of *Zea mays* root cells. Planta 165:383–391

Ramirez JM (ed) (1987) Redox functions of the eukaryotic plasma membrane, Gonsejo Superior de Investigaciones Cientificas, Madrid

Ramirez JM, Gallego GG, Serrano R (1984) Electron transfer constituents in plasma membrane fractions of *Avena sativa* and *Saccharomyces cerevisiae*. Plant Sci Lett 34:103–110

Römheld V, Marschner H (1983) Mechanism of iron uptake by peanut plants. I. Fe^{III} reduction, chelate splitting, and release of phenolics. Plant Physiol 71:949–954

Rubinstein B, Stern AI (1986) Relationship of transplasmalemma redox activity to proton and solute transport by roots of *Zea mays*. Plant Physiol 80:805–811

Rubinstein B, Stern AI, Stout RG (1984) Redox activity at the surface of oat root cells. Plant Physiol 76:386–391

Sandelius AS, Barr R, Crane FL, Morré DJ (1987) Redox reactions of plasma membranes isolated from soybean hypocotyls by phase partition. Plant Sci 48:1–10

Schonbaum GR, Bonner WD, Storey BT, Bahr JT (1971) Specific inhibition of the cyanide-insensitive respiratory pathway in plant mitochondria by hydroxamic acids. Plant Physiol 47:124–128

Sijmons PC, Bienfait HF (1983) Source of electrons for extracellular Fe^{3+} reduction in iron-deficient bean roots. Physiol Plant 59:409–415

Sijmons PC, van den Briel W, Bienfait HF (1984a) Cytosolic NADPH is the electron donor for extracellular Fe^{III} reduction in iron-deficient bean roots. Plant Physiol 75:219–221

Sijmons PC, Lanfermeijer FC, de Boer AH, Prins HBA, Bienfait HF (1984b) Depolarization of cell membrane potential during trans-plasma membrane electron transfer to extracellular electron acceptors in iron-deficient roots of *Phaseolus vulgaris* L. Plant Physiol 76:943–946

Spreen Brouwer K, van Valen T, Day DA, Lambers H (1985) Hydroxamate-stimulated O_2 uptake in roots of *Pisum sativum* and *Zea mays*, mediated by a peroxidase. Its consequences for respiration measurements. Plant Physiol 82:236–240

Steward FC, Mapes MO, Kent AE, Hosten RD (1964) Growth and organized development of cultured plant cells. Science 143:20–27

Tangerås A, Flatmark T, Backström D, Ehrenberg A (1980) Mitochondrial iron not bound in heme and iron sulfur centers. Biochim Biophys Acta 589:162–175

Thiel G, Kirst GO (1988) Transmembrane ferricyanide reduction and membrane properties in the euryhaline charophyte *Lamprothamnium papulosum*. J Exp Bot 39:641–654

Thom M, Maretzki A (1985) Evidence for a plasmalemma redox system in sugarcane. Plant Physiol 77:873–876

van der Plas LHW, Gude H, Wagner MJ (1987) Hydroxamate-activated peroxidases in potato tuber callus. Interaction with the determination of the cytochrome and the alternative pathways. Physiol Plant 70:35–45

Ward MR, Tischner R, Huffaker RC (1988) Inhibition of nitrate transport by anti-nitrate reductase IgG fragments and the identification of plasma membrane associated nitrate reductase in roots of barley seedlings. Plant Physiol 88:1141–1145

Widell S (1987) Membrane-bound blue light receptors – possible connection to blue light photomorphogenesis In: Senger H (ed) Blue light responses: phenomena and occurrence in plants and microorganisms, vol II. CRC, Boca Raton, FL, pp 89–98

Widell S, Larsson C (1981) Separation of presumptive plasma membranes from mitochondria by partition in an aqueous polymer two-phase system. Physiol Plant 51:368–374

Widell S, Larsson C (1983) Distribution of cytochrome b photoreductions mediated by endogenous photosensitizers or methylene blue in fractions from corn and cauliflower. Physiol Plant 57:196–202

Widell S, Caubergs RJ, Larsson C (1983) Spectral characterization of light-reducible cytochrome in a plasma membrane-enriched fraction and in other membranes from cauliflower inflorescences. Photochem Photobiol 38:95–98

Yamamato Y, Niki E, Eguchi J, Kaniya Y, Shimasaki H (1985) Oxidation of biological membranes and its inhibition. Free radical chain oxidation of erythrocyte ghost membranes by oxygen. Biochim Biophys Acta 819:29–36

Zambiano F, Fleischer S, Fleischer B (1975) Lipid composition of the Golgi apparatus of rat kidney and liver in comparison with other subcellular organelles. Biochim Biophys Acta 380:357–369

Chapter 6 Plasma Membrane ATPase

R. SERRANO[1]

1 An Historical Perspective on the Path to the Molecular Basis of the Plant Proton Plump

The existence of an ATP-driven proton pump in plant plasma membranes was suggested two decades ago from physiological studies on active transport (Poole 1978) and by measurement of electrical potentials (Spanswick 1981) across the plasma membrane of whole plant cells (Chap. 8). The pioneering work of Hodges et al. (1972) represented the first step toward the molecular characterization of this pump. These authors demonstrated the presence in partially purified plant plasma membranes of a K^+-stimulated ATPase distinct from the mitochondrial and chloroplast ATPases. During the last decade the proton-pumping activity of this enzyme has been demonstrated in isolated plasma membrane vesicles (Sze 1985, see Chap. 7) and the ATPase has been partially purified and reconstituted in liposomes catalyzing ATP-driven proton transport (Serrano 1983). These studies also demonstrated similarities between ATPase enzymes from plant and fungal plasma membrane, which together

[1]European Molecular Biology Laboratory, Postfach 10.2209, D-6900 Heidelberg, FRG

C. Larsson, I.M. Møller (Eds):
The Plant Plasma Membrane

constitute a novel group of ion-pumping ATPase (Serrano 1984a, 1985). Table 1 summarizes the properties of the two types of ion-pumping ATPases identified in biological membranes (Pedersen and Carafoli 1987a,b). The fungal and plant plasma membrane ATPases have the same structure and reaction mechanism as the other (E-P) ATPases involved in Na^+, K^+, and Ca^{2+} transport. On the other hand, they operate as electrogenic proton pumps, like most of the (F_0F_1) ATPases. However, the (F_0F_1) ATPases have different structures and reaction mechanisms.

During the last 3 years the amino acid sequence of many (E-P) ATPases has been deduced from the corresponding nucleotide sequence. These data indicate that all these enzymes evolved from a single ancestral pump and that the (F_0F_1) ATPases had an independent evolutionary origin (Serrano 1988). The partial sequences available for the plant plasma membrane ATPase show extensive similarities with plasma membrane ATPases from fungi.

This chapter summarizes our current knowledge about the plant plasma membrane ATPase. In addition, it is hoped that the proposal of specific models for the structure, mechanism, and regulation of the enzyme will stimulate future research. This review will focus on recent work with purified preparations of the solubilized enzyme. Previous work with membrane preparations has been reviewed by Hodges (1976), Leonard (1983), and Sze (1985) and will only be

Table 1. Properties of the two families of cation-pumping ATPases (after Pedersen and Carafoli 1987a,b, with modifications)

	(F_0F_1) ATPases	(E-P) ATPases
Membranes	Bacteria Mitochondria Thylakoids Vacuolar system	Bacteria Plasma membranes Endoplasmic reticulum
Cation pumped	H^+	H^+, H^+/K^+, Na^+/K^+, Ca^{2+}/H^+
Inhibitors (K_i)	Dicyclohexylcarbodiimide (10^{-6} M) Nitrate (10^{-2} M)	Vanadate (10^{-6} M)
Subunits	F_1 part: 3 catalytic and 3 others of 50–80 kD F_0 part: 6–12 proteolipids of 8–16 kD Other subunits in different enzymes	Catalytic of 70–140 kD Other subunits of 10–50 kD in different enzymes
Mechanism	No phosphorylated intermediate Cooperativity of catalytic subunits	Acyl-phosphate intermediate Monomer is active

referred to as a complement to information obtained with the purified enzyme. My earlier reviews (Serrano 1983, 1984a, 1985) preceded both the more recent information obtained with the purified enzyme and the new perspective opened by the elucidation of the primary structure of several (E-P) ATPases. Recent reviews by Sussman and Surowy (1987) and Serrano (1988) provide a similar coverage although slightly different approaches were emphasized.

2 The Variety of ATP-Hydrolyzing Activities in Plant Homogenates

The identification of the plant plasma membrane ATPase in crude fractions and even in purified plasma membranes is made difficult by the existence in plant homogenates of a plethora of enzymes capable of hydrolyzing ATP (Table 2). There are no specific inhibitors of the plasma membrane ATPase and only a combination of several inhibitors may provide a sufficiently specific assay for the enzyme. The phosphohydrolase activity measured at pH 6.5 in the presence of ATP, $MgSo_4$, KNO_3 (to inhibit vacuolar ATPase), azide (to inhibit mitochondrial and chloroplast ATPases), and molybdate (to inhibit acid phosphatase) should correspond to either plasma membrane ATPase or apyrase. Of these two, only the plasma membrane ATPase is sensitive to diethylstilbestrol and, in theory, this combination of inhibitors should provide a specific assay. However, in practice, this is only possible when the plasma membrane ATPase is relatively abundant. The efficiency of inhibition of interfering enzymes is never 100% and some of these are sensitive to diethylstilbestrol. As there are no inhibitors for apyrase, this enzyme may completely mask the plasma membrane ATPase in some plant tissues. Some of the interfering activities may be removed during purification of the plasma membrane or by extraction with salts and mild detergents which do not solubilize the plasma membrane ATPase.

The practice of considering as plasma membrane ATPase the part of the activity stimulated by K^+ (the so-called K^+-ATPase, Hodges et al. 1972) is not recommended. Many of the interfering activities, such as the mitochondrial ATPase, are also activated by K^+. Furthermore, the plasma membrane ATPase is not dependent on K^+ and is only slightly activated by this cation (see below).

Recent cytochemical studies indicate that the plasma membrane ATPase is particularly sensitive to inactivation by the fixatives glutaraldehyde and formaldehyde and by the lead utilized to precipitate phosphate (Katz et al. 1988). Therefore only unspecific phosphatase is usually visualized by these methods.

Table 2. ATP-hydrolyzing enzymes in plant homogenates (after Hollander 1971; Vara and Serrano 1981; Serrano 1983, 1985; Sze 1985). NDP and NTP indicate any nucleotide di- or triphosphate

Enzymes	pH optimum	Substrate	Cofactor	Activators	Inhibitors (concentration for > 90% inhibition)
Mitochondrial and thylakoid ATPases (EC 3.6.1.34)	8–9	ATP > GTP	$Mg^{2+} > Ca^{2+}$	HCO_3^- HSO_3^-	Oligomycin (1 μM, only mitochondria), azide (1 mM), nitrate (50 mM), dicyclohexylcarbodiimide (1 μM)
Vacuolar ATPase	7–8	ATP > GTP	$Mg^{2+} > Ca^{2+}$	Cl^-	Nitrate (20 mM), dicyclohexylcarbodiimide (10 μM) N-ethylmaleimide (10 μM) diethylstilbestrol (0.1 mM), erythrosin B (50 μM)
Plasma membrane H^+-ATPase (EC 3.6.1.35)	6–7	ATP ≫ NTP	Mg^{2+} (not Ca^{2+})	K^+	Dicyclohexylcarbodiimide (0.1 mM), diethylstilbestrol (0.1 mM), erythrosin B (50 μM), vanadate (50 μM)
Acid phosphatase (EC 3.1.3.2)	4–6	Phosphate esters	None	None	Molybdate (0.1 mM), vanadate (0.1 mM)
Apyrase (EC 3.6.1.5)	6–7	NTP and NDP	Ca^{2+} or Mg^{2+}	None	Unknown

3 Solubilization and Purification of the Plant Plasma Membrane ATPase

Several purification procedures have been described for the plant plasma membrane ATPase and Table 3 presents those which result in an active proton pump after reconstitution in liposomes. This is a requirement of any purified preparation, because proton pumping may have more requirements in terms of enzyme stability and subunit composition than uncoupled ATP hydrolysis.

In most of the procedures a plasma membrane-enriched fraction from young roots is initially prepared. Easily extracted proteins are subsequently removed by a combination of salt (KCl or KBr, 0.5 M) and mild detergents (Triton X-100, bile salts, octylglucoside), treatments which do not solubilize the plasma membrane ATPase. During this and following steps the ATPase is stabilized by high concentrations of glycerol (20–45%) and by dithioerythritol and EDTA. The latter two reagents seem to protect essential sulfhydryl group(s) of the enzyme against oxidation and heavy metals. A recent report suggests that optimum stability is obtained in the presence of ATP and at pH 6.7 (Palmgren et al. 1988).

Solubilization of the plasma membrane ATPase is effected by the strong zwitterionic detergents lysolecithin (egg yolk lysophosphatidylcholine) and zwittergent 3–14 (N-tetradecyl-N,N-dimethyl-3-ammonio-1-propanesulfonate),

Table 3. Purified preparations of plant plasma membrane ATPAse

Reference	Source	Purity of 100 kD band (%)	Specific activity μmol min^{-1} mg^{-1}	Yield mg (kg tissue)$^{-1}$
Vara and Serrano (1982)	Oat roots	20	2.7	3.0
Serrano (1984)	Oat roots	80	6	0.9
Cocucci and Marrè (1984) Cocucci et al. (1985)	Radish seedlings	? [a]	12	1.6
Kasamo (1986, 1987)	Mung bean roots	20	3[b]	?
Anthon and Spanswick (1986)	Tomato roots	80	10[c]	3.4

[a] Gels showed proteolytic degradation but more recent evidence indicates this was a problem of the electrophoresis (personal communication of MC Cocucci).

[b] The reported value (6) has been corrected by a factor of 0.5 because the assays were conducted at 38 °C.

[c] The reported value (17) has been corrected by a factor of 0.6 due to the underestimation of protein concentration resulting from the utilization of the Bio-Rad Protein Assay reagent with bovine serum albumin as standard.

either alone or in combination with bile salts. The solubilized ATPase exists as a large aggregate which is separated from other solubilized proteins by rate zonal centrifugation on a glycerol gradient (Fig. 1). Ammonium sulfate fractionation of the solubilized extract results in much cruder preparations. The procedure of Cocucci and Marrè (1984) is completely different from the others and this may be related to the nature of the starting material: whole seedlings instead of roots. The radish seedling ATPase is solubilized by cholate plus sulfate salts and then fractionated with ammonium sulfate.

Evaluation of the efficacy of the different purification procedures is complicated by detergent-induced activation and inactivation, phenomena which may even be compensatory in some cases. Detergent activation may be caused by unmasking of latent ATPase in sealed, right-side out vesicles (Larsson et al. 1984), but a direct effect on the enzyme cannot be discounted. Detergent inactivation may reflect delipidation of the ATPase (Serrano et al. 1988), which may be irreversible if it results in enzyme instability. Although the available information is limited, it seems that solubilization of the ATPase and rate zonal centrifugation may irreversibly inactivate the enzyme by 30–70%.

The best preparations (Serrano 1984b; Anthon and Spanswick 1986) contain a major polypeptide of 90–100 kD which accounts for about 80% of the protein in Coomassie-stained gels. Other polypeptides are present in variable and less than stoichiometric amounts and these therefore probably represent contaminants. The only rigorous demonstration that no other subunits are components of the plasma membrane H^+-ATPase has been made in the fungus

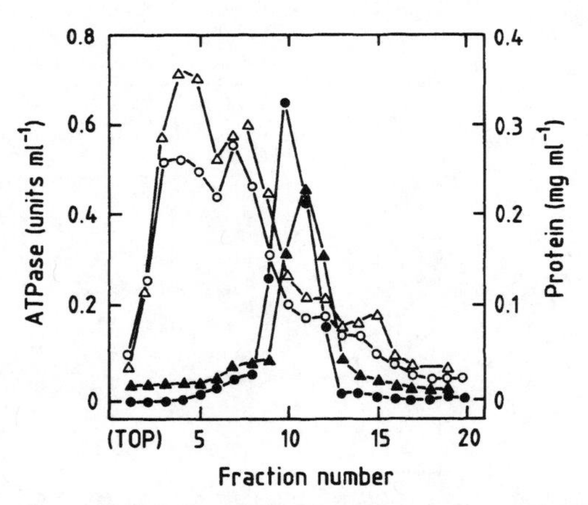

Fig. 1. Purification of the oat root plasma membrane ATPase by rate-zonal centrifugation in a glycerol gradient (Serrano 1984b). Oat root plasma membranes were extracted with Triton X-100 and KCl and the ATPase in the residue was solubilized with lysolecithin. Four ml of solubilized extract was applied to a glycerol gradient (34 ml, 25–50%) and centrifuged for 16 h at 2°C and 170 000 g in a vertical rotor. The distribution of ATPase activity (*closed symbols*) and protein (*open symbols*) for two different experiments (*circles* and *triangles*) is shown

Neurospora crassa (Scarborough and Addison 1984). Similar experiments should be performed with the plant enzyme. It is important to realize that special precautions need to be adopted during electrophoresis at ATPase preparations because the 100 kD polypeptide is aggregated after boiling in sodium dodecylsulfate. Boiling is used to inactivate endogenous proteases which could degrade the ATPase after denaturation in sodium dodecylsulfate. Therefore, proteases must be inactivated by an alternative procedure, such as the addition of protease inhibitors (Gallagher and Leonard 1987) or the precipitation of the samples with trichloroacetic acid prior to dissolution in sodium dodecylsulfate (Vara and Serrano 1982). These problems seem to explain the anomalously low molecular weight of the polypeptides observed after electrophoresis of the preparation of Cocucci and Marrè (1984; M.C. Cocucci, personal communication). Cross-linking of purified ATPase by Anthon and Spanswick (1986) suggests that the ATPase exists as a trimer. On the other hand, the target molecular size for radiation inactivation suggests a dimeric structure (Briskin et al. 1985). Future investigations are required to determine the oligomeric state of the plant ATPase by the application of a combination of techniques, including cross-linking of membrane-bound ATPase and sedimentation analysis of the solubilized enzyme. In the case of *Neurospora crassa*, monomers of the 100 kD polypeptide have been shown to be active (Goormaghtigh et al. 1986), but the enzyme may exist as an hexamer (Chadwick et al. 1987).

4 Enzymatic Properties of the Isolated Enzyme

4.1 Lipid Requirements

All the purified preparations require exogenous phospholipids for optimal activity, although the degree of stimulation and the specificity of the effect is variable. Preparations solubilized with lysolecithin (Serrano 1984b; Anthon and Spanswick 1986) are less stimulated by lipids than those made with zwittergent 3-14 and bile salts (Vara and Serrano 1982; Cocucci and Marrè 1984; Kasamo 1986), probably because the natural detergent satisfies the lipid requirements of the enzyme (see below). Studies on the lipid specificity of the ATPase have produced contradictory results. Cocucci and Marrè (1984) found that lysophosphatidylcholine was the best activating lipid (eightfold stimulation), while other lipids, including phosphatidylcholine, were much less effective. On the other hand, Kasamo and Nouchi (1987) found phosphatidylcholine and phosphatidylserine to be the best activators (two- to threefold stimulation), followed by lysophosphatidylcholine and phosphatidylglycerol. Reconstitution of the delipidated enzyme with phospholipids was effected by simple mixing under assay conditions. With some preparations of membrane enzymes the observed lipid specificity reflects requirements of the reconstitution procedure rather than requirements of the enzyme activity (Eytan 1982). The discrepancies between different ATPase preparations could therefore reflect particular requirements of the simple mixing reconstitution procedure. However, it cannot be discounted that ATPases from different sources could have different lipid specificities. A detailed study with the oat root ATPase (Serrano et al. 1988) indicates that the stimulation of the delipidated enzyme by lipids is relatively unspecific (Fig. 2). Even Triton X-100 activates partially, although maximal activity is obtained with lipids containing a zwitterionic polar head group, i.e., phosphorylcholine and phosphorylethanolamine. The same pattern of response is observed with phospholipids and with the corresponding lysophospholipids. As lysophospholipids are detergents which interact with membrane proteins without a requirement for reconstitution into vesicles, the observed specificity probably reflects the requirements of enzyme activity rather than reconstitution. This study also indicates that the plant plasma membrane ATPase does not require unsaturated hydrophobic chains for activity. This unspecific lipid requirement is also observed with the Ca^{2+}-ATPase from sarcoplasmic reticulum, which can be activated by detergents (Melgunov and

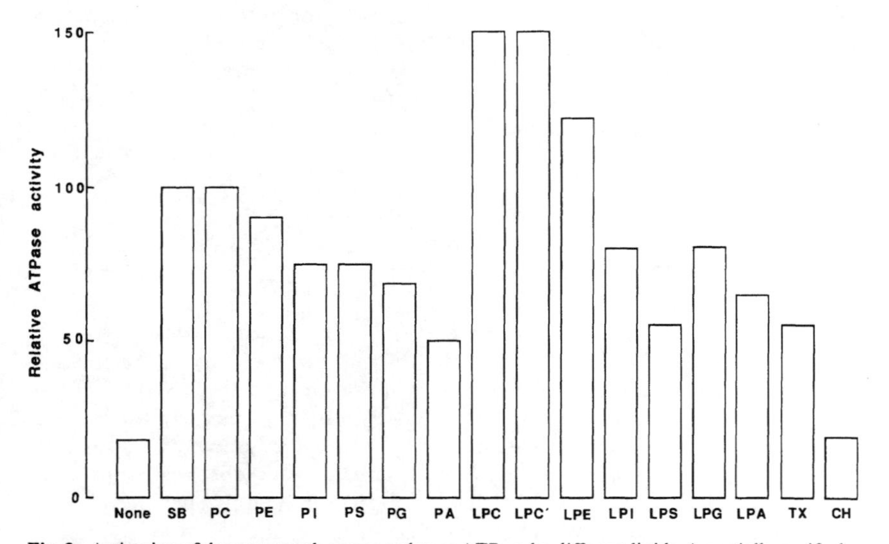

Fig. 2. Activation of the oat root plasma membrane ATPase by different lipids. A partially purified preparation was delipidated by cholate treatment and reconstituted with phospholipid vesicles and detergent micelles by mixing in the assay buffer (Serrano et al. 1988). The activity with soybean phospholipids (*SB*), taken as 100%, was 1.4 μmol min^{-1} (mg protein)$^{-1}$. The abbreviations for the other lipids are: *PC*, phosphatidylcholine from egg yolk; *PE* phosphatidylethanolamine; *PI* phosphatidylinositol from soybean; *PS* phosphatidylserine from bovine brain; *PG* phosphatidylglycerol from egg yolk; *PA* phosphatidic acid from egg yolk; *LPC* lysophosphatidylcholine from egg yolk, mostly palmitoyl and stearoyl; *LPC′* lysophosphatidylcholine, oleoyl; *LPE* lysophosphatidylethanolamine from egg yolk, mostly stearoyl and palmitoyl; *LPI* lysophosphatidylinositol from soybean, mostly palmitoyl and stearoyl; *LPS* lysophosphatidylserine from bovine brain, mostly palmitoyl and stearoyl; *LPG* lysophosphatidylglycerol from egg yolk, mostly palmitoyl and stearoyl; *LPA* lysophosphatidic acid, oleoyl; *TX* Triton X-100; *CH* cholate

Akimova 1980) and by all phospholipid species, although a slight preference for phosphatidylcholine exists (Bennet et al. 1978). On the other hand, the yeast plasma membrane H^+-ATPase (Serrano et al. 1988) and the Na^+/K^+-ATPase of animal cells (Kimelberg and Papahadjopoulos 1972; Roelofsen and Van Deenen 1973) require lipids with an unsaturated hydrophobic chain and a negatively charged polar head.

4.2 Kinetics

Detailed kinetic studies have not been performed with most purified enzymes. The results obtained with partially purified preparations indicate that the true substrate is the MgATP complex (Balke and Hodges 1975; Cocucci and Marrè 1984; Bennet et al. 1985). This substrate gives hyperbolic kinetics with K_m values of 0.3–0.7 mM. Excess of free Mg^{2+} is only slightly inhibitory ($K_i > 10$ mM) while excess of free ATP is a more effective inhibitor ($K_i \simeq 2$ mM). However, the last effect may reflect a requirement for small concentrations of free Mg^{2+} (10–100 μM), as demonstrated for the yeast plasma membrane ATPase (Ahlers 1984). ADP is a competitive inhibitor ($K_i \simeq 2$ mM) while P_i produces a less severe noncompetitive inhibition ($K_i \simeq 20$ mM).

Although Mg^{2+} is probably the physiological cofactor, it can be replaced by either Mn^{2+}, Co^{2+} or Zn^{2+}. Ca^{2+} is completely inactive and it is inhibitory in the presence of Mg^{2+} (0.2–0.4 mM Ca^{2+} produces 50% inhibition) (Vara and Serrano 1982; Cocucci and Marrè 1984; Kasamo 1986; Chap. 2).

The plasma membrane ATPase is moderately activated by K^+ at acidic pH, the effect disappearing at physiological pH (Fig. 3). The pH dependence of the

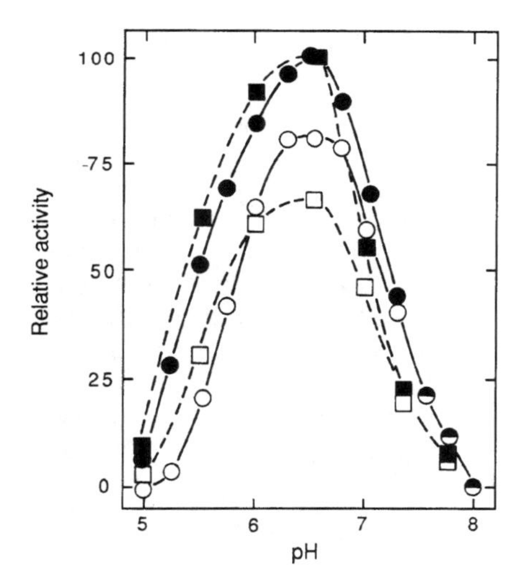

Fig. 3. Effect of pH and K^+ on the activity of the plant plasma membrane ATPase. *Circles*, tomato root ATPase (data from Anthon and Spanswick 1986); *squares*, oat root ATPase (data from Vara and Serrano 1982). *Closed symbols*, 50 mM KCl added to the basic assay medium; *open symbols*, no KCl added

ATPase suggests that the activity of the enzyme depends on the deprotonation of a group with pK_a 5.6–5.7 and on the protonation of another group with a pK_a of about 7.1, resulting in an optimal pH of 6.5. K^+ displaces the acidic pK_a to about 5.4 without affecting the neutral one. This selective effect on a critical acidic residue of the enzyme may explain the K^+ activation at acid pH. This residue could be a carboxylic group, the dissociation of which would be promoted by K^+. The critical residue, the neutral pK_a of which is unaffected by K^+, could be a histidine or a lysine with greatly altered pK_a. The effect of K^+ does not seem to be a simple salt effect because it shows some specificity ($K^+ > Rb^+ > Na^+ > Li^+$) and a half-maximal effect is obtained with only 1–4 mM K^+ (Vara and Serrano 1982; Kasamo 1986). Therefore, it is likely that the enzyme contains a K^+ binding site connected to a critical carboxylic group which must be dissociated for activity. On the other hand, the facts that K^+ activation is only observed at acidic pH and that it never amounts to more than a doubling of activity speaks against the enzyme being a K^+-dependent ATPase involved in K^+ transport. Enzymes showing this capability such as the animal Na^+/K^+-ATPase (Jørgensen 1982), the stomach H^+/K^+-ATPase (Sachs et al. 1982) and the bacterial K^+-ATPase (Furst and Solioz 1986) are largely dependent on K^+ for activity (more than tenfold activation at all pH values). The Ca^{2+}-ATPase of sarcoplasmic reticulum is slightly activated by K^+ without transporting it (Shigekawa and Wakabayashi 1985) and this seems also to be the case with the plant enzyme (see below).

The plasma membrane ATPase is highly specific for ATP. The activity with other nucleotide triphosphates, ADP, or phosphate monoesters is less than 5–10% of that with ATP (Vara and Serrano 1982; Cocucci and Marrè 1984).

Maximum ATPase activity is obtained at about 38 °C (Leonard and Hodges 1973; Cocucci and Marrè 1984) and there is about half as much activity at 30 °C.

4.3 Inhibitors

The most powerful inhibitors of the plant plasma membrane ATPase seem to be Cu^{2+} ($CuSO_4$) and mercurials (p-chloromercuribenzenesulfonate), which produce half-maximal inhibition at 2–5 μM and complete inhibition at 10–20 μM (Vara and Serrano 1982; Cocucci and Marrè 1984). This confirms the presence in the ATPase of an essential sulfhydryl group, as suggested by the protective effects of EDTA and dithiothreitol during purification (see above). At alkaline pH (> 8) the ATPase is inhibited by covalent modification with mM concentrations of N-ethylmaleimide. Both the labeling of the ATPase with radioactive sulfhydryl reagent and the inhibition of activity are prevented by MgATP and ADP, suggesting that the sensitive residue is in the active site (Katz and Sussman 1987). The same sulfhydryl residue may be responsible for the inhibition by heavy metals. All these inhibitors, however, are not specific for the plasma membrane ATPase and should have a general toxicity on whole cells. Other ATPases such as the mitochondrial and vacuolar ATPases also have essential sulfhydryl groups and N-ethylmaleimide is even more reactive with these enzymes than with the plasma membrane ATPase (Sze 1985).

Vanadate is also a very effective inhibitor of the plant plasma membrane ATPase, 50% inhibition being obtained with 2–10 μM and full inhibition with 50–100 μM (Vara and Serrano 1982; Cocucci and Marrè 1984; Kasamo 1986; Surowy and Sussman 1986). However, vanadate is a general inhibitor of phosphohydrolases which form a phosphorylated intermediate (Macara 1980) and it therefore inhibits acid phosphatases (Gallagher and Leonard 1982). A more important problem for "in vivo" studies is that vanadate may penetrate intact cells very slowly (Ball et al. 1987) and be reduced to noninhibitory vanadyl by cellular metabolism (Willsky et al. 1984).

Erythrosin B inhibits the plasma membrane ATPase very effectively, both "in vitro" (Cocucci and Marrè 1984) and "in vivo" (Cocucci and Marrè 1986; Ball et al. 1987). Half-maximal inhibition of the purified ATPase is obtained with only 0.3–1 μM erythrosin and full inhibition with 10–30 μM. The only known side effect is the inhibition of the vacuolar ATPase (Cocucci 1986).

Diethylstilbestrol is a weak inhibitor of the plasma membrane ATPase, requiring 20–40 μM for 50% inhibition and 100–200 μM for full inhibition (Vara and Serrano 1982; Cocucci and Marrè 1984). In addition, it inhibits mitochondrial and vacuolar ATPases (Pedersen and Carafoli 1987a) and in whole cell experiments it may affect the barrier properties of the plasma membrane (Borst-Pauwels et al. 1983).

The inhibition of the plasma membrane ATPase by dicyclohexylcarbodiimide results from a covalent reaction of the drug with the enzyme and is therefore time-dependent (Cid et al. 1987; Sussman and Surowy 1987; Oleski and Bennet 1987). Half-maximal inhibition is usually obtained in 10–30 min with 10–30 μM and full inhibition with 100–200 μM. Kinetic studies suggest that inhibition results from the covalent modification of a single site on the ATPase molecule but the stoichiometry of the reaction has not been determined by direct analysis in purified preparations. MgATP does not protect against the inhibition, suggesting that the sensitive residue is not part of the active site (Cid et al. 1987). The second-order rate constant of the reaction is about 500 M^{-1} min^{-1}, one or two orders of magnitude lower than those of the more sensitive mitochondrial and vacuolar ATPases. This inhibitor may also affect the barrier properties of the plasma membrane during experiments with whole cells (Borst-Pauwels et al. 1983).

The herbicide 2,2,2-trichloroethyl 3,4-dichlorocarbanilate (SW26) inhibits the plant plasma membrane ATPase completely at 100 μM, the half-maximal effect being observed at 8–14 μM (Blein et al. 1986, 1987). The mitochondrial ATPase is not affected but no information is available on its effect on the vacuolar ATPase. MgATP does not protect against SW26 inhibition, suggesting a binding site separate from the active site. This drug also inhibits the ATPase "in vivo" (Blein et al. 1986) but it is less effective than erythrosin (Ball et al. 1987).

As a general conclusion it can be stated that no completely specific inhibitors of the plant plasma membrane ATPase are available. Therefore, the effects of ATPase inhibitors on whole cells must be interpreted with caution and controls should be made for unspecific toxicity effects such as ATP depletion and disruption of the cell permeability barrier. Only a completely specific inhibitor without side effects, such as ouabain and its specific inhibition of the animal Na^+/K^+-ATPase (Jørgensen 1982), would help to unambiguously ascertain the physiological role of the plant plasma membrane ATPase in whole plant cells and tissues.

4.4 Phosphorylated Intermediate

As expected from its sensitivity to vanadate, the plant plasma membrane ATPase forms a phosphorylated intermediate (Briskin and Leonard 1982; Scalla et al. 1983; Vara and Serrano 1983). The demonstration of this intermediate in crude preparations is complicated by the presence of protein kinases, but in purified preparations the phosphoprotein formed from [γ-^{32}P]ATP has all the characteristics of a catalytic intermediate: (1) it is rapidly formed, with the maximum phosphorylation being obtained in less than 15 s at 0°C; (2) it turns over rapidly, as the addition of excess unlabeled ATP results in complete loss of label from the protein in less than 15 s at 0°C; (3) after denaturation, the phosphoprotein bond is sensitive to alkaline pH and to hydroxylamine, suggesting the presence of an acyl-phosphate bond rather than a hydroxyl-phosphate characteristic of protein kinase activity. The intermediate has been identified as a β-aspartyl-phosphate (Briskin and Poole 1983b) which is surrounded by the same amino acids that occur in other (E-P) ATPases (Walderhaug et al. 1985). The steady-state phosphorylation level is dependent on MgATP concentration and the same apparent K_m value can be calculated for the phosphorylation level and for the rate of ATP hydrolysis (Scalla et al. 1983; Vara and Serrano 1983). This suggests that the rate of P_i production is proportional to the level of intermediate, as predicted by the kinetic model developed by Amory et al. (1980) for the yeast plasma membrane ATPase. In this model the reversible binding of MgATP to the enzyme is followed by the irreversible phosphorylation of the ATPase (under initial conditions, where ADP is negligible) and irreversible hydrolysis of the intermediate:

$$\begin{array}{ccccccc} E & \overset{K_d}{\rightleftharpoons} & E{\cdot}MgATP & \overset{k_1}{\rightarrow} & E\text{-}P & \overset{k_2}{\rightarrow} & E \\ + & & & & + & & + \\ MgATP & & & & MgADP & & P_i \end{array}$$

The relationship $V = k_2$ (E-P) explains the proportionality between ATP hydrolysis and the level of intermediate. Assuming steady-state conditions, we have:

$$k_1\,(E{\cdot}MgATP) = k_2\,(E\text{-}P) \quad (1)$$

and from the equilibrium of MgATP binding and the conservation of enzyme forms:

$$K_d = \frac{(E)(MgATP)}{(E{\cdot}MgATP)} \tag{2}$$

$$(E_t) = (E) + (E{\cdot}MgATP) + (E\text{-}P). \tag{3}$$

The solution of this equation system is:

$$(E\text{-}P) = \frac{\frac{(E_t)k_1}{k_1+k_2}\ (MgATP)}{\frac{K_d k_2}{k_1+k_2} + (MgATP)}. \tag{4}$$

The saturating level of intermediate would be:

$$(E\text{-}P)_{max} = \frac{(E_t)\,k_1}{k_1+k_2}\ , \tag{5}$$

and the apparent K_m value for both the rate of P_i production and the formation of the intermediate:

$$K_m = \frac{(E_t)k_1}{k_1+k_2}. \tag{6}$$

The maximum level of intermediate in purified preparations of the ATPase is 0.6 nmol (mg protein)$^{-1}$, corresponding to the phosphorylation of about 10% of the 100 kD polypeptides (Serrano 1984b). This suggests that k_2 is nine times greater than k_1 and, therefore, that the hydrolysis of the intermediate is not rate-limiting. However, it is possible that the enzyme preparation contains inactive enzyme molecules which cannot be phosphorylated and, therefore, until a titration of active sites is available, no definitive interpretation can be reached for the partial phosphorylation of the enzyme under steady-state conditions. The estimated turnover of the phosphorylated intermediate is about 10 000 min^{-1}, corresponding to an overall ATPase turnover of 1000 min^{-1}.

The phosphorylation of the ATPase seems to be the only step of the catalytic cycle where Mg^{2+} is required because neither ATP binding nor hydrolysis of the intermediate are dependent on the presence of Mg^{2+} (Briskin and Poole 1983a). On the other hand, phospholipids seem to be required for dephosphorylation because delipidated ATPase has low ATPase activity but a normal level of phosphorylated intermediate (Vara and Serrano 1983). Actually, the addition of phospholipids (Vara and Serrano 1983) and of K^+ (Briskin and Leonard 1982; Scalla et al. 1983) slightly reduces the steady-state level of intermediate, suggesting activation of the dephosphorylation step. It must be stressed that K^+ has only minor effects on the hydrolysis of the phosphorylated intermediate of the plant plasma membrane ATPase, while it is essential for the hydrolysis of the intermediate in K^+-pumping enzymes (Jørgensen 1982; Sachs et al. 1982). This, again, speaks against the plant plasma membrane ATPase being a K^+ pump.

Dicyclohexylcarbodiimide (Scalla et al. 1983; Cid et al. 1987) and diethylstilbestrol (Scalla et al. 1983) inhibit to a similar extent both ATP hydrolysis and phosphorylation of the enzyme, suggesting that they mainly affect the phosphorylation step. On the other hand, vanadate inhibits ATPase activity more strongly than the phosphorylation of the enzyme (Scalla et al. 1983; Vara and Serrano 1983), suggesting that vanadate acts on the hydrolysis of the intermediate.

5 Transport Properties in Reconstituted Proteoliposomes

Purified preparations of the plant plasma membrane ATPase have been reconstituted into liposomes capable of ATP-driven proton transport. Either freeze-thaw sonication (Vara and Serrano 1982; Serrano 1984b; Anthon and Spanswick 1986), cholate dialysis (Cocucci et al. 1985), or octylglucoside dilution (Kasamo 1987) were employed and proton accumulation inside the vesicles was followed by fluorescence quenching (Vara and Serrano 1982; Serrano 1984b; Anthon and Spanswick 1986; Kasamo 1987) or by the change of absorbance (Cocucci et al. 1985) of acridine dyes. These results demonstrated the ability of the purified preparations to pump protons, suggesting that all the components necessary for this activity are present in the 100 Kd polypeptide (but see above, Sect. 3). In addition, the reconstituted systems indicated that the ATPase is an electrogenic proton pump, which is slightly activated by K^+ but not involved in the transport of this cation. The evidence is as follows (Fig. 4). The

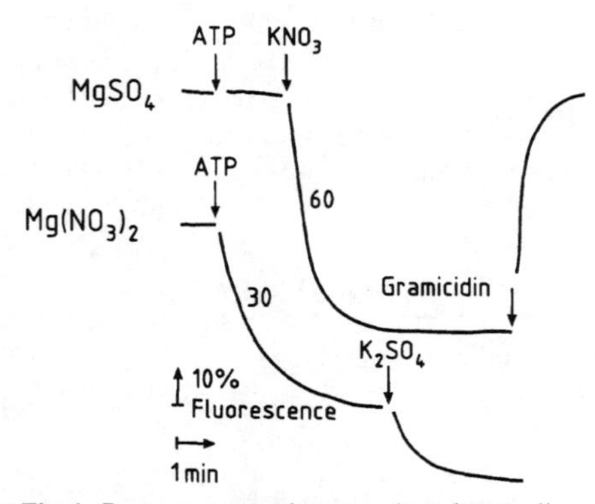

Fig. 4. Proton transport in reconstituted proteoliposomes containing the oat root plasma membrane ATPase (data from Vara and Serrano 1982). Proteoliposomes were reconstituted by freeze-thaw sonication in the presence of 25 mM $MgSO_4$ (*upper trace*) or 25 mM $Mg(NO_3)_2$ (*lower trace*) and proton transport was measured by the quenching of acridine dye fluorescence. Tris-ATP (1.25 mM), K_2SO_4 (25 mM), KNO_3 (50 mM), and gramicidin D (5 μg ml^{-1}) were added as indicated. The initial rate of quenching, expressed as percent of total fluorescence min^{-1}, is indicated on the *lines*

enzyme can translocate protons in the absence of K^+ if a permeable anion such as nitrate is present for electrical balance (lower trace). This indicates that proton transport is electrogenic and that the enzyme does not catalyze a H^+-K^+ exchange. On the other hand, K^+ slightly stimulates proton transport to the same extent as ATP hydrolysis (cf. upper and lower traces). The effect of externally added K^+ was instantaneous (lower trace), suggesting action on the external side of the vesicles, where ATP is hydrolyzed. In the H^+-K^+ exchange hypothesis the stimulation by K^+ should occur after diffusion to the internal side of the vesicle and a lag would be expected. Electrical balance in the absence of permeable anions can be obtained by high concentrations of K^+ in the presence of valinomycin (Vara and Serrano 1982). The requirement for a K^+ ionophore suggests that the permeability of the reconstituted vesicles to K^+ is low and therefore it is unlikely that K^+ uptake in the experiment of Fig. 4 occurred faster than the fluorescence response. These crucial experiments, together with studies conducted at the whole cell level (Serrano 1985), disprove previous suggestions of a K^+ pumping activity of the plant plasma membrane ATPase (Hodges 1976; Leonard 1983). The plant and fungal ATPases seem to pump only protons. K^+ transport probably occurs by voltage-gated channels (Schroeder et al. 1987). It must be stressed that the number of K^+ channels per cell (10–100, Schroeder et al. 1984) is much lower than the estimated number of ATPase molecules (10^5–10^6 per cell, Serrano 1988). Therefore, it is very unlikely that the two activities are present in the same protein. However, it has been reported that highly purified ATPase from the yeast *Schizosaccharomyces pombe* contains a voltage-sensitive K^+ channel (Villalobo 1982, 1984) and this possibility should be investigated with preparations of the plant ATPase.

The H^+/ATP stoichiometry of the plant plasma membrane ATPase has not been directly determined in reconstituted systems. Electrophysiological studies with whole cells suggest that the stoichiometry of the proton-pumping plasma membrane ATPase is 1 H^+/ATP in higher plants and 2 H^+/ATP in algae (Serrano 1985). The first value agrees with the stoichiometry of the fungal ATPase determined from electrophysiological measurements in whole cells (Warncke and Slayman 1980) and thermodynamic (Malpartida and Serrano 1981) and kinetic (Perlin et al. 1986) measurements in reconstituted proteoliposomes.

6 Regulation by Plant Hormones and Toxins

The plant plasma membrane ATPase seems to play a central role in plant physiology. The proton gradient generated by the enzyme drives the active transport of different molecules in and out of the cells by uniport, symport, and antiport mechanisms. In addition, the acidification of the cell wall resulting from ATPase activity causes the loosening required for cell enlargement. Therefore, cell nutrition and growth are controlled by the plasma membrane ATPase and the same is true for important phenomena at the whole organism level, such as phloem and xylem loading and unloading, changes in guard cell turgor and concomitant stomatal opening, breaking of seed dormancy, etc. Therefore, it is not surprising that the activity of the plasma membrane ATPase is modulated by many important regulatory factors known to plant physiologists: plant hormones, toxins produced by pathogens, light and stress conditions (Rayle and Cleland 1977; Marrè 1979; Serrano 1985; see Chaps. 7–9, 14 and 16).

Very little is known about the molecular mechanism of the regulation of the plasma membrane ATPase in plants. The clearest and most rapid effects are the activations produced by the fungal phytotoxin fusicoccin (Marrè 1979) and by the bacterial phytotoxin syringomycin (Bidwai and Takemoto 1987). Both toxins activate the ATPase in isolated plasma membrane vesicles (Rasi-Caldogno et al. 1986; Bidwai et al. 1987), suggesting that the membranes and the ATPase assay medium contain all the components necessary for the response. In the case of syringomycin there is evidence for the participation of a membrane-bound, Ca^{2+}-dependent protein kinase acting on the 100 kD ATPase polypeptide (Bidwai and Takemoto 1987). Fusicoccin also seems to activate protein kinases, but an effect on the phosphorylation of the ATPase itself has not been demonstrated (Tognoli and Colombo 1986).

The activation of the ATPase by auxin seems less direct, because protein synthesis is required and auxin does not usually activate the ATPase in isolated membranes (Rayle and Cleland 1977; Marrè 1979). The regulatory signals triggered by auxin in plant cells have been subjected to intensive investigation (Chaps. 9 and 16). Both cytosolic acidification (Felle et al. 1986) and phospholipase C-mediated hydrolysis of phosphoinositides (Ettlinger and Lehle 1988) have recently been proposed as immediate responses to auxin. Both conditions may activate protein kinase activity. A protein kinase stimulated by acidification has been described in plasma membrane vesicles from oat roots (Schaller and Sussman 1988a) and a Ca^{2+}-dependent protein kinase is present in plant tissues (Hetherington and Trewavas 1982). Artificially-induced internal acidification mimics auxin effects (Vesper and Evans 1979) and intracellular Ca^{2+} could be increased by inositol trisphosphate, as demonstrated in animal cells (Downes and Michell 1985). However, there is no direct evidence for these regulatory cascades in living plant cells.

A different regulatory response has been described for plant tissues subjected to different types of injury or stress (cutting, rubbing, chilling, heating). This causes ATPase inhibition and seems to correlate with increased phosphorylation of membrane proteins catalyzed by Ca^{2+}-dependent protein kinases (Zocchi et al. 1983; Zocchi 1985). Ca^{2+} seems to enter injured or shocked plant cells through channels opened on depolarization of the plasma membrane (Rincon and Hanson 1986).

Together these reports suggest that protein kinases mediate the modulation of plasma membrane ATPase activity by different regulatory factors. However, it must be stressed that there is no definitive evidence for any of the proposed regulatory mechanisms. The best-known example of regulation of ATPase activity by protein kinases is the modulation of Ca^{2+}-pumping ATPase in the cardioplasmic reticulum by cAMP and Ca^{2+}-dependent protein kinases (Katz 1981). In this case a low-molecular-weight polypeptide and not the ATPase is the target for the phosphorylation. This possibility should be kept in mind in future investigations on the effect of membrane phosphorylation on the activity of plant plasma membrane ATPase.

7 Structure and Mechanism

Very little is known of the plant plasma membrane ATPase at the molecular level. The sequence of some tryptic peptides (Walderhaug et al. 1985; Schaller and Sussman 1988b) and nucleotide sequences (Pardo and Serrano) indicate that the plant plasma membrane ATPase contains all the conserved regions of eukaryotic (E-P) ATPases and that it is closely related to fungal H^+-ATPases (Fig. 5). The immunological cross-reactivity between fungal and plant plasma membrane ATPases (Clement et al. 1986; Surowy and Sussman 1986) is consistent with this observation. Therefore the structure and mechanism of the plant plasma membrane ATPase can be discussed in the light of other better studied ATPases of the (E-P) family (Serrano 1988).

The proposed transmembrane structure and functional domains are shown in Fig. 6. There are eight hydrophobic stretches of about 20 amino acids which could form transmembrane α-helices. These, together with the membrane-flanking stretches, could constitute a transmembrane channel. However, it has been suggested that in the Na^+/K^+-ATPase the seventh hydrophobic stretch is not in the membrane and therefore that the polypeptide chain crosses the membrane only seven times (Ovchinnikov et al. 1987b). Conserved region 3, which contains the phosphorylated intermediate, would be located in the cytoplasmic opening of the channel, in an ideal position to couple ion transport to ATP hydrolysis. The coupling involves phosphorylation and dephosphorylation steps (see Sect. 4) and I have recently proposed that these steps are catalyzed by separate domains of the enzyme (Serrano 1988). Chemical modification studies with animal (E-P) ATPases (Farley and Faller 1985; Ohta et al. 1986; Ovchinnikov et al. 1987a) and site-directed mutagenesis of the yeast ATPase (Portillo and Serrano 1988) support the participation of conserved regions 4–7 in a kinase domain involved in ATP binding and the formation of the phosphorylated intermediate. Site-directed mutagenesis of the yeast enzyme (Portillo and Serrano 1988) indicates that at least conserved region 2 is part of a separate phosphatase domain which catalyzes the hydrolysis of the phosphorylated intermediate. The coupling mechanism seems to follow the principles enunciated by Hammes (1982) and Tanford (1982), as described by Serrano (1988; Fig. 7). The (E-P) ATPases seem to exist in two conformations which alternate during the catalytic cycle. In the E_1 conformation the kinase domain is active but the phosphatase domain is not and the proton binding site in the gate of the channel binds protons with high affinity from the cytoplasmic side of the membrane. In the E_2 conformation the phosphatase domain is active

```
                  1                  2                 3              4
            **-- -  *- ---*      *-- -****        ---*****-* -       ----
Ca      148 GDIVEIAVGDKVPAD  176 DQSILTGES    348 ICSDKTGTLTTN   514 KGAP
                 V
NaK     186 GDLVEVKGGDRIPAD  212 DNSSLTGES    371 ICSDKTGTLTQN   506 KGAP
                 I     V
HK      197 GDLVEMKGGDRVPAD  223 DNSSLTGES    382 ICSDKTGTLTQN   517 KGAP

K       129 GDIVLVEAGDIIPCD  154 DESAITGES    304 LLLDKTGTITLG   no present
              RLI R   KM T           V            IM      L Q
Hf      199 GDILQLEDGTVIPTD  226 DQSAITGES    375 LCSDKTGTLTKN   474 KGAP
                KVDE  I CA           L
Hld     173 GDLVKLASGSAVPAD  198 DEAALTGES    348 LCSDKTGTLTLN   445 KGAP
Hor         GDIVSIK              DQSGLTGES         CSDKTGTLTLN
Hat         GDIVSIKLGDIIPAD      DQSALTGES        LCSDKTGTLTLN       KGAP
```

```
           5             6                     7
          ----       --   *-***       - ----****-**-*-*--*-
Ca    580 DPPR   617 GIRVIMITGD   694 DEITAMTGDGVNDAPALKKAE

NaK   591 DPPR   608 GIKVIMVTGD   707 GAIVAVTGDGVNDSPALKKAD

HK    602 DPPR   619 GIRVIMVTGD   718 GAIVAVTGDGVNDSPALKKAD

K     no present 464 GIKTVMITGD   510 GRLVAMTGDGTNDAPALAQAD
                     N IP  L           KK I V   I    S  R T
Hf    534 DPPR   551 GLRVKMLTGD   626 GYLVAMTGDGVNDAPSLKKAD
                       SI
Hld   501 DPPR   518 GVDVKMITGD   597 GYTCAMTGDGVNDAPALKRAD
Hor                                    GIVGMTGDGVNDAPALK
Hat       DPPR       GVNVKMITGD       KHICGMTGDGVNDAPALKKAD
```

Fig. 5. Highly conserved regions in catalytic subunits of (E-P) ATPases. The *single-letter code* for amino acids is employed. *Asterisks* identify fully conserved amino acids and *dashes* correspond either to amino acids only conserved in the eukaryotic enzymes or replaced by similar amino acids. *Ca*: Ca^{2+}-ATPase from rabbit sarcoplasmic reticulum, the sequence of the slow twitch isoenzyme is shown, with changes in the fast twitch isoenzyme indicated below (Brandl et al. 1986); *NaK*: animal Na^{+}/K^{+}-ATPase, the sequence of the sheep kidney enzyme (Shull et al. 1985) is shown, with changes in pig kidney (Ovchinnikov et al. 1986), *Torpedo californica* (Kawakami et al. 1985) and rat brain forms (Shull et al. 1988) indicated below; *HK*: H^{+}/K^{+}-ATPase from rat stomach (Shull and Lingrel 1986); *K*: K^{+}-ATPase from bacteria, the sequence of the *Escherichia coli* enzyme (Hesse et al. 1984) is shown, with changes in *Streptococcus faecalis* (Solioz et al. 1987) indicated below; *Hf*: fungal H^{+}-ATPase, the sequence of the *Saccharomyces cerevisiae* enzyme (Serrano et al. 1986) is shown, with changes in *Neurospora crassa* (Addison 1986; Hager et al. 1986) and in *Schizosaccharomyces pombe* (Ghislain et al. 1987) indicated below; *Hld*: probably H^{+}-ATPase from *Leishmania donovani* (Meade et al. 1987); *Hor*: partial amino acid sequences from the oat root H^{+}-ATPase (Schaller and Sussman 1988b) completed with the tetrapeptide determined for the corn root enzyme (Walderhaug et al. 1985); *Hat*: amino acid sequences from the *Arabidopsis thaliana* H^{+}-ATPase (Pardo and Serrano 1989)

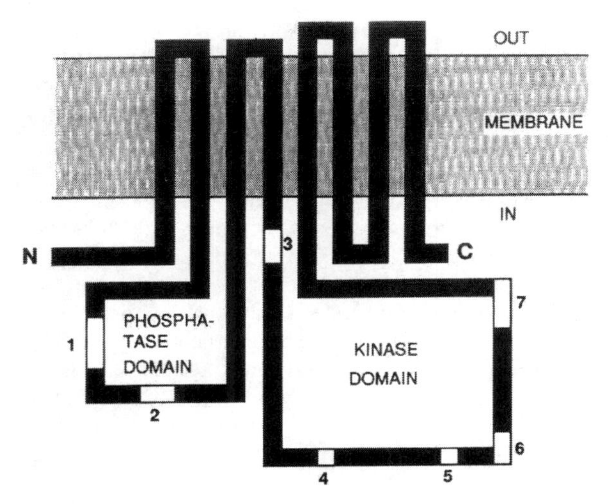

Fig. 6. Model for the transmembrane structure and functional domains of (E-P) ATPases. The seven highly conserved regions of Fig. 5 are indicated

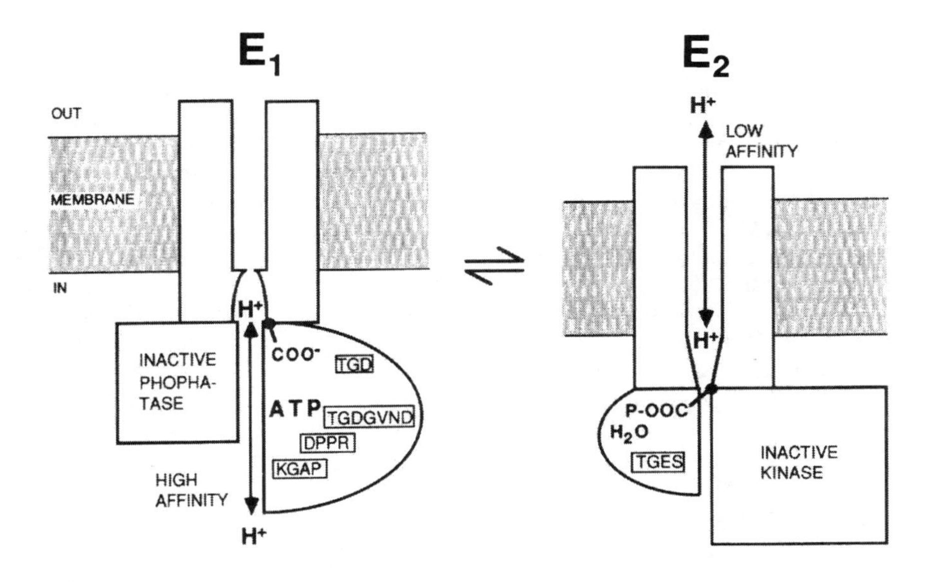

Fig. 7. Model for the mechanism of coupling ATP hydrolysis to proton transport in (E-P) ATPases. The conserved motifs from Fig. 5 implicated in catalysis are indicated with the *single-letter code* for amino acids

while the kinase domain is not and the proton binding site in the gate of the channel binds protons with low affinity from the external side of the membrane. Therefore, in order for the enzyme to complete its catalytic cycle it must alternate between both conformations and in so doing effects proton pumping. The conformational change seems to be triggered by the phosphorylation of conserved region 3 and it is easy to visualize that, as this region may be located in the gate of the channel, a concerted conformational change may modify both the affinity and sidedness of the channel, as well as the activity of the kinase and phosphatase domains. Most authors like to visualize this concerted change as the indirect, conformational coupling between distant transport and catalytic sites (Hammes 1982; Tanford 1982). This model seems to be supported by measurements on the Ca^{2+}-pumping ATPase of sarcoplasmic reticulum (Scott 1985; Teruel and Gomez-Fernandez 1986) which indicate that the ATP binding site is quite remote from the Ca^{2+} binding site (about 4 nm). On the other hand, Mitchell (1981) has suggested that the phosphorylated intermediate may move from a kinase catalytic site to a phosphatase catalytic site and in so doing "drag" the transported cation from the high-affinity, cytoplasmic site to the low-affinity, external site. The recent evidence for separate kinase and phosphatase domains (Portillo and Serrano 1988) would be consistent with this mechanism. However, the existence of independent domains is also compatible with indirect coupling between distant transport and catalytic sites. One point which does not fit with the measured distances described above is that all current models locate the phosphorylated intermediate close to the membrane, probably in the "stalk" of the enzyme protruding from the membrane, and the Ca^{2+} binding site also seems to be located in this region. Therefore, as the ATP must reach the amino acid forming the phosphorylated intermediate, it may be predicted that the ATP binding site should also be close to the membrane. Future studies should clarify these inconsistencies.

Very little is known about the E_1-E_2 conformational change, except that it affects the susceptibility of the enzymes to proteases (Addison and Scarborough 1982; Andersen et al. 1986; Jørgensen 1982; Perlin and Brown 1987; Serrano 1988); that it does not involve large changes in secondary structure (Hastings et al. 1986; Nakamoto and Inesi 1986; Hennessey and Scarborough 1988); and that the E_2 conformation is more exposed to the hydrophobic membrane phase (Jørgensen 1982; Andersen et al. 1986).

8 On the Existence of a Ca^{2+}-Pumping ATPase in Plant Plasma Membranes

Preparations containing plasma membrane vesicles from different plant tissues exhibit Ca^{2+}-pumping activity dependent on ATP and resistant to proton ionophores (Dieter and Marmé 1980; Rasi-Caldogno et al. 1982; Zocchi and Hanson 1983; Gräf and Weiler 1989). The fact that Ca^{2+} transport is not dependent on a proton gradient allows the rejection of a H^+-Ca^{2+} antiport mechanism driven by the proton-pumping ATPase. In contrast, this indirect coupling mechanism seems to be the only Ca^{2+} transport present in fungal plasma membrane vesicles (Stroobant and Scarborough 1979).

Ca^{2+} pumping in plant plasma membrane vesicles is stimulated by calmodulin (Dieter and Marmé 1980; Zocchi et al. 1983) and a Ca^{2+} and calmodulin-dependent ATPase has been identified in plant microsomal membranes (Dieter and Marmé 1981) and in highly purified plasma membrane vesicles (Robinson et al. 1988). Recent experimental evidence indicates that the plant Ca^{2+}-pumping ATPase catalyzes a Ca^{2+}/H^+ exchange (Rasi-Caldogno et al. 1987). All these properties suggest that the plant enzyme is very similar to the Ca^{2+}-pumping ATPase characterized in animal plasma membranes (Carafoli and Zurini 1982). Actually, recent work on the purification of the calmodulin-stimulated ATPase from maize coleoptiles indicates that the enzyme is a 140 kD polypeptide which shows immunological cross-reactivity with the 138 kD erythrocyte Ca^{2+} pump (Briars et al. 1988).

The measured Ca^{2+} pumping and (Ca^{2+} + calmodulin)-stimulated ATPase activities are very low, 1–4 nmol min^{-1} (mg protein)$^{-1}$ versus 0.2–1 μmol min^{-1} (mg protein)$^{-1}$ for the proton-pumping ATPase. Therefore, the Ca^{2+}-pumping ATPase seems to be a minor component of plant plasma membranes and this justifies the convention employed throughout this chapter of considering the proton-pumping ATPase as the predominant plasma membrane ATPase.

9 Perspectives

The study of plant plasma membrane H^+-ATPase is reaching the state where molecular and genetic perspectives are needed to complement the classical physiological and biochemical approaches. The proposed crucial role of the enzyme in plant physiology has to be documented by suitable mutants in which either ATPase activity or regulation is affected. The model plant *Arabidopsis thaliana* (Meyerowitz 1987) is the system of choice for this genetic approach. The molecular mechanism of proton pumping could be clarified by site-directed mutagenesis of critical amino acids of the enzyme. However, only the three-dimensional structure obtained from ATPase crystals could provide the intimate details of the proton pump. Finally, a combination of protein chemistry and directed mutagenesis could elucidate the role of phosphorylation in the regulation of ATPase activity.

References

Addison R (1986) Primary structure of the *Neurospora* plasma membrane H^+-ATPase deduced from the gene sequence. J Biol Chem 261:14896–14901

Addison R, Scarborough GA (1982) Conformational changes of the *Neurospora* plasma membrane H^+-ATPase during its catalytic cycle. J Biol Chem 257:10421–10426

Ahlers J (1984) Effect of different salts on the plasma membrane ATPase and on proton transport in yeast. Can J Biochem Cell Biol 62:998–1005

Amory A, Foury F, Goffeau A (1980) The purified plasma membrane ATPase of the yeast *Schizosaccharomyces pombe* forms a phosphorylated intermediate. J Biol Chem 255:9353–9357

Andersen JP, Vilsen B, Collins JH, Jørgensen PL (1986) Localization of E_1-E_2 conformational transitions of sarcoplasmic reticulum Ca-ATPase by tryptic cleavage and hydrophobic labeling. J Membr Biol 93:85–92

Anthon GE, Spanswick RM (1986) Purification and properties of the H^+-translocating ATPase from the plasma membrane of tomato roots. Plant Physiol 81:1080–1085

Balke NE, Hodges TK (1975) Plasma membrane adenosine triphosphatase of oat roots. Activation and inhibition by Mg and ATP. Plant Physiol 55:83–86

Ball JH, Williams L, Hall JL (1987) Effect of SW26 and erythrosin B on ATPase activity and related processes in *Ricinus* cotyledons and cucumber hypocotyls. Plant Sci 52:1–5

Bennet AB, O'Neill SD, Eilman M, Spanswick RM (1985) H^+-ATPase from storage tissue of *Beta vulgaris*. III Modulation of ATPase activity by reaction substrates and products. Plant Physiol 78:495–499

Bennet JP, Smith GA, Houslay MD, Hesketh TR, Metcalfe JC, Warren GB (1978) The phospholipid head group specificity of an ATP-dependent calcium pump. Biochim Biophys Acta 513:310–320

Bidwai AP, Takemoto JY (1987) Bacterial phytotoxin, syringomycin, induces a protein-kinase mediated phosphorylation of red beet plasma membrane polypeptides. Proc Natl Acad Sci USA 84:6755–6759

Bidwai AP, Zhang L, Bachmann RC, Takemoto JY (1987) Mechanism of action of *Pseudomonas syringae* phytotoxin, syringomycin. Stimulation of red beet plasma membrane ATPase activity. Plant Physiol 83:39–43

Blein JP, De Cherade X, Bergon M, Calmon JP, Scalla R (1986) Inhibition of adenosine triphosphatase activity from a plasma membrane fraction of *Acer pseudoplatanus* cells by 2,2,2-trichloroethyl 3,4-dichlorocarbanilate. Plant Physiol 80:782–785

Blein JP, Martinez J, Bergon M, Calmon JP, Scalla R (1987) Inhibition of adenosine triphosphatase activity from a plasma membrane fraction of *Acer pseudoplatanus* cells by carbanilate derivatives. Plant Physiol 83:469–471

Borst-Pauwels GWFH, Theuvenet APR, Stols ALH (1983) All-or-none interactions of inhibitors of the plasma membrane ATPase with *Saccharomyces cerevisiae*. Biochim Biophys Acta 732:186–192

Brandl CJ, Green NM, Korczak B, MacLennan DH (1986) Two Ca-ATPase genes: homologies and mechanistic implications of deduced amino acid sequence. Cell 44:597–607

Briars SA, Kessler F, Evans DE (1988) The calmodulin-stimulated ATPase of maize coleoptiles is a 140,000 Mr polypeptide. Planta 176:283–285

Briskin DP, Leonard RT (1982) Partial characterization of a phosphorylated intermediate associated with the plasma membrane ATPase of corn roots. Proc Natl Acad Sci USA 79:6922–6926

Briskin DP, Poole RJ (1983a) Role of magnesium in the plasma membrane ATPase of red beet. Plant Physiol 71:969–971

Briskin DP, Poole RJ (1983b) Evidence for a beta-aspartyl-phosphate residue in the phosphorylated intermediate of the red beet plasma membrane ATPase. Plant Physiol 72:1133–1135

Briskin DP, Thornley WR, Roti-Roti JL (1985) Target molecular size of the red beet plasma membrane ATPase. Plant Physiol 78:642–644

Carafoli E, Zurini M (1982) The Ca^{2+}-pumping ATPase of plasma membranes. Purification, reconstitution and properties. Biochim Biophys Acta 683:279–301

Chadwick CC, Goormaghtigh E, Scarborough GA (1987) A hexameric form of the *Neurospora crassa* plasma membrane H^+-ATPase. Arch Biochem Biophys 252:348–356

Cid A, Vara F, Serrano R (1987) Inhibition of the proton pumping ATPases of yeast and oat roots plasma membranes by dicyclohexylcarbodiimide. Arch Biochem Biophys 252:496–500

Clement JD, Ghislain M, Dufour JP, Scalla R (1986) Immunodetection of a 90 000-Mr polypeptide related to yeast plasma membrane ATPase in plasma membranes from maize shoots. Plant Sci 45:43–50

Cocucci MC (1986) Inhibition of plasma membrane and tonoplast ATPases by erythrosin B. Plant Sci 47:21–27
Cocucci MC, Marrè E (1984) Lysophosphatidylcholine-activated, vanadate inhibited, Mg-ATPase from radish microsomes. Biochim Biophys Acta 771:42–52
Cocucci MC, Marrè E (1986) Erythrosin B as an effective inhibitor of electrogenic H^+ extrusion. Plant Cell Environ 9:677–679
Cocucci MC, De Michelis MI, Pugliarello MC, Rasi-Caldogno F (1985) Reconstitution of proton pumping activity of a plasma membrane ATPase purified from radish. Plant Sci Lett 37:189–193
Dieter P, Marmé D (1980) Calmodulin activation of plant microsomal calcium uptake. Proc Natl Acad Sci USA 77:7311–7314
Dieter P, Marmé D (1981) A calmodulin-dependent, microsomal ATPase from corn (*Zea mays* L.). FEBS Lett 125:245–248
Downes CP, Michell RH (1985) Inositol phospholipid breakdown as a receptor controlled generator of second messengers. In: Cohen P, Houslay MD (eds) Molecular mechanisms of transmembrane signalling. Elsevier, Amsterdam, pp 3–56
Ettlinger C, Lehle L (1988) Auxin induces rapid changes in phosphatidylinositol metabolites. Nature 331:176–178
Eytan GD (1982) Use of liposomes for reconstitution of biological functions. Biochim Biophys Acta 694:185–202
Farley RA, Faller LD (1985) The amino acid sequence of an active site peptide from the H,K-ATPase of gastric mucosa. J Biol Chem 260:3899–3901
Felle H, Brummer B, Bertl A, Parish RW (1986) Indole-3-acetic acid and fusicoccin cause cytosolic acidification of corn coleoptile cells. Proc Natl Acad Sci USA 83:8992–8995
Furst P, Solioz M (1986) The vanadate-sensitive ATPase of *Streptococcus faecalis* pumps potassium in a reconstituted system. J Biol Chem 261:4302–4308
Gallagher SR, Leonard RT (1982) Effect of vanadate, molybdate and azide on membrane-associated ATPase and soluble phosphatase activities of corn roots. Plant Physiol 70:1335–1340
Gallagher SR, Leonard RT (1987) Electrophoretic characterization of a detergent-treated plasma membrane fraction from corn roots. Plant Physiol 83:265–271
Ghislain M, Schlesser A, Goffeau A (1987) Mutations of a conserved glycine modifies the vanadate sensitivity of the plasma membrane H^+-ATPase from *Schizosaccharomyces pombe*. J Biol Chem 262:17549–17555
Goormaghtigh E, Chadwick C, Scarborough GA (1986) Monomers of the *Neurospora* plasma membrane H^+-ATPase catalyze efficient proton translocation. J Biol Chem 261: 7466–7471
Gräf P, Weiler EW (1989) ATP-driven Ca^{2+} transport in sealed plasma membrane vesicles prepared by aqueous two-phase partitioning from leaves of *Commelina communis*. Physiol Plant 75:469–478
Hager KM, Mandala SM, Davenport JW, Speicher DW, Benz EJ, Slayman CW (1986) Amino acid sequence of the plasma membrane H^+-ATPase of *Neurospora crassa*: deduction from genomic and cDNA sequences. Proc Natl Acad Sci USA 83:7693–7697
Hammes GG (1982) Unifying concept for the coupling between ion pumping and ATP hydrolysis or synthesis. Proc Natl Acad Sci USA 79:6881–6884
Hastings DF, Reynold JA, Tanford C (1986) Circular dicroism of the two major conformational states of mammalian (Na + K) ATPase. Biochim Biophys Acta 860:566–569
Hennessey JP, Scarborough GA (1988) Secondary structure of the *Neurospora crassa* plasma membrane H^+-ATPase as estimated by circular dichroism. J Biol Chem 263:3123–3130
Hesse JE, Wieczorek L, Altendorf K, Reicin AS, Dorus E, Epstein W (1984) Sequence homology between two membrane transport ATPases, the Kdp-ATPase of *Escherichia coli* and the Ca-ATPase of sarcoplasmic reticulum. Proc Natl Acad Sci USA 81:4746–4750
Hetherington A, Trewavas A (1982) Calcium-dependent protein kinase in pea shoot membranes. FEBS Lett 145:67–71
Hodges TK (1976) ATPases associated with membranes of plant cells. In: Lüttge U, Pitman MG (eds) Encyclopedia of plant physiology, New Series, vol 2, Part A, Springer, Berlin Heidelberg New York, pp 260–283

Hodges TK, Leonard RT, Bracker CE, Keenan TW (1972) Purification of an ion-stimulated ATPase from plant roots: association with plasma membranes. Proc Natl Acad Sci USA 69:3307–3311

Hollander VP (1971) Acid phosphatases. In: Boyer PD (ed) The enzymes, 3rd edn, vol IV. Academic Press, New York, pp 449–498

Jørgensen PL (1982) Mechanism of the Na^+,K^+-pump. Protein structure and conformations of the pure (Na^+ + K^+)-ATPase. Biochim Biophys Acta 694:27–68

Kasamo K (1986) Purification and properties of the plasma membrane H^+-translocating adenosine triphosphatase of *Phaseolus mungo* L. roots. Plant Physiol 80:818–824

Kasamo K (1987) Reconstitution and characterization of H-translocating ATPase from the plasma membrane of *Phaseolus mungo* L. roots. Plant Cell Physiol 28:19–28

Kasamo K, Nouchi I (1987) The role of phospholipids in plasma membrane ATPase activity in *Vigna radiata* L. (mung bean) roots and hypocotyls. Plant Physiol 83:323–328

Katz AM (1981) Regulation of calcium transport in the cardiac sarcoplasmic reticulum by cyclic-AMP-dependent protein kinase. In: Rosen OM, Krebs EG (eds) Protein phosphorylation. Cold Spring Harbor Laboratory, Cold Spring Harbor, NY, pp 849–854

Katz DB, Sussman MR (1987) Inhibition and labeling of the plant plasma membrane H^+-ATPase with N-ethylmaleimide. Plant Physiol 83:977–981

Katz DB, Sussman MR, Mierzwa RJ, Evert RF (1988) Cytochemical localization of ATPase activity in oat root localizes a plasma membrane-associated soluble phosphatase, not the proton pump. Plant Physiol 86:841–847

Kawakami K, Noguchi S, Noda M, Takahashi H, Ohta T, Kawamura M, Nojima H, Nagano K, Hirose T, Inayama S, Hayashida H, Miyata T, Numa S (1985) Primary structure of the α-subunit of *Torpedo californica* (Na + K) ATPase deduced from cDNA sequence. Nature 316:733–736

Kimelberg HK, Papahadjopoulos D (1972) Phospholipid requirements for (Na + K) ATPase activity: head group specificity and fatty acid fluidity. Biochim Biophys Acta 282:277–292

Larsson C, Kjellbom P, Widell S, Lundborg T (1984) Sidedness of plant plasma membrane vesicles purified by partition in aqueous two-phase systems. FEBS Lett 171:271–276

Leonard RT (1983) Potassium transport and the plasma membrane ATPase in plants. In: Robb DA, Pierpoint WS (eds) Metals and micronutrients: uptake and utilization by plants. Academic Press, New York, pp 71–86

Leonard RT, Hodges TK (1973) Characterization of plasma membrane-associated ATPase activity of oat roots. Plant Physiol 52:6–12

Macara IG (1980) Vanadium, an element in search of a role. Trends Biochem Sci 5:92–94

Malpartida F, Serrano R (1981) Proton translocation catalyzed by the purified yeast plasma membrane ATPase reconstituted in liposomes. FEBS Lett 131:351–354

Marrè E (1979) Integration of solute transport in cereals. In: Laidman DL, Wyn Jones RG (eds) Recent advances in the biochemistry of cereals. Academic Press, New York, pp 3–25

Meade JC, Shaw J, Gallagher G, Lemaster S, Stringer JR (1987) Structure and expression of a tandem gene pair in *Leishmania donovani* that encodes a protein structurally homologous to eukaryotic cation-transporting ATPases. Mol Cell Biol 7:3937–3946

Melgunov VI, Akimova EI (1980) The dependence for reactivation of lipid-depleted Ca^{2+}-ATPase of sarcoplasmic reticulum by non-ionic detergents on their hydrophile/lipophile balance. FEBS Lett 121:235–238

Meyerowitz EM (1987) *Arabidopsis thaliana.* Annu. Rev Genet 21:93–111

Mitchell P (1981) Bioenergetic aspects of unity in biochemistry: evolution of the concept of ligand conduction in chemical, osmotic and chemiosmotic reaction mechanisms. In: Semenza G (ed) Of oxygen, fuels and living matter. Part 1. John Wiley, New York, chap 1

Nakamoto RK, Inesi G (1986) Retention of ellipticity between enzymatic states of the Ca-ATPase of sarcoplasmic reticulum. FEBS Lett 194:258–262

Ohta T, Nagano K, Yoshida M (1986) The active site structure of Na^+/K^+-transporting ATPase: location of the 5′-(p-fluorosulfonyl)benzoyladenosine binding site and soluble peptides released by trypsin. Proc Natl Acad Sci USA 83:2071–2075

Oleski NA, Bennett AB (1987) H^+-ATPase activity from storage tissue of *Beta vulgaris.* IV. N,N′-dicyclohexylcarbodiimide binding and inhibition of the plasma membrane H^+-ATPase. Plant Physiol 83:569–572

Ovchinnikov YA, Modyanov NN, Broude NE, Petrukhin KE, Grishin AV, Arzamazova NM, Aldanova NA, Monastyrskaya GS, Sverdlov ED (1986) Pig kidney Na,K-ATPase. Primary structure and spatial organization. FEBS Lett 201:237–245

Ovchinnikov YA, Dzhandzugazyan KN, Lutsenko SV, Mustayev AA, Modyanov NN (1987a) Affinity modification of E_1-form of Na^+,K^+-ATPase revealed Asp-710 in the catalytic site. FEBS Lett 217:111–116

Ovchinnikov YA, Arzamazova NM, Arystarkhova EA, Gevondyan NM, Aldanova NA, Modyanov NN (1987b) Detailed structural analysis of exposed domains of membrane-bound Na^+,K^+-ATPase. A model of transmembrane arrangement. FEBS Lett 217:269–274

Palmgren MG, Sommarin M, Jørgensen PL (1988) Substrate stabilization of lysophosphatidylcholine-solubilized plasma membrane H^+-ATPase from oat roots. Physiol Plant 74:20–25

Pardo JM, Serrano R (1989) Structure of a plasma membrane H^+-ATPase gene from the plant *Arabidopsis thaliana*. J Biol Chem 264:8557–8562

Pedersen PL, Carafoli E (1987a) Ion motive ATPases. I. Ubiquity, properties and significance to cell function. Trends Biochem Sci 12:146–150

Pedersen PL, Carafoli E (1987b) Ion motive ATPases. II. Energy coupling and work output. Trends Biochem Sci 12:186–189

Perlin DS, Brown CL (1987) Identification of structurally distinct catalytic intermediates of the H^+-ATPase from yeast plasma membranes. J Biol Chem 262:6788–6794

Perlin DS, San Francisco MJD, Slayman CW, Rosen BP (1986) Proton-ATP stoichiometry of proton pumps from *Neurospora crassa* and *Escherichia coli*. Arch Biochem Biophys 248:53–61

Poole RJ (1978) Energy coupling for membrane transport. Annu Rev Plant Physiol 29:437–460

Portillo F, Serrano R (1988) Dissection of functional domains of the yeast proton-pumping ATPase by directed mutagenesis. EMBO J 7:1793–1798

Rasi-Caldogno F, De Michelis MI, Pugliarello MC (1982) Active transport of Ca^{2+} in membrane vesicles from pea. Evidence for a H^+/Ca^{2+} antiport. Biochim Biophys Acta 693:287–295

Rasi-Caldogno F, De Michelis MI, Pugliarello MC, Marrè E (1986) H^+-pumping driven by the plasma membrane ATPase in membrane vesicles from radish: stimulation by fusicoccin. Plant Physiol 82:121–125

Rasi-Caldogno F, Pugliarello MC, De Michelis MI (1987) The Ca^{2+}-transport ATPase of plant plasma membrane catalyzes an nH^+/Ca^{2+} exchange. Plant Physiol 83:994–1000

Rayle DL, Cleland RE (1977) Control of plant cell enlargement by hydrogen ions. Curr Top Dev Biol 11:187–214

Rincon M, Hanson JB (1986) Controls on calcium ion fluxes in injured or shocked corn root cells: importance of proton pumping and cell membrane potential. Physiol Plant 67:576–583

Robinson C, Larsson C, Buckhout TJ (1988) Identification of a calmodulin-stimulated (Ca^{2+} + Mg^{2+})-ATPase in a plasma membrane fraction isolated from maize (*Zea mays*) leaves. Physiol Plant 72:177–184

Roelofsen B, Van Deenen LLM (1973) Lipid requirement of membrane-bound ATPase. Studies on human erythrocyte ghosts. Eur J Biochem 40:245–257

Sachs G, Faller LD, Rabon E (1982) Proton/hydroxyl transport in gastric and intestinal epithelia. J Membr Biol 64:123–133

Scalla R, Amory A, Rigaud J, Goffeau A (1983) Phosphorylated intermediate of a transport ATPase and activity of protein kinase in membranes from corn roots. Eur J Biochem 132:525–530

Scarborough GA, Addison R (1984) On the subunit composition of the *Neurospora* plasma membrane H^+-ATPase. J Biol Chem 259:9109–9114

Schaller GE, Sussman MR (1988a) Phosphorylation of the plasma membrane H^+-ATPase of oat roots by a calcium-stimulated protein kinase. Planta 173:509–518

Schaller GE, Sussman MR (1988b) Isolation and sequence of tryptic peptides from the proton-pumping ATPase of the oat root plasma membrane. Plant Physiol 86:512–516

Schroeder JI, Hedrich R, Fernandez JM (1984) Potassium-selective single channels in guard cell protoplasts of *Vicia faba*. Nature 312:361–362

Schroeder JI, Raschke K, Neher E (1987) Voltage-sensitive K^+ channels in guard cell protoplasts. Proc Natl Acad Sci USA 84:4107–4112

Scott TL (1985) Distances between the functional sites of the ($Ca^{2+} + Mg^{2+}$)-ATPase of sarcoplasmic reticulum. J Biol Chem 260:14421–14423
Serrano R (1983) Purification and reconstitution of the proton-pumping ATPase of fungal and plant plasma membranes. Arch Biochem Biophys 227:1–8
Serrano R (1984a) Plasma membrane ATPase of fungi and plants as a novel type of proton pump. Curr Top Cell Regul 23:87–126
Serrano R (1984b) Purification of the proton pumping ATPase from plant plasma membranes. Biochem Biophys Res Commun 121:735–740
Serrano R (1985) Plasma membrane ATPase of plants and fungi. CRC, Boca Raton, FL
Serrano R (1988) Structure and function of proton translocating ATPase in plasma membranes of plants and fungi. Biochim Biophys Acta 947:1–28
Serrano R, Kielland-Brandt MC, Fink GR (1986) Yeast plasma membrane ATPase is essential for growth and has homology with (Na + K), K and Ca ATPases. Nature 319:689–693
Serrano R, Montesinos C, Sanchez J (1988) Lipid requirements of the plasma membrane ATPases from oat roots and yeast. Plant Sci 56:117–122
Shigekawa M, Wakabayashi S (1985) Sidedness of K^+ activation of calcium transport in the reconstituted sarcoplasmic reticulum calcium pump. J Biol Chem 260:11679–11687
Shull GE, Lingrel JB (1986) Molecular cloning of the rat stomach (H + K) ATPase. J Biol Chem 261:16788–16791
Shull GE, Schwartz A, Lingrel JB (1985) Amino-acid sequence of the catalytic subunit of the (Na + K) ATPase deduced from a complementary DNA. Nature 316:691–695
Shull GE, Greeb J, Lingrel JB (1988) Molecular cloning of three distinct forms of the Na,K-ATPase α-subunit from rat brain. Biochemistry 25:8125–8132
Solioz M, Mathews S, Furst P (1987) Cloning of the K-ATPase of *Strepcoccus faecalis*. Structural and evolutionary implications of its homology to the KdpB-protein of *Escherichia coli*. J Biol Chem 262:7358–7362
Spanswick RM (1981) Electrogenic ion pumps. Annu Rev Plant Physiol 32:267–289
Stroobant P, Scarborough GA (1979) Active transport of calcium in *Neurospora* plasma membrane vesicles. Proc Natl Acad Sci USA 76:3102–3107
Surowy TK, Sussman MR (1986) Immunological cross-reactivity and inhibitor sensitivities of the plasma membrane H^+-ATPase from plants and fungi. Biochim Biophys Acta 848:24–34
Sussman MR, Surowy TK (1987) Physiology and molecular biology of membrane ATPases. Oxf Surv Plant Mol Cell Biol 4:47–69
Sze H (1985) H^+-translocating ATPase: advances using membrane vesicles. Annu Rev Plant Physiol 36:175–208
Tanford C (1982) Simple model for the chemical potential change of a transported ion in active transport. Proc Natl Acad Sci USA 79:2882–2884
Teruel JA, Gomez-Fernandez JC (1986) Distances between the functional sites of sarcoplasmic reticulum ($Ca^{2+} + Mg^{2+}$)-ATPase and the lipid/water interface. Biochim Biophys Acta 863:178–184
Tognoli L, Colombo R (1986) Protein phosphorylation in intact cultured sycamore (*Acer pseudoplatanus*) cells and its response to fusicoccin. Biochem J 235:45–58
Vara F, Serrano R (1981) Purification and characterization of a membrane-bound ATP-diphosphohydrolase from *Cicer arietinum* (chick-pea) roots. Biochem J 197:637–643
Vara F, Serrano R (1982) Partial purification and properties of the proton-translocating ATPase of plant plasma membranes. J Biol Chem 257:12826–12830
Vara F, Serrano R (1983) Phosphorylated intermediate of the ATPase of plant plasma membranes. J Biol Chem 258:5334–5336
Vesper MJ, Evans ML (1979) Nonhormonal induction of H^+ efflux from plant tissues and its correlation with growth. Proc Natl Acad Sci USA 76:6366–6370
Villalobo A (1982) Potassium transport coupled to ATP hydrolysis in reconstituted proteoliposomes of yeast plasma membrane ATPase. J Biol Chem 257:1824–1828
Villalobo A (1984) Energy-dependent H^+ and K^+ translocation by reconstituted yeast plasma membrane ATPase. Can J Biochem Cell Biol 62:865–877

Walderhaug MO, Post RL, Saccomani G, Leonard RT, Briskin DP (1985) Structural relatedness of three ion-transport adenosine triphosphatases around their sites of phosphorylation. J Biol Chem 260:3852–3859
Warncke J, Slayman CL (1980) Metabolic modulation of stoichiometry in a proton pump. Biochim Biophys Acta 591:224–229
Willsky GR, White DA, McCabe BC (1984) Metabolism of added orthovanadate to vanadyl and high-molecular-weight vanadates by *Saccharomyces cerevisiae.* J Biol Chem 259:13273–13281
Zocchi G (1985) Phosphorylation-dephosphorylation of membrane proteins controls the microsomal H^+-ATPase activity of corn roots. Plant Sci 40:153–159
Zocchi G (1988) Separation of membrane vesicles from maize roots having different calcium transport activities. Plant Sci 54:103–108
Zocchi G, Hanson JB (1983) Calcium transport and ATPase activity in a microsomal vesicle fraction from corn roots. Plant Cell Environ 6:203–209
Zocchi G, Rogers SA, Hanson JB (1983) Inhibition of proton pumping in corn roots is associated with increased phosphorylation of membrane proteins. Plant Sci Lett 31:215–221

Chapter 7 Transport in Plasma Membrane Vesicles – Approaches and Perspectives

D.P. BRISKIN[1]

1 Introduction

Preparations of isolated vesicles represent an ideal system for the study of transport processes associated with native membranes. With this approach, it is possible to characterize single transport systems without the complications associated with examining transport in the intact cell. Since the vesicles are not associated with the metabolic machinery of the cell, solute transport can be studied in the absence of metabolic transformation or compartmentation and the energy supply to drive transport (e.g., ATP) can be added exogenously.

[1]Department of Agronomy, University of Illinois, 1102 S. Goodwin Avenue, Urbana, IL 61801, USA

Abbreviations: BTP, bis-tris propane; CCCP, carbonylcyanide m-chlorophenylhydrazone; DES, diethylstilbestrol; DCCD, N,N′-dicyclohexylcarbodiimide; EPR, electron paramagnetic resonance; PMSF, phenylmethylsulfonyl fluoride.

C. Larsson, I.M. Møller (Eds):
The Plant Plasma Membrane

Therefore, it becomes possible to exert a level of control over a transport experiment which not only includes the ability to determine the conditions on both the *cis* and *trans* side of the membrane but also the magnitude and nature of the driving force for transport. In addition, it becomes a much easier task to address questions regarding the mechanism of energy coupling to the transport of a solute since ionophores can be used without concerns about effects upon the energy source for transport. Finally, the use of membrane vesicles promotes the biochemical understanding of transport processes since transport systems are much more accessible and can be treated as actual membrane proteins. The utility of this system as a means to directly investigate transport systems is shown by its widespread application to the study of transport systems in animal, fungal, and bacterial cells (see Turner 1983; Stein 1986 for reviews). The initial demonstration of the usefulness of membrane vesicles as a means to study bacterial transport was carried out by Kaback in the 1960's (Kaback 1974a,b; 1983).

More recently, this approach has been applied to transport systems associated with the membranes of higher plant cells (see Sze 1985 for review). Since the initial development of methodology for the preparation of transport-competent vesicles from plant cells by Sze in 1980, a number of investigators have used this approach to examine transport processes associated with plasma membrane (e.g., De Michelis and Spanswick 1986; Giannini et al. 1987a,b; Giannini et al. 1988b), tonoplast (e.g., Bennett and Spanswick 1983; Poole et al. 1984; Briskin et al. 1985; Schumaker and Sze 1985; Blumwald and Poole 1985b; Blumwald and Poole 1986), and endoplasmic reticulum (e.g., Buckhout 1983, 1984; Bush and Sze 1986; Giannini et al. 1988a) vesicles. Of these particular plant cell membranes, major progress on the use of isolated vesicles to study plasma membrane-associated transport processes has only been achieved within the last few years. It is the overall intent of this chapter to discuss recent advances in the methodology to produce transport-competent plasma membrane vesicles and their use in the study of transport systems at the plant cell surface. In addition, the general experimental aspects of obtaining quantitative data for transport studies using isolated vesicles will be presented. However, prior to addressing these topics, it will be necessary to present a brief discussion of energy coupling for transport at the plasma membrane in relation to the use of isolated vesicles for transport studies.

2 Energy Coupling for Plasma Membrane Transport: Cells vs Vesicles

The ability of plant cells to actively acidify their immediate exterior and establish a pH difference between the cytoplasm and cell wall space has been known for quite some time (Poole 1982 and references therein). With the emergence of Mitchell's *Chemiosmotic Hypothesis* (Mitchell 1976, 1985 for review), it became apparent that this process represented part of the scheme for energy coupling to solute transport at the plant cell surface. This hypothesis suggests that a proton electrochemical gradient (often denoted as $\Delta \mu_{H^+}$) established across a biological membrane can serve as an interconvertable energy source for the energization of solute translocation. In addition to extruding H^+ to the cell exterior, plant cells also maintain a significant, negative-interior electrical potential difference (often denoted as $\Delta \Psi$) across the plasma membrane (Sze 1985 and references therein). These two components of the proton electrochemical gradient (ΔpH and $\Delta \Psi$) are produced as the result of "primary" transport processes where there is a direct linkage of a scalar chemical reaction to electrogenic H^+ translocation. From studies utilizing whole plant tissues (e.g., Mercier and Poole 1980), reconstituted plasma membrane ATPase (e.g., Vara and Serrano 1982; O'Neill and Spanswick 1984; Singh et al. 1987; Chap. 6) and more recently, isolated plasma membrane vesicles (e.g., De Michelis and Spanswick 1986; Giannini et al. 1987a), it appears that this primary H^+ transport reflects the activity of an H^+-translocating ATPase (Fig. 1). This enzyme couples the exergonic reaction of ATP hydrolysis to the vectorial movement of H^+ to the cell exterior. For a more extensive discussion of the properties of the plant plasma membrane H^+-translocating ATPase, readers should refer to Chapter 6. To accommodate the variety of other solutes which are transported at the cell surface, the plasma membrane is proposed to contain separate transport systems (i.e., symports, antiports, uniports) which can utilize the ΔpH or $\Delta \Psi$ components of the proton electrochemical gradient to drive solute movement. In addition, some transport systems such as the H^+/K^+ symport, proposed to exist at the plasma membrane of *Neurospora*, would have the capacity to utilize both of these components of the proton electrochemical gradient (Rodriguez-Navarro et al. 1986). This coupling of the electrochemical gradient of one "solute" (H^+) to the movement of another solute is referred to as secondary transport and would utilize separate transport proteins associated with the plasma membrane (Gunn 1980; Sze 1985).

What is apparent from this discussion is that energy coupling for transport at the plant plasma membrane represents a vectorial process where the specific orientation of transport systems with respect to the driving forces for transport is an important consideration. However, the use of isolated membrane vesicles necessitates the homogenization of plant cells which causes membrane vesicles to form which are either right side-out or inside-out (Fig. 1). Frequently, this results in a membrane preparation which contains a mixture of vesicles of both orientations. In studies where transport will be energized via the plasma membrane H^+-ATPase, this does not pose a serious problem because the ATP substrate will generally not permeate the membrane (Van Thieman and Postma 1973). Thus, only inside-out plasma membrane vesicles, which contain the ATP hydrolytic site on the outside, will be active in the production of a proton electrochemical gradient. As long as the vesicle population is not so skewed that there is an extremely low number of inside-out plasma membrane vesicles, this approach can be used in the study of both primary and secondary transport systems. What becomes more difficult, however, is the use of artificially imposed pH or electrical gradients in the study of secondary transport systems using a mixed population of vesicle orientations. While the orientation of the vesicles

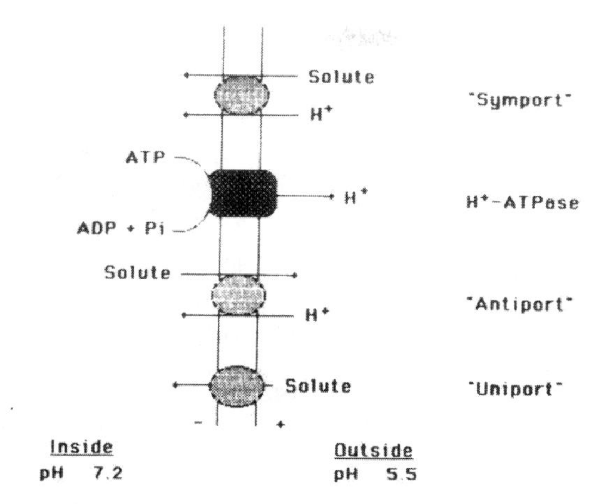

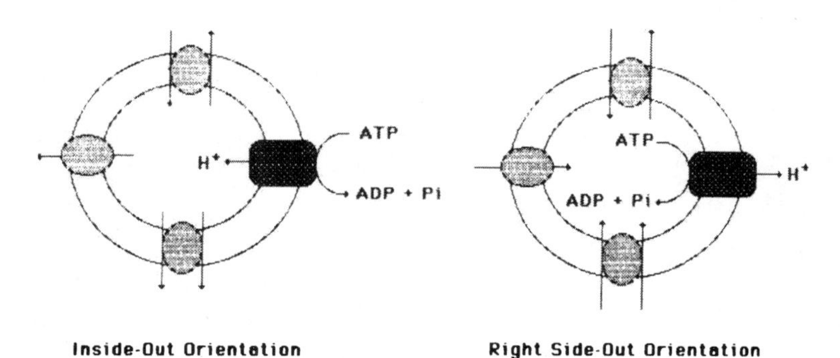

Fig. 1. Energy coupling for solute transport at the plasma membrane of plant cells: arrangement of the proton pump and carriers in the native membrane and in isolated plasma membrane vesicles

has been ignored in some studies of this type (Schumaker and Sze 1985; Blumwald and Poole 1986), the more desirable situation would involve the use of membrane vesicles enriched for one orientation or another. With bacterial cells (Kaback 1974a,b) and erythrocytes (Steck 1974), this has been achieved by modification of the conditions of homogenization to produce predominantly right side-out or inside-out vesicles. For higher plant plasma membrane vesicles, the separation of inside-out and right side-out vesicles was achieved only recently (Canut et al. 1987; Larsson et al. 1988; Chap. 3). This lack of preparations of vesicles with well-defined polarity has required alternative approaches which will be discussed in a subsequent section.

3 Quantitative Measurement of Proton Electrochemical Gradients and Solute Fluxes in Isolated Vesicles

The use of isolated membrane vesicles in transport experiments requires the application of experimental approaches much different than those utilized for transport experiments with either intact plant tissues, isolated plant cells, or intact protoplasts. This difference in approach is largely due to the small size of the vesicles themselves. Whereas an intact plant protoplast might have a diameter of about 30 μm (Mettler and Leonard 1979), isolated membrane vesicles generally have diameters in the range of 0.2 μm. Because of this small size it becomes much more convenient to monitor pH gradients and electrical potential differences across the membrane indirectly using molecular probes. While radioisotopes are generally used for measuring solute fluxes in vesicles, their quantitation can become more difficult than analogous use in whole cells because of the small internal volume of the vesicles and the tendency for some solutes to rapidly leak out during vesicle collection on filters. The following approaches are generally used in the quantitation of pH gradients, electrical gradients, and solute fluxes during vesicle transport experiments.

3.1 Measurement of ΔpH

In most studies with vesicles from higher plant cells, weak permeant bases have been used as probes for the measurement of the pH gradient established across the vesicle membrane. As most studies have involved transport experiments where the interior of the vesicle is acidified either using the H^+-ATPase or "pH jumps", the use of basic probes (as opposed to weak acid probes) is favored because the probe molecules accumulate within the vesicles in response to ΔpH (see below). These probes are either radioactively labeled, fluorescent, colored, or bear a group which produces an electron paramagnetic resonance (EPR) signal. While the means by which these probes "report" the pH gradient may differ (i.e., radiolabel, fluorescence quench, absorbance change, EPR signal change), the basic principle behind the probe response to ΔpH is somewhat the same for each probe type. The basic principles for use of these probes in the measurement of ΔpH in membrane vesicles are outlined in Fig. 2.

All of the probes generally used for the measurement of ΔpH are similar in being somewhat hydrophobic and containing an amine group as a weak base (Deamer et al. 1972; Lee and Forte 1978; Bennett and Spanswick 1983; Melhorn and Packer 1984). In the unprotonated form, the probe is assumed to have the capacity for free movement across the membrane while protonation of the basic amine group prevents free transmembrane movement. Under these conditions, the probe will distribute across the membrane in accordance to the pH difference between the interior and exterior of the vesicle as described by the following equation (Deamer et al. 1972; Lee and Forte 1978):

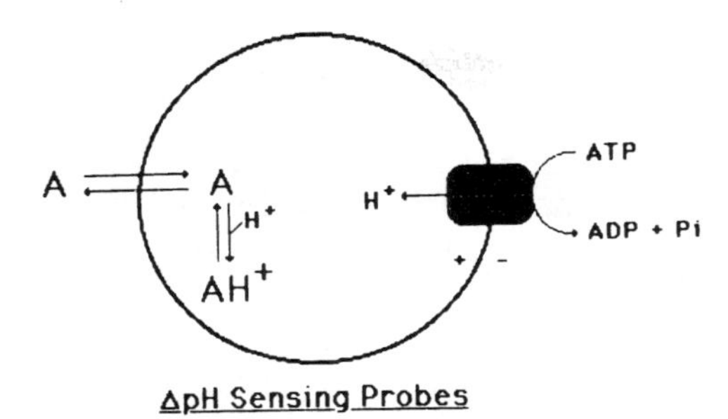

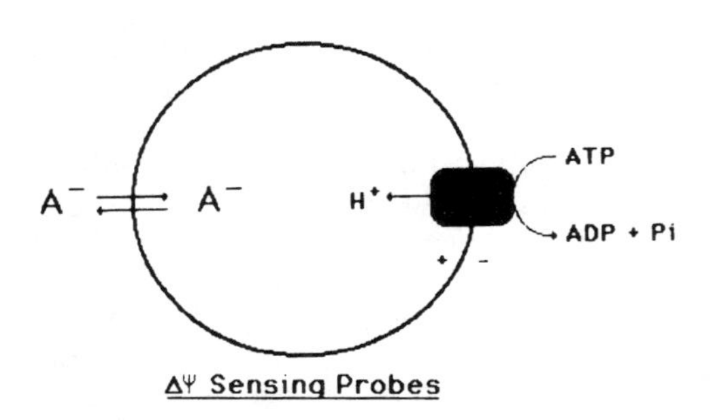

Fig. 2. Mechanism for the use of ΔpH and ΔΨ sensing probes in isolated membrane vesicles

$$pH_o - pH_i = \log (A_i/A_o) + \log (V_o/V_i), \quad (1)$$

where A refers to the total amount of probe inside (A_i) or outside the vesicle (A_o), and V refers to the volume of the vesicle interior (V_i) or external solution (V_o). When the interior pH of the vesicle is decreased relative to the exterior solution, as during ATP-driven H^+ transport in inside-out plasma membrane vesicles (see above), the probe will be accumulated in the interior of the vesicles as indicated by Eq. (1). The removal (or decrease) of the pH gradient would cause a corresponding efflux of the probe from the vesicles into the external solution. This represents the basic principle behind the probe response to ΔpH and what differs for each of the probe types is the means by which the partitioning of probe between the vesicle interior and exterior solution is measured.

In the use of ^{14}C-methylamine, a radioactive probe for ΔpH measurement, the accumulation of radiolabel within the vesicles is determined by collecting the vesicles on Millipore-type filters (Sze 1985). Radioactivity associated with the filters can then be determined by liquid scintillation spectroscopy. Thus, this method is discontinuous in that a number of separate filtrations need to be carried out and plotted to establish a profile for H^+ transport.

When fluorescent probes such as 9-aminoacridine or acridine orange are used in the measurement of Δ pH, the partitioning of the probe between the vesicles and exterior solution can be monitored by changes in probe fluorescence. While the precise explanation for the fluorescence change is not entirely understood, the most commonly accepted mechanism involves "stacking" of the planar probe molecules during accumulation in the vesicles which results in a self-quenching of probe fluorescence. As shown by Deamer et al. (1972) the relationship between the quenching of probe fluorescence during Δ pH-driven accumulation and the magnitude of the pH gradient can be described by the following equation:

$$\Delta pH = \log[\%Q/(100 - \%Q)] + \log(V_o/V_i), \quad (2)$$

where %Q refers to the percent reduction of probe fluorescence during accumulation within the vesicles. Therefore, this approach provides a continuous measurement of the pH difference where a decrease in probe fluorescence indicates internal acidification of the vesicles. Due to the common availability of instruments for measuring fluorescence, this method has been widely used for measuring Δ pH in membrane vesicles isolated from plant cells. It should be noted, however, that 9-aminoacridine can also exhibit fluorescence changes in response to the membrane surface potential (Møller et al. 1984). Therefore, when used for the determination of Δ pH, measurements with this probe should be conducted in the presence of relatively high concentrations of cations. In addition, recent work by Pope and Leigh (1988) has indicated that the use of acridine orange as a fluorescent Δ pH probe with certain anions (e.g., NO_3^-) can result in an artifactual increase in membrane permeability. These observations demonstrate that some caution must be exercised in the use of these probes as a means to monitor vesicle Δ pH.

Another optical method for measuring vesicle Δ pH in plant vesicles has involved the measurement of the absorbance change of probes such as acridine orange (Rasi-Caldogno et al. 1985) or neutral red (Hager et al. 1980). Since protonation of these probes would render them positively charged, it is probable that the partitioning of probe to the vesicle interior would operate as discussed above. However, protonation of the probe molecule also results in a change in the absorption spectrum which can be used as a means to monitor the vesicle pH gradient. While determinations of Δ pH using absorbance measurements for acridine orange have been carried out at either one wavelength (e.g., Giannini et al. 1988c) or as the difference of absorbance at two wavelengths (e.g., Rasi-Caldogno et al. 1985), Hager et al. (1980) pointed out that for neutral red a comparison of the entire probe spectra may be necessary for an accurate evaluation of Δ pH.

In the use of EPR for measuring Δ pH in isolated vesicles, the EPR signal associated with the nitroxide portion of the probe molecule is used as a means to determine the partitioning of the probe between the vesicle interior and the external medium (Melhorn et al. 1982; Melhorn and Packer 1984). The probe used for this application is an amine derivative of the nitroxide spin probe "Tempo" (2,2,6,6-tetramethylpiperidine-N-oxyl) referred to as "Tempamine" (4-amino-2,2,6,6-tetramethylpiperidine-N-oxyl). This method is based upon using a nonpermeant paramagnetic quenching agent which serves to broaden the signal of the nitroxide spin label in the external medium. Compounds used for this purpose include ferricyanide and Mn/EDTA (Melhorn et al. 1982; Poole et al. 1985). With the addition of a nonpermeant, signal-broadening agent, the only signal which can be measured is that which is retained inside the vesicles. Thus, like the other methods presented above, this can provide information on the amount of weak base probe which is accumulated in response to vesicle acidification.

A major point of debate in the use of weak base probes for ΔpH measurement is the appropriate means to express the data when presented in tabular form. The typical time course for vesicle acidification in the presence of ATP involves an initial rapid (often linear) phase of ΔpH production which slows to reach a steady-state pH gradient. This steady-state gradient represents the balance between ATP-driven H^+ influx and the leakage of H^+ out of the vesicle. In addition, thermodynamic "limitations" on the H^+ pump might also contribute to this slowing of ΔpH production. In the reduction of data from numerous plots for presentation in tabular form, some laboratories (e.g., Bennett and Spanswick 1983; De Michelis and Spanswick 1986) have expressed H^+ transport in terms of the initial rate of vesicle acidification while other laboratories (e.g., Giannini et al. 1987a,b) have also expressed H^+ transport in terms of the steady-state pH gradient determined after the addition of a proton-conducting ionophore. While the former tends to reflect rates of ATP-driven H^+ pumping and the latter tends to reflect not only H^+ pumping but also the degree of leakiness of the vesicles, the two parameters generally show strong relative agreement when being used to compare specific properties of H^+ transport (i.e., inhibitor sensitivity, pH optimum, etc.). When possible, the best approach would be to present both parameters (see subsequent sections).

3.2 Measurement of $\Delta\Psi$

As with the methods to measure ΔpH described in the previous section, the approach to measuring the electrical potential difference ($\Delta\Psi$) relies upon one general principle for all the methods used, but several different ways to report the response of the probe to the vesicle $\Delta\Psi$. As discussed above, most studies have involved situations where a positive-interior membrane potential has been produced either by the action of H^+-ATPases or the imposition of K^+-diffusion potentials using K^+ plus valinomycin. The probes generally used to measure $\Delta\Psi$ under these conditions are lipophilic anions negatively charged but hydrophobic enough to migrate across the membrane (Fig. 2). These probes will equilibrate across the membrane according to the Nernst equation:

$$\Delta\Psi = (RT/zF)\ln(A_o/A_i), \quad (3)$$

where R is the gas constant, T is the absolute temperature, z is the valency of the anion, and F is the Faraday constant. As in Eq. (1), A refers to the concentration of the probe outside (A_o) and inside (A_i) the vesicle. With a positive interior membrane potential, the lipophilic anion probes will accumulate in the vesicles and radioisotope, optical (absorbance and fluorescence) and EPR approaches have been used to quantify the degree of accumulation of the $\Delta\Psi$-sensing probes. As these approaches were discussed above and similar principles are involved, a more abbreviated presentation will follow.

A common radioactive probe used to measure interior-positive $\Delta\Psi$ is radiolabeled thiocyanate ($S^{14}CN^-$). While a variety of methods can be used to remove external label so that internal label can be determined (e.g., centrifugation, columns, flow dialysis), filtration methods have been preferred because of the shorter sampling time (Sze 1985 and references therein). Optical probes such as oxonol V and oxonol VI have found wide use in the measurement of vesicle membrane potential (e.g., Bennett and Spanswick 1983; Blumwald and Poole 1985a; Briskin et al. 1985) and have the advantage that events can be measured rapidly and continuously. In response to a positive interior $\Delta\Psi$, these probes partition across the membrane and exhibit a $\Delta\Psi$-dependent quenching of probe fluorescence (Oxonol V) or change in absorbance (Oxonol VI) due to binding on membrane sites (Waggoner 1979). Lipophilic anion probes containing the nitroxide spin label as derivatives of tetraphenyl phosphonium have also been used in the measurement of $\Delta\Psi$, although their slower equilibration across the membrane may place some constraints upon use for rapid responses (Melhorn and Packer 1984).

3.3 Measurement of Solute Fluxes

A number of methods have been developed for the measurement of radiolabeled solute flux in isolated membrane vesicles. General methods for determining solute movement in vesicles have involved flow dialysis (e.g., Ramos and Kaback 1977), removal of charged solutes from external medium by ion-exchange columns (Furst and Solioz 1986), and collection of vesicles on Millipore-type filters (e.g., Giannini et al. 1988b). This last technique has been the one most commonly used in the measurement of radiolabeled solute flux with plant membrane vesicles. The basic procedure for this method is as follows. Membrane vesicles are suspended in a medium containing radiolabeled solute and all other ligands as needed for a transport experiment. After an appropriate period of time, a stop solution is added and the vesicles are collected on a Millipore-type filter. The filter is subsequently washed and then radioactivity associated with the filter is determined by liquid scintillation spectroscopy.

To conduct efflux studies, this procedure can be used with vesicles where the radiolabeled solute has been preloaded inside the vesicle. Our recent studies have suggested that a "slow freeze/thwa" method can be effectively used to load solutes inside plasma membrane vesicles (Giannini and Briskin 1988). In this method, membrane vesicles are combined with the solute to be preloaded and slowly frozen in a -20°C freezer. After thawing at room temperature and removal of nonincorporated label by either centrifugation (Giannini and Briskin 1988) or column chromatography (Giannini et al. 1987b), the vesicles with internal label can be used in studies of efflux. As radiolabeled solute may begin to leak from the vesicles during the removal of external label, some caution must be exercised in the application of these approaches. For situations where background labeling of filters can be reduced to an acceptable level by the washing solution, it may be more desirable to simply dilute the external radiolabel during the efflux assay (e.g., Ruiz-Cristin and Briskin 1988).

As a general rule, the rapid filtration technique must be optimized for each solute which is investigated in transport experiments. This optimization includes determining the appropriate solution composition for uptake or efflux experiments as well as the appropriate stop solution used to terminate transport reactions. In addition, the composition and volume of the filter washing solution as well as the number of washes need to be optimized for each solute used. For example, it may be desirable to include unlabeled solute to exchange with labeled solute which has adsorbed on the filter or surface of the vesicles. However, this must be approached with care because it is also possible that such an addition could cause transmembrane exchange of internally accumulated label. For a complete discussion of the optimization of vesicle filtration assays, readers are encouraged to consult the review by Turner (1983).

4 Preparation of Transport-Competent Plasma Membrane Vesicles from Plant Tissues

The ability to produce membrane preparations containing substantial amounts of plasma membrane vesicles which are competent in transport is a relatively recent advance. For many plant tissues, it has been possible to readily recover transport-competent vesicles derived from the tonoplast (e.g., Dupont et al. 1982; Bennett and Spanswick 1983; Churchill and Sze 1983; Bennett et al. 1984; Poole et al. 1984) or endoplasmic reticulum (e.g., Buckhout 1983, 1984; Bush and Sze 1986) while it has proven much more difficult to routinely recover transport-competent plasma membrane vesicles in substantial quantity. When microsomal membrane fractions have been prepared from many plant tissues and centrifuged on linear sucrose gradients, what has often been observed is that although a substantial amount of plasma membrane is present (as indicated by marker enzyme activities), the major peak of transport-competent vesicles (as assayed by ATP-dependent H^+-transport) are derived from the tonoplast (Dupont et al. 1982; Churchill et al. 1983; Lew and Spanswick 1984). Since any measurement of transport capacity requires the vesicle to be "sealed" to the extent that solute gradients can be produced, this problem may be related to the greater tendency of previous plant tissue fractionation methods to yield plasma membrane vesicles which are more "leaky" to solutes (Sze 1985 and references therein). This hypothesis was supported in initial studies to measure transport activity with plasma membrane-enriched fractions produced by such methods

(Sze and Hodges 1976; DP Briskin, unpublished results). Perlin and Spanswick (1982) found that maize root plasma membrane vesicles, produced by differential and sucrose density gradient centrifugation, were leaky to H^+ as indicated by their inability to maintain an imposed pH difference (acid-interior). In addition, Lew et al. (1985) demonstrated a pronounced difference in H^+ permeability for plasma membrane and tonoplast vesicles produced from zucchini fruit. Based upon monitoring the collapse of preformed pH gradients in these vesicles, these workers found half-times which differed as much as eight fold. To some extent, these permeability differences might also reflect a natural difference in passive H^+-conductance between these two membrane types.

An alternative explanation for the difficulties in obtaining transport-competent plasma membrane vesicles may relate to the sidedness produced following cell rupture. If the detection of transport-competent vesicles is based upon assaying ATP-dependent H^+-transport, then plasma membrane vesicle preparations, which are predominantly right side-out, will demonstrate little transport activity. As discussed below, this has been the situation when plasma membrane-enriched fractions have been produced by the phase partitioning method.

More recently, what has become apparent is that a number of factors must be taken into consideration in attempts to isolate substantial amounts of transport-competent plasma membrane vesicles. These will be considered in the next section of this chapter with the emphasis on work carried out in our laboratory.

4.1 Choice of Plant Tissue

The choice of plant tissue to be used in the production of membrane vesicles can have a major bearing on the recovery of transport-competent plasma membrane vesicles. This was dramatically shown in studies by Rasi-Caldogno et al. (1985). While microsomal membrane fractions produced from most plant tissues have tended to yield predominantly transport-competent tonoplast vesicles (although other membranes are present), these workers found that a microsomal membrane fraction produced from dark-grown seedlings of radish contained transport-competent vesicles derived predominantly from the plasma membrane! This was shown by the correspondence of the properties of H^+-transport activity with those of the plasma membrane ATPase and the insensitivity of H^+-transport to inhibitors of mitochondrial or tonoplast H^+-ATPase. While this appears to be the only reported situation where a specific plant or tissue type allowed a selective production of transport-competent plasma membrane vesicles, it is apparent from the literature that the relative amounts of transport-competent plasma membrane versus tonoplast (or endoplasmic reticulum) vesicles may be somewhat tissue-dependent (Hager and Helme 1981; Dupont et al. 1982; Sze 1982; Churchill and Sze 1983; Lew and Spanswick 1984). As suggested by Sze (1983) differences in reported recoveries of transport-competent plasma membrane versus tonoplast vesicles could also result from the developmental stage of the tissue used in vesicle isolations or the inherent tendency of the membranes to reseal in the appropriate orientation (inside-out plasma membrane or right side-out tonoplast) upon tissue homogenization so that transport activity can be measured.

4.2 Homogenization Conditions

The conditions of homogenization, including the composition of the medium used in tissue disruption as well as forms of tissue pretreatment prior to homogenization or washing of membrane fractions, can have a major effect upon the recovery of transport-competent plasma membrane vesicles. In some cases, modification of the chemical composition of the medium used to homogenize the tissue can increase the overall proportion of transport-competent plasma membrane vesicles in relation to tonoplast vesicles. An example of this effect was shown in studies on maize root membrane vesicles carried out by De Michelis and Spanswick (1986). These workers varied the composition of the homogenization medium and then determined the recovery of transport-competent plasma membrane or tonoplast vesicles in a microsomal membrane fraction. They found that the addition of 10% glycerol enhanced the recovery of transport-competent plasma membrane vesicles while the addition of 10% methanol or 5% polyvinylpyrrolidone enhanced the recovery of transport-competent tonoplast vesicles. Other than the usual components present in homogenization media (i.e., osmoticum, buffer, chelators, reductants), a survey of papers which reported substantial recoveries of transport-competent plasma membrane vesicles (Rasi-Caldogno et al. 1985; De Michelis and Spanswick 1986; Giannini et al. 1987a; Dupont et al. 1988) also indicated the common addition of phenymethylsulfonyl fluoride (PMSF) as a protease inhibitor to the homogenization medium. This would suggest that proteolytic degradation of plasma membrane vesicles may have been a potential problem during earlier unsuccessful attempts at obtaining transport-competent vesicles.

Work from our laboratory has shown a rather interesting and pronounced effect of adding moderate concentrations of monovalent salts to the homo genization medium used to grind plant tissue (Giannini et al. 1987a). As a pretreatment prior to homogenization, this medium was vacuum infiltrated into the tissue. The effect of adding monovalent salts on the recovery of transport-competent plasma membrane and tonoplast vesicles in microsomal membrane fractions from red beet storage tissue is shown in Fig. 3. The recovery of each vesicle type was monitored by assaying the sensitivity of ATP-driven H^+-transport to characteristic inhibitors of either the plasma membrane or tonoplast ATPase. The assay for ATP-driven H^+-transport was based upon the quenching of quinacrine fluorescence where nitrate inhibition would indicate H^+-transporting tonoplast vesicles and vanadate inhibition would indicate H^+-transporting plasma membrane vesicles. The addition of gramicidin D, a

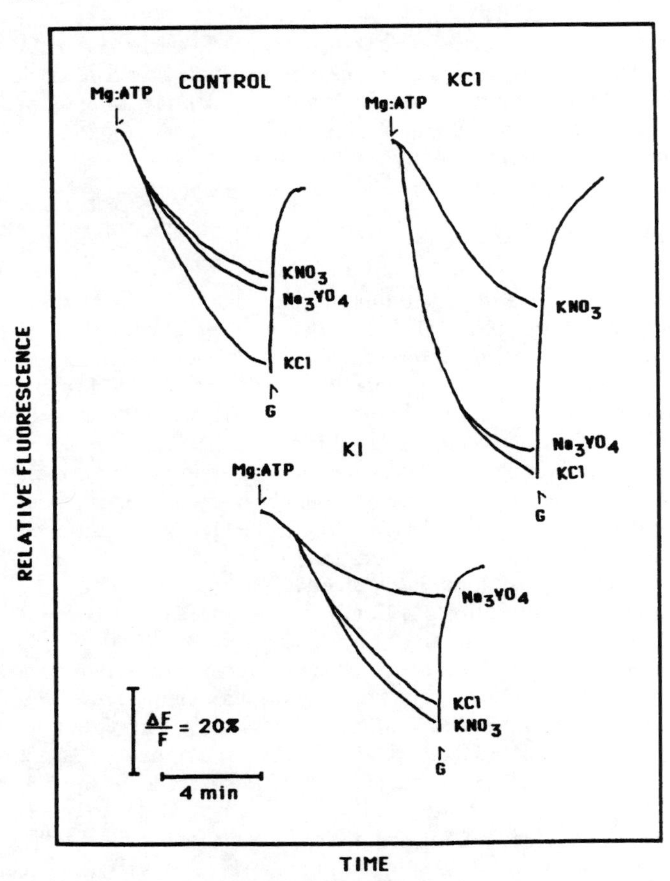

Fig. 3. Quenching of quinacrine fluorescence by microsomal membrane fractions isolated in homogenization media containing either no added monovalent salts, 0.25 M KCl, or 0.25 mM KI. Proton transport was measured in the presence of 250 mM sorbitol, 3.75 mM ATP (BTP salt, pH 6.5), 3.75 mM $MgSO_4$, 5 μM quinacrine, 25 mM BTP/MES (pH 6.5), and 100 mM KCl. When nitrate was present, the KCl in the transport assay was replaced by an equal concentration of KNO_3. When Na_3VO_4 was present, the final concentration was 100 μM and 100 mM KCl was present. Fluorescence quenching was reversed by the addition of 5 μM gramicidin D (*G*) (Data from Giannini et al. 1987a; reprinted with permission from Archives of Biochemistry and Biophysics, Academic Press)

channel-forming ionophore, caused a recovery in fluorescence for each case which indicated that the reduction in fluorescence corresponded to the production of an acid-interior pH gradient.

When membrane vesicles were produced in a medium containing the usual components and protectants, but in the absence of added monovalent salts,

ATP-driven H^+-transport associated with the microsomal pellet demonstrated sensitivity to both nitrate and vanadate which would indicate a recovery of transport-competent plasma membrane and tonoplast vesicles (Fig. 3). Since previous studies on red beet storage tissue (Bennett et al. 1984; Poole et al. 1984) had indicated only a low recovery of transport-competent plasma membrane vesicles, it is apparent that the modifications to the homogenization medium made in these more recent studies (Giannini et al. 1987a) improved the recovery of transport-competent plasma membrane vesicles. However, when 0.25 M KI was included in the homogenization medium in addition to the other components, only transport-competent plasma membrane vesicles were recovered as shown by sensitivity to vanadate but not nitrate. In fact, nitrate consistently stimulated the rate and extent of ATP-driven H^+-transport. In subsequent studies this was shown to occur because this anion could effectively reduce the vesicle membrane potential (Giannini and Briskin 1987). Finally, the replacement of KI in the homogenization medium with an identical concentration of KCl resulted in a shift in the relative proportion of transport-competent vesicle types. In this case, transport-competent vesicles were mainly derived from the tonoplast as shown by the enhancement of nitrate inhibition of H^+-transport when compared to the plot for vesicles isolated in the absence of monovalent salts.

This change in the relative proportion of transport-competent plasma membrane versus tonoplast vesicles recovered by salt treatment during tissue homogenization was further shown when these vesicle types were resolved by centrifugation on linear sucrose gradients. As shown in Fig. 4, it was apparent that the major effect of the KI treatment was to reduce the recovery of H^+-transporting tonoplast vesicles associated with the microsomal membrane preparation. Since both tonoplast (NO_3^--inhibited) and plasma membrane (VO_4^{3-}-inhibited) ATPase activities were still present and migrated to their usual low and high density regions of the gradient, it would appear that the KI treatment might serve to somehow "uncouple" tonoplast ATPase activity from H^+-transport. It should be noted that this figure contains the combined data from separate linear gradients for KCl (solid symbols) and KI (open symbols) treated membranes, so that differences in the location of the peak activities do not reflect changes in isopycnic density due to salt treatment. Rather, these differences are related to differences in the sucrose gradient profile between these two gradients.

While the mechanism by which moderate concentrations of KI promote the selective production of transport-competent plasma membrane vesicles during homogenization is uncertain, the practical use of this approach is apparent. With the use of a KI-containing homogenization buffer and conventional cell fractionation methods (differential and density gradient centrifugation), a preparation of transport-competent plasma membrane vesicles can be produced which is relatively free of transport-competent tonoplast vesicles. It should also be pointed out that our subsequent work has shown this method to be widely applicable to a number of plant tissues including

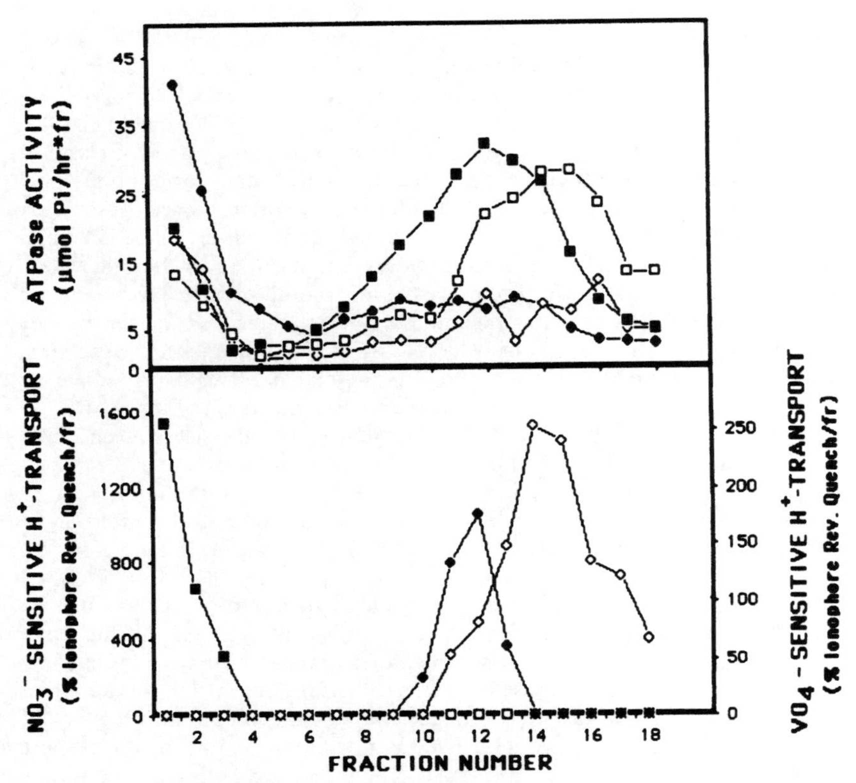

Fig. 4. Distribution of inhibitor-sensitive components of ATPase activity and proton transport when microsomal membrane fractions prepared in the presence of 0.25 M KI (*open symbols*: ◇,□) or 0.25 M KCl (*solid symbols*: ◆,■) were centrifuged in a linear sucrose density gradient. The sucrose gradients were centrifuged at 100 000 g for 3 h and then fractionated into 18 equal fractions. Vanadate-sensitive (□,■) and nitrate-sensitive (◇,◆) components of ATPase activity were assayed in the fractions as described in Giannini et al. (1987a; *top panel*). Proton transport sensitive to either 100 μM Na_3VO_4 (◇,◆) or 100 mM KNO_3 (□,■) was assayed as described in Fig. 3 and the level of ionophore-reversible fluorescence quenching was plotted (*bottom panel*). Inhibitor-sensitive components represent the difference in ionophore-reversible quenching between assays carried out in the presence of KCl and the assay with inhibitor (After Giannini et al. 1987a; reprinted with permission from Archives of Biochemistry and Biophysics, Academic Press)

maize roots, soybean roots, and cultured cells (Giannini et al. 1987a). Of the tissues tested, red beet storage tissue has the distinct advantage that large quantities of transport-competent plasma membrane vesicles can be isolated (in the presence of KI) from the bulky storage tissue. Most recently, we have also used this method to directly isolate transport-competent plasma membrane vesicles from the fungal pathogen, *Phytophthora megasperma* (Giannini et al. 1988c).

4.3 Resolution of Plasma Membrane Vesicles from Microsomal Membrane Fractions

It is generally desirable to further purify plasma membrane vesicles present in microsomal membrane fractions from other cellular membranes. At present, two major approaches have been used to achieve enrichment for plasma membrane vesicles. As shown in the previous section, one approach involves centrifugation on density gradients. While sucrose (Giannini et al. 1987a; Dupont et al. 1988) or dextran (Churchill and Sze 1983) has been used for this purpose, most results have indicated better resolution of membranes with the former density medium. Based upon results obtained from linear sucrose gradients, discontinuous gradients can be constructed to recover transport-competent plasma membrane vesicles on a gradient interface. For plasma membrane vesicles produced from red beet storage tissue, optimal recovery occurs at a 26%/38% (w/w) sucrose density interface (Giannini et al. 1987a).

The other approach which has been used to recover plasma membrane vesicles involves phase partitioning of microsomal membranes in mixtures of dextran-polyethylene glycol (Briskin et al. 1987; Larsson et al. 1987). In contrast to separations based upon bouyant density, this method resolves membranes according to surface properties and generally produces membrane fractions with a much higher degree of purity (ca. 95% plasma membrane) than those produced using density gradients (ca. 50% plasma membrane) (Larsson et al. 1987 and references therein). However, a major problem with this approach has been that these high-quality fractions mainly contain plasma membrane vesicles which are present in a right side-out orientation (Larsson et al. 1984; Briskin et al. 1987). Until most recently, this has prevented detailed studies of the plasma membrane H^+-ATPase in native membranes and other secondary transport processes driven by the $\Delta\mu_{H^+}$ produced by this enzyme. Modifications to the phase partitioning procedure developed by Larsson et al. (1988) now allow for the production of fractions enriched in vesicles with either sidedness. With this method one fraction can be obtained containing ca. 90% inside-out vesicles and another containing ca. 90% right side-out vesicles. These estimates of the proportions of inside-out and right side-out vesicles were based on ATPase latency (addition of Triton X-100) and H^+-transport activity. This modification will greatly increase the utility of the phase partitioning method for the production of plasma membrane vesicles which can be used in studies of solute transport.

5 Transport Systems Studied Using Plasma Membrane Vesicles

Since the development of methodology to obtain reasonable quantities of transport-competent plasma membrane vesicles represents a relatively recent event, reports on the use of this system for the study of plasma membrane-associated transport systems have been fairly limited. Most studies to date have focused on characterizing the properties of the plasma membrane H^+-translocating ATPase (Rasi-Caldogno et al. 1985; De Michelis and Spanswick 1986; Giannini et al. 1987a; Dupont et al. 1988). In part, this work has served to demonstrate that the investigators have, in fact, isolated sealed, transport-competent plasma membrane vesicles. In this section, the use of plasma membrane vesicles in transport studies will be discussed. Again, emphasis will be placed on results obtained from studies carried out in our laboratory.

5.1 Plasma Membrane H^+-Translocating ATPase

Proton transport mediated by the plasma membrane H^+-ATPase has been examined in vesicles isolated from maize (De Michelis and Spanswick 1986), radish (Rasi-Caldogno et al. 1985), barley (Dupont et al. 1988), and red beet (Giannini and Briskin 1987; Giannini et al. 1987a). For the present discussion, data will be presented from our studies with plasma membrane vesicles isolated from red beet in the presence of KI and further enriched by separation on sucrose density gradients. However, it should be pointed out that this transport enzyme (ATPase activity) demonstrates a number of characteristic properties consistently observed in plasma membrane fractions from a variety of species (Leonard 1984 and references therein) and these properties are the ones generally examined in any study attempting to demonstrate the presence of transport-competent plasma membrane vesicles. Thus, while data will only be presented from studies on plasma membrane vesicles from red beets, the type of results obtained are fairly similar to those observed for the other plant species mentioned above.

A characteristic property of the H^+-ATPase, associated with the plasma membrane, is a pH optimum for activity at about 6.5 (Leonard 1984 and references therein; Chaps. 2 and 6). Figure 5 clearly shows a similar pH optimum for H^+ transport determined by monitoring quinacrine fluorescence quenching. This optimum was identical when expressed either in terms of the initial rate of fluorescence quenching or the total level of ionophore-reversible quenching. Proton transport demonstrated a high specificity for ATP as shown by the lack of transport activity with other phosphorylated compounds as substrates, similar to that observed for ATP hydrolytic activity mediated by the plasma membrane ATPase (Leonard 1984; Giannini et al. 1987a). In contrast to that observed for tonoplast vesicles (Rea and Poole 1985; Blumwald 1987), no H^+ transport could be supported by the addition of pyrophosphate, indicating the lack of a proton-pumping pyrophosphatase in these preparations. Proton transport in the plasma membrane vesicles was insensitive to KNO_3 and NaN_3 which are inhibitors of tonoplast and mitochondrial ATPase activities, respectively (Poole et al. 1984 and references therein). However, H^+ transport was inhibited by N,N′-dicyclohexylcarbodiimide (DCCD), diethylstilbestrol (DES), and orthovanadate (Na_3VO_4) (Table 1). These compounds are inhibitors of ATP hydrolytic activity associated with the plasma membrane ATPase

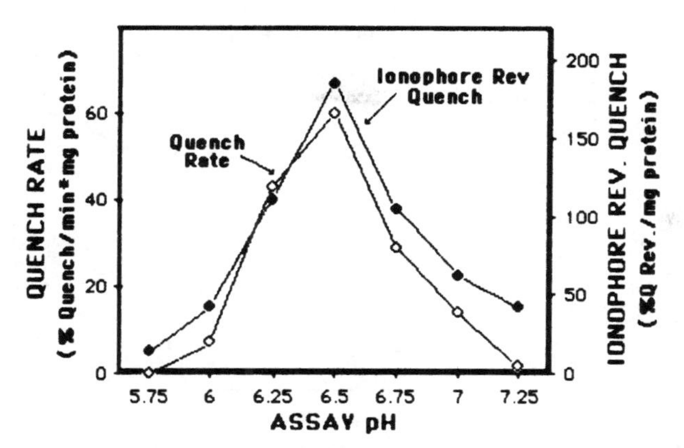

Fig. 5. Effect of assay pH on proton transport in plasma membrane vesicles from red beet storage tissue. Proton transport was measured as a function of assay pH where both the initial rate of fluorescence quenching and the ionophore-reversible quench are plotted (Data from Giannini et al. 1987a; reprinted with permission from Archives of Biochemistry and Biophysics, Academic Press)

Table 1. Effect of various inhibitors on proton transport activity associated with plasma membrane vesicles from red beet storage tissue[a]

Treatment	Quench rate[b] (%Q min^{-1} mg^{-1})	Ionophore-rev. quench (ionophore-rev. Q mg^{-1})
Control	61 (100)	181 (100)
+50 mM KNO_3	65 (106)	202 (111)
+100 μM Na_3VO_4	16 (26)	36 (20)
+1 mM NaN_3	65 (106)	232 (128)
+50 μM DCCD	16 (26)	15 (8)
+30 μM DES	18 (29)	20 (11)
+10 mM BTP/F	36 (59)	80 (44)

[a] Data adapted from Giannini et al. (1987a). Reprinted with permission from Archives of Biochemistry and Biophysics (Academic Press).
[b] Values in parentheses represent the percentage of the control assay carried out in the presence of 50 mM KCl.

(Leonard 1984 and references therein; Chap. 6). A final similarity to that observed for plasma membrane ATPase activity regards the kinetics of H^+ transport at varying Mg:ATP concentration. As with ATP hydrolytic activity (Leonard 1984), the kinetic profile demonstrates simple saturation kinetics with a K_m around 0.6 mM (Giannini et al. 1987a).

One property which differs substantially from that observed for ATP hydrolytic activity regards the effect of anions. The plasma membrane ATPase is often referred to as a monovalent cation-stimulated ATPase (Briskin and Poole 1983; Leonard 1984; Chap. 6). However, when H^+-transport activity was

Table 2. Anion effects upon ΔpH and ΔΨ in plasma membrane vesicles from red beet storage tissue[a]

Treatment	ΔpH		ΔΨ
	Initial rate (%Q min^{-1} mg^{-1})	Ionophore reversible (Q mg^{-1})	Ionophore reversible (Q mg^{-1})
Control[b]	12	25	1120
KI	136	525	250
KNO_3	102	475	312
KBr	94	438	462
$KClO_3$	75	425	436
KCl	56	212	888
K_2SO_4	36	120	920

[a] Data from Giannini and Briskin (1987). Reprinted with permission from Plant Physiology.
[b] Control assay was carried out in the presence of 250 mM sorbitol, 3.75 mM ATP (BTP salt, pH 6.5), 3.75 mM $MgSO_4$, 25 mM BTP/MES (pH 6.5), and 100 μg of membrane protein. Acid-interior pH gradients were measured in the presence of 2.5 μM quinacrine while interior-positive membrane potentials were measured in the presence of 15 μM Oxonol V, over a 4-min period. When potassium salts of anions were tested, the anion concentration was 50 mM.

measured for this enzyme in sealed, inside-out plasma membrane vesicles, anions also had substantial effects (Table 2). When the corresponding effects of various anions (as K^+ salts) on both ΔpH and ΔΨ were examined, it was found that the effect of anions was primarily related to their relative ability to reduce the electrical membrane potential and increase the pH gradient. Therefore, while cations such as K^+ appear to stimulate ATPase activity through a direct effect on the enzyme (Leonard 1984 and references therein), the effects of anions may be more indirect via a modification of the conditions for H^+ transport (Chap. 6).

At this point, our studies (and others) on the plasma membrane H^+-ATPase in native vesicles are preliminary and only involve correlations of properties previously observed to be characteristic of the ATP hydrolytic activity of this enzyme. Clearly, this system will be quite useful for future work on aspects of the enzyme (e.g., lipid regulation of activity, hormone regulation, etc.) which require measuring both ATPase activity and H^+-transport properties in the native membrane environment (see Chap. 6).

5.2 Plasma Membrane Ca^{2+}-Translocating ATPase

Currently, there is great interest in understanding the role of Ca^{2+} in the regulation of plant cell metabolism. As in animal cells, it appears that plant cells maintain low cytoplasmic levels of free Ca^{2+} through various transport processes and there is evidence for a role of this cation in signal transduction (Hepler and Wayne 1986; Chap. 9). In response to the appropriate signal, it is proposed that cytoplasmic Ca^{2+} levels would transiently increase and this would modulate the activities of other enzymes such as protein kinases (Poovaiah and Reddy 1987). Therefore, an important aspect of this process is represented by the transport activities involved in maintaining low cytoplasmic levels of Ca^{2+} since these are responsible for keeping the system "poised" for response. While a secondary H^+/Ca^{2+} antiport system has been demonstrated for tonoplast vesicles (Schumaker and Sze 1985; Blumwald and Poole 1986) and a primary Ca^{2+}-translocating ATPase demonstrated for endoplasmic reticulum vesicles (Buckhout 1983, 1984; Bush and Sze 1986; Giannini et al. 1988a), a system for Ca^{2+} transport at the plasma membrane has not been conclusively demonstrated until most recently (Giannini et al. 1988b). By analogy with that observed for animal cells (Itano and Penniston 1985 and references therein), it has been proposed that plant cells should contain a primary Ca^{2+}-translocating ATPase at the plasma membrane which would be responsible for driving Ca^{2+} efflux (Hepler and Wayne 1986). Until most recently, the only evidence for such a system was based upon studies with crude microsomal membrane fractions (Dieter and Marmé 1980, 1981) where the activity of the Ca^{2+}-translocating ATPase from the endoplasmic reticulum could not be ruled out.

In recent studies, we have characterized Ca^{2+} transport using a fraction enriched for transport-competent plasma membrane vesicles obtained from red beet storage tissue (Giannini et al. 1988b). As this membrane fraction did not contain substantial levels of marker enzyme activity for endoplasmic reticulum, Ca^{2+} transport could be analyzed in the absence of the analogous endoplasmic reticulum-associated system. Using the rapid filtration assay, it was shown that these vesicles carry out ATP-driven uptake of $^{45}Ca^{2+}$ (Fig. 6). Uptake in the absence of ATP was negligible and when the Ca^{2+} ionophore, A23187, was added to vesicles that had accumulated $^{45}Ca^{2+}$, the radiolabel rapidly leaked back into the external medium. Uptake of $^{45}Ca^{2+}$ was only slightly affected by the proton ionophore carbonylcyanide m-chlorophenylhydrazone (CCCP) but stimulated by gramicidin D. Since both of these compounds would be expected to completely eliminate Ca^{2+} movement driven by $\Delta\mu_{H^+}$ (i.e., H^+/Ca^{2+} antiport), this result would suggest that the coupling of Ca^{2+} transport to ATP utilization was direct via a Ca^{2+}-translocating ATPase. The slight inhibition by CCCP is thought to represent some direct effect of this compound upon the activity of the Ca^{2+}-translocating ATPase (see Giannini et al. 1988b for discussion). The proposal for a direct coupling of ATP utilization to Ca^{2+} transport was also consistent with the observations that the imposition of an acid-interior ΔpH ("pH jump") could not drive $^{45}Ca^{2+}$ uptake and that the addition of Ca^{2+} to vesicles with an acid-interior ΔpH did not enhance the release of H^+ (Giannini et al. 1988b). These approaches have been used to demonstrate the existence of a secondary H^+/Ca^{2+} antiport associated with the tonoplast (Schumaker and Sze 1985; Blumwald and Poole 1986). Since the addition of ATP would only "energize" inside-out plasma membrane vesicles, this uptake of $^{45}Ca^{2+}$ would represent the activity of a Ca^{2+}-transport system responsible for driving Ca^{2+} efflux from the cell as originally proposed.

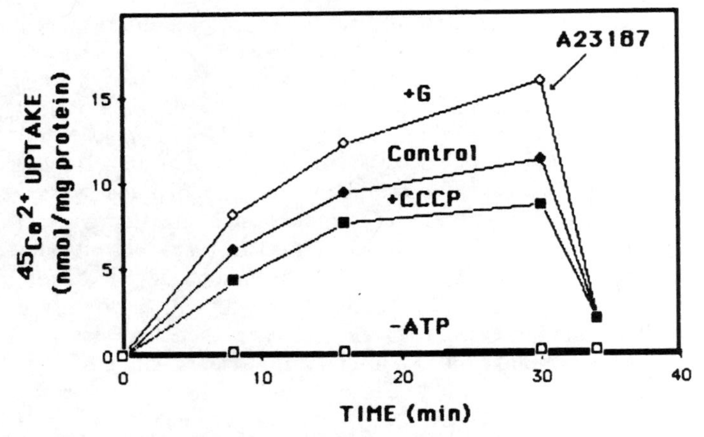

Fig. 6. Time course of $^{45}Ca^{2+}$ uptake by plasma membrane vesicles from red beet storage tissue. The uptake of $^{45}Ca^{2+}$ was carried out in the presence of 250 mM sorbitol, 3.75 mM ATP (BTP salt, pH 7.0), 3.75 mM $MgSO_4$, 100 mM KNO_3, 25 mM BTP/MES (pH 7.0), 0.4 mM NaN_3, and 10 μM $CaCl_2$ (containing 90 kBq $^{45}Ca^{2+}$). As indicated, 0.3 μg ml^{-1} of the Ca^{2+} ionophore A23187 was added to release calcium gradients produced by ATP-dependent uptake. Gramicidin D (*G*) and *CCCP*, when added, were present at 5 and 10 μM, respectively (Data from Giannini et al. 1988b; reprinted with permission from Plant Physiology)

Further characterization of the plasma membrane Ca^{2+}-translocating ATPase indicated that, like the plasma membrane H^+-ATPase, this activity was inhibited by vanadate, DES, and DCCD (Giannini et al. 1988b). The activity differed, however, in having some capacity to utilize GTP as a substrate for driving $^{45}Ca^{2+}$ transport. This ability to use GTP in the energization of $^{45}Ca^{2+}$ transport also differed from the endoplasmic reticulum Ca^{2+}-ATPase of red beet which is specific for ATP (Giannini et al. 1988a). The kinetics of $^{45}Ca^{2+}$ uptake as a function of Ca^{2+} concentration showed a simple saturable profile and consistent with its proposed role in maintaining low cytoplasmic Ca^{2+} levels, the K_m was in the low micromolar range (Giannini et al. 1988b).

Studies by Robinson et al. (1988) have also shown the presence of Ca^{2+}-ATPase activity in a highly purified plasma membrane fraction from maize leaves produced by phase partitioning. This ATPase activity was stimulated by low concentrations of Ca^{2+} (less than 100 μM) and calmodulin. Although no measurement of Ca^{2+} transport was conducted in this study, the similarity of this ATPase to the erythrocyte Ca^{2+}-ATPase could imply a transport function for this enzyme (see Robinson et al. 1988 for discussion). Further evidence for a plasma membrane-associated Ca^{2+}-ATPase with properties similar to the erythrocyte Ca^{2+} pump was found in studies by Briars et al. (1989). In their work, a Ca^{2+}-ATPase purified from maize membranes by detergent solubilization and chromatography on a calmodulin affinity column cross-reacted on a Western blot with antibodies prepared against erythrocyte Ca^{2+}-ATPase. The cross-reacting band on the blot had a molecular weight of about 140 kD which is similar to the molecular weight of the purified erythrocyte enzyme (Itano and Penniston 1985 and references therein).

5.3 Redox Systems Associated with Plasma Membrane Vesicles

While it is often assumed that the primary energy source for driving solute transport at the plasma membrane is ATP (Leonard 1984 and references therein), several laboratories (Federico and Giartosio 1983; Rubinstein et al. 1984; Marrè et al. 1988; Chap. 5) have presented evidence that a plasma membrane-associated redox system might have a role in establishing or regulating the driving force for transport. It has been proposed that oxidation of NAD(P)H on the plasma membrane and electron transfer to oxygen at the cell surface could result in proton extrusion to the cell exterior and/or the production of a membrane potential (Møller and Lin 1986 and references therein). Alternatively, these electron transport systems could modulate the $\Delta\mu_{H^+}$ through effects upon the action of the plasma membrane H^+-ATPase (Rubinstein and Stern 1986).

To further examine these possibilities, we have carried out studies using plasma membrane vesicles selectively isolated from red beet storage tissue using sucrose gradient centrifugation (Giannini and Briskin 1988). Consistent with the above-mentioned previous reports in the literature, the plasma membrane vesicles from red beet oxidized NAD(P)H. The activity was increased about ten fold when ferricyanide was used as an electron acceptor and appeared to prefer NADH over NADPH as the substrate for oxidation. To determine whether NADH oxidation could produce a transmembrane $\Delta\mu_{H^+}$, assays were carried out in which the production of an acid-interior ΔpH was monitored by quinacrine fluorescence and the production of an interior-positive $\Delta\Psi$ was monitored by Oxonol V fluorescence. In these experiments, NADH was either added externally to the vesicles alone or to vesicles which had been preloaded with ferricyanide by a freeze/thaw method (Giannini and Briskin 1988 for details). This experimental arrangement would set up the situation for transmembrane electron flux to either oxygen or ferricyanide, as electron acceptor, in inside-out plasma membrane vesicles.

When the vesicles were incubated with NADH, or NADH was added to vesicles preloaded with ferricyanide, the production of a ΔpH or $\Delta\Psi$ could not be detected (Giannini and Briskin 1988). The addition of NADH to vesicles either alone or preloaded with ferricyanide also had no effect upon the ability of the plasma membrane H^+-ATPase to mediate ATP-dependent ΔpH or $\Delta\Psi$ formation. Finally, when ferricyanide reduction was monitored spectrophotometrically in the preloaded vesicles, the addition of NADH to the vesicle exterior did not result in the reduction of ferricyanide present inside the vesicles and reduction of this electron acceptor could only be observed when the vesicles were made leaky by the addition of 0.05% Triton X-100.

Taken together, these results would suggest that although the red beet plasma membrane vesicles have the capacity for NAD(P)H oxidation, this process does not appear to involve transmembrane electron flux, the production of a $\Delta\mu_{H^+}$, or the modulation of the plasma membrane H^+-ATPase (see Giannini and Briskin 1988 for discussion). However, these results are in conflict with more recent electrophysiological studies suggesting transmembrane electron flux, where ferricyanide reduction by *Elodea* leaf cells was associated with an acidification of the external medium, a 0.2 to 0.3 unit decrease in cytoplasmic pH, depolarization of the membrane electrical potential, and a net efflux of K^+ (Marré et al. 1988). In addition, earlier studies by Sijmons et al. (1984) demonstrated a depolarization of the membrane potential with trans-plasma membrane electron transfer to ferricyanide in iron-deficient roots from bean (*Phaseolus vulgaris* L.). Clearly, these conflicts reflect the need for further research in this area. A more detailed discussion of plasma membrane-associated redox systems can be found in Chapter 5.

5.4 Secondary Transport Systems Driven by $\Delta\mu_{H^+}$

In addition to their use in the study of primary ATP-driven transport systems, the use of transport-competent plasma membrane vesicles in the study of $\Delta\mu_{H^+}$-driven secondary transport systems is emerging. Bush and Langston-Unkefer (1988) have used vesicles isolated from zucchini to examine an H^+/amino acid symport associated with the plasma membrane. Using washed microsomal membrane fractions, these workers could demonstrate uptake of radiolabeled amino acid into vesicles in response to an artificially imposed acid-exterior ΔpH. Amino acid uptake was inhibited by the proton ionophore, CCCP, and by pretreatment of the vesicles with DCCD. The effect of this latter inhibitor is significant in being consistent with that generally observed for H^+-driven transport systems (Bush and Langston-Unkefer 1988 and references therein). This system appeared to have the capacity to mediate H^+ cotransport with several amino acids including alanine, leucine, glutamate and glutamine. Evidence that this system was associated with the plasma membrane was shown by the comigration of ΔpH-driven alanine uptake with plasma membrane marker enzymes in fractions produced by centrifuging the microsomal membranes on a linear sucrose density gradient.

Using plasma membrane vesicles from red beet storage tissue, preliminary studies in our laboratory have focused on the transport of K^+ and NO_3^- with this system. In the case of K^+, there has been substantial debate as to whether this cation is transported into plant cells by a secondary $\Delta\mu_{H^+}$-driven process or by direct, primary coupling to ATP utilization with the plasma membrane ATPase acting in H^+/K^+ exchange (Briskin 1987 and references therein; Chap. 6). To examine this question, we characterized ATP-driven K^+ transport in plasma membrane vesicles using $^{86}Rb^+$ as a radiotracer analog (Giannini et al. 1987b). After preloading the radiolabel into the vesicles, efflux was examined using the filtration method. Radiolabel efflux was stimulated by ATP but only partially inhibited (about 50%) by the addition of CCCP. Since the addition of CCCP completely collapsed the vesicle Δ pH and $\Delta \Psi$, this result would argue that at least some portion of the ATP-driven $^{86}Rb^+$ transport might be directly coupled to ATP utilization. These results would also indicate that a portion of the transport is also coupled to $\Delta \mu_{H^+}$ by either a $\Delta \Psi$-driven K^+ uniport (Briskin 1987) or H^+/K^+ symport (Rodriguez-Navarro et al. 1986). We have also carried out initial studies to examine the mechanism of NO_3^- transport using $^{36}ClO_3^-$ as a radiotracer analog (Ruiz-Cristin and Briskin 1988). In this case, it was possible to demonstrate ATP-dependent uptake of $^{36}ClO_3^-$ occurring through a $\Delta \Psi$-driven NO_3^- ($^{36}ClO_3^-$) uniport which would correspond to a system for NO_3^- efflux from the plant cell.

With the current availability of transport-competent plasma membrane vesicles from several plant species, it is anticipated that this approach will be extensively used in future studies on secondary transport processes associated with the plasma membrane. This will provide much needed information on the biochemical mechanisms responsible for mediating mineral nutrient and metabolite transport at the plant cell surface.

6 Perspectives

In this chapter, the use of transport-competent membrane vesicles as an experimental tool in the study of plasma membrane transport processes was presented. Despite initial difficulties in obtaining transport-competent plasma membrane vesicles, it is apparent that with recent advances in cell fractionation methodology this approach will find more widespread application in the study of the biochemistry of plasma membrane transport processes. Together with modern approaches in molecular biology, this will lead to a detailed understanding of the structure, mechanism, and regulation of transport systems which govern the flow of solutes at the plant cell surface.

Acknowledgments. The research reported in this chapter from the author's laboratory was supported by USDA competitive grant 86-CRCR-1-1977, an individual research award from the McKnight Foundation, and funds from the University of Illinois Experiment Station.

References

Bennett AB, Spanswick RM (1983) Optical measurement of ΔpH and ΔΨ in corn root membrane vesicles: kinetic analysis of Cl^- effects on a proton translocating ATPase. J Membr Biol 71:95–107

Bennett AB, O'Neill SD, Spanswick RM (1984) H^+-ATPase from storage tissue of *Beta vulgaris* I. Identification and characterization of an anion sensitive H^+-ATPase. Plant Physiol 74:538–544

Blumwald E (1987) Tonoplast vesicles as a tool in the study of ion transport at the plant vacuole. Physiol Plant 69:731–734

Blumwald E, Poole RJ (1985a) Nitrate storage and retrieval in *Beta vulgaris* L.: effects of nitrate and chloride on proton gradients in tonoplast vesicles. Proc Natl Acad Sci USA 83:3683–3687
Blumwald E, Poole RJ (1985b) Na^+/H^+ antiport in isolated tonoplast vesicles from storage tissues of *Beta vulgaris*. Plant Physiol 78:163–167
Blumwald E, Poole RJ (1986) Kinetics of Ca^{2+}/H^+ antiport in isolated tonoplast vesicles from storage tissue of *Beta vulgaris* L. Plant Physiol 80:727–731
Briars SA, Kessler F, Evans DE (1989) The calmodulin-stimulated ATPase of maize coleoptiles is a 140,000 Mr polypeptide. Planta 176:283–285
Briskin DP (1987) Plasma membrane H^+-transporting ATPase: role in potassium ion transport? Physiol Plant 68:159–163
Briskin DP, Poole RJ (1983) Characterization of a K^+-stimulated adenosine triphosphatase associated with the plasma membrane of red beet. Plant Physiol 71:350–355
Briskin DP, Thornley WR, Wyse RE (1985) Membrane transport in isolated vesicles from sugarbeet taproot. II. Evidence for a sucrose/H^+-antiport. Plant Physiol 78:871–875
Briskin DP, Leonard RT, Hodges TK (1987) Isolation of plasma membrane: membrane markers and general principles. Methods Enzymol 148:542–558
Buckhout TJ (1983) ATP-dependent Ca^{2+}-transport in endoplasmic reticulum isolated from roots of *Lepidium sativum* L. Planta 159:84–90
Buckhout TJ (1984) Characterization of Ca^{2+}-transport in purified endoplasmic reticulum membranes from *Lepidium sativum* L. roots. Plant Physiol 76:962–967
Bush DR, Langston-Unkefer PJ (1988) Amino acid transport into membrane vesicles isolated from zucchini: evidence for a H^+/amino acid symport in the plasmalemma. Plant Physiol 88:487–490
Bush DR, Sze H (1986) Calcium transport in tonoplast and endoplasmic reticulum vesicles isolated from cultured carrot cells. Plant Physiol 80:549–555
Canut H, Brightman AO, Boudet AM, Morré DJ (1987) Determination of sidedness of plasma membrane and tonoplast vesicles isolated from plant stems. In: Leaver C, Sze H (eds) Plant membranes: structure, function and biogenesis. Alan R Liss, pp 141–159
Churchill KA, Sze H (1983) Anion sensitive, H^+-pumping ATPase in membrane vesicles from oat roots. Plant Physiol 71:610–617
Churchill KA, Holloway B, Sze H (1983) Separation of two types of electrogenic H^+-pumping ATPases from oat roots. Plant Physiol 73:921–928
Deamer DW, Prince RC, Crofts AR (1972) The response of fluorescent amines to pH gradients across liposome membranes. Biochim Biophys Acta 274:323–335
De Michelis MI, Spanswick RM (1986) H^+-pumping driven by the vanadate-ATPase in membrane vesicles from corn roots. Plant Physiol 81:542–547
Dieter P, Marmé D (1980) Calmodulin activation of plant microsomal Ca^{2+} uptake. Proc Natl Acad Sci USA 77:7311–7314
Dieter P, Marmé D (1981) A calmodulin-dependent, microsomal ATPase from corn (*Zea mays* L.). FEBS Lett 125:245–248
Dupont FM, Bennett AB, Spanswick RM (1982) Localization of a proton-translocating ATPase on sucrose gradients. Plant Physiol 70:1115–1119
Dupont FM, Tanaka CK, Hurkman WJ (1988) Separation and immunological characterization of membrane fractions from barley roots. Plant Physiol 86:717–724
Federico R, Giartosio CE (1983) A trans-plasma membrane electron transport system in maize roots. Plant Physiol 73:182–184
Furst P, Solioz M (1986) The vanadate-sensitive ATPase of *Streptococcus faecalis* pumps potassium in a reconstituted system. J Biol Chem 261:4302–4308
Giannini JL, Briskin DP (1987) Proton transport in plasma membrane and tonoplast vesicles from red beet (*Beta vulgaris* L.) storage tissue. A comparative study of ion effects on ΔpH and $\Delta\Psi$. Plant Physiol 84:613–618
Giannini JL, Briskin DP (1988) Pyridine nucleotide oxidation by a plasma membrane fraction from red beet (*Beta vulgaris* L.) storage tissue. Arch Biochem Biophys 260:653–660
Giannini JL, Gildensoph LH, Briskin DP (1987a) Selective production of sealed plasma membrane vesicles from red beet (*Beta vulgaris* L.) storage tissue. Arch Biochem Biophys 254:621–630

Giannini JL, Gildensoph LH, Ruiz-Cristin J, Briskin DP (1987b) Isolation and characterization of sealed plasma membrane vesicles from red beet (*Beta vulgaris* L.) storage tissue. Plant Physiol Suppl 83:55

Giannini JL, Gildensoph LH, Reynolds-Niesman I, Briskin DP (1988a) Calcium transport in sealed vesicles from red beet (*Beta vulgaris* L.) storage tissue. I. Characterization of a Ca^{2+}-pumping ATPase associated with the endoplasmic reticulum. Plant Physiol 85:1129–1136

Giannini JL, Ruiz-Cristin J, Briskin DP (1988b) Calcium transport in sealed vesicles from red beet (*Beta vulgaris* L.) storage tissue. II. Characterization of $^{45}Ca^{2+}$ uptake into plasma membrane vesicles. Plant Physiol 85:1137–1142

Giannini JL, Holt JS, Briskin DP (1988c) Isolation of sealed plasma membrane vesicles from *Phytophthora megasperma* f. sp. glycinea. I. Characterization of proton pumping and ATPase activity. Arch Biochem Biophys 265:337–345

Gunn RB (1980) Co- and counter transport mechanisms in cell membranes. Annu Rev Physiol 42:249–259

Hager A, Helme M (1981) Properties of an ATP fueled, Cl^--dependent proton pump localized in membranes of microsomal vesicles from maize coleoptiles. Z Naturforsch 36c:927–937

Hager A, Frenzel R, Laible D (1980) ATP-dependent proton transport into vesicles of microsomal membranes of *Zea mays* coleoptiles. Z Naturforsch 35c:783–793

Hepler PK, Wayne RO (1986) Calcium and plant development. Annu Rev Plant Physiol 36:397–439

Itano T, Penniston JT (1985) Ca^{2+}-pumping ATPase of plasma membranes. In: Cheung A (ed) Calmodulin antagonists and cellular physiology. Academic Press, New York, pp 335–345

Kaback HR (1974a) Transport in isolated bacterial vesicles. Methods Enzymol 31:698–709

Kaback HR (1974b) Transport studies in bacterial membrane vesicles. Science 186:882–892

Kaback HR (1983) The lac carrier protein in *Escherichia coli*. J Membr Biol 76:95–112

Larsson C, Kjellbom P, Widell S, Lundborg T (1984) Sidedness of plant plasma membrane vesicles purified by partitioning in aqueous two-phase systems. FEBS Lett 171:271–276

Larsson C, Widell S, Kjellbom P (1987) Preparation of high-purity plasma membranes. Methods Enzymol 148:558–568

Larsson C, Widell S, Sommarin M (1988) Inside-out plant plasma membrane vesicles of high purity obtained by aqueous two-phase partitioning. FEBS Lett 229:289–292

Lee HC, Forte JG (1978) A study of H^+ transport in gastric microsomal vesicles using fluorescent probes. Biochim Biophys Acta 508:339–359

Leonard RT (1984) Membrane associated ATPases and nutrient absorption by roots. In: Tinker PB, Läuchli A (eds) Advances in plant nutrition, Vol I. Praeger Scientific, New York, pp 209–240

Lew RR, Spanswick RM (1984) Proton-pumping activities of soybean (*Glycine max* L.) root microsomes: localization and sensitivity to nitrate and vanadate. Plant Sci Lett 36:187–193

Lew RR, Bushunow N, Spanswick RM (1985) ATP-dependent proton-pumping activities of zucchini fruit microsomes. A study of tonoplast and plasma membrane activities. Biochim Biophys Acta 821:341–347

Marrè MT, Moroni A, Albergoni FG, Marrè E (1988) Plasmalemma redox activity and H^+ extrusion. I. Activation of the H^+-pump by ferricyanide-induced potential depolarization and cytoplasm acidification. Plant Physiol 87:25–29

Melhorn RJ, Packer L (1984) Bioenergetic studies of cells with spin probes. Ann NY Acad Sci 414:180–189

Melhorn RJ, Candu P, Packer L (1982) Measurements of volumes and electrochemical gradients with spin probes in membrane vesicles. Method Enzymol 88:752–761

Mercier J, Poole RJ (1980) Electrogenic pump activity in red beet: its relation to ATP levels and cation influx. J Membr Biol 55:165–174

Mettler IJ, Leonard RT (1979) Ion transport in isolated protoplasts from tobacco suspension cells. I. General characteristics. Plant Physiol 63:183–190

Mitchell PM (1976) Vectorial chemistry and the molecular mechanism of chemiosmotic coupling: power transmission by proticity. Biochem Soc Trans 4:399–430

Mitchell PM (1985) The correlation of chemical and osmotic forces in biochemistry. J Biochem 97:1–18

Møller I, Lin W (1986) Membrane-bound NAD(P)H dehydrogenases in higher plant cells. Annu Rev Plant Physiol 37:309–334
Møller IM, Lundborg T, Bérczi A (1984) The negative surface charge density of plasmalemma vesicles from wheat and oat roots. FEBS Lett 167:181–185
O'Neill SD, Spanswick RM (1984) Solubilization and reconstitution of a vanadate-sensitive H^+-ATPase from the plasma membrane of *Beta vulgaris*. J Membr Biol 79:245–256
Perlin DS, Spanswick RM (1982) Isolation and assay of corn root membrane vesicles with reduced proton permeability. Biochim Biophys Acta 690:178–186
Poole RJ (1982) Electrogenic transport at the plasma membrane of plant cells. In: Martinosi AN (ed) Membranes and transport, vol 2. Plenum Publishing, New York, pp 651–655
Poole RJ, Briskin DP, Kratky Z, Johnstone RM (1984) Density gradient localization of plasma membrane and tonoplast from storage tissue of growing and dormant red beet: characterization of proton transport and ATPase in tonoplast vesicles. Plant Physiol 74:549–556
Poole RJ, Melhorn RJ, Packer L (1985) A study of transport in tonoplast vesicles using spin-labelled probes. In: Marin B (ed) Biochemistry and function of vacuolar adenosine triphosphatase in fungi and plants. Springer, Berlin, Heidelberg New York Tokyo, pp 114–118
Poovaiah BW, Reddy ASN (1987) Calcium messenger system in plants. CRC Crit Rev Plant Sci 6:47–103
Pope AJ, Leigh RA (1988) Dissipation of pH gradients in tonoplast vesicles and liposomes by mixtures of acridine orange and anions. Implications of the use of acridine orange as a ΔpH probe. Plant Physiol 86:1315–1322
Ramos S, Kaback HR (1977) The electrochemical proton gradient in *Escherichia coli* membrane vesicles. Biochemistry 16:848–854
Rasi-Caldogno F, Pugliarello MC, De Michelis MI (1985) Electrogenic transport of protons driven by the plasma membrane ATPase in membrane vesicles from radish. Biochemical characterization. Plant Physiol 77:200–205
Rea PA, Poole RJ (1985) Proton-translocating inorganic pyrophosphatase in red beet (*Beta vulgaris* L.) tonoplast vesicles. Plant Physiol 77:46–52
Robinson C, Larsson C, Buckhout TJ (1988) Identification of a calmodulin-stimulated (Ca^{2+} + Mg^{2+})-ATPase in a plasma membrane fraction isolated from maize (*Zea mays*) leaves. Physiol Plant 72:177–184
Rodriguez-Navarro A, Blatt MR, Slayman CL (1986) A potassium-proton symport in *Neurospora crassa*. J Gen Physiol 87:649–674
Rubinstein B, Stern AI (1986) Relationship of transplasmalemma redox activity to proton and solute transport by roots of *Zea mays*. Plant Physiol 80:805–811
Rubinstein B, Stern AI, Stout RG (1984) Redox activity at the surface of oat root cells. Plant Physiol 76:386–391
Ruiz-Cristin J, Briskin DP (1988) Nitrate transport in plasma membrane vesicles from red beet (*Beta vulgaris* L.) storage tissue. Plant Physiol Suppl 86:79
Schumaker KS, Sze H (1985) A Ca^{2+}/H^+ antiport system driven by the proton electrochemical gradient of a tonoplast H^+-ATPase from oat roots. Plant Physiol 79:1111–1117
Sijmons PC, Lanfermeijer FC, De Boer AH, Prins HBA, Bienfait HF (1984) Depolarization of cell membrane potential during trans-plasma membrane electron transfer to extracellular electron acceptors in iron-deficient roots of *Phaseolus vulgaris* L. Plant Physiol 76:943–946
Singh SP, Kesav BVS, Briskin DP (1987) Reconstitution and rapid partial purification of the red beet plasma membrane ATPase. Physiol Plant 69:617–626
Steck TL (1974) Preparation of impermeable inside-out and right-side-out vesicles from erythrocyte membranes. Methods Membr Biol 2:245–281
Stein WD (1986) Transport and diffusion across cell membranes. Academic Press, Orlando, FL
Sze H (1980) Nigericin-stimulated ATPase activity in microsomal vesicles of tobacco callus. Proc Natl Acad Sci USA 77:5904–5908
Sze H (1982) Characterization of nigericin-stimulated ATPase from sealed vesicles of tobacco callus. Plant Physiol 70:495–505

Sze H (1983) H^+-pumping ATPase in membrane vesicles of tobacco callus: sensitivity to vanadate and K^+. Biochim Biophys Acta 732:586–594
Sze H (1985) H^+-translocating ATPases: advances using membrane vesicles. Annu Rev Plant Physiol 36:175–208
Sze H, Hodges TK (1976) Characterization of passive ion transport in plasma membrane of oat roots. Plant Physiol 58:304–308
Turner RJ (1983) Quantitative studies of co-transport systems: models and vesicles. J Membr Biol 76:1–15
Van Thieman B, Postma PW (1973) Coupling between energy conservation and active transport of serine in *E. coli*. Biochim Biophys Acta 323:429–440
Vara F, Serrano R (1982) Partial purification and properties of the proton-translocating ATPase of plant plasma membranes. J Biol Chem 252:5334–5336
Waggoner AS (1979) Dye indicators of membrane potential. Annu Rev Biophys Bioeng 8:47–68

Chapter 8 Electrophysiology of the Plasma Membrane of Higher Plant Cells: New Insights from Patch-Clamp Studies

R. HEDRICH[1], H. STOECKEL[2], and K. TAKEDA[3]

1 Introduction

There can be little argument as to the fundamental importance of ion transport mechanisms for the physiology of plant cells. While excitable electrical behavior was first observed in plant cells about a century ago (e.g., Sanderson 1888), the underlying mechanisms responsible for this behavior are only now being directly studied at the molecular level. Ion channels are integral transmembrane

[1]Universität Göttingen, Pflanzenphysiologisches Institut, D-3400 Göttingen, FRG
[2]Université Louis Pasteur de Strasbourg, Laboratoire de Biologie Cellulaire Végétale-CNRS UA1182, F-67083 Strasbourg, France
[3]Université Louis Pasteur de Strasbourg, Laboratoire de Pharmacologie Cellulaire et Moléculaire-CNRS UA600, BP. 24, F-67401 Illkirch, France

Abbreviations: FV, fast vacuole; SV, slow vacuole; TEA, triethanolamine.

C. Larsson, I.M. Møller (Eds):
The Plant Plasma Membrane

proteins, which when open allow the movement of ions and some nonelectrolytes down their electrochemical gradients (for review, Hille 1984; Catterall 1988). Although ionic currents in plant cell membranes were among the first to be studied in detail (e.g., Michaelis 1925; Cole and Curtis 1938), by comparison with their animal cell counterparts the electrophysiological characterization of plant ion channels has been somewhat slower. This has been due to problems specific to plant cells, such as the presence of the cell wall, having the plasma membrane and vacuolar membrane in series, and the relatively small cytoplasmic compartment. The latter is especially a problem in higher plants.

These difficulties are rapidly being overcome with the application of the patch-clamp technique, pioneered by Neher and Sakmann (1976) at Göttingen, to a variety of plant cell membranes (Hedrich and Schroeder 1989). The patch-clamp technique is a revolutionary electrophysiological method allowing high resolution recording of ionic currents from biological membranes, both at the single channel level and from whole cells or organelles (Hamill et al. 1981). Furthermore, ionic currents resulting from electrogenic ion pumping activity have proved amenable to direct study at the single cell level using the patch-clamp technique. Our intention here is to review new insights into ion channel function and ion transport in plasma membranes from isolated protoplasts of higher plant cells as revealed following the application of patch-clamp methods starting on 1984 (Moran et al. 1984; Schroeder et al. 1984).

2 Electrophysiological Measurements on Biological Membranes: Advantages and Limitations

2.1 Extracellular Recordings

Extracellularly recorded potentials (or currents) are a reflection of transmembrane ion transport being resolved as current flowing across an extracellular series resistor. Although these types of measurements are relatively easy to make, they suffer from lack of resolution, both at the absolute signal amplitude level and spatially. A large body of earlier work on whole plant tissues using extracellular recording electrodes concerned the stimulation and propagation of action potentials (e.g., Umrath 1937).

A more recent example of this approach has been the studies from L.F. Jaffe's group on sea urchin eggs and plant pollen using the vibrating probe. The vibrating probe technique is based on the rapid, extracellular measurement of potential at two different locations close to one another. Any spatial difference in potential is attributed to the presence of net ionic current, being the sum of currents arising from membrane conductances (e.g., for H^+, K^+, Ca^{2+}, Cl^-). These currents result in a characteristic topology of the electric field along single cells or whole tissues. The results obtained using this approach have led to the suggestion that ionic currents are required as triggers for the establishment of cellular polarity or developmental responses in some cells (for review, Jaffe and

Nuccitelli 1977). It has been shown that changes in the intensity of external electric fields precede changes in cell polarity, e.g., causing unequal cell division, and that certain types of tissue growth are associated with transcellular current flow. Furthermore, cellular morphology can be altered by externally applied electromagnetic fields.

Nevertheless, the vibrating probe technique is limited because the contribution of individual ion species (e.g., K^+ fluxes) to the overall measured current is not easily distinguishable, and because direct access to the source of ionic currents (the membrane) is constrained by the physical size of the probe. Due to the highly charged nature of plant cell walls, ion concentrations are much higher close to the wall, and ionic currents here may differ considerably from those recorded in the bulk solution.

2.2 Intracellular Membrane Potentials and Transmembrane Currents

If the ionic composition on the two sides of a biological (or artificial) membrane differs, and if the membrane is semipermeable, then an electrical potential difference will exist across the membrane, the size of which is a function of the membrane conductance for the permeant ion species. This is the transmembrane potential, the difference between the intra- and extracellular potentials. It can be measured directly using fine intracellular glass microelectrodes or more indirectly with potential-sensitive dyes. Our understanding of much of the biophysics of ion channel function is based on studies where intracellular *potentials* were measured. Subsequently, with the application of voltage-clamp techniques, the measurement of transmembrane ionic *current* became possible and much of what we know today concerning the electrical behavior of biological membranes is based on observations made using these methods (for review, Hille 1984). The best example of the use of these techniques is the characterization of the ionic basis (depolarization-activated Na^+ and K^+ conductances) for the action potential in the squid giant axon by Cole, Hodgkin, Huxley, and Katz roughly 40 years ago (for review, Hodgkin 1964).

In animal cells, following the macroscopic description of cellular ionic currents and their activating mechanisms (e.g., changes in membrane voltage, the presence of an agonist agent such as a neurotransmitter, or an intracellular second messenger like Ca^{2+}), much effort was expended in trying to determine the elementary characteristics of individual ion channels (Hille 1984). At that time (the 1970's), noise analysis was the method of choice and was used to study the fluctuations in current arising from varying numbers of open channels in the population present in a single cell. This allowed indirect estimation of the single channel conductance and the average mean open time for both agonist-activated and voltage-dependent ion channels (for review, Neher and Stevens 1977).

Excitable membranes and action potentials are also known in plant cells. For example, action potentials were investigated in giant algae many years ago

(Cole and Curtis 1938) and were also found in higher plants, such as the mechanically sensitive plant, *Mimosa*, and the Venus fly trap (for review, Sibaoka 1966). In contrast to animal cells where Na^+, K^+, and Ca^{2+} are the main charge carriers involved in the generation of action potentials, in plant cells K^+, Cl^-, and Ca^{2+} are the most important permeant ion species.

Ion channels are now accepted as the main transporters for ion movement down electrochemical gradients (Hille 1984). Questions of current interest concern notably the regulation and modulation of ion channel activity by circulating factors, internal Ca^{2+}, or other second messengers generated following, for example, the induction of phosphoinositol metabolism or kinase-mediated protein phosphorylation (Shiina et al. 1988). Physiologists gained access to the cytoplasm of squid giant axons by internal perfusion techniques (for review, Hodgkin 1964), thus defining exactly the composition of the solutions on both sides of the membrane. This condition proved critical in the later investigation of nonchannel ion transport mechanisms, for example, the Na^+/K^+-ATPase or the Na^+/Ca^{2+} exchanger. A similar strategy of internal perfusion was adopted for giant algal cells. Using "artificial cytoplasmic" solutions, the basic components of ion transport through membranes of giant algae and their regulation was similarly investigated (Tazawa 1964; Tazawa et al. 1987). However, the use of both internal perfusion and the axial wire voltage-clamp techniques was for practical reasons limited to these "giant" cell types.

Conventional voltage-clamp techniques, especially the axial wire type, but also the two-microelectrode voltage-clamp technique (Takeuchi and Takeuchi 1959) are limited to reasonably large cells ($\geq 30\ \mu m$). Otherwise, the inevitable leakage conductances associated with intracellular microelectrode impalements are so large with respect to the inherent membrane conductances as to make virtually impossible the reliable interpretation of voltage-clamp data from small cells.

For plant cells, several additional constraints arise which are all unfavorable for intracellular microelectrode work of adequate technical quality, especially where multiple penetrations of a single cell are required for voltage-clamp measurements.

Firstly, to study intact cells, blunt low-resistance microelectrodes must be used in order to penetrate the cell wall, thus leading unavoidably to high leakage conductances. These relatively large tip diameter electrodes are sometimes subject to pressure artifacts when used to study cells maintaining significant turgor pressure. Further, they may give rise to uncontrolled shifts in intracellular ion concentrations, and therefore poorly defined equilibrium potentials, due to excessive leakage into the cell of the electrolyte (often 3 M KCl) used to fill the microelectrodes. If, instead, low concentrations of electrolyte are used to fill the microelectrodes, an unacceptably high electrode resistance might be the probable result.

Secondly, plant cells are normally 5–60 μm in diameter, are often connected by cell-cell junctions (plasmodesmata) in situ, and usually contain a large central vacuole. Thus, voltage-clamp studies of plant tissue in situ, where electrical coupling between cells is present, may be impossible due to lack of spatial control of voltage. Adequate temporal control of voltage seems less an issue compared to the case in animal cells, as current time courses in plant cells are typically about 3 or 4 orders of magnitude slower.

Thirdly, another very serious problem is the precise location of the intracellular microelectrode, in other words whether the electrode tip after impalement is in the cytoplasm or inside the vacuole. This problem is particularly acute in higher plant cells, where the vacuole makes up most of the cellular volume and the cytoplasm is just a thin layer between the vacuolar membrane and the plasma membrane. The measured intracellular potentials are then not always easy to interpret with confidence.

Finally, because the plasma membrane and the vacuolar membrane are in series, bipolar gating phenomena may arise due to ionic currents across both membranes (as has been found for gap junctions; Spray et al. 1981). This difficulty is compounded in voltage-clamp studies because upon depolarization, the plasma and vacuolar membranes rectify in opposite directions (Sect. 3.3).

All of these complications have contributed to the relatively slow advancement in the understanding of electrical phenomena in plant cells compared to their animal counterparts. However, many of these limitations have been circumvented in recent years following the development of the patch-clamp technique (Neher and Sakmann 1976; Neher et al. 1978; Hamill et al. 1981) and its application to the study of isolated plant protoplasts and vacuoles (Takeda et al. 1985; Hedrich et al. 1987, 1988; Satter and Moran 1988).

2.3 The Patch-Clamp Technique

Although the GΩ seal patch-clamp technique has been developed over the past 10 years, one of the most important theoretical considerations for improving the signal to noise ratio in recordings of membrane currents was understood many years ago (Strickholm 1961; Frank and Tauc 1964). The critical idea was to measure current from a small, electrically isolated membrane patch, having a very high impedance. Earlier efforts in establishing this high impedance included the use of coaxial recording electrodes, with the outer electrode being filled with sucrose. The initial reports from the Göttingen group describing the recording of single-channel ionic currents from biological membranes relied on electrically isolating a membrane patch by pressing a heat-polished glass microelectrode onto a clean (enzyme-treated) membrane surface (Neher and Sakmann 1976; Neher et al. 1978). Neher, Sakmann, and colleagues subsequently refined this approach, with the notable discovery that slight negative pressure inside the patch pipette resulted in seal resistances (electrical isolation) in the GΩ (10^9 Ω) range (Hamill et al. 1981), instead of MΩ (10^6 Ω) as previously obtained (Neher et al. 1978). Thus, a basic requirement for the successful application of the patch-clamp technique to plant cells is a clean membrane surface. The development of methods permitting the removal of the cell wall, thereby allowing isolated protoplasts to be obtained (for review, Pilet 1985), has been instrumental to patch-clamp studies on higher plant cells. Over the last few years, such measurements have been obtained from protoplasts, cytoplasts (evacuolated protoplasts), vacuoles, or membrane fractions incorporated into giant liposomes (Fig. 1A-D).

We give here a brief, nontechnical description of some important aspects of the patch-clamp technique. More detailed accounts of both the theoretical and practical aspects of this methodology are available elsewhere (Hamill et al. 1981; Sakmann and Neher 1983; Rae and Levis 1984; Smith et al. 1985; Kado et al. 1986). Patch pipettes made from glass capillaries using a two-stage pulling method typically have resistances of 1-10 MΩ when filled with standard physiological saline solutions. Pipette glass of different composition and varying inner and outer diameters (having different electrical properties) have been used. The choice of glass type is often made empirically, with the success rate of seal formation being a key factor, but critical is also the desired recording band width, which determines the effective temporal resolution. Pipettes are often treated with a hydrophobic coating like Sylgard (Dow Corning), Sigmacote (Sigma Chemicals), or electrical insulating solutions in order to minimize pipette capacitance, thereby maximizing the recording band width. Some of these treatments require that the pipette tip be subsequently heat polished in order to obtain successful GΩ seals, but this step has been found to be optional by many workers.

The filled pipette is gently pressed onto the cell surface using a high-quality micromanipulator under visual control (e.g., using an inverted microscope with magnification of 300-400 ×), and a slight negative pressure (5-10 cm of water) is applied to the pipette (most often by mouth suction). A large drop in baseline current noise is rapidly apparent after formation of a GΩ seal resistance. This arises because the voltage gain of the current-to-voltage converter circuit in the patch-clamp amplifier decreases with increasing resistance of the current source impedance. Procedures facilitating high-resistance seal formation on guard cell protoplasts have been described (e.g., Schroeder et al. 1984; Raschke and Hedrich 1989). For protoplasts, if high-resistance seals are difficult to obtain, it may be worthwhile verifying cytochemically (e.g., Nagata and Takebe 1970) that the membrane is free of cellulose and other cell wall constituents (see Fig. 1E). Varying the pipette resistance to find the optimal pipette tip size may be necessary (usually, whole cell pipettes have lower resistances than those used for single-channel recording). Filtering all solutions (with a 0.22 μm Millipore filter) before use is also recommended. Some workers maintain positive pressure on the pipette as it enters the bath solution to minimize contamination of the pipette tip surface as it passes through the air/water interface. Pipettes are used only once.

Passage from the cell-attached (or on-cell) patch, GΩ seal configuration to the tight-seal, whole-cell recording mode can be accomplished by increasing the negative pressure in the patch pipette or by brief, high voltage shocks. Because

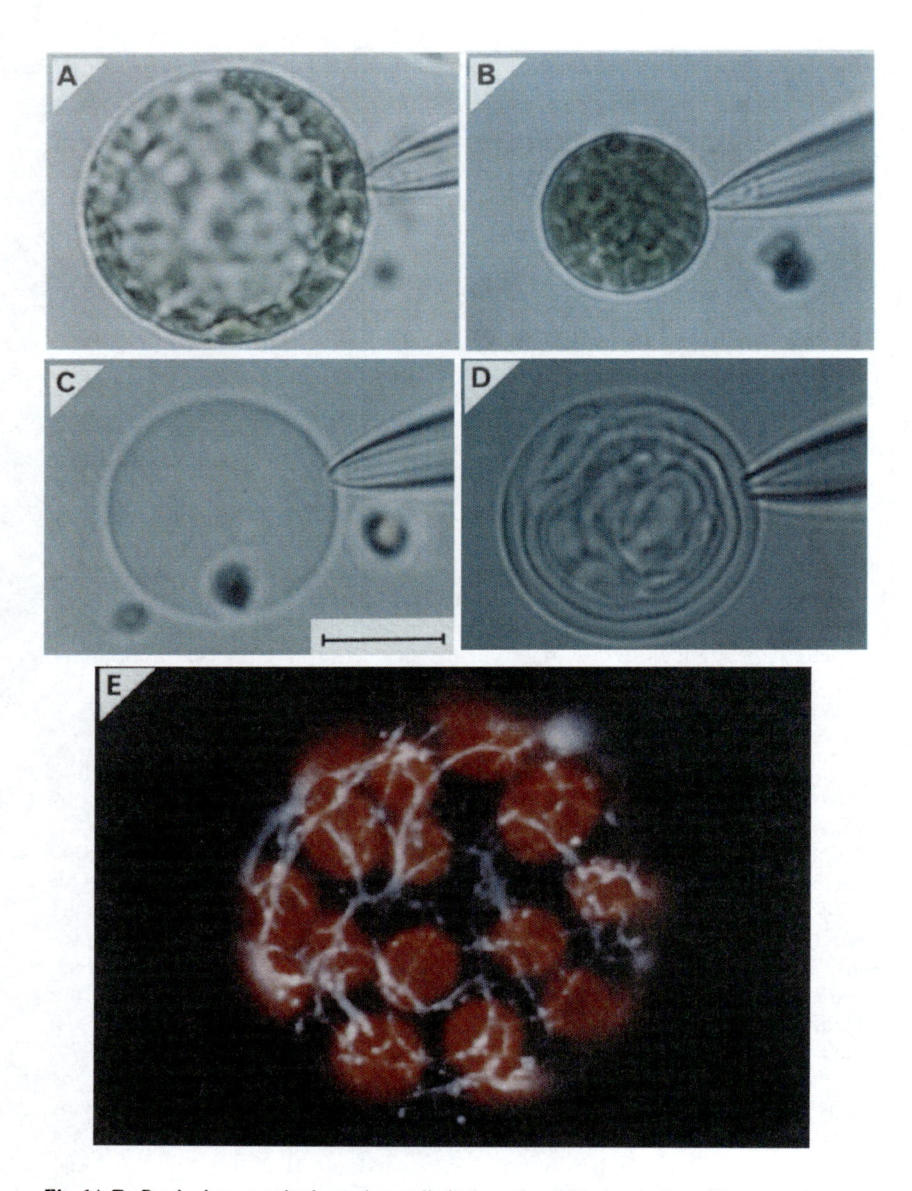

Fig. 1A-E. Patch-clamp methods can be applied to a variety of isolated plant cell preparations: **A** protoplast; **B** cytoplast; **C** vacuole from barley mesophyll cell; **D** giant liposome containing purified and reconstituted channel proteins from spinach chloroplast outer envelope. In each panel, a patch-clamp pipette is sealed onto the membrane surface of the specimen. *Bar* = 10 μm. **E** Isolated tobacco protoplast fluorescently labeled for cell wall components: the white fiber-like strands (cellulose) are stained with calcofluor white. (The *red spots* are chloroplasts which autofluoresce) (P. Müller and H. Barbier Brygoo, unpublished)

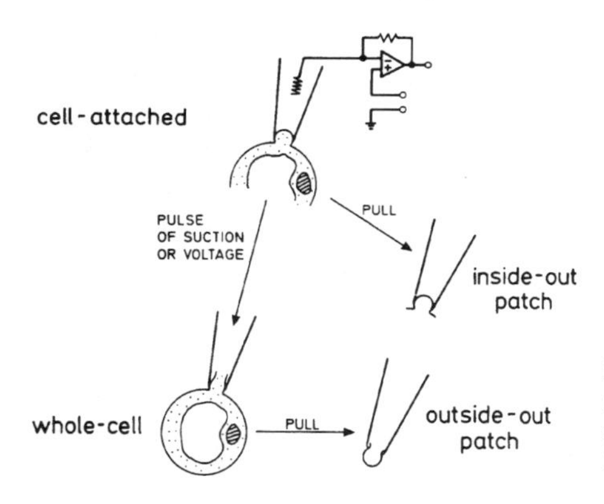

Fig. 2. The four principal modes of recording using the patch-clamp technique. See text for details (after Hamill et al. 1981)

the GΩ seal resistance is also mechanically very stable, it is possible to excise membrane patches (which are thus cell-free) by gently removing the electrode from the cell surface. Two different recording configurations result (Fig. 2). Starting from the cell-attached mode, inside-out patches are made, where the cytoplasmic membrane surface faces the bath and the extracellular membrane surface faces the recording pipette interior. Outside-out (or right-side out) patches result when the electrode is pulled away starting from the whole-cell recording configuration. Note that it is now possible to control precisely the ionic milieu on both sides of the membrane.

2.4 Electrical Properties of the Plasma Membrane

Information at the microscopic or single-channel current level can be obtained in recordings from cell-attached or cell-free (either inside-out or outside-out) membrane patches (for review, Sakmann and Neher 1983). Of interest are the elementary single-channel conductance, the ion selectivity (relative permeabilities) of the channel, the current-voltage relation for the open channel (in case of rectification), and the kinetic (open/closed) behavior of the channel. Recordings made in the whole-cell (or organelle, in the case of vacuoles) configuration are macroscopic, that is, they are measures of current flowing across the entire cell membrane and arise from the ensemble behavior of a population of channels. It is often easier (and faster) to characterize membrane conductance properties at the whole-cell level, leaving single-channel work for questions otherwise unanswerable. Also, an effective internal perfusion of the cell with the pipette solution occurs in whole-cell recording, allowing control of solution composition on both sides of the membrane. However, this sometimes leads to "run-down" of the observed currents, perhaps due to dilution (or washout) of critical intracellular factors. For example, voltage-activated Ca^{2+} currents in animal cells are usually subject to run-down with time over tens of minutes, although this may be counteracted by the presence of ATP, ATP-generating systems, high-capacity Ca^{2+} and pH buffers and/or protease inhibitors like leupeptin. On the other hand, information such as single-channel conductance or mean open lifetime is obtained in the most direct fashion from single-channel recordings. State diagrams for open/closed kinetic channel models generally are easiest determined using single-channel analysis. Rapid kinetics, which are often found for open-channel blocking effects, are best studied in detail at the single-channel level.

3 Ion Transport Through the Plasma Membrane

To study ion fluxes across the plasma membrane, motor tissues provide the ideal material, because ion uptake and release processes can be triggered separately. A change in the direction of ion transport in guard cells or pulvinar cells can be caused by physiological stimuli such as light or hormones (Raschke 1979; Satter et al. 1988). Of particular interest to plant physiologists are the mechanisms responsible for activation of ion channels and for the eventual regulation of this ion transport. To date, ion channels whose probability of opening is sensitive to changes in transmembrane voltage have been most commonly observed, although some channels are also dependent on the internal Ca^{2+} concentration, while others are mechanically sensitive. Voltage-dependent ion channels underlying action potential activity in excitable cells are also of obvious interest.

3.1 Ion Channels

Patch-clamp techniques allow the separate measurement of ionic currents through the plasma membrane and currents through the vacuolar membrane. An increasing number of ion channels in the plasma membrane of higher plant cells is currently being discovered. Ion channels can be classified by the ion species able to permeate the open channel (e.g., K^+, Cl^-, Ca^{2+}, and nonspecific cation channels). Other critical factors permitting characterization of different channel types include the mechanism of activation (e.g., membrane depolarization or hyperpolarization and/or sensitivity to internal Ca^{2+} concentration).

3.1.1 K^+ Channels

Following the initial characterization of K^+ channels in guard cell protoplasts (Schroeder et al. 1984), the presence of a variety of K^+ channels has been reported in the plasma membrane of various higher plant tissues (Moran et al. 1984, 1988; Iijima and Hagiwara 1987; Schauf and Wilson 1987a,b; Bush et al. 1988; Hosoi et al. 1988; Stoeckel and Takeda 1989b). K^+ transport across the plasma membrane of plant cells is closely linked to tissue- and cell-specific processes related to plant growth and development (e.g., Jones 1973; Satter et al. 1974; Lüttge and Pitman 1976; Raschke 1979). These recently described K^+ channels have received much interest due to their probable involvement in K^+ transport in plants.

To date, K^+ channels have been characterized in detail in guard cell protoplasts (Schroeder et al. 1984, 1987; Schroeder 1988). These K^+ channels are strongly regulated by the membrane potential, giving rise to K^+ efflux upon activation by membrane depolarization (with a threshold of about -40 mV) or K^+ influx upon hyperpolarization beyond about -100 mV. Outward K^+ current is carried by a few hundred K^+ channels in the plasma membrane of a guard cell (Schroeder et al. 1987). External Ba^{2+}, Cs^+, and TEA^+ (triethanolamine) are generally effective blockers of both outward and inward current flow through K^+ channels, while internal Cs^+, Na^+, and TEA^+ block outward K^+ efflux. Examples of voltage-gated K^+ currents are shown at the whole-cell level for protoplasts from a tumoral tobacco cell line (Fig. 3A) and from *Mimosa* pulvinus (Fig. 3C). Note that while both cells present outward rectifying properties at depolarized potentials, the kinetics of channel activation are faster for *Mimosa*, perhaps reflecting the importance of this current for action potential repolarization in these excitable motor cells. An example of outward single-channel K^+ currents at different patch potentials is illustrated in Fig. 4.

The K^+ channels found in various higher plant protoplasts have properties very similar to the outward K^+ conductance in algal cells (for review, Tazawa et al. 1987). For instance, the voltage dependence and alkali-metal ion selectivity are very similar in *Nitella* (Sokolik and Yurin 1986) and guard cells from *Vicia faba* (Schroeder 1988). Cytoplasmic droplets from giant algae have provided a favorable preparation for the recording of single-channel currents (Krawczyk 1978; Lühring 1986; Bertl and Gradmann 1987; Homblé et al. 1987; Laver and Walker 1987; Bertl et al. 1988). A large conductance (150–200 pS) K^+ channel was found in *Chara* and *Acetabularia* (Lühring 1986; Bertl and Gradmann 1987; Laver and Walker 1987; cf. Krawczyk 1978; Homblé et al. 1987). The membrane enclosing these droplets has been reported to be tonoplastic in origin (e.g., Lühring 1986), although this point remains controversial. Thus, the actual function of these channels is for the moment unclear. Finally, a small conductance (10 pS) K^+ channel was observed in the plasma membrane of slime mold (Müller et al. 1986).

One physiologically significant role of K^+ channels in plant cells is K^+ uptake and release as shown for motor cells (Schroeder et al. 1984, 1987; Moran et al. 1988) and algae (Tazawa et al. 1987). The magnitude of K^+ fluxes through K^+ channels in guard cells can account for the K^+ fluxes that occur during stomatal movement (Outlaw 1983; Schroeder et al. 1987). Properties of K^+ channels in guard cell protoplasts as characterized by patch-clamp studies (Schroeder 1988) agree with in situ K^+ fluxes observed from guard cells embedded in their original environment of the epidermis (Outlaw 1983). Moran, Satter, and colleagues have found that TEA^+ at mM concentrations blocks K^+ channels in pulvini and reduces leaf movements, supporting the importance of these channels for turgor regulation (Moran et al. 1988). An

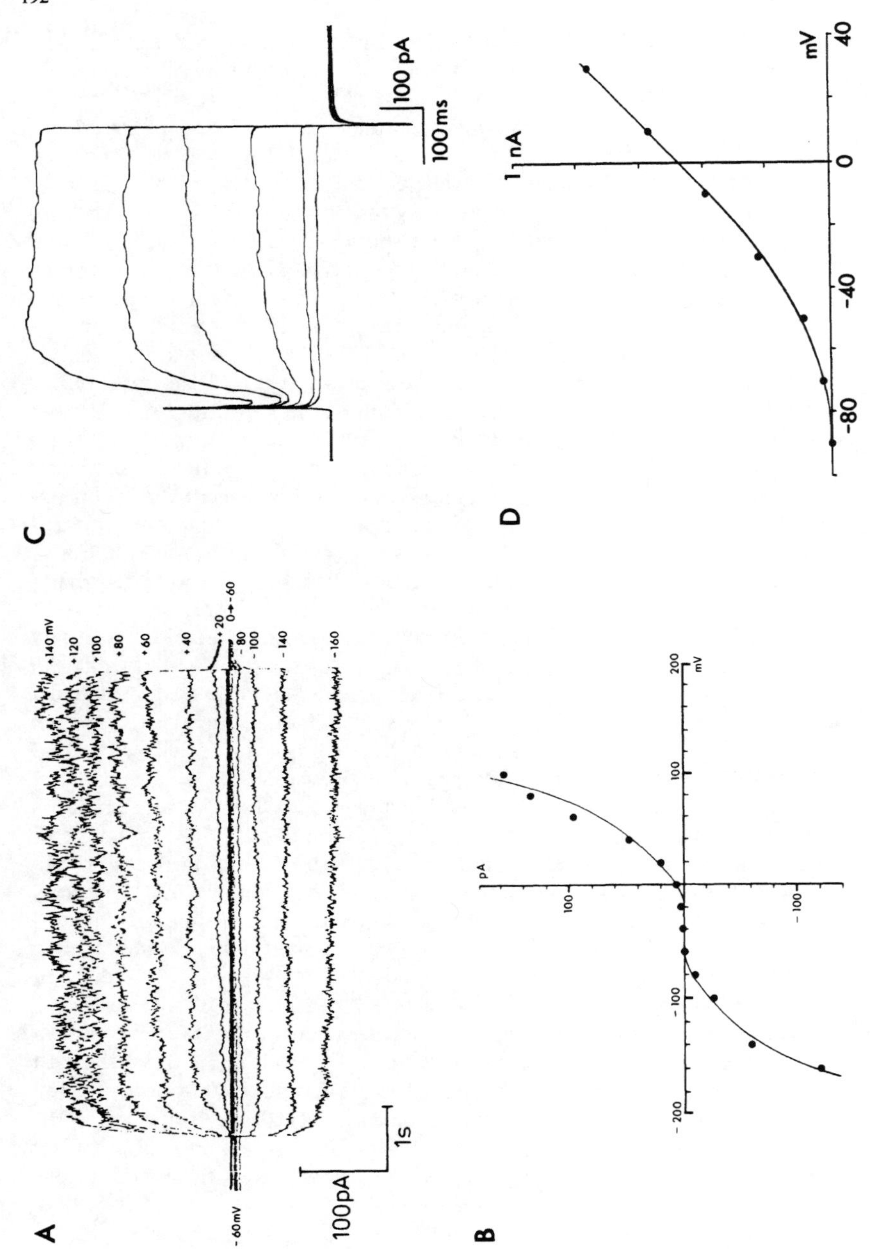
A
-60 mV
+140 mV
+120
+100
+80
+60
+40
+20
0→-60
-80
-100
-140
-160
100pA
1s
B
pA
100
-100
-200
100
200
mV
C
100 pA
100 ms
D
1 nA
mV
40
0
-40
-80

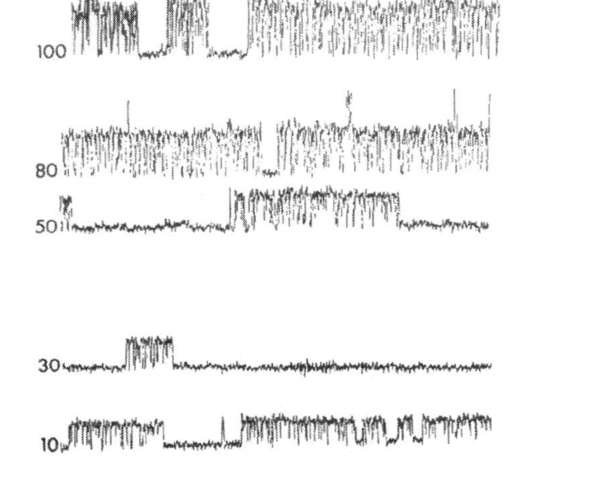

Fig. 4. Single-channel K^+ currents from a barley cytoplast at maintained voltages (in mV) recorded in the cell-attached configuration with a pipette solution containing 100 mM KCl and a bath solution containing 10 mM KCl. Potentials are given on the *left*, in mV relative to the resting potential of the plasma membrane (R. Hedrich, unpublished results)

interesting problem is the mechanism(s) responsible for activating or modulating the activity of these channels, as significant current is generated only at membrane potentials that are quite different from the normal resting potential.

Apart from contributing to volume/turgor regulation, voltage-activated, outwardly rectifying K^+ conductances probably play a predominant role in the repolarization of plant action potentials in both algae and higher plant cells (Sibaoka 1966; Findlay and Hope 1976; Simons 1981). Therefore, these outward K^+-current channels, in conjunction with the Cl^- and Ca^{2+} permeabilities responsible for regenerative inward current flow (Findlay 1961; Mullins 1962; Beilby 1982; Lunevsky et al. 1983), may form the ionic basis for

Fig. 3. A Whole-cell K^+ currents from a tumoral tobacco cell protoplast under voltage-clamp. The bath solution was 10 mM K-glutamate and the pipette contained 100 mM K-glutamate. Outward currents are shown upwards. Test pulses were given from −160 to 140 mV in steps of 20 mV; holding potential −60 mV **B** Current-voltage relations for the K^+ currents shown in **A**. (**A** and **B**, R. Hedrich, unpublished results). **C** Depolarization-activated, outwardly rectifying, whole-cell K^+ currents from a pulvinar protoplast from *Mimosa*. Voltage-clamp pulses were from −70 to 30 mV in steps of 20 mV; holding potential, −90 mV. The pipette contained (mM): KCl 150, $MgCl_2$ 2, mannitol 11, HEPES-KOH 10, pH 7.2, while the bath solution contained (mM): $CaCl_2$ 8, mannitol 600, MES-KOH 25, pH 5.5. Currents are shown nonleak subtracted. **D** Current-voltage relations for the currents shown in **C** (**C** and **D** after Stoeckel and Takeda 1989b)

excitability in plants. To date (see Sect. 3.1.3), definitive demonstrations of voltage-gated, inward, single-channel currents underlying action potential generation have so far not been reported.

Overall, it may be concluded that K^+ channels represent a major pathway for K^+ uptake/efflux in guard cells, and possibly in plant cells in general.

3.1.2 Ca^{2+}-Activated, Nonselective Cation Channels

In naturally occurring protoplasts from young endosperm cells, a Ca^{2+}-activated, voltage-dependent, nonselective cation channel has been identified (Fig. 5). Native protoplasts can be obtained directly from young *Haemanthus* or *Clivia* endosperm without the usual enzymatic digestion and can maintain mitotic division once isolated. When activated, this channel would give rise to an outward current of K^+ (Stoeckel and Takeda 1989a). However, external TEA^+ and Ba^{2+} were ineffective in blocking outward K^+ flow, and both internal Cs^+ and Na^+ were permeant, all uncharacteristic of typical K^+ channels. While being nonselective among the monovalent alkali metal cations, these channels were essentially impermeable to anions. It may be that the Ca^{2+} dependence of the channels is related in some way to critical Ca^{2+}-sensitive steps in mitosis. However, no correlation was found between channel activity and any particular mitotic stage. Another possibility is that these Ca^{2+}-dependent channels may be related to the specific function of endosperm as a storage tissue. Currents were outwardly rectifying, and voltage-dependent, with increasing depolarization giving rise to higher probabilities of channels being open. The significance of voltage-dependent gating remains unclear in these nonexcitable cells.

3.1.3 Cl^- Channels

Evidence for Cl^- channels has been reported in protoplasts from suspension-cultured *Asclepias tuberosa* (Schauf and Wilson 1987a). These channels were characterized by a strong voltage dependence, large-channel conductance, and inhibition by Zn^{2+}. The functional role of these channels is uncertain, although it may be related to osmoregulatory processes. Single Cl^- channel currents have also been described in the plasma membrane of *Chara* (Coleman 1986), but a

Fig. 5. A Outward single-channel currents recorded from an inside-out patch (see Fig. 2) from *Haemanthus* endosperm at maintained, depolarized patch membrane potentials. The bath solution (in contact with the cytoplasmic surface) contained (mM): NaCl 100, $MgCl_2$ 2, $CaCl_2$ 0.1, HEPES-NaOH 5, pH 7.2, while the pipette solution (in contact with the external surface) contained (mM): KCl 25, $MgCl_2$ 2, $CaCl_2$ 2, mannitol 22, MES-KOH 5, pH 5.5. **B** Current-voltage relations for the currents illustrated in **A**. The slope conductance was 33 pS. **C** Outward currents recorded as in **A** for a different inside-out patch, except that repetitive depolarizing steps (60 mV in amplitude from a holding potential of −10 mV; *uppermost trace*) were given (as illustrated in the *middle 9 traces*), and that the bath solution contained Ba^{2+} instead of Ca^{2+} and K^+ instead of Na^+. The *lowermost trace* is the averaged current for 137 test pulses. *Dotted lines* indicate the zero current level (After Stoeckel and Takeda 1989a)

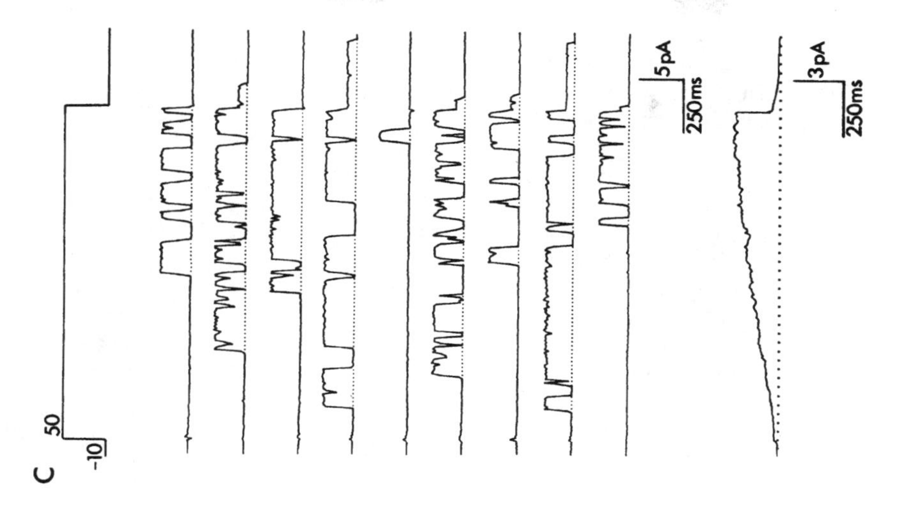
C
50
-10
5pA
250ms
3pA
250ms

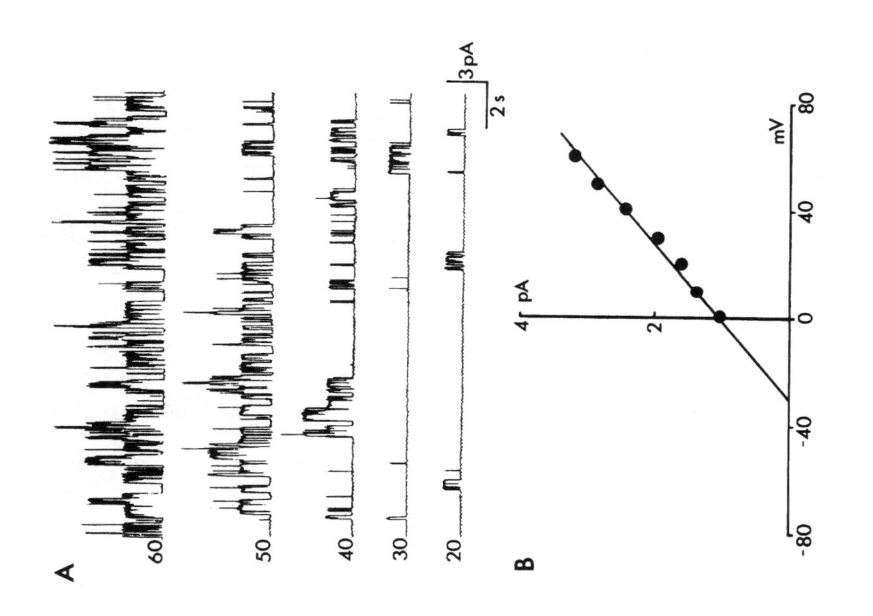
A
60
50
40
30
20
3pA
2 s
B
4 pA
2
mV
-80
-40
0
40
80

detailed characterization of these channels was not made. It will be of great interest to demonstrate convincingly that they are the basis for the regenerative Cl^- current (e.g., Beilby 1982; Lunevsky et al. 1983) underlying the action potential in these excitable cells.

3.1.4 Stretch-Activated Channels

Channels activated by mechanically stretching the plasma membrane have been found in protoplasts of cultured tobacco cells (Edwards and Pickard 1987; Falke et al. 1988), aleurone cells (R. Hedrich and D.S. Bush, unpublished results), guard cells (Schroeder 1988), in yeast, and in the outer membrane of *E. coli* (Saimi et al. 1988). Stretch is usually applied by maintaining negative or positive pressure on the patch pipette interior. In all cases, these channels were rather nonselective. It has been suggested that stretch-activated channels may function as both turgor sensors and mechanosensors (Edwards and Pickard 1987; Schroeder and Hedrich 1989), perhaps involved in down-regulation of the osmotic potential.

3.2 Proton Pumps

The patch-clamp technique allows the investigation of ion pump dynamics by direct measurement of the electrogenic current (Assmann et al. 1985; Hedrich et al. 1986). These were among the first direct measurements of pump current at the single cell level and emphasize the suitability of plant cells for studies using patch-clamp methods. Subsequently, new insights into pump activation by blue and red light (Assmann et al. 1985; Schroeder 1988; Serrano et al. 1988) and by the hormone-like compound fusicoccin (Hedrich et al. 1987; Serrano et al. 1988) have been obtained. The activation of these pumps results in outward current flow and, under current clamp, leads to plasma membrane hyperpolarization.

In guard cells, low flux rates of blue light and high flux rates of red light activate plasma membrane H^+ pumps by different mechanisms (Assmann et al. 1985; Serrano et al. 1988). Blue-light activation of H^+ pumping shows a marked delay of approximately 30 s with respect to the initial exposure to light (Assmann et al. 1985). In contrast, exposure to red light activates H^+ pumps without a measurable delay (Serrano et al. 1988). Both mechanisms require cytoplasmic substrates other than MgATP for maximal pump activity (Schroeder 1988; Serrano et al. 1988). The identification of these substrates, together with the regulatory pathways involved, should be a fruitful subject for further investigation.

An obvious future direction of research is to study H^+-ATPase pump currents responsible for the resting membrane potential in many cells (Serrano 1988; Chap. 6) at the single cell level using the whole-cell, patch-clamp mode. In certain cells with resting potentials much more hyperpolarized than the K^+

equilibrium potential (presumably due to pump activity), the presence of an inwardly rectifying K^+ conductance, as described in guard cells (Schroeder 1988), would lead to K^+ uptake. Thus, during stomatal opening in guard cells the pump is activated, extruding protons and giving rise to hyperpolarized membrane potentials favoring K^+ influx.

For cells where both proton pumps and inwardly rectifying K^+ channels (which are open at hyperpolarized potentials) exist, the question of net charge transfer arises, as this should determine the eventual steady-state resting potential. A clearer understanding of the probable role of ion channels as an alternative to ion transport (Schroeder et al. 1987; Schroeder 1988) mediated by proposed, "active", voltage-dependent K^+/H^+ symporters, as suggested in *Neurospora* (Blatt and Slayman 1987), seems a likely outcome of the further application of patch-clamp techniques.

Other types of electrogenic ion exchangers, cotransporters, or carriers, possibly involving nonelectrolytes, may also be amenable to investigation with whole-cell recording, provided that the current density (which is proportional to the number of transporter molecules per surface area) is large enough.

3.3 Comparison of the Electrical Properties of the Plasma Membrane and the Vacuolar Membrane

Ion fluxes across the vacuolar membrane contribute to regulating turgor, volume, and osmolarity. Additionally, vacuoles serve as metabolic compartments, maintaining internal nutrient homeostasis via e.g., the accumulation and mobilization of sugars, organic and inorgamic ions, and the activity of lytic enzymes. Direct, whole-cell, patch-clamp measurements of voltage-dependent ionic currents and currents resulting from activation of H^+-ATPase pumps have been described recently (for review, Hedrich et al. 1987, 1988). Single-channel ionic currents have also been characterized, e.g., the first direct demonstration in plants of channel gating by cytoplasmic Ca^{2+} was for the vacuolar membrane (Hedrich and Neher 1987; Wada et al. 1987).

3.3.1 Ion Channels

Inwardly rectifying whole-vacuole currents are evoked by hyperpolarizing voltage pulses (Hedrich et al. 1986, 1988; Coyaud et al. 1987). These currents are carried by slow vacuolar (SV) channels, so called because of their slow activation kinetics following a voltage step to negative values inside the vacuole (Hedrich and Neher 1987). Typical single-channel conductances are in the 60–80 pS range. The SV currents display two distinctive properties: firstly, they are activated at high cytoplasmic Ca^{2+} concentrations ($<$ 1–10 μM; Hedrich and Neher 1987) and secondly, they are relatively nonselective (Coyaud et al. 1987; Colombo et al. 1988; Kolb et al. 1987; Wada et al. 1987). Cl^- channel blockers like Zn^{2+} and the stilbene derivatives DIDS and SITS inhibit SV currents

(Hedrich and Kurkdjian 1988). The SV channels would appear to be a pathway for anion efflux and for equilibration of cations across the vacuolar membrane. A smaller conductance, 30 pS, fast vacuolar (FV; fast activation kinetics) channel was observed mainly at low concentrations of free cytoplasmic Ca^{2+} ($<$ 300 nM) (Hedrich and Neher 1987). These channels may be responsible for anion uptake in the resting state.

3.3.2 Pump Currents

Activation of vacuolar proton pumping resulted when MgATP was added to the extravacuolar ("cytoplasmic") solution and an inwardly directed current was measured (Hedrich et al. 1986; Coyaud et al. 1987). This current depolarizes the vacuole and is presumably responsible for the positive internal vacuolar resting potential. A K_m for MgATP in the order of 0.5–0.8 mM has been found (Hedrich et al. 1988). The pump was capable of pumping against a $10^4 \times$ proton gradient (pH 3.5 in the vacuole and 7.5 externally). Note that while the direction of pump current flow across the vacuolar membrane (inward into the vacuole) is opposite to that found for the plasma membrane proton pump (out of the cell), the activity of both pumps would contribute to a hyperpolarization of the cytoplasmic compartment. As for the plasma membrane, a balance between pump activity and current flow through ion channels resulting from vacuolar potentials set by the pump and/or rises in the concentration of free Ca^{2+} in the cytoplasm would determine the direction and magnitude of net ion transport across the vacuolar membrane.

4 Concluding Remarks

Patch-clamp studies over the past few years have provided direct information about the existence of ion channels in membranes of both higher plant cells and algae. There can be little doubt of the enormous potential offered by patch-clamp methods for the study of membrane electrical phenomena in plant cells. Together with the use of isolated cells or cell culture systems, many of the technical uncertainties that previously hindered electrophysiological studies in plants have now been overcome.

Ion channels have been found in the plasma membrane of all higher plant cells so far studied and thus may constitute a ubiquitous component of all plant cells. To date, K^+ channels have been the most commonly observed channel type, and thus probably play prominent roles in K^+ uptake/efflux in general, and certainly in action potential repolarization in excitable cells. Future investigations may lead to the detailed characterization of other channels and of tissue-specific subtypes of the channel classes discussed here. The importance of Ca^{2+} in many of the fundamental physiological processes in plant cells (e.g., for review, Gilroy et al. 1987; Poovaiah and Reddy 1987; Chaps. 9, 12, 14, and 16) highlights the interest in characterizing Ca^{2+} channels. Mutants may provide suitable systems to study plant or tissue specificity and control of channel density. Potential problems are the differences between observations made on isolated cell preparations (protoplasts and vacuoles) and those made on intact tissues.

The second phase of study of ion channels and ion transport in plant cell membranes should involve more than just demonstrating and describing that channels exist. The physiological role of channels, together with the mechanisms underlying activation and regulation of channel activity, will be obvious targets for investigation. These questions are for the moment largely unanswered, especially for nonexcitable plant cells. Isolation and purification of membrane proteins responsible for ion transport (pumps or channels), followed by functional reconstitution into artificial lipid bilayers or liposomes, should also contribute to answering these kinds of questions.

Another area of interest will be the search for ligand-gated (or receptor-operated) channels, which to date are unknown in plant cells. Nevertheless, the presence of specific, high-affinity receptors for certain hormones like auxin and abscisic acid (Hornberg and Weiler 1984) and the observations of auxin-stimulated phosphoinositol turnover (Ettlinger and Lehle 1988) and protein phosphorylation (Ranjeva and Boudet 1987) suggest that, like in animal cells, specific membrane receptors may be coupled through G-proteins (Blum et al. 1988) to phospholipases or cyclases. By analogy, the net effect of this hypothetical cascade of cellular signal transduction mechanisms may be changes in internal Ca^{2+} levels (Schumaker and Sze 1987) and/or the phosphorylation of key effector proteins. Thus, the existence of second messenger-regulated or Ca^{2+}-sensitive channels (Hedrich and Neher 1987; Wada et al. 1987; Stoeckel and Takeda 1989a) may be a general phenomenon in plant cells, as found for their animal counterparts. Similarly, modulation of normally voltage-dependent channels by plant hormones or other circulating factors may be widespread and of significance, especially in view of the number of developmental processes affected by these substances (see also Chaps. 9 and 12).

The role of transcellular currents in the determination of cell polarity or tissue development may be more easily elucidated, following the characterization of the underlying channels. The relative ease with which pump currents have been studied in plant cells should encourage further investigation of these and other carrier-mediated ion transporters.

References

Assmann SM, Simoncini L, Schroeder JI (1985) Blue light activates electrogenic ion pumping in guard cell protoplasts of *Vicia faba*. Nature 318:285–287

Beilby MJ (1982) Cl^- channels in *Chara*. Philos Trans R Soc Lond B 299:435–455

Bertl A, Gradmann D (1987) Current-voltage relationships of potassium channels in the plasmalemma of *Acetabularia*. J Membr Biol 99:41–49

Bertl A, Klieber HG, Gradmann D (1988) Slow kinetics of a potassium channel in *Acetabularia*. J Membr Biol 102:141–152

Blatt MR, Slayman CL (1987) Role of "active" potassium transport in the regulation of cytoplasmic pH by nonanimal cells. Proc Natl Acad Sci USA 84:2737–2741

Blum W, Hinsch K-D, Schulz G, Weiler EW (1988) Identification of GTP-binding proteins in the plasma membrane of higher plants. Biochem Biophys Res Commun 156:954–959

Bush DS, Hedrich R, Schroeder JI, Jones RL (1988) Channel-mediated K^+ flux in barley aleurone protoplasts. Planta 176:368–377

Catterall WA (1988) Structure and function of voltage-sensitive ion channels. Science 242:50–61

Cole KS, Curtis HJ (1938) Electric impedance of *Nitella* during activity. J Gen Physiol 22:37–64

Coleman HA (1986) Cl^- currents in *Chara* – a patch clamp study. J Membr Biol 93:55–61

Colombo R, Cerana R, Lado P, Peres A (1988) Voltage-dependent channels permeable to K^+ and Na^+ in the membrane of *Acer pseudoplatanus* vacuoles. J Membr Biol 103:227–236

Coyaud L, Kurkdjian A, Kado R, Hedrich R (1987) Ion channels and ATP-driven pumps involved in ion transport across the tonoplast of sugarbeet vacuoles. Biochim Biophys Acta 902:263–268

Edwards KL, Pickard BG (1987) Detection and transduction of physical stimuli in plants. In: Wagner E, Greppin H, Biller B (eds) The cell surface in signal transduction. NATO ASI Series H12. Springer, Berlin Heidelberg New York Tokyo, pp 41–66

Ettlinger C, Lehle L (1988) Auxin induces rapid changes in phosphatidylinositol metabolites. Nature 331:176–178

Falke LC, Edwards KL, Pickard BG, Misler S (1988) A stretch-activated anion channel in tobacco protoplasts. FEBS Lett 237:141–144

Findlay GP (1961) Voltage-clamp experiments with *Nitella*. Nature 191:812–814

Findlay GP, Hope AB (1976) Electrical properties of plant cells: methods and findings. In: Lüttge U, Pitman MG (eds) Encyclopedia of plant physiology, new series, Vol. 2, Part A. Transport in plants. Springer, Berlin Heidelberg New York, pp 53–92

Frank E, Tauc L (1964) Voltage clamp studies of molluscan neuron membrane properties. In: Hoffmann JF (ed) The cellular functions of membrane transport. Prentice Hall, Englewoods Cliff NJ, pp 26–51

Gilroy S, Blowers DP, Trewavas AJ (1987) Calcium: a regulation system emerges in plant cells. Development 100:181–184

Hamill OP, Marty A, Neher E, Sakmann B, Sigworth FJ (1981) Improved patch clamp techniques for high resolution current recording from cells and cell-free membrane patches. Pflügers Arch 391:85–100

Hedrich R, Kurkdjian A (1988) Characterization of an anion-permeable channel from sugar beet vacuoles: effect of inhibitors. EMBO J 7:3661–3666

Hedrich R, Neher E (1987) Cytoplasmic calcium regulates voltage-dependent ion channels in plant vacuoles. Nature 329:833–836

Hedrich R, Schroeder JI (1989) The physiology of ion channels and electrogenic pumps in higher plants. Annu Rev Plant Physiol Plant Mol Biol 40:539–569

Hedrich R, Flügge UI, Fernandez JM (1986) Patch-clamp studies of ion transport in isolated plant vacuoles. FEBS Lett 204:228–232

Hedrich R, Schroeder JI, Fernandez JM (1987) Patch-clamp studies on higher plant cells: a perspective. Trends Biochem Sci 12:49–52

Hedrich R, Barbier-Brygoo H. Felle H, Flügge UI, Maathuis FJM, Marx S, Prins HBA, Raschke K, Schnabl H, Schroeder JI, Struve I, Taiz L, Ziegler P (1988) General mechanisms for solute transport across the tonoplast of plant vacuoles: a patch-clamp survey of ion channels and proton pumps. Bot Acta 101:7–13

Hille B (1984) Ionic channels of excitable membranes. Sinauer, Sunderland

Hodgkin AL (1964) The conduction of the nervous impulse. Ch C Thomas, Springfield, MA

Homblé F, Ferrier JM, Dainty J (1987) Voltage-dependent K^+-channels in protoplasmic droplets of *Chara corallina*. Plant Physiol 83:53–57

Hornberg C, Weiler EW (1984) High-affinity binding sites for abscisic acid on the plasmalemma of *Vicia faba* guard cells. Nature 370:321–324

Hosoi S, Iino M, Shimazaki K (1988) Outward-rectifying K^+ channels in stomatal guard cell protoplasts. Plant Cell Physiol 29:907–911

Iijima T, Hagiwara S (1987) Voltage dependent K^+ channels in protoplasts of trap-lobe cells of *Dionea muscipula*. J Membr Biol 100:73–81
Jaffe LF, Nuccitelli R (1977) Electrical controls of development. Annu Rev Biophys Bioeng 6:445–476
Jones RL (1973) Gibberellic acid and ion release from barley aleurone tissue. Plant Physiol 52:303–308
Kado RT, Kurkdjian A, Takeda K (1986) Transport mechanisms in plant cell membranes: an application for the patch clamp technique. Physiol Vég 24:227–244
Kolb HA, Köhler K, Martinoia E (1987) Single potassium channels in membranes of isolated mesophyll barley vacuoles. J Membr Biol 95:163–169
Krawczyk S (1978) Ionic channel formation in a living cell membrane. Nature 273:56–57
Laver DR, Walker NA (1987) Steady-state voltage dependent gating and conduction kinetics of single K^+ channels in the membrane of cytoplasmic drops of *Chara australis*. J Membr Biol 100:31–42
Lühring HE (1986) Recording of single K^+ channels in the membrane of cytoplasmic drop of *Chara australis*. Protoplasma 133:19–28
Lunevsky VZ, Zherelova OM, Vostrikov IY, Berestovsky GN (1983) Excitation of *Characeae* cell membranes as a result of activation of calcium and chloride channels. J Membr Biol 72:43–58
Lüttge U, Pitman MG (eds) (1976) Transport in plants. Vols. I, II & III. Springer, Berlin Heidelberg New York
Michaelis L (1925) Contribution to the theory of permeability of membranes for electrolytes. J Gen Physiol 8:33–59
Moran N, Ehrenstein G, Iwasa K, Bare C, Mischke C (1984) Ion channels in plasmalemma of wheat protoplast. Science 226:835–838
Moran N, Ehrenstein G, Iwasa K, Mischke C, Bare C, Satter RL (1988) Potassium channels in motor cells of *Samanea saman*: a patch clamp study. Plant Physiol 88:643–648
Müller U, Malchow D, Hartung K (1986) Single ion channels in the slime mold *Dictyostelium discoideum*. Biochim Biophys Acta 857:287–290
Mullins LJ (1962) Efflux of chloride ions during the action potential of *Nitella*. Nature 196:986–987
Nagata T, Takebe I (1970) Cell wall regeneration and cell division in isolated tobacco mesophyll protoplasts. Planta 92:301–308
Neher E, Sakmann B (1976) Single-channel currents recorded from membrane of denervated frog muscle fibres. Nature 260:799–802
Neher E, Stevens CF (1977) Conductance fluctuations and ionic pores in membranes. Annu Rev Biophys Bioeng 6:345–381
Neher E, Sakmann B, Steinbach JH (1978) The extracellular patch clamp: a method for resolving currents through individual open channels in biological membranes. Pflügers Arch 375:219–228
Outlaw WH (1983) Current concepts on the role of potassium in stomatal movements. Physiol Plant 59:302–311
Pilet PE (ed) (1985) The physiological properties of plant protoplasts. Springer, Berlin Heidelberg New York Tokyo
Poovaiah BW, Reddy ASN (1987) Calcium messenger systems in plants. CRC Crit Rev Plant Sci 6:47–103
Rae JL, Levis RA (1984) Patch voltage clamp of lens epithelial cells: theory and practise. Mol Physiol 6:115–162
Ranjeva R, Boudet AM (1987) Phosphorylation of proteins in plants: regulatory effects and potential in stimulus/response coupling. Annu Rev Plant Physiol 38:73–93
Raschke K (1979) Movements of stomata. In: Haupt, W, Feinleib E (eds) Encyclopedia of plant physiology, new series. Vol. 7. Springer, Berlin Heidelberg New York, pp 383–441
Raschke K, Hedrich R (1989) Patch-clamp measurements on isolated guard cell protoplasts and vacuoles. Methods Enzymol 174, in press
Saimi Y, Martinac B, Gustin MC, Culbertson MR, Adler J, Kung C (1988) Ion channels in *Paramecium*, yeast and *Escherichia coli*. Trends Biochem Sci 13:304–309
Sakmann B, Neher E (eds) (1983) Single channel recording. Plenum, New York

Sanderson JB (1888) On the electromotive properties of the leaf of *Dionaea* in the excited and unexcited states. Philos Trans R Soc Lond B 179:417–449

Satter RL, Moran N (1988) Ionic channels in plant cell membranes. Physiol Plant 72:816–820

Satter RL, Geballe GT, Applewhite PB, Galston AW (1974) Potassium flux of leaf movements in *Samanea saman*. I. Rhythmic movement. J Gen Physiol 64:413–430

Satter RL, Morse MJ, Lee Y, Crain RC, Coté GG, Moran N (1988) Light- and clock-controlled leaflet movements in *Samanea saman*: a physiological, biophysical and biochemical analysis. Bot Acta 101:205–213

Schauf CL, Wilson KJ (1987a) Properties of single K^+ and Cl^- channels in *Asclepias tuberosa* protoplasts. Plant Physiol 85:413–441

Schauf CL, Wilson KJ (1987b) Effect of abscisic acid on K^+ channels in *Vicia faba* guard cell protoplasts. Biochem Biophys Res Commun 145:284–290

Schroeder JI (1988) K^+ transport properties of K^+ channels in the plasma membrane of *Vicia faba* guard cells. J Gen Physiol 92:667–684

Schroeder JI, Hedrich R (1989) A model for the concerted action of ion transport mechanisms across guard cell membranes. Trends Biochem Sci 14:187–192

Schroeder JI, Hedrich R, Fernandez JM (1984) Potassium-selective single channels in guard cell protoplasts. Nature 312:361–362

Schroeder JI, Raschke K, Neher E (1987) Voltage dependence of K^+ channels in guard cell protoplasts. Proc Natl Acad Sci USA 84:4108–4112

Schumaker K, Sze H (1987) Inositol 1,4,5-triphosphate releases Ca^{2+} from vacuolar membrane vesicles of oat roots. J Biol Chem 262:3944–3946

Serrano EE, Zeiger E, Hagiwara S (1988) Red light stimulates an electrogenic proton pump in *Vicia faba* guard cell protoplasts. Proc Natl Acad Sci USA 85:436–440

Serrano R (1988) Structure and function of proton translocating ATPase in plasma membranes of plants and fungi. Biochim Biophys Acta 947:1–28

Shiina T, Wayne R, Lim Tung HY, Tazawa M (1988) Possible involvement of protein phosphorylation/dephosphorylation in the modulation of Ca^{2+} channel in tonoplast-free cells of *Nitellopsis*. J Membr Biol 102:255–264

Sibaoka T (1966) Action potentials in plant organs. Symp Soc Exp Biol 20:49–74

Simons PJ (1981) The role of electricity in plant movements. New Phytol 87:11–37

Smith TG, Lecar H, Redman SJ, Gage PW (eds) (1985) Voltage and patch clamping with microelectrodes. Williams & Wilkins, Baltimore

Sokolik AI, Yurin VM (1986) Potassium channels in plasmalemma of *Nitella* cells at rest. J Membr Biol 89:9–22

Spray DC, Harris AL, Bennett MVL (1981) Equilibrium properties of a voltage-dependent junctional conductance. J Gen Physiol 77:77–93

Stoeckel H, Takeda K (1989a) Calcium-activated, voltage-dependent, nonselective cation currents in endosperm plasma membrane from higher plants. Proc R Soc Lond B, in press

Stoeckel H, Takeda K (1989b) Voltage-activated, delayed rectifier K^+ current from pulvinar protoplasts of *Mimosa pudica*. Pflügers Arch, in press

Strickholm A (1961) Impedance of a small electrically isolated area of the muscle cell surface. J Gen Physiol 44:1073–1088

Takeda K, Kurkdjian A, Kado RT (1985) Ionic channels, ion transport and plant cell membranes: potential applications of the patch-clamp technique. Protoplasma 127:147–162

Takeuchi A, Takeuchi N (1959) Active phase of frog's end-plate potential. J Neurophysiol 22:395–411

Tazawa M (1964) Studies on *Nitella* having artificial sap. I. Replacement of the cell sap with artificial solutions. Plant Cell Physiol 5:33–43

Tazawa M, Shimmen T, Mimura T (1987) Membrane control in the *Characeae*. Annu Rev Plant Physiol 38:95–117

Umrath K (1937) Der Errungsvorgang bei höheren Pflanzen. Ergeb Biol 14:1–142

Wada Y, Ohsumi Y, Tanifuji M, Kasai M, Anraku Y (1987) Vacuolar ion channel of the yeast, *Saccharomyces cerevisiae*. J Biol Chem 262:17260–17263

Chapter 9 Signal Sensing and Signal Transduction Across the Plasma Membrane

S. GILROY and A. TREWAVAS[1]

[1]Department of Botany, University of Edinburgh, King's Buildings, Mayfield Road, Edinburgh EH9 3JH, U.K.

Abbreviations: ABA, abscisic acid; DAG, diacylglycerol; IP_3, inositol-1,4,5-trisphosphate; P_{fr}, active phytochrome; P_r, inactive phytochrome; PI, phosphatidylinositol; PIP, phosphatidylinositol-4-phosphate; PIP_2, phosphatidylinositol-4,5-bisphosphate; SH, sulfhydryl group.

C. Larsson, I.M. Møller (Eds):
The Plant Plasma Membrane

1 Introduction

This topic is so broad that we have had to severely constrain what is included if the chapter is not to become imponderably large. The plasma membrane, the outer skin of the cell, delimits the boundary between the highly organized and enzyme-rich cytosol and the enzyme-poor and less organized cell wall; any molecular traffic, in or out of the cell must cross the plasma membrane. This traffic may be sensed at the plasma membrane and converted to an entirely new set of information, or it may be modified in transit, or it may pass unhindered. The latter traffic is small and not considered here. Discussion is limited to what is termed signal sensing and stimulus-response coupling although both are difficult to define exactly. Ion flux phenomena in general are not considered here (see Chap. 8) although such fluxes are clearly important information conveyed to the plant cell across the plasma membrane. Signalling by the cell to the cell wall is a relevant phenomenon to cell development and an example of information conveyed from inside to outside. But it has been excluded for reasons of space and is dealt with elsewhere (Chaps. 11, 14, and 15).

The notion that the plasma membrane is the important sensing organelle of the plant cell is an hypothesis derived from animal studies. We must be wary of falling into the all too common trap of viewing the plant as a green animal. Hard evidence for signal transduction, as distinct from signal conveyance, across the plant plasma membrane is very limited. Even when present we can expect differences from animals.

Green plants maintain continued embryogenesis troughout their life cycle. Much of the signalling to which plants are subjected invokes changes in growth and development. Adequate models for these phenomena are not freely available from other organisms. Thus the unravelling of signal transduction processes in plants should provide concepts of remarkable molecular variety and novelty. The challenge is there – we can only hope there are enough sufficiently skilled to accept and master it.

2 There Must be Differences in the Way Animals and Plants Sense and Transduce Signals

2.1 Animals Have an Elaborate Surface Recognition System

Most higher animals undergo their critical development in the protected environment of an egg or uterus. Tissues and cell types are usually specified during this period and further development outside this protected environment involves growth in size of already specified tissues; a quantitative rather than a qualitative phenomenon. Some limited tissue changes can accompany sexual maturation. Early animal embryogenesis involves extensive cell movement and recognition. Cells directly interact with each other and specifically sense cellular partners through interlocking proteins on the cell surface. Protein-protein interaction is a 'sine qua non' of animal embryogenesis.

Recognition via surface proteins is at its most complex in immune phenomena. But this could only evolve because the capability was already inherently present in animal cells as a result of embryogenic requirements. For the same reason it is not surprising that many mammalian hormones and growth factors are proteins; or that sensing of these and other signals use an eleborate surface sensing and transduction phenomenon based on proteins The potential originating from embryogenic mechanisms was already present.

Animal cells are bathed in a fluid the composition of which is carefully regulated and in the case of mammals and birds is maintained at constant temperature. Uncontrolled variation in composition or character induced by a changing environment is thus avoided. Furthermore, many cells and tissues are highly specialized and the variety of function is thus reduced. Plasticity is confined to behaviour and is produced by coordinating different groups of cells using nervous and hormonal information. However, for many specialized cells one or a few transduction sequences are probably all that is necessary since the variety of function is so constrained. A muscle cell can only contract for example.

2.2 Individual Plant Cells Sense a Much Greater Variety of Signals than Animals

The sessile plant presents a very different picture to the animal. Although there is a period of embryological development to form the seed, which again occurs in a protected environment, most embryogenesis occurs in meristems which continue development throughout the life cycle. Environmental experience and variation change the character of development resulting in plasticity in development (Trewavas 1986b). The shapes of leaves and stems vary as a response to the influence of environmental change; the numbers of leaves, buds, root branches etc. are highly variable.

There are few specialized cell types. Lack of specialization of cell type is essential to survive inevitable predation. Individual plant cells have many more compartments than animal cells and these require a more complex intracellular coordination (Trewavas 1986b). Many individual plant cells retain a multiplicity of functions and respond to a variety of signals. In animals these functions are usually distributed amongst different cells. In contrast to animals, there is a weakly controlled and compositionally variable vascular system. Cells are always separated by a wall and signals between cells may have to travel over a variable distance as wall thickness varies. The continued presence of a wall obviates a requirement for cell-to-cell interaction through interlocking plasma membrane proteins.

Individual cells have to integrate the great variety of variable external information which they express in a coordinated response. The *Fucus* zygote can sense some 14 different environmental parameters any one of which can be used to align the axis of polarity; stomatal cells respond to light (intensity, quality and photoperiod are distinguished), temperature, water vapour pressure deficit, CO_2, air pollutants, plant pathogen-produced chemicals, abscisic acid (ABA), cytokinin, K^+; the growth of rice coleoptile cells is sensitive to blue and red light, temperature, gravity, auxin, gibberellin, ABA, CO_2, O_2, ethylene, wounding, pathogens, sucrose, amino acids, pH, Ca^{2+}, unknown effects of submergence and long lists of laboratory chemicals, pathogen-derived products and so on. The sheer variety of inductive signals is common to all forms of plant development which have been thoroughly investigated; the list of inductive chemicals can be bewildering both in character and scope (Trewavas et al. 1984; Trewavas 1986a,b). The plant cell has to integrate many different signals which arrive at the same time into a coherent response. Usually the effects of such simultaneous signals are not additive although there are exceptions.

2.3 There May be Many Transduction Sequences in Plant Cells

Morphologically there are only two patterns to rice coleoptile growth, submerged growth and aerobic growth. The submerged characteristics are evidenced by a thinner, longer coleoptile, full of aerenchyma (Trewavas 1987a), but biochemically there are probably many different types of coleoptile cell development (Trewavas 1986a). In the absence of sucrose the cell walls are thinner; plants grown on different media show compositional variation reflecting their different biochemistries. Morphology deceptively hides a great deal of molecular variation. The same is true for seed dormancy breakage which can be induced by respiratory inhibitors, SH-group reagents, acids, growth substances, electron acceptors and many others. It is very unlikely that a uniform biochemistry can underlie the apparent uniform morphological response of dormancy breakage (Trewavas 1987b), flowering induction (Bernier 1986) and other aspects of plant development (Trewavas 1986a; Trewavas and Jennings 1986).

From this we can conclude that there may be many signal transduction sequences in plant cells which issue from a requirement to respond to many signals. At first sight it is unlikely that the same signal transduction sequence will induce phytoalexin formation, change leaf morphology, initiate de-etiolation, or induce aerenchyma formation, to name but a few.

2.4 Could Time Variations in Sensing Reduce the Requirement for Many Transduction Sequences?

The developing *Fucus* zygote responds to many different parameters which can specify polarity but it is not equally sensitive to these parameters throughout initial development (some 16–18 h) (Jaffe 1969). There are narrow temporal windows of sensitivity which in an individual zygote may last no more than 1 h. Sensitivity to applied electrical fields substantially precedes that of light for example. Instead of integrating different phenomena at the same time, it is possible to conceive of a temporal spacing of sensing of stimuli. The same transduction sequence could then be used, but at different times it occurs in a different biochemical context eliciting a different response (Alkon and Rasmussen 1988).

Definitive evidence to support such a view is certainly lacking in higher plants but there are indications. These derive from easily made observations on individual cells, such as guard cells or germinating pollen grains. Simple observations show considerable individual cell variation in the sensitivity of response to a uniform stimulus. Some cells respond quickly, others slowly, showing temporal variation in sensitivity. Pericycle cells show sporadic responses to applied auxin such that branch roots develop anywhere between 1–4 days after treatment (Blakeley et al. 1972). Other tissue responses in ripening or amylase formation suggest that plant cells are partially synchronized by exogenous treatment with growth regulators (Trewavas 1987b). This suggests that cells are not temporally uniform in their sensitivity to applied chemicals. Again some respond early, some later. But the effect of high levels of applied growth regulators is to reduce the variation in temporal spacing and synchronize the response.

A similar stochastic response to a uniform stimulus has been noted for dividing cells (Smith and Martin 1973). Likewise cytosol Ca^{2+} measurements in animal cells (Tsien and Poenie 1986) have shown similar stochastic behaviour. Cells can respond sporadically with the responsive number increasing with time. Other examples of stochastic temporal response in mammalian cell development have been detailed (Maclean and Hall 1987). In the slime mould *Dictyostelium* a differential response is produced if the cells are in different phases of the cell cycle when they receive a stimulus (Newell et al. 1987). It should be obvious from this that we have much to learn concerning temporal aspects of signal transduction. If there is temporal sensitivity variation in the cells composing a tissue some will always be capable of responding. The proportion that do respond may simply in turn specify the size of the subsequent tissue response in a specified time period.

2.5 Plant Tissues and Cells May Have Elaborated Their Vectorial Sensing Systems

If animal cells have elaborated their surface protein sensing system during evolution, a consequence of its requirement during embryogenesis, is there a similar amplification of a particular character in plant cells? Since plants are anchored and plant cells fixed, it is perhaps the sensing of the direction (the vector) of the incoming information, which has been greatly elaborated.

Vectorial sensing would have to be carried out by a fixed component of the cell, the most likely is the plasma membrane. Furthermore, there would have to be transduction mechanisms, a means of assessing differences between parts of the plasma membrane of the same cell and probably some sort of counting character, an adding up of numbers of signal molecules arriving within a time period.

The vectorial transport of auxin, well-known tropic phenomena and evident polarity in all aspects of plant growth and form are adequate testimony to these vectorial sensing characteristics. This sensing capability is presumably initiated when cells are first formed in the meristem. The direction from which signals arrive can determine subsequent development. Nitrate entering pericycle cells from the outside induces cell division and branch root formation (Hackett 1972; Drew et al. 1973). But pericycle cells, which must experience some leakage of nitrate from the xylem higher up the root but in the opposite direction, do not respond. Branch roots will form in a layer of root soil where water transit is from outside to inside but not in the root in other drier layers where water transit across the pericycle must presumably reverse if the other root cells are to remain viable.

Studies on *Fucus* have shown that it is a discrete localization of channels, pumps and other enzymes in the plasma membrane that determine inherent polarity (Quatrano 1978). In cells in which polarity to all outside information is not completely fixed, a change in the direction of normal incoming information from the environment or the vascular system or treatments with chemicals that disaggregate clusters of channels, for example, may disorganize planes of division and growth (Trewavas et al. 1984). Polarity is a labile phenomenon in plants (plant tissue culture is the obvious example) and can be altered. If cells grow with respect to a defined direction for incoming growth materials from the vascular system, experimentally altering only that direction may have profound and unforeseen consequences. To re-establish organization, the direction of information impinging on cells may be crucial. Much of plant development may be stabilized by electrical polarities which in turn result from the position and numbers of cells contributing to the endogenous electrical field itself (Trewavas et al. 1984).

3 To Which Stimuli do Plant Cells Respond?

3.1 Wherein Lies the Specificity of Response?

A stimulus must be perceived and transduced to initiate a response. Tables 1 and 2 are a compilation of some of the stimuli to which plant cells respond, however, it is not comprehensive. Also included are molecules supposedly involved in transduction sequences.

Clear problems arise immediately. If all the transduction sequences involve Ca^{2+} and/or H^+, then the specificity inherent in the responses of cells has to be explained. Alternatively, these transduction sequences might simply and non-specifically prepare cells for change. Or they may enhance general flux rates or act as simple signal amplifiers with the specificity of information interpreted by some alternative means.

The sequence, stimulus – perception (receptor) – transduction – response, should be appreciated, but is purely conceptual and unlike real events. The metabolic system of cells is a complicated network and such simple ordering is very unlikely. The application of stimulus and observation of response are separable in time in that the response must follow the stimulus but there can be great flexibility in the kinetics of change of individual metabolic components (Trewavas 1986a). Perception, transduction and response are aggregate events and it is unlikely that any stimulus will have just one point of impact on the metabolic network. Furthermore, the effect of any one stimulus occurs in the context of all others. Any response which arises is the result of an integration of the complexity of all information reaching the cell from outside along with the current developmental state of the metabolic network.

3.2 The Detection of Response Requires the Construction of Suitable Assays for the Response

The data in Table 1 indicate the remarkable sensitivity which can be exhibited by plant cells to outside information. Some of this may result from the production of specialized cells but this in turn should derive from a potential present in most plant cells. Evolution has merely amplified a pre-existing potential to produce these very sensitive tissues. The detection of this potential will depend on the construction of assays for it.

Thus plant tendrils are viewed as being very sensitive to touch because there is a rapid morphological and very visible response. But it is now known that most plant cells respond to touch with much less visible but no less important metabolic and cellular changes (Jaffe and Telewski 1983). The inductive effects of fungal elicitors could only emerge when an assay for phytoalexins was constructed. In turn, the inductive effects of these carbohydrate elicitors may reflect interference in the means whereby all plant cells sense information about their own cell walls. Action potentials are often regarded only in the context of

Table 1. Sensitivity of plant cells to external stimuli[a]

Stimulus	Possible receptor	Possible location of receptor	Transduction may involve	Example of response	Threshold sensitivity	Reference
Red light	Phytochrome	Plasma membrane	Ca^{2+} membrane potential	Photomorphogenesis	6×10^4 photons m^{-2} s^{-1} (100 pW m^{-2} s^{-1})	Song (1984)
Blue light	Cryptochrome	Plasma membrane	Ca^{2+}	Phototropism	4.4×10^{13} photons m^{-2} s^{-1} (10 mW m^{-2} s^{-1})	Dennison (1979)
Nitrate and phosphate	Nitrate reductase (or nitrate or phosphate transporter)	Cytosol, plasma membrane	H^+	Root branching	10 μmol NO_3^- 10–50 μM P_i	Hackett (1972) Drew et al. (1973)
Gravity	Statolith (or microfilament)	Amyloplast	Ca^{2+} (electrical change)	Gravitropism	240 *g* s	Johnson (1965)
Water	Turgor pressure	Plasma membrane (stretch channel)	Electrical change	ABA accumulation	500 kPa reduction from ambient	Pierce and Raschke (1980)
				Hydrotropism	0.2% cm^{-1}	Hooker (1915)
CO_2	Carboxylating enzyme	Soluble	H^+	Stomatal closure	10% change in ambient CO_2	Raschke (1979)
O_2	Oxidases, NADH oxidase	Cytosol, plasma membrane	H^+	Promotion of growth, abscission	15% change in ambient O_2	Ishizawa and Esashi (1985) Addicott (1982)
Touch	Plasma membrane (stretch channel)	Plasma membrane	Ca^{2+}	Tendril coiling	25 μg	Paturi (1974)
Tension	Plasma membrane (cytoskeleton)	Plasma membrane	Ca^{2+}	Reaction wood (cell division)	50 mg	Wilson and Archer (1977)
Temperature	Plasma membrane (action potential)	Plasma membrane	H^+/Ca^{2+} (CO_2 can mimic effects sometimes)	Crocus flower opening	0.2°C	Crombie (1962)
Electrical field	Plasma membrane	Plasma membrane	Ca^{2+}	Growth change	1 μA base to tip coleoptile or 0.3 mV/cell	Schrank (1959) Jaffe and Nuccitelli (1977)
Boron	–	–	–	Wall thickness	2×10^{-7} M	Spurr (1957)

[a]The table summarizes external stimuli known to affect plant cells with suggestions of molecules involved in transduction, possible receptor and suggested localization of receptor. The list is not comprehensive.

Table 2. Sensitivity of plant cells to internally generated stimuli[a]

Stimulus	Possible receptor	Possible location of receptor	Transduction may involve	Example of response	Threshold sensitivity	Reference
Action potential	Plasma membrane	Plasma membrane	Ca^{2+}	Leaf movement	1 Action potential	Satter (1979)
Electrical field	Plasma membrane	Plasma membrane	Ca^{2+}	Tropic bending	–	Sievers and Hensel (1982)
Sucrose	–	Plasma membrane	H^+	Vascular tissue formation	10–20 mM	Jeffs and Northcote (1967)
Cell wall fragments (oligo-saccharides)	Glycoprotein	Plasma membrane	–	Vegetative bud formation	1 nM	Tranh Thanh Van et al. (1985)
Ca^{2+}	Calmodulin	Plasma membrane and soluble	Ca^{2+}	Secretion	10 μM (external) 10 μM (internal)	Stickler et al. (1981) Hepler and Wayne (1985)
pH	Wall-loosening enzymes	Cell wall	–	Cell extension	0.2 pH unit change from ambient	Rayle (1973)
Growth regulators (e.g.' abscisic acid)	Proteins	Plasma membrane	Ca^{2+}	Stomatal closure	10 nM	Tucker and Mansfield (1971) Hornberg and Weiler (1984)
Fatty acids	–	Plasma membrane	Ca^{2+}	Amylase synthesis inhibition	100 μM	Buller et al. (1976)
Galactose	–	–		Growth inhibition	1 mM	Knudson (1916)
Amino acids	Amino acid porter	Plasma membrane	H^+	Various morphological changes	200 μM	Carr (1966)
Turgor pressure	Plasma membrane	Plasma membrane	Electrical change (stretch channel)	Action potential	10% Change in turgor	Zimmermann (1977) Zimmermann and Beckers (1978)

[a] The table lists internally generated stimuli which can affect plant cells with examples of response and threshold sensitivities where known. The table is not comprehensive.

plant movements but recent data suggest that they may initiate profound (but invisible) metabolic changes in protein synthesis in response to injury (Davies 1987). Again, action potentials are the more extreme version of electrical signalling which probably occurs between all plant cells.

The data in Tables 1 and 2 indicate then the wealth of incoming information which is sensed by all plant cells. They are probably all sensitive to incoming ions, gases and chemicals, as well as to electrical, light and mechanical stimuli. Visible growth or turgor changes are only one aspect of change in response to stimulus which occurs in certain instances. Other more subtle (but morphologically invisible) metabolic events are no less important but help integrate the response of the whole plant. Responses may depend on the kinetics of the stimulus as well as its direction. There are many instances known where mature non-growing cells directly influence response in others (e.g. Brown and Wightman 1953).

4 Is There Specific Recognition of External Information at the Plant Plasma Membrane?

Signal transduction sequences usually start with recognition at the plasma membrane. For animal-derived models a specific protein receptor located in the plasma membrane is assumed to initiate the sequence. This section will briefly consider the evidence for specific recognition in plant cells. It is self-evident that the plasma membrane will see a stimulus first. It is less evident that it may see it very specifically.

4.1 Cell Surface Glycoproteins are Believed to be Involved in Recognition Phenomena in Fucus, Pollen/Stigma and Legume Root/Rhizobium Interactions

The plant plasma membrane contains large numbers of glycoproteins. It is thought that these glycoproteins can interact with each other via lock and key type mechanisms to mediate specific recognition phenomena (Heslop-Harrison 1978). The interactions are believed to occur via the carbohydrate moieties of the glycoproteins.

There is certainly evidence that glycoproteins containing fucose and mannose are involved in sperm/egg interactions in *Fucus* (Callow et al. 1982). However, it is unknown whether this interaction is sufficiently specific to explain the species discrimination inherent in fertilization or whether it may be a simple method for adherence of sperm to egg.

There is substantial evidence implicating pollen and pistil glycoproteins in pollen/stigma interactions and self-incompatibility (Knox et al. 1986). However, the stigma surface can bind non-specifically and, although genetic evidence points to a single S gene explaining incompatibility, the role of stylar S-associated glycoproteins is still uncertain. Inhibition of pollen tube growth is affected by many genes involved in pistil physiology and pollen tube growth, although the role of the S gene might alone be responsible for self-recognition (Knox et al. 1986).

There is evidently considerable specificity amongst rhizobial species for their legume hosts (Chap. 15). Although there is good evidence that glycoproteins are involved in adhesion of *Rhizobium* to root hairs, whether this adhesion is species-specific is very uncertain. Bauer (1981) comments that *Rhizobium* adheres non-specifically to many surfaces and that the role of glycoproteins in specific recognition should not be overemphasized.

In these three examples the function of glycoproteins may simply be for surface adherence; a phenomenon which may not necessarily lead to signal sensing or transduction but clearly is a necessary part of the biological response.

4.2 Complex Carbohydrates Sensed at the Plasma Membrane May be Involved in Defence Mechanisms and Development

Oligosaccharides derived from fungal and plant cell walls can elicit phytoalexin, callose and chitinase synthesis in plant cells (Chap. 14; Callow 1982; Kauss 1987). It is believed that these signals are sensed and transduced at the plasma membrane (Ebel and Grisebach 1988). Furthermore, since interactions of pathological fungi and plants are often host-specific (Daly 1984; De Wit et al. 1987) and there is inherently great potential for specificity in the information content of these oligosaccharides, it is thought that specific recognition can occur at the plasma membrane (Callow 1982). However, phytoalexin production and callose and chitinase synthesis can be induced in plant cells not normally infected by the fungus (Daly 1984). Furthermore, they can be induced by a variety of molecules like polymixin, poly-L-ornithine, digitonin, acylcarnitine and chitosan which are chemically unrelated to most fungal elicitors (Kauss 1987). The synthesis of callose is carried out by a plasma membrane-localized glucan synthase (EC 2.4.1.34) and can also be induced by simple pressure on the plasma membrane or bending of plant tissues (Kauss 1987). Although fungal elicitors are sensed at the plasma membrane, poly-L-lysine treatment leads to rapid increases in ion leakage suggesting that relatively nonspecific perturbations or stress can initiate defence reactions.

In appropriate tissues, cell wall-derived oligosaccharides can profoundly modify bud, flower and root development (Tranh Thanh Van et al. 1985). However, the same tissues respond to a great variety of experimental perturbations (e.g. container shape, Tranh Thanh Van 1981) leading to the conclusion again that there may be a lack of specificity in plasma membrane recognition.

4.3 Are Light Signals Recognized by Phytochrome in the Plasma Membrane?

It is thought that photobiologically relevant phytochrome is located in the plasma membrane (Hepler and Wayne 1985). The evidence for this view derives from experiments using polarized red light on *Mougeotia* (Haupt 1972) and from rapid red light-induced changes in the surface charge of root tips (Tanada 1968). Conversion of inactive phytochrome (P_r) to active phytochrome (P_{fr}) leads to pelletability of phytochrome believed to result from attachment to membranes (Quail 1982).

However, Speth et al. (1986) reported an exacting study of phytochrome distribution using immune localizaton. They found that P_r was diffusely distributed through cells and that exposure to red light caused the formation of aggregates located in the vicinity of the vacuole. Since green plants have only 1% of the total phytochrome of etiolated plants (Pratt 1983) the observation suggests that the pelletability or aggregation of phytochrome is a preparation for degradation. Experiments attempting to demonstrate membrane localization in higher plant cells using polarized red light are as yet unconvincing or negative (Song 1984). Definitive evidence for an in vivo interaction of phytochrome and membranes is simply lacking and phytochrome may normally be a soluble protein. In mitigation, as indicated later, it may be the case that there is always a great excess of receptor molecules; the biologically relevant phytochrome may be a tiny and easily missed proportion of the total.

4.4 Are Growth Substances Sensed at the Plasma Membrane?

Auxin, gibberellin, cytokinin, ethylene and ABA are often assumed to be sensed at the plasma membrane. These notions derive from mammalian hormone models. Hornberg and Weiler (1984) have presented the only concrete evidence for surface receptors for any growth substance (ABA). The effects of ABA can be detected in patch-clamp experiments using guard cell membranes supporting this view (Schauf and Wilson 1987). Auxin is reported to have effects on membrane potential which again might imply plasma membrane reception (Schroeder et al. 1987).

But against this must be set the fact that for mammalian hormones, surface receptors are found only for hydrophilic hormones; hydrophobic hormones have their receptors inside the cell. Plant growth substances are amphipathic and penetrate membranes easily. The key identification of an ABA surface receptor by Hornberg and Weiler (1984) has not been confirmed by other independent workers. An extensive search through many ABA-sensitive systems failed to reveal anything comparable (Smart et al. 1987). Furthermore, Lea and Collins (1979) have suggested ABA may form channels in membranes without a specific protein attachment. The only well-characterized binding protein for auxin may be located in the endoplasmic reticulum [see Venis (1985) who discusses other binding data]. Löbler and Klämbt (1985) have claimed to observe inhibition of coleoptile growth by exogenously added auxin-binding protein antibodies. This might suggest a plasma membrane locale.

Every stage of plant development which can be influenced by growth regulators can also be induced and influenced by a wide variety of chemicals such as reducing agents, dyes, minerals, organic solvents and amino acids

(Trewavas 1986b). Such evidence does not suggest a requirement for specific recognition of plant growth regulators at the plasma membrane but only that cells have the capability of sensing general perturbations at the cell surface and of acting accordingly. If plant growth regulators have very specific effects, the origin of these must be sought elsewhere.

4.5 Many Stimuli May Non-Specifically Alter Ion Flux Activity of the Plasma Membrane

The remaining stimuli listed in Tables 1 and 2 are likely to have an influence on the cell surface. Electrical fields cannot pass the insulating lipid bilayer; touch or deformation of cells should modify plasma membrane structure; turgor pressure changes are believed to alter the electrical character of the plasma membrane (Zimmermann 1977; Zimmermann and Beckers 1978); action potentials are a plasma membrane phenomenon; sucrose and other carbohydrates and nitrates may have specific transporting proteins and their transport may be coupled to proton transport. The plant plasma membrane is an excitable structure with Ca^{2+}, K^+ and Cl^- channels capable of responding to membrane potential, surface charges (changed by available ions and pH), and deformation (stretching). Alteration of channel activity may be a general, but non-specific, way in which signals are transduced across the plasma membrane but with resultant metabolic changes (Chap. 8). At present, specific surface recognition at the plant plasma membrane has simply not been established.

5 Stimulus-Response Coupling by Ca^{2+}

5.1 The Basic Model

Figure 1 describes the model that lies at the heart of much of the work on plant cell signal transduction. We have taken as an example one of the best-studied plant signal transduction mechanisms with Ca^{2+} as the second messenger and protein phosphorylation as one of the activities modified by changes at the second messenger level. However, the basic outline could equally well be applied to other second messengers (if and when they emerge) and responsive elements (which we know to exist in the case of the Ca^{2+} system).

The model then, is that a primary stimulus (for example light, growth substances or gravity) interacts with a specific receptor at a cellular membrane (usually assumed to be the plasma membrane). This interaction causes an increase in the concentration of free Ca^{2+} in the cytoplasm either by influx at the plasma membrane or Ca^{2+} release from intracellular stores. This influx amplifies the original stimulus-receptor signal that induced it. Increased levels of cytoplasmic Ca^{2+} then modulate a variety of Ca^{2+}-dependent cellular processes, such as protein phosphorylation either directly, or by forming a complex with other, regulatory proteins, such as calmodulin (Fig. 1).

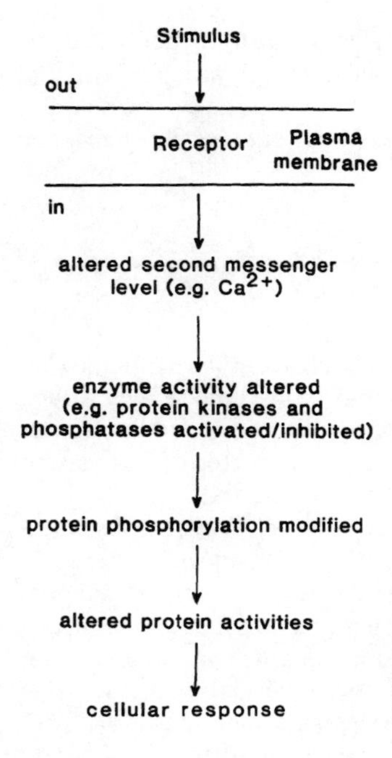

Fig. 1. Signal transduction across the plasma membrane. Although stimulus-evoked changes in the second messenger (Ca^{2+}) concentration have been shown to affect protein phosphorylation, giving rise to the physiological response, other enzyme activities and protein functions will be altered in parallel

5.2 Evidence that Ca^{2+} is Involved in Signal Transduction

A wide range of plant physiological processes are affected by treatments thought to modulate intracellular Ca^{2+} levels: Ca^{2+} ionophores have been used to alter cytoplasmic Ca^{2+} levels, whilst Ca^{2+} fluxes have been blocked with putative Ca^{2+}-channel antagonists or reduction in extracellular Ca^{2+} (Hepler and Wayne 1985). Also, changes in 'membrane-bound' Ca^{2+} and $^{45}Ca^{2+}$ fluxes have been correlated with physiological events (Hepler and Wayne 1985). Such investigations have identified a possible role for Ca^{2+} and/or Ca^{2+}-calmodulin for example in: gravitropism, stomatal responses, mitosis, polarized growth, cytoplasmic streaming, secretion, membrane fusion, moss bud induction and several phytochrome-related events. This list is by no means exhaustive. However, whether the Ca^{2+} requirement of these systems reflects its role as a second messenger in signal transduction is poorly understood.

5.3 The Concentration of Free Ca^{2+} in the Cytoplasm is Maintained at a Low Concentration

For Ca^{2+} to act as an intracellular regulator plants must closely regulate their cytoplasmic Ca^{2+} concentration, holding it well below the cytotoxic levels found extracellularly (10^{-3}M) (Hepler and Wayne 1985). However, direct measurement of the resting cytoplasmic Ca^{2+} level, and demonstration that this level is modulated by stimuli, has proved technically difficult. When possible these measurements have shown a basal Ca^{2+} level of $< 10^{-6}$ M in algal cells of *Chara, Nitella, Nitellopsis* and *Fucus*, pollen tubes and higher plant cell protoplasts (Williamson and Ashley 1982; Brownlee and Wood 1986; Gilroy et al. 1986, 1987b, 1989; Miller and Sanders 1987; Nobiling and Reiss 1987; Bush and Jones 1988). The basal Ca^{2+} level has also been observed to change and such changes correlate with physiological processes. A sustained, elevated (10^{-6} M), but spatially localized gradient of Ca^{2+} has been observed in the tip-growing cells of pollen tubes (Nobiling and Reiss 1987) or *Fucus* (Brownlee and Wood 1986) and transient changes have been measured in mitotic cells of lily and *Tradescantia* (Keith et al. 1985; Hepler and Callaham 1987), photosynthesizing *Nitellopsis* cells (Miller and Sanders 1987), osmotically regulating *Lamprothamnium* cells (Okazaki et al. 1987) and *Chara* cells undergoing an action potential (Williamson and Ashley 1982). So, in accordance with the signal transduction scheme presented in Fig. 1, plant cells do seem to closely regulate their cytoplasmic Ca^{2+} levels and modulation of this level do correlate with physiological processes and a fairly non-specific stimulus such as light.

5.4 Organelles Help Maintain a Low Cytoplasmic Concentration of Free Ca^{2+}

Fluxes in Ca^{2+} are likely to arise from modulations in either active Ca^{2+}-transporting systems or Ca^{2+} channels in cellular membranes. Mitochondria, chloroplasts and vacuoles have been demonstrated to actively accumulate Ca^{2+}, often to 10^{-3} M, either by an ATP-dependent process or via nH^+/Ca^{2+} antiporter (Moore and Åkerman 1984; Gilroy et al. 1987a). These systems generally have an affinity for Ca^{2+} of less than 10^{-6} M and may represent low-affinity, high-capacity storage sites. However, they may also contribute to changes in cytoplasmic Ca^{2+}, as shown by its depletion on illumination of *Nitellopsis* (Miller and Sanders 1987), which may represent Ca^{2+} uptake by chloroplasts. The endoplasmic reticulum and plasma membrane probably possess Ca^{2+}-transporting ATPases which have a ten-fold higher affinity for Ca^{2+} and may be the primary sites at which the basal Ca^{2+} level is set (Moore and Åkerman 1984; Gilroy et al. 1987a). Stimuli may also alter Ca^{2+} fluxes through Ca^{2+}-selective ion channels which have been tentatively identified at the plasma and vacuolar membranes (Hetherington and Trewavas 1984; Andrejauskas et al. 1985). The activity of these components of the Ca^{2+} regulatory system have been indirectly shown to respond to physiologically important stimuli, for example light increases both Ca^{2+} influx into plant cell protoplasts and modulates Ca^{2+} uptake by mitochondria and chloroplasts (Moore and Åkerman 1984; Roux et al. 1986). Evidence of the direct modulation of a Ca^{2+} channel or pump by a stimulus has yet to be reported.

5.5 Calmodulin is a Primary Ca^{2+} Receptor

The plasma membrane, endoplasmic reticulum and possibly tonoplast Ca^{2+} ATPases may be stimulated by the Ca^{2+}-dependent regulatory protein, calmodulin (Moore and Åkerman 1984; Fukumoto and Venis 1986; Briars et al. 1988; Robinson et al. 1988). Both soluble and membrane-bound calmodulin have been detected in a range of plants (Biro et al. 1984; Allan and Trewavas 1985). Calmodulin is highly conserved, spinach calmodulin differing by only 13 of 148 residues from the bovine form (Roberts et al. 1986) and plant calmodulin can often replace animal calmodulin in enzyme activation. A conformational change is induced in a calmodulin molecule when it binds three or four Ca^{2+}. The Ca^{2+}-calmodulin combination forms a ternary complex with enzymes generally causing their activation. Several Ca^{2+}-calmodulin stimulated enzymes have been found in plants. The best characterized of these enzymes are NAD^+ kinase (EC 2.7.1.23) (Anderson and Cormier 1978; Allan and Trewavas 1985), quinate:NAD oxidoreductase (EC 1.1.1.24) (Graziana et al. 1984) and ATPases (Briars et al. 1988; Robinson et al. 1988) which may represent Ca^{2+}-transporting activities of plasma, vacuolar and endoplasmic reticulum membranes (Gilroy et al. 1987a), and numerous soluble and membrane-bound protein kinases (Ranjeva and Boudet 1987). These enzymes generally have binding constants for Ca^{2+}-calmodulin of 10^{-8}–10^{-6} M or less. Cellular calmodulin concentrations are thought to be about 10^{-6} M.

5.6 Ca^{2+}- and Calmodulin-Dependent Protein Kinase

Though plant protein kinases phosphorylate endogenous proteins both in vitro and in vivo (Ranjeva and Boudet 1987) in general their substrates have not been identified. The only identified substrates of these Ca^{2+}- and Ca^{2+}- calmodulin-stimulated protein kinases are the quinate oxidoreductase of light-grown carrot cells (Graziana et al. 1983), the small subunit of ribulose-1,5-bisphosphate carboxylase (EC.4.1.1.39; Muto and Shimogawara 1985), and protein kinases themselves (Blowers and Trewavas 1987; Harmon et al. 1987). Interestingly, autophosphorylation of the membrane-bound protein kinase alters its activity towards other substrates and relieves its calmodulin stimulation (Blowers and Trewavas 1987). The general ignorance of substrates makes definite conclusions about the role of protein kinases in signal transduction difficult to make. However, theoretically at least, they are attractive as intermediates in signal transduction. Protein kinases could both prolong and further amplify a stimulus-provoked second messenger change, which as we have seen may be transient in the case of Ca^{2+}. Kinases also represent a site at which several stimuli could be integrated into the modulation/phosphorylation of similar proteins (Blowers and Trewavas 1988). The activity of animal Ca^{2+} channels is modified (inhibited) by phosphorylation (Nastainczyk et al. 1987). Thus, a Ca^{2+}-calmodulin-dependent protein kinase could modify the Ca^{2+} regulatory system that originally led to its activation. The relevance of this feedback activity in plants awaits the isolation and characterization of the plant Ca^{2+}-channel proteins. In many cases the sensitivity of animal cell protein kinase activation to second messengers is enhanced by its 'pairing' with an antagonistic phosphatase that is inhibited by those conditions that stimulated the kinase (Koshland 1987; Ranjeva and Boudet 1987). To date Ca^{2+}-calmodulin-inhibited protein phosphatases have not been reported in plant cells but may represent a profitable area for future research.

Although it is the most completely characterized Ca^{2+} receptor, calmodulin is not the only Ca^{2+}-binding protein in plants. Other Ca^{2+}-binding proteins have been noted for example in phloem (Sabnis and McEuen 1986) and carrot cells (Ranjeva et al. 1986).

6 The Role of Inositol Phospholipids in Signal Transduction

Recently much interest has centered on a possible role for inositol phospholipids in the transduction of stimuli across the plant plasma membrane. In animal cells it is now established that primary stimuli, such as hormones and neurotransmitters, bind to surface receptors on the plasma membrane and activate a polyphosphatidylinositol-specific phospholipase C (possibly transduced through a G-protein) (Berridge 1987; Pfaffmann et al. 1987). In animals, G-proteins are membrane-associated proteins that bind and hydrolyze GTP. The GTP-G protein complex acts as an intermediate between a hormone-receptor complex and membrane-bound enzyme (Gilman 1987). cAMP and cGMP phosphodiesterase are the two best-characterized animal enzymes modulated by G-proteins; although the currently held view is that cAMP does not act as an intracellular messenger in plants the way it does in animal cells. The stimulation of phosphoinositide hydrolysis and direct modulation of ion channel activity by G-proteins is still under investigation (Gilman 1987). GTP-binding proteins have been identified in bulk extracts of *Lemna paucicostata* (Hasunuma and Funadera 1987) and zucchini membranes (Drøbak et al. 1988).

Phospholipase C hydrolyzes phosphatidylinositol-4,5-bisphosphate (PIP_2), a lipid assumed to be associated with the inner face of the plasma membrane, to release diacylglycerol (DAG) and inositol-1,4,5-trisphosphate (IP_3) (Fig. 2). DAG then activates a Ca^{2+}- and phospholipid-dependent protein kinase (kinase C) in the plasma membrane, whereas the soluble IP_3 induces release of Ca^{2+} from intracellular stores such as the endoplasmic reticulum (Berridge 1987). The IP_3-induced Ca^{2+} increase would then activate the Ca^{2+}- and Ca^{2+}-calmodulin-dependent processes such as protein phosphorylation. Although the precise details of inositol metabolism may vary between plants and animals (Loewus and Loewus 1983) the essential features shown in Fig. 2 have been assumed to be closely related. What evidence is there to support a phosphoinositide-mediated signal transduction pathway at the plant plasma

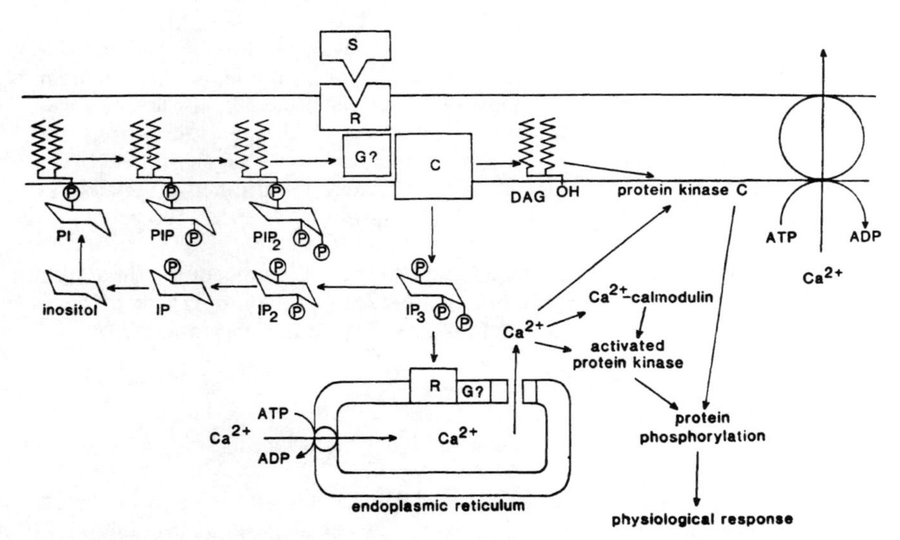

Fig. 2. Stimulus evoked turnover of phosphatidylinositol at the plasma membrane, leading to the release of intracellular Ca^{2+}. Protein phosphorylation has been taken as an example of processes known to be modulated by changes in cytoplasmic [Ca^{2+}]. See text for further details. Abbreviations: *S* stimulus; *R* receptor; *G* G-protein; *C* phospholipase C; *DAG* diacylglycerol; *IP_3* inositol-1,4,5-trisphosphate; IP_2 inositol-1,4-bisphosphate; *IP* inositol-1-phosphate; *PI* phosphatidylinositol; *PIP* phosphatidylinositol-4-phosphate; *PIP_2* phosphatidylinositol-4,5-bisphosphate. Other inositol lipids thought to have regulatory activity in animal cells (Berridge 1987) have been omitted for the sake of clarity. *G?* reflects the uncertainty of the role of G-proteins in the stimulation of phosphatidylinositol hydrolysis or gating of ion channels

membrane? Most of the reaction products (PI, PIP, PIP_2, IP_3 and DAG) have been observed in plants (Boss and Massel 1985; Connett and Hanke 1986; Heim and Wagner 1986; Sandelius and Sommarin 1986; Cote et al. 1987; Sandelius and Morré 1987) as has a phosphatidyl-inositol kinase and a PIP kinase (Sandelius and Sommarin 1986; Sommarin and Sandelius 1988). A Ca^{2+}-dependent and plasma membrane-bound PIP_2-specific phospholipase C has been identified (Melin et al. 1987). However, to date, a stimulus-evoked turnover of inositol phospholipids has only been reported on the addition of auxin to cells of *Catharanthus roseus* (Ettlinger and Lehle 1988) or isolated soybean membranes (Morré et al. 1984).

IP_3 has been shown to release $^{45}Ca^{2+}$ from isolated crude microsomal (Drøbak and Ferguson 1985) and tonoplast membranes (Schumaker and Sze 1987). However, the sensitivity and reproducibility of this release is unknown. Detectable increases in cytoplasmic Ca^{2+} levels in electro-permeabilized carrot cell protoplasts treated with up to 100 μM IP_3 have not been observed (Gilroy et al. 1989). In animal cells, phospholipase C also produces DAG from PIP_2 hydrolysis and DAG goes on to activate protein kinase C (EC 2.7.1.37). DAG has been identified as a component of plant lipids (Connett and Hanke 1986) and several investigators have observed a Ca^{2+}- and phospholipid-activated protein kinase which is antigenically similar to animal protein kinase C (Schäfer et al. 1985; Elliot and Skinner 1986; Oláh and Kiss 1986). However, a stimulus-evoked modulation of the activity of a putative kinase C has yet to be reported in plants, and needless to say, its substrates are unknown.

Phosphorylation of plant proteins on tyrosine residues has been observed (Ranjeva and Boudet 1987), although it is a very minor component of total protein phosphorylation. Tyrosine phosphorylation of animal proteins is thought to be developmentally important. The animal protein kinase C is tyrosine phoshorylated.

Thus, plant cells do possess many of the elements of the inositol phospholipid transduction pathway found in animal cells (Fig. 2). However, the evidence is too limited to date to identify these elements as a signal-transduction pathway. Although it is interesting to note that mitotic progression in *Tradescantia* root hairs, a system fairly well characterized as being Ca^{2+}-dependent (Hepler and Callaham 1987), is inhibited by Li^{+} (Tong-Ling and Wolniak 1987). Li^{+} is thought to block inositol phosphate turnover and so inhibit PIP_2 production (Fig. 2). This mitotic inhibition is reversed by both inositol and Ca^{2+} (Tong-Ling and Wolniak 1987). Tucker (1988) has also reported that cell-cell communication via plasmodesmata may be inhibited by IP_2 and IP_3 in staminal hairs of *Setcreasea purpurea*.

7 Physiological Characteristics of Sensory Systems that can be Explained by Signal Transduction Involving Ca^{2+}

Before the molecular details of signal transduction were being unravelled, physiological examination demonstrated that signal transduction must possess the properties of amplification, adaptation, thresholds, sensitivity modulation and discrimination of signals or gradients within a single cell. Using largely examples from phytochrome, the next section examines how signal transduction involving Ca^{2+} can in part explain some of these phenomena. Regardless of the arguments over phytochrome localization, conversion to P_{fr} does initiate electrical changes within a short period. These in turn are the result of changes in channel activity or will modify them. Any of the stimuli presented in Table 1 and 2 could be used, but more information is available on phytochrome.

7.1 Amplification of Weak Signals via Ca^{2+}

When lettuce seeds are irradiated within 24 h of imbibition by weak red light, dormancy can be broken. We can make estimates of the energy input from the data of Blaauw et al. (1976). About 20 nJ red light energy/seed is needed to break dormancy. The force required to break the endosperm layer (Abeles 1986), the energy output, is about 0.5 J seed. This is sufficient to lift a 20 g weight. The difference between energy input and output is about seven orders of magnitude. Thus, amplification of the energy in the signal is required. An even greater amplification occurs in a thigmotropic stimulus on a tendril. A 250 μg weight is the threshold sensitivity to initiate coiling (lower figures have been quoted). The maximum force that can be used to coil the tendril (the output) is turgor pressure. Although the calculations have to be very approximate the difference is (1) input force, 10^{-4} N m^{-2}, (2) output force, 8×10^5 N m^{-2}.

There are numerous ways in which amplification can occur in biological systems (Koshland et al. 1982; Koshland 1987). Most involve changing the conformation of a protein that turns on or turns off a processing system either by allosteric interaction or covalent modification. In the system shown in Fig. 1, opening of one Ca^{2+} channel theoretically could admit thousands of Ca^{2+} ions. In turn, each Ca^{2+} ion could activate a calmodulin-dependent protein kinase, which in turn could phosphorylate many millions of enzymes, channel proteins and cytoskeletal proteins. Whether amplification actually occurs depends on the concentrations of metabolites and the binding constants but the potential is certainly there. This method of amplification is known as amplitude amplification. An equivalent amplification could also result from the sudden appearance of large amounts of calmodulin. This is termed sensitivity amplification for reasons which will be apparent later. Both forms of amplification have been observed to occur in animal systems and there is certainly evidence to suggest that both could occur in plants. Variations of one order of magnitude in calmodulin concentration have been observed in different cells of the pea root (Allan and Trewavas 1985) for example.

In fact, the amplification available from a Ca^{2+}-based system is too large. A single sensing molecule, e.g., at 10^{-10}–10^{-9} M, could activate ten channels, each transporting 10^4 atoms of Ca^{2+} and a protein kinase could activate 10^4 molecules of enzyme (Koshland et al. 1982). Thus the product could end up at something over 1 M which is biologically unreasonable. Therefore, it is necessary to damp the amplification. This is accomplished by receptor degradation, a plasma membrane-localized Ca^{2+}-ATPase to reduce Ca^{2+} concentrations, other internal stores for cytosol Ca^{2+}, inactivation of channels by phosphorylation, binding of calmodulin to inhibitory proteins and finally protein phosphatases.

7.2 Threshold Phenomena May be Controlled by Bistable Ca^{2+}-Regulated Protein Kinases

Figure 3 shows two dose-response curves for lettuce seed germination responding to applied red light (Blaauw et al. 1976). It compares thermodormant seeds to those incubated in far-red light. An important feature in these dose-response curves, reflecting a very narrow threshold range, is the effective all-or-none character of seed germination. Some indication of the narrowness of the threshold is derived from the steepness of the response curves; germination

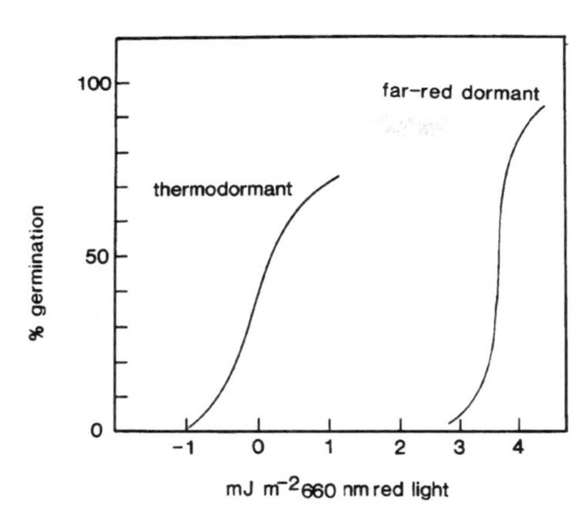

Fig. 3. Dose response curves for induction of germination of lettuce seeds by red light (660 nm). Dormancy was produced by exposure to 37°C or by exposure to far-red light. Data calculated from Blaauw et al. (1976). The scale along the x-axis is a $\log_{10}$ scale

can increase from 15–60% with a doubling of light intensity. Below the threshold for an individual seed, red light must be absorbed but is ineffective. Only when the threshold is passed does germination commence. Thus the increase in light intensity to pass the threshold and lose dormancy may be no more than 10–15% for an individual seed. How many cells need to be switched from dormancy to germination for the whole seed to germinate is an unanswered but intriguing question (Trewavas 1987b).

Discussion of the possible molecular mechanisms of thresholds can be found in Lewis et al. (1977), Lisman (1985) and Koshland (1987). Some sort of bistable enzymes are needed. Membrane-bound autophosphorylating protein kinases (and phosphatases) with a supply of ATP have properties commensurate with the required bistability, as shown by Lisman (1985). A Ca^{2+}-dependent protein kinase in which Ca^{2+} activates autophosphorylation and autophosphorylation modifies kinase activity would show a very steep switch-like character. Koshland (1987) has pointed out that a protein kinase which is saturated with its protein substrate would also show zero-order, ultrasensitive responses. Autophosphorylating protein kinases should always be saturated, since the substrate, the autophosphorylation site, which is on the enzyme itself, is always bound to the catalytic site. A plant protein kinase which has the required 'Lisman' characteristics, autophosphorylation with subsequent modification of catalytic activity, has been described by us (Blowers and Trewavas 1987). This enzyme has combined kinase and phosphatase activity, is Ca^{2+} dependent and is found in cell fractions rich in plasma membrane.

7.3 Sensory Adaptation can Occur Through Channel Phosphorylation and Variation in Calmodulin Level

A feature common to all sensory systems is that they undergo sensory adaptation to a continuous signal so as to maintain the sensing system within the most sensitive area. Figure 3 shows an example of sensory adaptation with phytochrome. Seeds maintained under far-red conditions become very much less sensitive to red light needing some four orders of magnitude higher light intensity to initiate a response. The probable reason for this is simply inhibition of phytochrome synthesis; green plants possess only 1% of the phytochrome of an etiolated plant (Pratt 1983). The lettuce seed interprets the far-red light as indicative of sunlight.

The phenomenon of sensory adaptation is best characterized in bacterial chemotaxis. Here, methylation of chemoreceptors changes the binding constants enabling the bacterial cell to retain high sensitivity to gradients at higher concentrations of the chemotactic agent (Koshland et al. 1982). Phosphorylation of other sensory receptors (rhodopsin, acetyl choline receptor and a variety of hormone receptors) may occur for the same reason. But phosphorylation is also known to inactivate Ca^{2+}-channel proteins (Reuter 1986). This will reduce the sensitivity of cells to external Ca^{2+} by limiting the Ca^{2+} input in response to an environmental stimulus. 'Fatigue' in the production of plant action potentials to a constant stimulus is well known and could result from progressive channel inactivation.

But an additional mechanism which alters the sensitivity of a cell to a signal derives from variations in receptor level and is an important property deriving from 'spare' or 'excess' receptors. This concept can best be explained by reference to Fig. 4 (Cox 1986). These data involve a three-component system, Ca^{2+}, calmodulin and phosphodiesterase (EC 3.1.16.1). It measures phosphodiesterase activation against free Ca^{2+} concentration at designated cal-

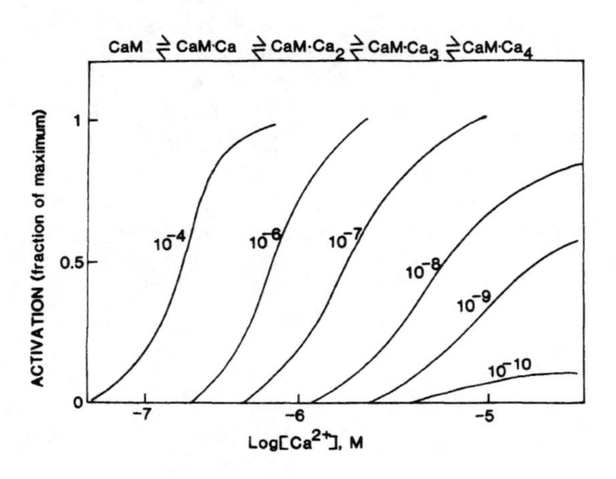

Fig. 4. Variation in phosphodiesterase activation (as fraction of maximum) with different Ca^{2+} and calmodulin concentrations. Data calculated from Cox (1986). The *numbers* by each curve are concentrations of calmodulin in M

modulin concentrations. As the calmodulin concentration is decreased, the activation level can be maintained by increasing the exogenously applied Ca^{2+}. The sensitivity to Ca^{2+} is thus altered by changing the concentration of calmodulin, the Ca^{2+} receptor. This property arises from the binding constant of Ca^{2+} to calmodulin (K_D about 10^{-5} M and the Ca^{2+}/calmodulin complex to phosphodiesterase (K_D about 10^{-9}–10^{-8} M). At any one time only a tiny proportion of the calmodulin is bound to form the active complex. Thus, there is always an apparent excess of or spare calmodulin. It is not redundant calmodulin, however, since only in the presence of this apparent excess of calmodulin can the sensitivity be modified by changing the concentration of calmodulin. The cellular concentration of calmodulin is 10^{-6}–10^{-5} M. Only a tiny proportion is directly bound to other receptor proteins at any one time.

Figure 3 can now be more clearly explained if the concept of spare phytochrome (excess light receptors) holds. Only a tiny proportion of P_{fr} is ever bound to other protiens, and is thus biologically meaningful. As the total level of phytochrome decreases, an increased light flux is necessary to produce the same amount of active bound phytochrome P_{fr}.

Similar conclusions apply equally to hormones and quite probably plant growth substances (Venis 1985; Hollenberg and Goren 1986). The responses to plant growth substance are usually dominated by chemical equilibrium thinking, i.e. a fixed binding affinity for a growth substance with its receptor. Since the sensitivity of plant tissues to plant growth regulators frequently changes (Trewavas 1987a), a similar type of mechanism to the above can be proposed and 'spare' receptors may be responsible.

7.4 Signal Destruction by Noise May be Mitigated by Ca^{2+} Oscillations

One of the basic problems faced in any biological signalling system is destruction by noise, that is random fluctuations in the chemical apparatus sensing and interpreting the signals. Similar problems were faced by experimentalists attempting to measure the tiny electrical fields present around living organisms. Electrical equipment is normally too noisy to permit the detection of tiny variations in electrical fields. The problem was solved by allowing the signal to oscillate, a characteristic not shared by noise (Nuccitelli and Jaffe 1975).

Perhaps for the same reason it has been found that after perturbation there is an oscillation of cytoplasmic Ca^{2+} rather than a single changed level (Berridge 1987). This oscillation should greatly improve the accuracy of detection of low levels of Ca^{2+} against a variable background. Many plant nutational movements may occur for a similar reason. The continued oscillation diminishes sensory adaptation and improves the sensing of fine changes in the environment.

The origin of Ca^{2+} oscillations may be found in simple negative feedback phenomena. Entry of Ca^{2+} activates protein kinases which in turn phosphorylate and inactivate channels. Coupled with a phosphatase, oscillation of cytosol Ca^{2+} should then occur. Different signals and strength of signals may then produce oscillations of different wavelength; a rich source of information to the cell.

7.5 Detection of Gradients Across Plant Cells can Occur Through Local Clustering of Channels and Pimps

The developing *Fucus* zygote can detect gradients across itself in a number of environmental variables, e.g., light, temperature, electricity, ions, pH, auxin, pressure, flow etc. (Jaffe 1969). The growing zygote responds to these gradients by aligning the future axis of polarity. The mechanism by which this is believed to occur is by a differentiation of proteins of the plasma membrane, a separation of channels and pump, which lead to selective Ca^{2+} entry at the future growth pole. The process is believed to be autocatalytic, to require amplification of a labile polarity and to lead to the development of an asymmetric electrical field in which positive charge (in part carried by Ca^{2+}) enters the growth pole (Jaffe 1986). This field is much to small to influence other zygotes but that generated by other growing asymmetric structures such as pollen tubes is considerably larger and probably able to influence other cells nearby. The growing pollen tube also has a higher concentration of Ca^{2+} in the cytoplasm of the tip (Jaffe 1986; Nobiling and Reiss 1987). The *Fucus* zygote is an excellent model system explaining how gradients are sensed in other plant cells.

The mechanism of channel and pump separation is not understood, although the involvement of the cytoskeleton suggests a mechanism involving contractile elements or the stabilization of labile patches by attached microfilaments (Chap. 12; Trewavas 1982). However, because Ca^{2+} is a relatively immobile ion in the cytoplasm, the entry point acquires a net positive charge which may cause the electrophoretic movement of proteins either in the plasma membrane or cytoplasm. Alternatively, the higher Ca^{2+} level will accelerate Ca^{2+}-dependent reactions or lead to microtubule and microfilament orientation.

8 Final Comments

We have tried to indicate the present level of understanding of signal perception and transduction across the plant plasma membrane. Evidence concerning specific signal perception at the plant plasma membrane is weak. We have tried to indicate reasons for doubting a strict analogy to animal cells and have suggested instead that specific recognition may occur inside the cell. In that case the Ca^{2+} relationships of the plant plasma membrane have to be explained.

The best analogy may be that of a valve in a radio set. Whatever signal is picked up is amplified, but the specificity, the specific information, is still retained. Ca^{2+} may act as a metabolic amplifier; the specificity of the signal may reside in the unique spectrum of metabolic events each molecule will affect. Fixing on a sequence, receptor-transduction-response gives no hint of other specific metabolic changes which must underlie the apparent specificity of response. We have tried to indicate ways out of this impasse; temporal variation in sensitivity and differing metabolic environments or contexts in which each signal is interpreted. Signal perception may also use a variety of receptors and specific speeds of oscillation of the transduction molecule, Ca^{2+}, itself. There are certainly a variety of challenging possibilities as various as the signals to which plant cells are sensitive.

Acknowledgements. Unpublished work was supported by the A.F.R.C.

References

Abeles FB (1986) Role of ethylene in *Lactuca sativa* cv. Grand Rapids seed germination. Plant Physiol 81:780–787

Addicott FT (1982) Abscission. University of California Press, Berkeley

Alkon DL, Rasmussen M (1988) A spatial-temporal model of cell activation. Science 239:998–1004

Allan E, Trewavas AJ (1985) Quantitative changes in calmodulin and NAD kinase during early cell development in the root apex of *Pisum sativum*. Planta 165:493–501

Anderson SM, Cormier MJ (1978) Calcium dependent regulation of NAD kinase in higher plants. Biochem Biophys Res Commun 84:595–599

Andrejauskas E, Mertel R, Marmé D (1985) Specific binding of the calcium antagonist ^{3}H-verapamil to membrane fractions from plants. J Biol Chem 260:5411–5414

Bauer WD (1981) Infection of legumes by rhizobia. Annu Rev Plant Physiol 32:407–449

Bernier G (1986) The flowering process as an example of plastid development. In: Jennings D, Trewavas AJ (eds) Plasticity in plants. Company of Biologists. Society for Experimental Biology, pp 257–287

Berridge JB (1987) Inositol triphosphate and diacylglycerol: two interacting second messengers. Annu Rev Biochem 56:159–193

Biro RL, Daye S, Serlin B, Terry ME, Datta N, Sopory BK, Roux S (1984) Characterisation of oat calmodulin and radioimmunoassay of its subcellular localisation. Plant Physiol 75:382–391

Blaauw OH, Blaauw-Jansen G, Elgersma O (1976) Determination of hit numbers from dose response curves for phytochrome control of seed germination. Acta Bot Neerl 25:341–348

Blakeley LM, Rodaway SJ, Hollen LB, Croker SG (1972) Control and kinetics of branch root formation in cultured root segments of *Haplopappus*. Plant Physiol 50:35–42

Blowers DP, Trewavas AJ (1987) Autophosphorylation of plasma membrane bound calcium/calmodulin dependent protein kinase from pea seedlings. Biochem Biophys Res Commun 143:691–696

Blowers D, Trewavas AJ (1988) Second messengers: their existence and relationship to protein kinases. In: Boss W, Morré DJ (eds) Second messengers in plant growth and development. Alan R Liss, New York, NY, pp 1–28

Boss WF, Massel MO (1985) Polyphosphoinositides are present in plant tissue culture cells. Biochem Biophys Res Commun 132:1018–1023

Briars SA, Kessler F, Evans DE (1988) The calmodulin-stimulated ATPase of maize coleoptiles is a 140,000 M_r polypeptide. Planta 176:283–285

Brown R, Wightman F (1953) The effect of mature tissue on division in the meristem of the root. J Exp Bot 3:253–263

Brownlee C, Wood JW (1986) A gradient of cytoplasmic free calcium in growing rhizoid cells of *Fucus serratus*. Nature 320:624–626

Buller DC, Parker W, Reid JSG (1976) Short chain fatty acids as inhibitors of gibberellin induced amylolysis in barley endosperm. Nature 260:169–170
Bush DS, Jones RL (1988) Measurement of cytoplasmic calcium in aleurone protoplasts using Indo-1 and Fura-2. Cell Calcium 8:455–472
Callow JA (1982) Molecular aspects of fungal infection. In: Smith H, Grierson D (eds) The molecular biology of plant development. Blackwells, Edinburgh, pp 467–496
Callow ME, Evans LV, Callow JA (1982) Fucus. In: Smith H, Grierson D (eds) The molecular biology of plant development. Blackwells, Edinburgh, pp 159–185
Carr DJ (1966) Metabolic and hormonal regulation of growth and development. In: Cutter EG (ed) Trends in plant morphogenesis. Plenum, New York, NY, pp 253–283
Connett RJA, Hanke DE (1986) Breakdown of phosphatidylinositol in soybean callus. Planta 169:216–221
Cote GG, Morse MJ, Crain RC, Salter RL (1987) Isolation of soluble metabolites of the phosphatidylinositol cycle from *Samanea saman*. Plant Cell Rep 6:352–355
Cox JA (1986) Calcium-calmodulin interaction and cellular function. J Cardiovasc Pharmacol 8 (Suppl 6):545–551
Crombie WML (1962) Thermonasky. In: Bunning E (ed) Encyclopedia of plant physiology, Vol XVII/2. Springer, Berlin Heidelberg New York, pp 15–28
Daly JM (1984) The role of recognition in plant disease. Annu Rev Phytopathol 22:273–307
Davies E (1987) Plant responses to wounding. In: Davies DD (ed) The biochemistry of plants. A comprehensive treatise, Vol. 12, Physiology of metabolism. Academic Press, New York NY, pp 243–264
Dennison DG (1979) Phototropism. In: Haupt W, Feinleib ME (eds) Encyclopedia of plant physiology new series, Vol 7. Physiology of movements. Springer, Berlin Heidelberg New York, pp 506–566
De Wit PJGM, Hofman AE, Velthuis CM, Toma IMJ (1987) Specificity of active defense responses in plant fungus interactions: tomato-*Cladosporium fulvum*, a case study. Plant Physiol Biochem 25:345–351
Drew MC, Saker R, Ashley TW (1973) Nutrient supply and the growth of the seminal root system in barley. J Exp Bot 24:1189–1202
Drøbak BK, Ferguson IB (1985) Release of Ca^{2+} from plant hypocotyl microsomes by inositol 1,4,5-triphosphate. Biochem Biophys Res Commun 130:1241–1246
Drøbak BK, Allan EF, Comesford JG, Roberts K, Dawson AP (1988) Presence of guanine nucleotide-binding proteins in a plant hypocotyl microsomal fraction. Biochem Biophys Res Commun 150:899–903
Ebel J, Grisebach H (1988) Defense strategies of soybean against the fungus *Phytophthora megasperma,* a molecular analysis. Trends Biochem Sci 13:23–27
Elliott DC, Skinner JD (1986) Calcium-dependent, phospholipid activated protein kinase in plants. Phytochemistry 25:39–44
Ettlinger C, Lehle L (1988) Auxin induces rapid changes in phosphatidylinositol metabolites. Nature 331:176–178
Fukumoto M, Venis MA (1986) ATP-dependent calcium transport in tonoplast vesicles from apple fruits. Plant Cell Physiol 27:491–497
Gilman AG (1987) G-proteins: tranducers of receptor generated signals. Annu Rev Biochem 56:615–649
Gilroy S, Hughes WA, Trewavas AJ (1986) The measurement of intracellular calcium levels in protoplasts from higher plant cells. FEBS Lett 199:217–221
Gilroy S, Blowers DP, Trewavas AJ (1987a) Calcium: a regulation system emerges in plant cells. Development 100:181–184
Gilroy S, Hughes WA, Trewavas AJ (1987b) Calmodulin antagonists increase free cytosolic calcium in plant protoplasts in vivo. FEBS Lett 212:133–137
Gilroy S, Hughes WA, Trewavas AJ (1989) A comparison between Quin-2 and Aequorin as indicators of cytoplasmic calcium levels in higher plant cell protoplasts. Plant Physiol 90:482–491

Graziana A, Ranjeva R, Boudet AM (1983) Provoked changes in intracellular Ca^{2+} controlled protein phosphorylation and activity of quinate NAD^{+} oxidoreductase in carrot cells. FEBS Lett 156:325–328
Graziana A, Dillenschneider M, Ranjeva R (1984) A calcium binding protein is a regulatory subunit of QORase from dark grown carrot cells. Biochem Biophys Res Commun 125:774–783
Hackett C (1972) A method of applying nutrients locally to roots under controlled conditions and some morphological effects of locally applied nitrate on the branching of wheat roots. Aust J Biol Sci 25:1169–1180
Harmon AC, Putnam-Evans C, Cormier MJ (1987) A calcium-dependent but calmodulin-independent protein kinase from soybean. Plant Physiol 83:830–837
Hasunuma K, Funadera K (1987) GTP-binding protein(s) in green plant *Lemna paucicostata*. Biochem Biophys Res Commun 143:908–912
Haupt W (1972) Localisation of phytochrome within the cell. In: Mitrakos K, Shropshire W (eds) Phytochrome. Academic Press, London, pp 223–243
Heim S, Wagner KG (1986) Evidence of phosphorylated phosphatidylinositols in the growth cycle of suspension cultured plant cells. Biochem Biophys Res Commun 134:1175–1181
Hepler PK, Callaham DA (1987) Free calcium increases during anaphase in stamen hair cells of *Tradescantia*. J Cell Biol 105:2137–2145
Hepler PK, Wayne RO (1985) Calcium and plant development. Annu Rev Plant Physiol 36:397–439
Heslop-Harrison J (1978) Cellular recognition systems in plants. E Arnold, London, pp 1–58
Hetherington AM, Trewavas AJ (1984) Binding of a nitrendipine, a calcium channel blocker, to pea shoot membranes. Plant Sci Lett 35:109–113
Hollenberg MD, Goren HJ (1986) Ligand-receptor interactions at the cell surface. In: Poste G, Crooke ST (eds) Mechanisms of receptor regulation. Plenum, New York, NY, pp 323–373
Hooker HD (1915) Hydrotropism in roots of *Lupinus albus*. Ann Bot 29:265–283
Hornberg C, Weiler EW (1984) High affinity binding sites for abscisic acid on the plasma membrane of *Vicia faba* guard cells. Nature 310:321–322
Ishizawa K, Esashi Y (1985) Gaseous factors involved in the enhanced elongation of rice coleoptiles under water. Plant Cell Environ 7:239–245
Jaffe LF (1969) Localisation in the developing *Fucus* egg and the general role of localising currents. Adv Morphogen 7:295–328
Jaffe LF (1986) Calcium and morphogenetic fields. In: Calcium and the cell. Ciba Foundation Symposium 122. Wiley Interscience, New York, NY, pp 271–281
Jaffe LF, Nuccitelli R (1977) Electrical controls of development. Annu Rev Biophys Bioeng 6:445–476
Jaffe MJ, Telewski FW (1983) Thigmomorphogenesis: callose and ethylene in the hardening of mechanically stressed plants. In: Timmerman BN, Steelink C, Loewus FA (eds) Recent advances in phytochemistry, Vol 18. Plenum, New York, NY, pp 79–95
Jeffs RA, Northcote DH (1967) The influence of IAA and sugar on the pattern of induced differentiation in plant tissue culture. J Cell Sci 2:77–88
Johnson A (1965) Investigations of the reciprocity rule by means of geotropic and geoelectric measurements. Physiol Plant 18:945–967
Kauss H (1987) Some aspects of calcium dependent regulation in plant metabolism. Annu Rev Plant Physiol 38:47–72
Keith CH, Raten R, Maxfield FR, Bajer A, Selanski ML (1985) Local cytoplasmic calcium gradients in living mitotic cells. Nature 316:848–850
Knox RB, Williams EG, Dumas C (1986) Pollen, pistil and reproductive function in crop plants. In: Janich J (ed) Plant breeding reviews, Vol 4. Avi Publishing, Westport, CN, pp 9–79
Knudson L (1916) Influence of certain carbohydrates on green plants. Cornell Univ Agric Exp Stat Mem 9:747–813
Koshland DE (1987) Switches, thresholds and ultrasensitivity. Trends Biochem Sci 12:225–229
Koshland DE, Goldbeter A, Stock JB (1982) Amplification and adaptation in regulatory and sensory systems. Science 217:220–225

Lea EJA, Collins JC (1979) The effect of the plant hormone abscisic acid on lipid bilayer membranes. New Phytol 82:11–18

Lewis J, Slack JMW, Wolpert L (1977) Thresholds in development. J Theor Biol 65:579–590

Lisman JE (1985) A mechanism for memory storage insensitive to molecular turnover: a bistable autophosphorylating kinase. Proc Natl Acad Sci USA 82:3055–3057

Löbler M. Klämbt D (1985) Localisation of a putative auxin receptor. J Biol Chem 260:9854–9860

Loewus FA, Loewus MW (1983) Myo-inositol: its biosynthesis and metabolism. Annu Rev Plant Physiol 34:137–161

Maclean N, Hall BK (1987) Cell commitment and differentiation. Cambridge University Press, Cambridge, pp 1–244

Melin P-M, Sommarin M, Sandelius AS, Jergil B (1987) Identification of Ca^{2+}-stimulated polyphosphoinositide phospholipase C in isolated plant plasma membranes. FEBS Lett 223: 87–91

Miller AJ, Sanders D (1987) Depletion of cytosolic free calcium induced by photosynthesis. Nature 326:397–400

Moore AL, Åkerman KEO (1984) Calcium and plant organelles. Plant Cell Environ 7:423–429

Morré DJ, Gripshover B, Monroe A, Morré JT (1984) Phosphatidylinositol turnover in isolated soybean membranes stimulated by the synthetic growth hormone 2,4-dinitro-phenoxyacetic acid. J Biol Chem 259:15364–15368

Muto S, Shimogawara K (1985) Calcium and phospholipid-dependent phosphorylation of ribulose-1,5-bisphosphate carboxylase/oxygenase small subunit by a chloroplast envelope-bound protein kinase in situ. FEBS Lett 193:88–92

Nastainczyk W, Rohrkasten A, Sieber M, Rudolph C, Schachtele C, Marmé D, Hofmann F (1987) Phosphorylation of the purified receptor for calcium channel blockers by cAMP kinase and protein kinase C. Eur J Biochem 169:137–142

Newell PC, Europe-Finner GN, Small NV (1987) Signal transduction during amoebal chemotaxis of *Dictyostelium discoideum*. Microbiol Sci 4:5–11

Nobiling R, Reiss H-D (1987) Quantitative analysis of calcium gradients and activity in growing pollen tubes of *Lilium longiflorum*. Protoplasma 139:20–24

Nuccitelli R, Jaffe LF (1975) The pulse current generated by developing fucoid eggs. J Cell Biol 64:636–643

Okazaki Y, Yoshimoto Y, Hirarnote Y, Tazawa M (1987) Turgor regulation and cytoplasmic free Ca^{2+} in the alga *Lamprothanium*. Protoplasma 140:67–71

Olah Z, Kiss Z (1986) Occurrence of lipid and phorbol ester activated protein kinase in wheat cells. FEBS Lett 195:33–37

Paturi FZ (1974) Nature, mother of invention. Pelican, Penguin Books, London, pp 1–156

Pfaffmann H, Hartmann E, Brightman AO, Morré DJ (1987) Phosphatidylinositol specific phospholipase C of plant stems. Plant Physiol 85:1151–1155

Pierce M, Raschke K (1980) Correlation between loss of turgor and accumulation of abscisic acid in detached leaves. Planta 148:174–182

Pratt LH (1983) Phytochrome: the protein moiety. Annu Rev Plant Physiol 33:557–582

Quail PH (1982) Intracellular localisation of phytochrome. In: Helene C, Charbier M, Montenay Garestien Th, Lanstriat S (eds) Trends in photobiology. Plenum, New York, NY, pp 485–500

Quatrano KS (1978) Development of cell polarity. Annu Rev Plant Physiol 29:487–510

Ranjeva R, Boudet AM (1987) Phosphorylation of proteins in plants: regulatory effects and potential involvement in stimulus/response coupling. Annu Rev Plant Physiol 38:73–93

Ranjeva R, Graziana A, Dillenschneider M, Charpentean M, Boudet AM (1986) A novel plant calciprotein as transient subunit of enzymes. In: Trewavas AJ (ed) Molecular and cellular aspects of calcium in plant development. Plenum, New York, NY, pp 41–48

Raschke K (1979) Movements using turgor mechanisms. In: Haupt W, Feinleib ME (eds) Encyclopedia of plant physiology, new series, Vol 7. Physiology of movements. Springer, Berlin Heidelberg New York, pp 383–441

Rayle DL (1973) Auxin induced hydrogen ion secretion in *Avena* coleoptiles and its implications. Planta 114:63–73

Reuter HW (1986) Voltage dependent mechanisms for raising intracellular free calcium concentration via calcium channels. In: Cheung WY (ed) Calcium and the cell. Ciba Foundation Symposium 122. Wiley Interscience, London, pp 5–23
Roberts DM, Lokas TJ, Harrington HM, Watterson DM (1986) Molecular mechanisms of calmodulin action. In: Trewavas AJ (ed) Molecular and cellular aspects of calcium in plant development. Plenum, New York, NY, pp 11–18
Robinson C, Larsson C, Buckhout TJ (1988) Identification of a calmodulin-stimulated (Ca^{2+} + Mg^{2+})-ATPase in a plasma membrane fraction isolated from maize (*Zea mays*) leaves. Physiol Plant 72:177–184
Roux SJ, Wayne RO, Datta N (1986) Role of calcium ions in phytochrome responses: an update. Physiol Plant 66:344–348
Sabnis DD, McEuen AR (1986) Calcium and calcium binding proteins in phloem. In: Trewavas AJ (ed) Molecular and cellular aspects of calcium in plant development. Plenum, New York, pp 33–40
Sandelius AS, Morré DJ (1987) Characteristics of a phosphatidylinositol exchange activity of soybean microsomes. Plant Physiol 84:1022–1027
Sandelius AS, Sommarin M (1986) Phosphorylation of phosphatidylinositols in isolated plant membranes. FEBS Lett 201:282–286
Satter RL (1979) Leaf movements and tendril curling. In: Haupt W, Feinleib ME (eds) Encyclopedia of plant physiology, new series, Vol. 7, Physiology of movement. Springer, Berlin Heidelberg New York, pp 442–484
Schäfer A, Bygrave R, Matzenaver S, Marmé D (1985) Identification of a calcium- and phospholipid-dependent kinase in plant tissue. FEBS Lett 208:25–29
Schauf CL, Wilson KJ (1987) Effect of abscisic acid on K^+ channels in *Vicia faba* guard cell protoplasts. Biochem Biophys Res Commun 145:284–290
Schrank AR (1959) Electronasty and electrotropism. In: Bunning E (ed) Encyclopedia of plant physiology, Vol. 17, Springer, Berlin Göttingen Heidelberg, pp 148–168
Schroeder JI, Raschke K, Neher E (1987) Voltage dependence of K^+ channels in guard cell protoplasts. Proc Natl Acad Sci USA 84:4108–4112
Schumaker S, Sze H (1987) Inositol-1,4,5-P_3 releases Ca^{2+} from vacuolar membrane residues of oat roots. J Biol Chem 262:3944–3946
Sievers A, Hensel W (1982) Nature of graviperception. In: Wareing PF (ed) Plant growth substances. Academic Press, London, pp 497–506
Smart C, Longlang J, Trewavas AJ (1987) The turion: a biological probe for the molecular action of abscisic acid. In: Fox JE, Jacobs M (eds) Molecular biology of plant growth control. Alan R Liss, New York, NY, pp 345–359
Smith JA, Martin L (1973) Do cells cycle? Proc Natl Acad Sci USA 70:1263–1267
Sommarin M, Sandelius AS (1988) Phosphatydylinositol and phosphatidylinositol-phosphate kinases in plant plasma membranes. Biochim Biophys Acta 958:268–278
Song PG (1984) Phytochrome. In: Wilkins MB (ed) Advanced plant physiology. Pitman, Bath, pp 354–375
Speth V, Otto V, Schäfer E (1986) Intracellular localisation of phytochrome in oat coleoptiles by electron microscopy. Planta 168:299–304
Spurr AR (1957) The effect of boron on cell wall structure in celery. Am J Bot 44:637–650
Stickler L, Penel C, Greppin H (1981) Calcium requirement for the secretion of peroxidases by plant cell suspensions. J Cell Sci 48:354–353
Tanada T (1968) A rapid photoreversible response of barley root tips in the presence of 3-indoleacetic acid. Proc Natl Acad Sci USA 59:376–380
Tong-Ling LC, Wolniak SM (1987) Lithium induces cell plate dispersion during cytokinesis in *Tradescantia*. Protoplasm 141:56–63
Tranh Thanh Van K (1981) Control of morphogenesis in in vitro cultures. Annu Rev Plant Physiol 32:291–313
Tranh Thanh Van K, Toubarb P, Cousson A, Darvill AG, Gollin DJ, Chelf P, Albersheim P (1985) Manipulation of the morphogenetic pathways of tobacco explants by oligosaccharins. Nature 314:615–617

Trewavas AJ (1982) Possible control points in plant development. In: Smith H, Grierson D (eds) The molecular biology of plant development. Blackwells, Edinburgh, pp 7–28
Trewavas AJ (1986a) Understanding the control of development and the role of growth substances. Aust J Plant Physiol 13:447–457
Trewavas AJ (1986b) Resource allocation under poor growth conditions. A Major role for growth substances in development plasticity. In: Jennings DH, Trewavas AJ (eds) Plasticity in plants. Company of Biologists, Cambridge, pp 31–37
Trewavas AJ (1987a) Sensitivity and sensory adaptation in growth substance responses. In: Lenton JR, Jackson MB, Atkin RK (eds) Hormone action in plant development. A critical appraisal. Butterworths, London, p 19–37
Trewavas AJ (1987b) Timing and memory processes in seed embryo dormancy – a conceptual paradigm for plant development questions. Bioessays 6:87–92
Trewavas AJ, Jennings D (1986) Introduction. In: Jennings D, Trewavas AJ (eds) Plasticity in plants. Company of Biologists, Cambridge, pp 1–4
Trewavas AJ, Kelly P, Sexton R (1984) Polarity, calcium and abscission: molecular bases for developmental plasticity in plants. J Embryol Exp Morphol 83 (suppl):179–195
Tsien RY, Poenie M (1986) Fluorescence ratio imaging: a new window on intracellular signalling. Trends Biochem Sci 11:450–455
Tucker EB (1988) Inositol bisphosphate and inositol trisphosphate inhibit cell-to-cell passage of carboxyfluorescein in staminal hairs of *Setcreasea purpurea*. Planta 174:358–363
Tucker DJ, Mansfield TA (1971) A simple bioassay for detecting antitranspirant activity of naturally-occurring compounds such as abscisic acid. Planta 98:157–163
Venis M (1985) Hormone binding sites in plants. Longman, New York, NY, pp 1–190
Williamson RE, Ashley CC (1982) Free Ca^{2+} and cytoplasmic streaming in the alga *Chara*. Nature 296:647–651
Wilson BF, Archer RR (1977) Reaction wood: induction and mechanical action. Annu Rev Plant Physiol 28:23–43
Zimmermann U (1977) Cell turgor pressure regulation and turgor pressure-mediated transport processes. Symp Soc Exp Biol 31:117–154
Zimmermann U, Beckers F (1978) Generation of action potentials in *Chara corollina* by turgor pressure changes. Planta 138:173–179

Chapter 10 Coated Pits

D.G. Robinson and S. Hillmer[1]

1 Types of Membrane Coating

Over the last decade it has become increasingly clear that the cytoplasmic surface of a number of membranes is coated. Based on their appearance in thin section one can distinguish between at least three different types of coating (Fig. 1). Common to all of these types, however, is their discrete nature: they never cover extensive portions of a membrane and they are usually restricted to regions where vesiculation is occurring or is about to occur.

The most frequently occurring type is the clathrin coat which consists of a group of 10–20 nm spikes or bristles jutting out at right angles to the membrane surface. When, in the literature, reference is made to coated vesicles/coated pits it is this kind of coating which the author usually has in mind. Coats of this type were first recognized in the 1960s and can safely be claimed to occur ubiqui-

[1]Abteilung Cytologie, Pflanzenphysiologisches Institut und Botanischer Garten, Universität Göttingen, Untere Karspüle 2, D-3400 Göttingen, FRG

C. Larsson, I.M. Møller (Eds):
The Plant Plasma Membrane

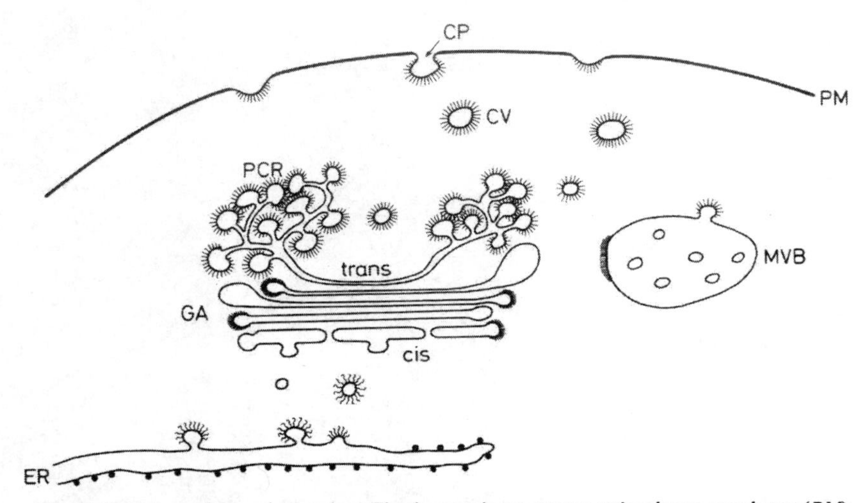

Fig. 1. Coated membranes in the plant. The three major coat types at the plasma membrane (*PM*), Golgi apparatus (*GA*) with *cis* (forming) and *trans* (maturing) faces, and endoplasmic reticulum (*ER*) are depicted according to their appearance in thin section. *CP* coated pit; *CV* coated vesicle; *MVB* multivesicular body; *PCR* partially coated reticulum

tously among eukaryotic cells (for a historical review, see Ockleford and Whyte 1980). In confirmation of morphological evidence, immunocytochemical data obtained on animal cells has shown that clathrin coats are principally present at the plasma membrane (Anderson et al. 1978; Wehland et al. 1982), and at the trans pole of the Golgi apparatus (Griffiths et al. 1981; Croze et al. 1982; Orci et al. 1985).

Although the corresponding immunocytochemistry has not yet been performed, their similar appearance in thin section allows one to predict with some degree of confidence that the same distribution of clathrin coats occurs in plant cells. There is, however, some controversy as to the relationship between the Golgi apparatus and a putative clathrin-coated structure in plant cells, known as the partially coated reticulum (Fig. 1). Whereas some workers (Pesacreta and Lucas 1985; Tanchak et al. 1988) have found direct connections between the Golgi apparatus and the partially coated reticulum to be both infrequent and difficult to discern, the present authors (Hillmer et al. 1988) have been able to clearly depict such connections. In addition, we postulated that free-lying, partially coated reticula develop from the sloughed-off trans-most cisternae of the Golgi apparatus.

A second type of membrane coating has been found associated with the membranes of the Golgi apparatus of animal cells. This coating appears, in section, to be more densely packed than the clathrin-type coat and does not extend into the cytosol as far as the latter. It is located on the surface of budding vesicles at the tubular periphery of medial and cis cisternae and is antigenically different from clathrin (Orci et al. 1986). This type of coating has not yet been recognized in the Golgi apparatus of plant cells but a closer inspection of published micrographs (e.g., Fig. 9B in Robinson 1984; Fig. 2A in Coleman et al. 1987) suggests that it may have been confused in the past with the clathrin type of coated membrane.

The third major type of membrane coating is characteristic of budding profiles at the endoplasmic reticulum and of the so-called transition vesicles, which are formed as a result of the vesiculation process occurring at this organelle. In section the coating on these membranes differs from that of clathrin in that it has a fuzzy or hairy rather than a bristle or spike-like appearance. This kind of coating has been observed frequently in both animal (e.g., Mollenhauer et al. 1976) and plant (e.g., Gunning and Steer 1975; Hübner et al. 1985) cells, but is particularly well preserved in algae (e.g., Zhang and Robinson 1986) and ciliates (e.g., Franke et al. 1971), probably as a result of better fixation.

2 Clathrin Coats

A number of different proteins constitute the clathrin coat referred to above (for a recent review see Robinson and Depta 1988). These proteins are relatively easy to dissociate from the surface of the membrane, e.g., by 2 M urea or 0.5 M Tris-HCI treatments, and spontaneously reassemble into so-called cages upon removal of the dissociating agent through dialysis (e.g., Woodward and Roth 1978). Such cages possess a polygonal architecture identical to that present in the original clathrin coat (Figs. 2 and 3c-e). Responsible for this unique morpholgy is the packing geometry of the subunits, known as triskelions (Figs. 2a and 3). Although the length of the polygon edges appears to be constant in the various eukaryotic cell types (around 15 nm; Heuser and Kirchhausen 1985; Coleman et al. 1987), the total leg length of triskelions from fungal (Mueller and Branton

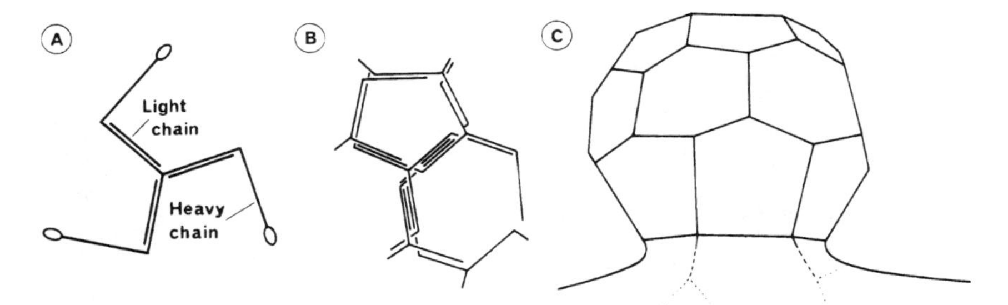

Fig. 2A-C. Diagrammatic representation of clathrin coating. **A** The basic building block, the triskelion; **B** hexagonal and pentagonal units form as a result of the particular way in which triskelions assemble with one another; **C** polygonal architecture at the surface of a coated pit

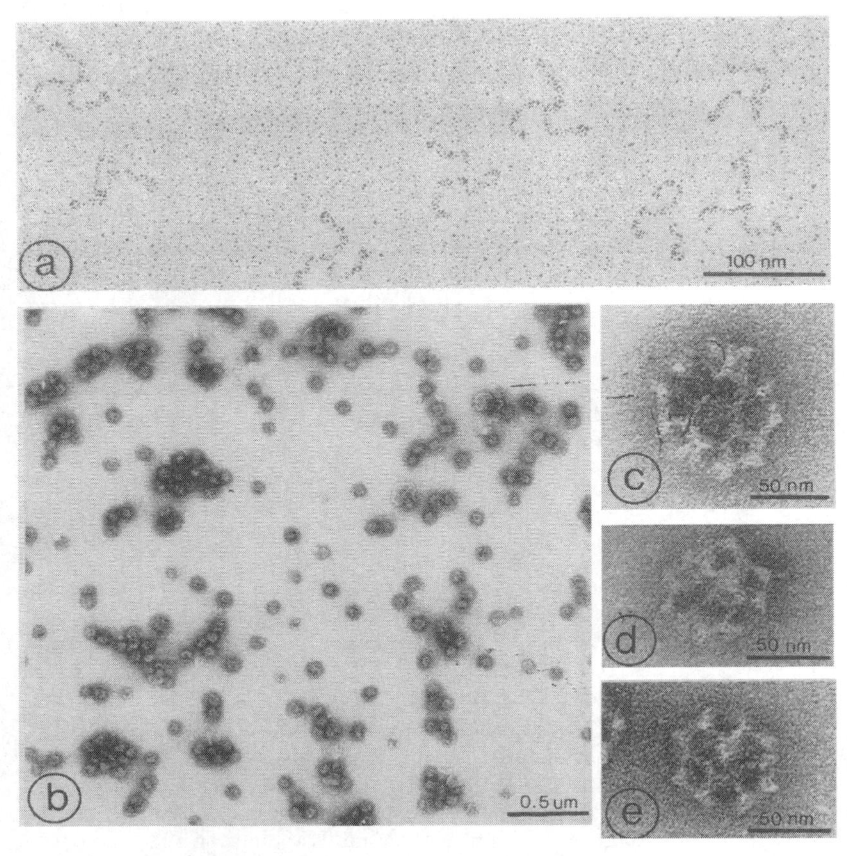

Fig. 3. a Triskelions isolated from bovine brain coated vesicles (micrograph courtesy of E. Ungewickell). **b** Coated vesicles isolated from bean leaves according to the procedure described in Depta et al. (1987b); negative stain preparation. **c-e** Clathrin cages from zucchini hypocotyl tissue

1984) and plant sources (Coleman et al. 1987) is some 10–20% greater than their animal counterparts.

Each triskelion is a hexamer consisting of three heavy chain and three light chain clathrin polypeptides. In animals the molecular mass of the former is 180 kD whereas in plants and fungi it is some 10 kD larger (Mersey et al. 1982; Mueller and Branton 1984; Depta and Robinson 1986; Coleman et al. 1987). Two different light chain polypeptides, which in animal cells are coded for by different genes (Jackson et al. 1987), are present in each triskelion, bound near the joints in the triskelion legs (Blank and Brodsky 1987). Their molecular masses differ according to the tissue type investigated (Brodsky and Parham 1983); in the most frequently used mammalian tissue, the brain, the light chains are 33 and 36 kD.

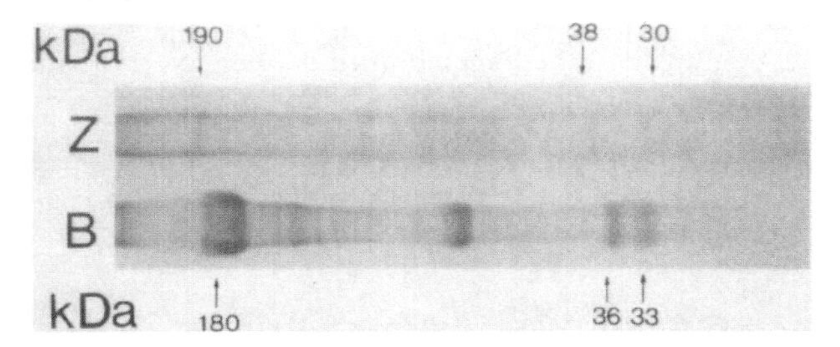

Fig. 4. SDS-polyacrylamide gel electrophoresis of coated vesicle fractions obtained from bovine brain (*B*) and zucchini hypocotyls (*Z*) stained with the carbocyanine dye "stains-all". Ca^{2+}-binding proteins (the light chains) stain blue, other proteins, including clathrin heavy chain, stain red. Molecular masses are indicated accordingly (micrograph courtesy of K. Balusek)

On the basis of cross-reactivity with an antibody prepared against clathrin heavy chain from brain, Cole et al. (1987) and Coleman et al. (1987) have proposed that the light chains of plant clathrin have molecular masses of 57 and 60 kD. This claim has recently been challenged by Balusek et al. (1988) who have put to use the facts that clathrin light chains from brain are heat-stable (Lisanti et al. 1982) and Ca^{2+}-binding (Mooibroek et al. 1987) proteins. Thus, a brief heat treatment of a plant coated vesicle preparation allowed the separation of two polypeptides of 30 and 38 kD, which stained blue with the carbocyanine "stains-all" dye (Fig. 4; Campbell et al. 1983).

Work on coated vesicles from brain (Vigers et al. 1986) strongly suggests that the triskelion framework is attached to the membrane via a group of proteins of 110–100 and 50 kD. These, together with another protein of 180 kD, also become dissociated from the membrane by 2 M urea treatment (see Table 1 in Robinson and Depta 1988 for details). Because of their promotive effect on triskelion polymerization (Zaremba and Keen 1983; Ahle and Ungewickell 1986; Keen 1987) these polypeptides have been given the collective name "the assembly polypeptides". The identification of such polypeptides in plant coated membranes has not yet been possible. However, based on the results of Coleman et al. (1987) and Wiedenhoeft et al. (1988), who have claimed to have obtained cage formation in vitro from dissociated plant coat proteins, it is justifiable to assume that assembly polypeptides will also be present in plant coated vesicles.

3 Coated Pits

Clathrin-coated membranes in eukaryotes can be artificially subdivided into two groups: coated pits, which sensu stricto are invaginations of the plasma membrane, and coated vesicles which are derived from the latter through budding. It is also possible to apply the term coated pit in a wider sense to vesiculating coated regions of endomembranes, e.g., those at the trans face of the

Golgi apparatus, since the lumen of these membranes is topographically the same as the cell exterior. This, however, is rarely seen in the literature; the term coated pit is restricted to the plasma membrane.

3.1 Methods of Visualization

Coated pits can be seen in cross-section (e.g., Fig. 5c; see also Ryser 1979; Van der Valk and Fowke 1981) but one has to search for such profiles in thin sections of plant cells. They are much easier to visualize when the plasma membrane is exposed in surface view. This is possible by scanning sections for tangential cuts between cytoplasm and cell wall (e.g., Fig 5a; see also Doohan and Palevitz 1980), but it is more gratifying when one of the following surface-visualizing techniques is employed.

Historically, the first of these is the protoplast-bursting method, introduced into plant cell biology by Marchant and Hines (1979). It involves attaching protoplasts to formvar-coated EM grids which have been treated with poly-L-lysine. The grid plus adhering protoplasts is then immersed in distilled H_2O which bursts open the protoplasts leaving part of the plasma membrane still attached to the grid. After the residual cytoplasm is washed out a drop of uranyl acetate is added to the grid allowing structures on the cytoplasmic face of the plasma membrane to be visualized in negative contrast (Fig. 5d). With this method coated pits have been demonstrated in protoplasts prepared from suspension-cultured tobacco cells (Van der Valk et al. 1980; Van der Valk and Fowke 1981), onion guard cells (Doohan and Palevitz 1980), tobacco mesophyll (Fowke et al. 1983), and bean mesophyll (Joachim and Robinson 1984).

Two other methods have proved most useful in visualizing coated pits in the plasma membrane of plants. One is the dry-cleaving procedure, originally introduced by Traas (1984) as a method to study the cytoskeleton (see Chap. 12). This technique can be applied to both protoplasts and whole cells and entails fixation, partial dehydration in ethanol, and critical-point-drying. The material can be allowed to adhere to poly-L-lysine/formvar grids either before or after fixation. The cells are then cleaved by inverting the grids, gently pressing them onto adhesive (e.g., Scotch) tape, and ripping them off again. As with the protoplast bursting method, parts of the plasma membrane remain attached to the grid which, because of the heavy metals present (due to osmification, often

Fig. 5a-d. Coated pits (indicated by *arrowheads*) in the plasma membrane from various cell types, visualized by different methods. **a** Grazing surface section of a staminal hair of *Tradescantia*. The cells were quick-frozen and freeze-substituted before sectioning (micrograph courtesy of P.K. Hepler). **b** Dry-cleaving preparation of a protoplast of *Nicotiana plumbaginifolia* (micrograph courtesy of J. Derksen). **c** Cross-section of a coated pit in a suspension-cultured carrot cell. Conventional fixation, dehydration, and embedment. **d** Negatively stained portion of the plasma membrane of a bean leaf protoplast; protoplast-bursting method

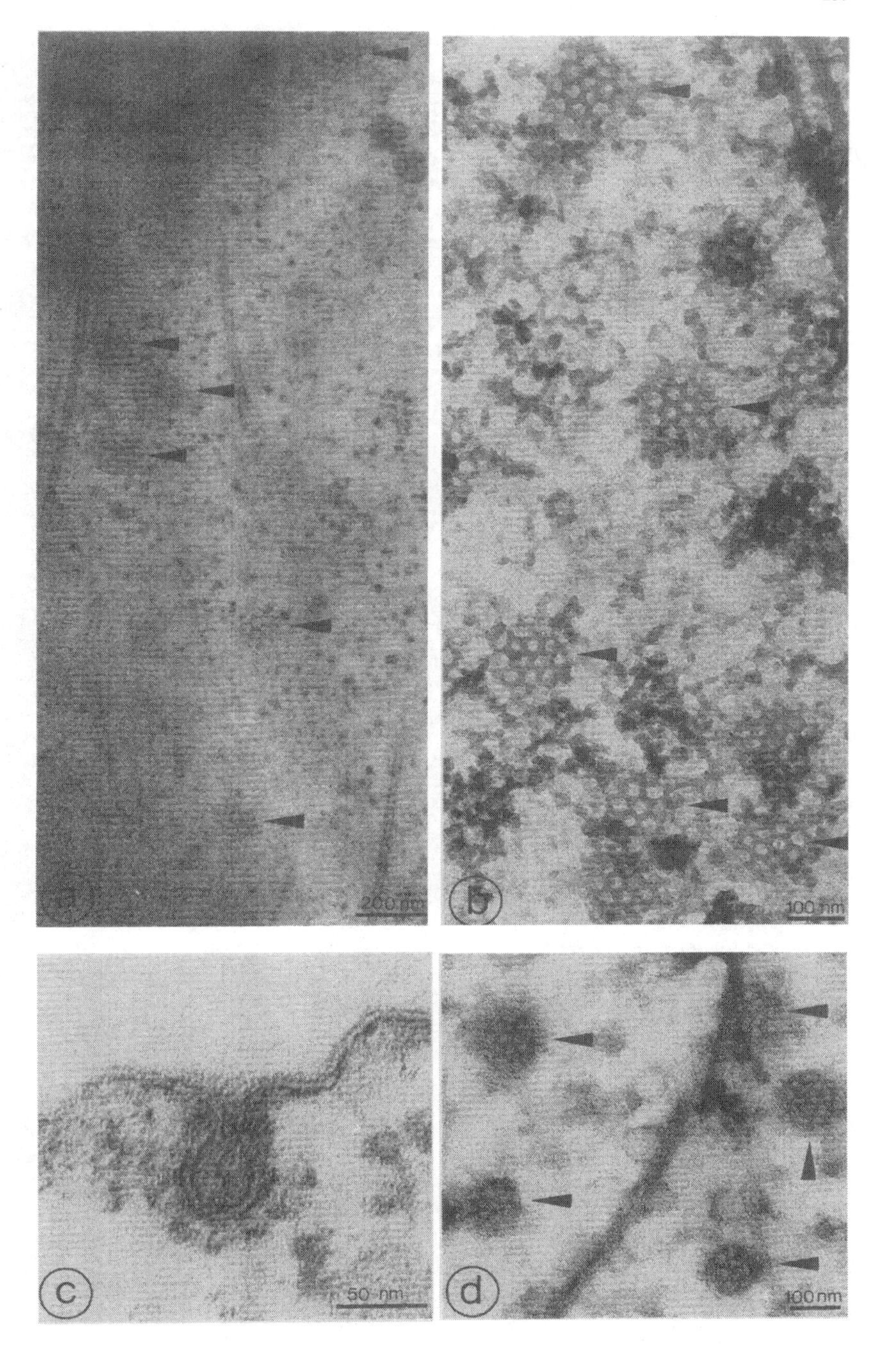
a
200 nm
b
100 nm
c
50 nm
d
100 nm

in the presence of tannic acid; en bloc staining with uranyl acetate), show quite spectacularly the polygonal architecture of the coated pit (Fig. 5b).

In contrast to the protoplast-bursting method, dry cleaving can also be applied to whole cells, however, the cell wall often must be partially digested prior to chemical fixation. Thus, in addition to protoplasts (Emons and Traas 1986), a variety of root hairs and root cells (Emons and Traas 1986), seed hairs (Quader et al. 1986, 1987), and suspension-cultured carrot cells (Hawes 1985) have been successfully processed by this procedure.

Probably the most spectacular visualization of coated pits is achieved by the technique of rapid freezing and deep etching. This method has been put to use primarily on animal cells, especially by Heuser (Heuser et al. 1979; Heuser 1980; Heuser and Reese 1981; Heuser et al. 1987). At the moment only Hawes has been able to successfully apply this method to plant cells (Hawes and Martin 1986; Coleman et al. 1987).

Although the above methods can give rise to different values for the width of the polygon edges, the center-center spacing of the polygons is relatively constant with a mean value of 23 nm. This is similar to values reported for the clathrin coats of animal cells (e.g., Heuser and Kirchhausen 1985), once again implying a common triskelion packing in all eukaryotes. The various surface techniques also allow the development of the coated pit to be visualized. That the triskelions assemble into polygons before the plasma membrane begins to invaginate is convincingly demonstrated in the micrographs of Emons and Traas (1986) and Van der Valk and Fowke (1981). These planar triskelion assemblies appear to consist only of hexagons suggesting that the formation of pentagons is related to the creation of the pit, as originally envisaged for coated pits in animal cells by Kanaseki and Kadota (1969).

3.2 Density and Numbers

There have been few measurements of coated pit density in plants, but some generalizations seem nevertheless possible. Thus, rapidly growing tobacco cells in suspension culture possess almost ten times more coated pits than corresponding leaf tissue cells (Van der Falk and Fowke 1981; Fowke et al. 1983). Meristematic cells do not appear to have higher densities of coated pits than elongating cells in the root cortex, and growing root hairs have higher densities than mature ones (Emons and Traas 1986). Interestingly, in tip-growing cells, e.g., root hairs, coated pit density is highest at the very tip (Robertson and Lyttleton 1982; Emons and Traas 1986). Based on these data and taking a cell diameter of 35 μm, Van der Valk and Fowke (1981) have calculated that protoplasts of suspension-cultured tobacco cells must have around 35 000 coated pits per cell. Assuming a similar cell surface area, the results of Emons and Traas (1986) suggest that tissue cells in higher plants probably have on the average of 5000–10 000 coated pits per cell.

4 Coated Pits and Endocytosis

4.1 Plasma Membrane Turnover

Assuming only vesiculation events are involved, plasma membrane turnover in plants can be defined as being "a consequence of secretion, which incorporates new membrane into the cell surface, and endocytosis which internalizes surface membrane" (Steer 1988). Clearly in nonexpanding cells endocytosis must counteract exocytosis in order to maintain a constant membrane area. This need not necessarily mean that endocytosis is less active in elongating cells where the plasma membrane is expanding. Indeed, calculations made recently by Phillips et al. (1988) indicate that, in epidermal cells of oat coleoptiles, two to three times more membrane is inserted into the plasma membrane than is needed to fulfill the demands made by cell elongation. However, measurements of secretory vesicle size and estimates of their rate of production from inhibitor studies (Kristen and Lockhausen 1983; Shannon and Steer 1984) do suggest that the role of plasma membrane turnover is much more important in nongrowing, glandular cells than in elongating cells. One might perhaps quibble with the actual values obtained in such studies, but they leave no doubts as to the necessity for endocytosis to occur in plant cells.

As far as coated pits are concerned one wants to know whether they are sufficient in terms of number and formation to counterbalance exocytotic input into the plasma membrane. Or are other, noncoated, endocytotic vesicles required for the process of membrane internalization in plants? The results and calculation of Emons and Traas (1986) suggest that coated pits, and the coated vesicles derived from them, might indeed account solely for membrane retrieval. Based on their measurements of coated pit density and the assumption that each coated pit is internalized within 60 s of its formation (see below), Emons and Traas (1986) estimated that an area equivalent to the whole of the plasma membrane can be internalized by coated pits within 20–40 min for rapidly growing and actively secreting cells. This is in the same order of magnitude as the values of exocytotic input given by Kristen and Lockhausen (1983) and Shannon and Steer (1984).

4.2 Studies on Protoplasts

A number of studies (Joachim and Robinson 1984; Tanchak et al. 1984; Hillmer et al. 1986) have established that endocytosis via coated pits does take place in protoplasts. The rationale behind such experiments is simple: electron dense markers (e.g., cationic ferritin, colloidal gold conjugates) are added to the protoplast suspension and their uptake is followed by electron microscopy of thin sections prepared from samples fixed after various incubation periods.

A description of the endocytotic pathway is beyond the scope of this chapter but mention of the organelles which eventually receive these extracellular markers should be made. Partially coated reticulum, Golgi cisternae, and multivesicular bodies (Fig. 1) are the initial targets for endocytosed markers, followed by the large vacuole after longer incubation periods (Tanchak and Fowke 1987). Important in this context is the speed with which such markers enter these endomembrane compartments. Tanchak et al. (1984, 1988) reported that cationic ferritin can be detected in the partially coated reticulum within 2 min of application of this marker to the protoplast suspension. This means that the enclosure of this substance in a coated pit and the vesiculation of the latter into an endocytotic coated vesicle as well as the fusion of the coated vesicle with the membranes of the Golgi apparatus must take place within 1–2 min.

4.3 Studies on Walled Cells

4.3.1 The Turgor Problem

Because plant cells usually possess a wall and a vacuole, the osmotic properties of such a system involve a parameter which does not exist in either animal cells or protoplasts. This is turgor pressure, which is essentially an internal hydrostatic pressure directed against the cell wall. Hence, for endocytosis to occur in turgid plant cells pressures of several atmospheres (up to 1 MPa) have to be overcome. To make the constraints placed on endocytosis in plants more obvious to the reader it would be the same as if invaginations of rubber from the inner tube of a well-pumped-up car tire were occurring. It is therefore understandable that, until recently, plant physiologists were not even prepared to consider the possibility of endocytosis taking place in plants.

Cram (1980) has calculated the maximum energy which plant cells can make available by respiration or photosynthesis and compared it with the work involved in creating a solute-containing vesicle of volume, V, against a pressure difference of δP across the plasma membrane. He concluded that it is energetically impossible for endocytosis to be the principal mechanism responsible for the uptake of the major osmotica into plant cells. In addition, he considered that the water taken up from the extracellular medium as a result of endocytosis would lead to turgor increases which the plant cell could not dissipate quickly enough. Nevertheless, Cram did concede "that a low rate of pinocytosis might occur in plants, providing a mechanism, possibly specific, for the uptake of macromolecules."

Raven (1987) also performed calculations on the energetics of endocytosis. He showed that, even in very large cells, when coated vesicles are the agents of endocytosis, the energy required for plasma membrane internalization is only a fraction of that which the cell expends on cell wall synthesis. A similar conclusion was reached recently by Saxton and Breidenbach (1988). They calculated the energy required to internalize the entire plasma membrane via coated vesicles of diameter 100 nm, in a cell of diameter 50 μm with an average density of 1 g cm^{-3} against a turgor pressure of 0.6 MPa. Assuming a turnover time for the plasma membrane of 10 min (see above), the energy required is 2 $\mu W\ g^{-1}$. This is considerably less than that which plant cells have at their disposal (0.27–2 $mW\ g^{-1}$; Cram 1980). Saxton and Breidenbach (1988) also showed that, due to the small size of endocytotic coated vesicles, the increase in turgor pressure which a turgid plant cell would have to accomodate as a result of one complete turnover of its plasma membrane is at the most 3 kPa. This would represent only a minor fluctuation in turgor.

Also emphasizing the importance of coated vesicle size in endocytosis, but coming to a more restrictive conclusion as to the occurrence of this event in plant cells, is the paper by Gradmann and Robinson (1989). In contrast to Cram (1980) and Saxton and Breidenbach (1988), these authors have concentrated on the thermodynamics of coated pit/coated vesicle formation. Based on the amount of energy known to be required to dissociate a clathrin cage in vitro (3–4 mol ATP per triskelion; Braell et al. 1984), Gradmann and Robinson (1989) have shown that the critical energy for endocytotic vesicle formation in turgid cells exceeds by far the available free energy for triskelion assembly. In terms of the conservation of free energy, endocytosis could then only occur in cells with very low turgor, e.g., root hairs or endosperm or cotyledon cells in developing seeds which act as sinks for phloem transport and are essentially bathed in a sucrose solution.

4.3.2 The Problem of Wall Permeability

The presence of a cell wall means that the plasma membrane of plant cells is not as freely accessible as its equivalent in animal cells. The question is how much of a barrier it really is to potential high-molecular-weight substances, irrespective of whether they are internalized as bound ligand or by fluid phase endocytosis.

There is no doubt that the cell wall can retard the passage of some secreted polysaccharides (e.g., Paull and Jones 1976) or proteins (e.g., Gubler et al. 1987) but, on the other hand, a number of hydrolytic enzymes are secreted through the wall of suspension-cultured cells (e.g., Wink 1984). Moreover, proteins can be centrifuged out of the cell wall space of tissue segments (Morrow and Jones 1986). Thus, it would appear that the cell wall is not entirely impervious to large molecules.

There is some discrepancy as to the actual limits to the porosity of the plant cell wall. According to Carpita et al. (1979), who used solutions of polyethylene glycol (0.4–6 kD) and dextran (2–10 kD), the average pore size of the plant cell wall lies around 4 nm in diameter, thereby just allowing the passage of a polyethylene glycol molecule of 1.6 kD. The authors state that this is equivalent to a globular protein of a 17 kD. This value has been challenged by Tepfer and Taylor (1981) on the grounds that polyethylene glycol solutions cause cell walls to shrink, thereby giving rise to smaller pore sizes. They have shown by gel filtration using columns packed with cell wall fragments from bean hypocotyls that globular proteins up to 60 kD can traverse cell walls. Similarly, Gogarten

(1988), who recently performed experiments with intact cells of suspension-cultured *Chenopodium rubrum*, determined that the exclusion limit of this plant cell wall lies between 30 and 40 kD. Experiments by Baron-Epel et al. (1988) on trans-wall transport in suspension-cultured soybean cells using fluorescent dextran and protein derivatives of differing Stokes radii confirm these estimates. They have set the upper size limit for unrestricted transport at 3.3 nm radius and have shown that molecules with Stokes radii greater than 4.6 nm cannot traverse the wall. Interestingly, one notes that the German biochemicals company Serva (Heidelberg, FRG) have now marketed a gel filtration substance called VESIPOR which is no more than a cell wall preparation from a suspension-cultured plant. In their product information (Serva 1988) it is claimed that polyethylene glycol molecules of 6–8 kD and proteins up to 45 kD can easily pass through the cell walls, but larger proteins as e.g., albumin (67 kD), are held back.

Results obtained on yeast cells are also contradictory. Experiments with polyethylene glycol (Scherrer et al. 1974) suggest that the walls of yeast are even more impermeable since polyethylene glycol molecules larger than 0.7 kD cannot pass through them. Yet it has been reported that α-amylase (molecular mass in excess of 45 kD) can be both taken up (Makarow 1985) and secreted (Rothstein et al. 1984) from yeast cells. In conclusion it would seem that cell walls are, in general, easily permeable to a whole range of proteins up to about 50 kD. Depending on their shape, even larger molecules may traverse the wall.

4.3.3 The Experimental Demonstration of Endocytosis in Walled Plant Cells

Direct evidence for endocytosis in plant cells with walls has been difficult to obtain. Hübner et al. (1985) incubated the roots of intact germinating maize seedlings in solutions of heavy metal salts and were able to detect electron-dense deposits in coated pits, coated vesicles, partially coated reticulum, multivesicular bodies, and Golgi cisternae (Fig. 6a-d). These results were interpreted in terms of endocytosis, although it was not possible to determine the sequence of events involved.

In another paper (Hillmer et al. 1989) endocytosis is implied rather than proven. Here the uptake of the non-permeating fluorochrome lucifer yellow into suspension-cultured carrot cells was followed with the help of the confocal optics of a laser scanning microscope. Lucifer yellow had already been successfully used by Riezman (1985) to demonstrate endocytosis in yeast, and in agreement with his results, this fluorochrome also eventually accumulates in the vacuolar system of carrot cells (Fig. 7a). Microinjection experiments by other

Fig. 6. Organelles involved in endocytosis in maize root cap cells. The roots of maize seedlings were incubated for 1 h in 1 mM $Pb(NO_3)_2$ before fixing. Electron-dense deposits of lead are indicated by *arrows*. **a** Coated pits in the plasma membrane; **b** coated vesicle; **c** Golgi dictyosome (*t* trans; *c* cis face); **d** multivesicular body (*MVB*) and partially coated reticulum (*PCR*)

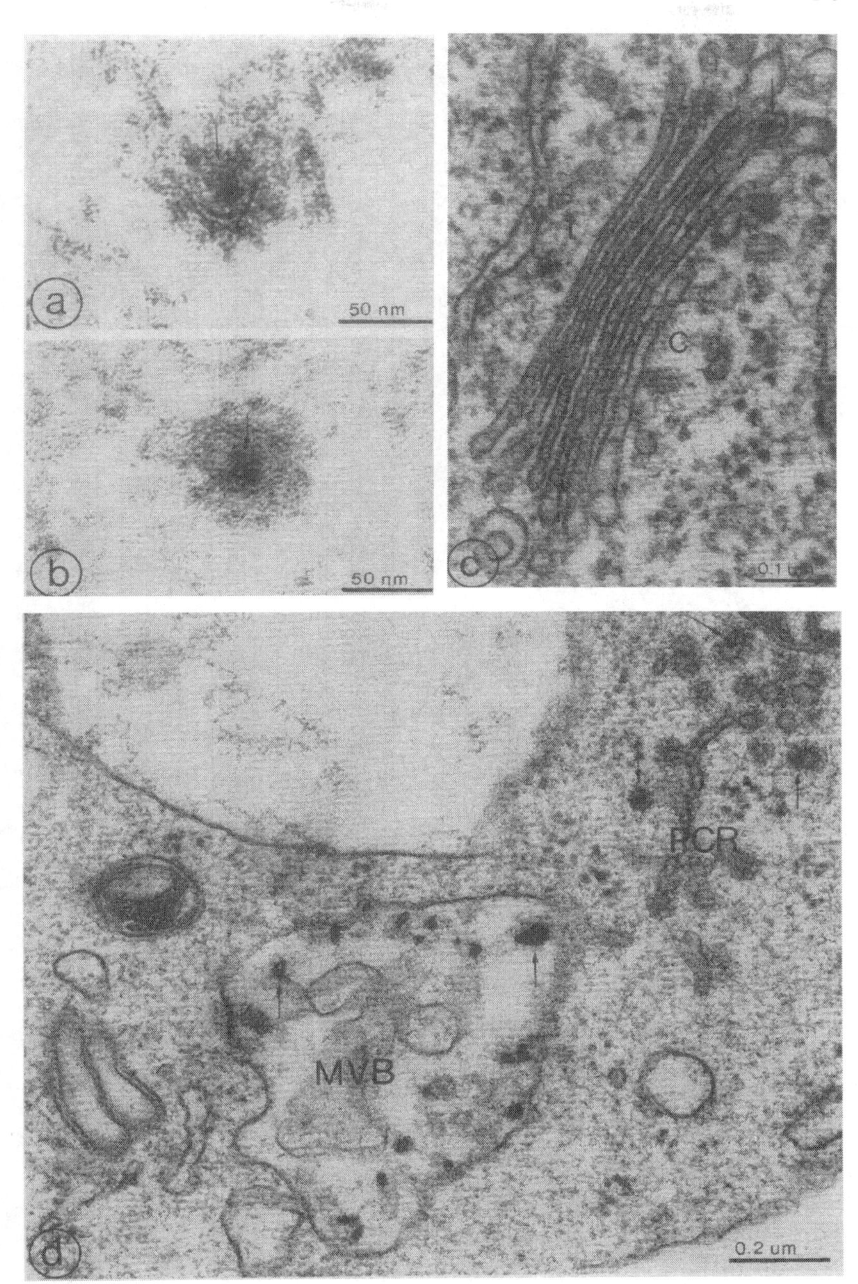
a
50 nm
b
50 nm
t
C
c
MVB
PCR
d
0.2 um

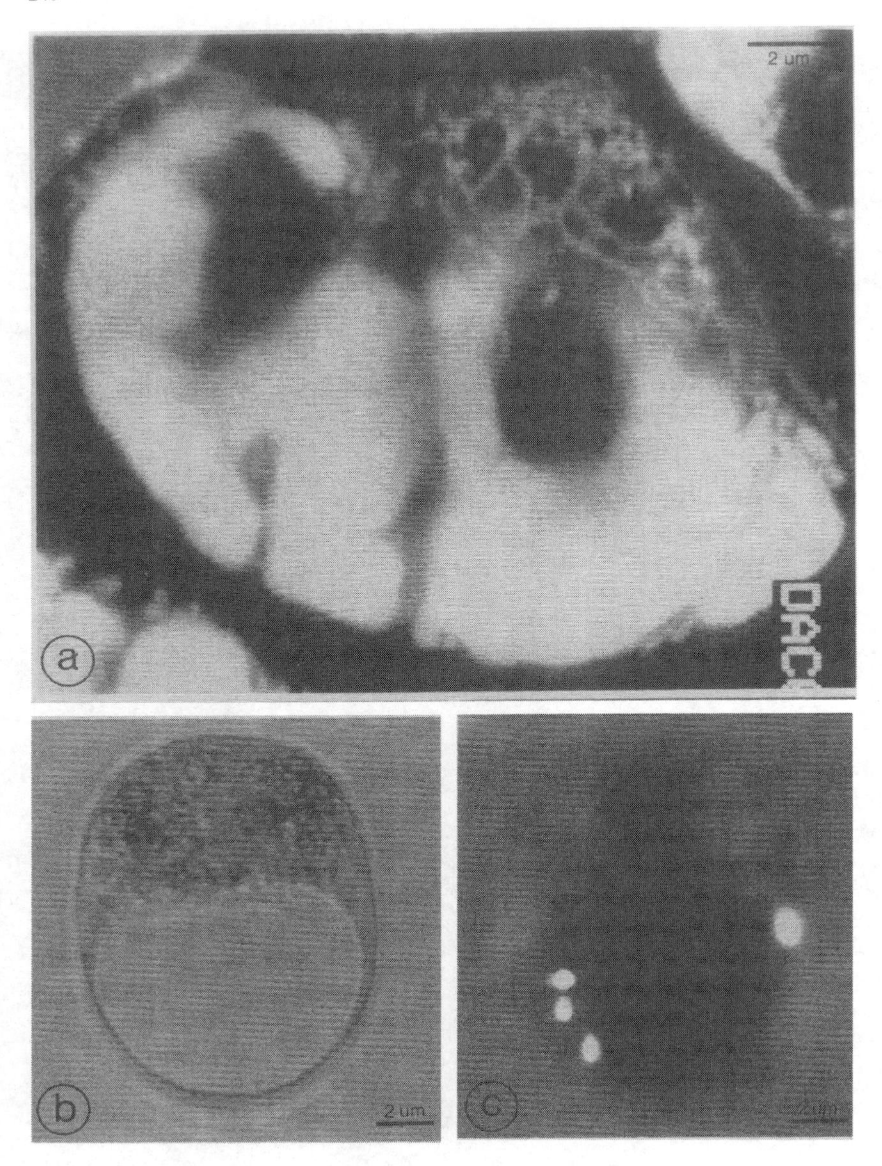

Fig. 7. a Endocytosis of lucifer yellow in a suspension-cultured carrot cell as visualized by confocal laser scanning microscopy. Cells were incubated for 18 h in lucifer yellow which seemed to have no deleterious effects. The fluorochrome is concentrated in the vacuole and not in the cytoplasm. Note, however, the ramifying, tubular, part-granular system of fluorescing elements in the cortical cytoplasm. **b** Phase contrast image of a carrot protoplast after 18 h incubation in lucifer yellow. **c** Nonconfocal fluorescence image of Fig. 6 b. Note that the central vacuole does not fluoresce (micrographs courtesy of M. Robert-Nicoud)

workers (Steinbiss and Stabel 1983; Palevitz and Hepler 1985) have shown that the tonoplast is also impermeable to this dye, strongly suggesting that the entry of lucifer yellow into the vacuole does indeed occur through the fusion of an endocytotic vesicle. Interestingly, in comparable uptake experiments with protoplasts (Fig. 7b,c), the central vacuole does not receive the dye. Instead, fluorescence is restricted to punctate structures in the peripheral cytoplasm. The reason for this difference is not yet clear but one should note that in contrast, Tanchak and Fowke (1987) were able to detect small amounts of endocytosed cationic ferritin in the central vacuole but only in long-term uptake experiments with soybean protoplasts.

4.3.4 Receptor-Mediated Endocytosis in Plants?

The participation of coated pits in receptor-mediated endocytosis in animal cells is now well-established (Steinman et al. 1983; Willingham and Pastan 1984). A large number of receptors have now been characterized from animal cells and the nature of the ligand-receptor interaction is known in many cases (Goldstein et al. 1985). As far as plants are concerned there has not yet been a single demonstration of receptor-mediated endocytosis. The ligands which are internalized by animal cells are physiologically irrelevant as far as plants are concerned. For those substances which might be potential ligands in plants there is insufficient knowledge about their receptors. Nevertheless, progress is being made in this direction (see also Chaps. 9 and 16).

One of the most likely candidates for receptor-mediated endocytosis in plants is the elicitor-stimulated defense response (Darvill and Albersheim 1984; Chaps. 14 and 16). The elicitors are either of fungal or bacterial origin and encompass branched, short-chain carbohydrates, glycoproteins, and even lipids. Their presence at the plant surface results in the synthesis of a number of toxic compounds collectively termed phytoalexins, in the secretion of a number of cell wall compounds, and in the synthesis of several proteases. In addition to such exogenous elicitors, evidence is accumulating for the hormone-like action of cell wall fragments, so-called oligosaccharins (Darvill and Albersheim 1984).

Suspension-cultured cells can respond to fungal elicitors very quickly (in less than 10 min; Young and Kauss 1983; Low and Heinstein 1986), and the response may well involve sudden fluxes of ions, e.g., Ca^{2+} (Köhle et al. 1985). As judged from studies with pH-sensitive fluorochromes (Low and Heinstein 1986; Apostol et al. 1987) the plasma membrane is clearly implicated as the initial target for elicitor binding, indeed an integral plasma membrane protein has now been identified as the receptor in question (P.S. Low, personal communication). In contrast, the receptor for the fungal elicitor associated with wheat stem rust infection (Kogel et al. 1985) has turned out to be a galactolipid (Kogel et al. 1984). In these studies binding of the lectin soybean agglutinin, which is known to prevent the elicitor response, was used as an assay for the presence of the receptor. It is therefore possible that the lipid environment around the protein, rather than the receptor protein itself, is being recognized.

A number of studies on the characterization of the receptor for the fungal phytotoxin, fusicoccin, which is known to stimulate proton extrusion at the plant plasma membrane (Marrè 1979), have also been performed. According to Feyerabend and Weiler (1988), this protein is thermolabile, can be degraded by trypsin, and is localized at the outer surface of the plasma membrane. Evidence exists from subcellular fractionation that fusicoccin specifically binds to the plasma membrane (Dohrmann et al. 1977; Pesci et al. 1979), and attempts to purify the receptor have been undertaken (Pesci et al. 1979; Stout and Cleland 1980; De Boer et al. 1989). The receptor can be solubilized through detergent treatment and has been purified by gel filtration followed by affinity chromatography. It appears to be a trimeric protein with subunits of 28 and 31 kD occurring in the ratio of 1:2. Preliminary autoradiographic experiments with ^{3}H-fusicoccin have been performed in an attempt to determine whether the fusicoccin-binding sites are internalized (Aducci et al. 1985). Although the quality of the micrographs leaves much to be desired, it can be seen that the binding sites are not restricted to the plasma membrane. Clearly of great potential use in future internalization studies will be the use of monoclonal antibodies against fusicoccin (Feyerabend and Weiler 1987).

The identification of a receptor for the major plant hormone, auxin, has occupied the efforts of a number of researchers for well over a decade (see review by Cross 1985). The situation with this potential ligand is made more complex by the fact that not only does auxin have more than one binding site, it can also traverse membranes in its uncharged form (review by Hertel 1987; Chap. 9). Both membrane-associated (Löbler and Klämbt 1985a; Venis 1987) and cytoplasmically located (Libbenga et al. 1987) auxin-binding proteins have been isolated from plant homogenates. The native binding protein appears to be a dimer with subunits of 40–45 kD. Antibodies have been prepared against the membrane-bound binding protein (Löbler and Klämbt 1985a) and have been shown to prevent auxin-stimulated growth when added exogenously to coleoptile segments (Löbler and Klämbt 1985b), suggesting that the auxin receptor is indeed localized at the plasma membrane.

Of the other plant hormones evidence for membrane-bound receptors for abscisic acid (Hornberg and Weiler 1984) and ethylene (Hall et al. 1987) is at hand, but it is unclear which membrane(s) are involved. On the other hand, the gibberellins do not appear to have a membrane-bound receptor (e.g., Liu and Srivastava 1987).

5 Coated Vesicles as Products of Coated Pits

5.1 Methods of Isolation; Purity of Fractions

Coated vesicles are the immediate vesiculation product of coated pits. Their isolation and characterization are clearly preliminary steps toward understanding their function. Because of their size (less than 100 nm in diameter), they can almost be treated as virus particles in terms of isolation. Thus, the major steps in the purification of coated vesicles from plant sources are as follows:

1. Centrifugation of the homogenate at 40 000–50 000 g for 30–60 min in order to remove large organelles and microsomal membranes;
2. Centrifugation of the postmicrosomal supernatant at 125 000 g for 60–90 min to pull down a crude coated vesicle fraction;
3. Treatment of this fraction with RNAse to remove ribosomal contamination;
4. Further centrifugations (e.g., in Ficoll/D_2O) in order to separate the coated vesicles from residual membrane contaminants and, in the case of green tissue, ribulose bisphosphate carboxylase (EC 4.1.1.39) as well.

With the exception of Griffing et al. (1986), who have isolated coated vesicles directly by centrifuging homogenates through a sucrose density gradient in a vertical rotor, coated vesicle fractions have invariably been prepared as above (Mersey et al. 1985; Depta and Robinson 1986; Coleman et al. 1987; Depta et al. 1987b; Wiedenhoeft et al. 1988). Other possible methods, such as electrophoresis (Rubinstein et al. 1981) or immunoaffinity chromatography (Merisko et al. 1982), which have been applied successfully to animal cells, have not yet been attempted on plant extracts.

Several problems have to be overcome in isolating coated vesicles from plant sources which are not so important when working with animal tissue. These problems relate to the vacuole and entail:

1. Combating endogeneous proteolytic activity (by including in the homogenizing medium an inhibitor, e.g., phenylmethylsulfonyl fluoride or chymostatin);
2. Preventing polyphenol oxidase (EC 1.10.3.1) activity (by the inclusion of thiourea or metabisulfite);
3. Overcoming the consequences of vacuolar dilution (homogenate volumes of 2 l or more are not uncommon).

The latter problem necessitates having access to more than one ultracentrifuge in order to perform the isolation in a reasonable time period, and also means that the recovery of coated vesicles from plant sources is at least one order of magnitude poorer than from an equivalent amount of animal tissue (usually brain) (Robinson et al. 1987). This, of course, severely limits experimentation.

The degree of enrichment in the coated vesicle preparations previously isolated from plant tissues differs somewhat. The fractions obtained by Mersey et al. (1982) and Coleman et al. (1987) are considerably contaminated with fragments of other membranes. The 30% contamination which Depta and Robinson (1986) observed in a coated vesicle fraction from suspension-cultured carrot cells was ascribed, in part, to stripped (i.e., previously coated) vesicles. The purest fractions so far obtained appear to be from bean leaves (Depta et al. 1987b) with less than 5% contamination. This tissue has also allowed the recognition of two different size classes of coated vesicles: one having a diameter of at least 85 nm and the other with a smaller diameter (70–80 nm).

5.2 Properties

Plant coated vesicle fractions have been tested for the presence of plasma membrane marker enzymes, based on the premise that coated vesicles derived from coated pits should have some characteristics in common with the plasma membrane. Depta et al. (1987a) have shown that one of the standard plasma membrane markers, glucan synthase II (EC 2.1.3.34; Quail 1979; Robinson 1985; Chap. 2), is enriched in plant coated vesicle fractions. The product is highly crystalline 1,3-β-glucan (callose), and the characteristics of the enzyme activity (synergistic stimulation through Ca^{2+} and spermine) is identical to that previously recorded for the plasma membrane enzyme (Kauss and Jeblick 1986; Fink et al. 1987; see also Chap. 14).

Doubts that the glucan synthase II activity measured in coated vesicle fractions may be due to contaminating plasma membrane fragments have been partially sustained. On the one hand, plasma membrane from zucchini hypocotyls, isolated by phase partitioning, does not coequilibrate with zucchini coated vesicles in Ficoll/D_2O gradients (H. Depta, D. Wagenseil, M. Lützelschwab, W. Michalke, D.G. Robinson, unpublished observations). Nor does an antibody, prepared against the 100 kD subunit of the plasma membrane H^+-ATPase from corn coleoptiles cross-react with the zucchini coated vesicle fraction, although it does cross-react strongly with zucchini plasma membrane preparations (Fichmann et al. 1989). On the other hand, it has been possible to separate glucan synthase II activity from the coated vesicles in a zucchini coated vesicle fraction. Since these contaminating membranes lack plasma membrane H^+-ATPase their origin remains unclear. However, it is possible that they may represent an H^+-ATPase deficient domain of the plasma membrane.

Despite the fact that highly purified coated vesicle fractions from zucchini are without 1,3-β-glucan synthase there are, nevertheless reports implicating the participation of coated vesicles in the wound healing response. Thus, O'Neil and La Claire (1984) and Foissner (1988) reported that the mechanical wounding of giant algal cells results in the appearance of large numbers of coated membranes at the cell surface where wound polysaccharides are being released. A similar situation exists in the liverwort *Riella* where wound callose can be specifically induced by the Ca^{2+} chelator, chlorotetracycline (Grotha 1986; D. Lerchl et al. 1989).

Acknowledgments. We thank Bernd Rauffeisen for preparing the line drawings. Work on coated pits and coated vesicles from our own laboratory was supported by the Deutsche Forschungsgemeinschaft.

References

Aducci P, Ballio A, Autuori F (1985) Fusicoccin binding sites: an autoradiographic study. Phytopathol Mediterr 24:294–295

Ahle S, Ungewickell E (1986) Purification and properties of a new clathrin assembly protein. EMBO J 5:3143–3149

Anderson RGW, Vasile E, Mello RT, Brown MS, Goldstein JL (1978) Immunocytochemical visualization of coated pits and vesicles in human fibroblasts: relation to low density lipoprotein receptor distribution. Cell 15:919–933

Apostol I, Low PS, Heinstein P, Stipanovic RD, Altman DW (1987) Inhibition of elicitor-induced phytoalexin formation in cotton and soybean cells by citrate. Plant Physiol 84:1276–1280

Balusek K, Depta H, Robinson DG (1988) Two polypeptides (30 and 38 kDa) in plant coated vesicles with light chain properties. Protoplasma 146:174–176

Baron-Epel O, Gharyal PK, Schindler M (1988) Pectins as mediators of wall porosity in soybean cells. Planta 175:389–395

Blank GS, Brodsky FM (1987) Clathrin assembly involves a light chain binding region. J Cell Biol 105:2011–2019

Braell WA, Schlossman P, Schmid SL, Rothman JE (1984) Dissociation of clathrin coats coupled to the hydrolysis of ATP: role of an uncoating ATPase. J Cell Biol 99:734–741

Brodsky FM, Parham P (1983) Polymorphism in clathrin light chains from different tissues. J Mol Biol 167:197–204

Campbell KP, MacLennan DH, Jørgensen AO (1983) Staining of the Ca^{2+}-binding proteins calsequestrin, calmodulin, troponin C, and S-100 with the cationic carbocyanine dye "stains-all". J Biol Chem 258:11267–11273

Carpita N, Sabularse D, Montezinos D, Delmer DP (1979) Determination of the pore size of cell walls of living plant cells. Science 205:1144–1147

Cole L, Coleman JOD, Evans DE, Hawes CR, Horsley D (1987) Antibodies to brain clathrin recognise plant coated vesicles. Plant Cell Rep 6:227–230

Coleman JOD, Evans DE, Hawes CR, Horsley D, Cole L (1987) Structure and molecular organization of higher plant vesicles. J Cell Sci 88:33–45

Cram WJ (1980) Pinocytosis in plants. New Phytol 84:1–17

Cross JW (1985) Auxin action: the search for the receptor. Plant Cell Environ 8:351–359

Croze EM, Morré DJ, Morré DM, Kartenbeck J, Franke WW (1982) Distribution of clathrin and spiny-coated vesicles on membranes within mature Golgi apparatus elements of mouse liver. Eur J Cell Biol 28:130–138

Darvill AG, Albersheim P (1984) Phytoalexins and their elicitors – a defense against microbial infection in plants. Annu Rev Plant Physiol 35:243–275

De Boer AH, Watson BA, Cleland RE (1989) Purification and identification of the fusicoccin binding protein from oat root plasma membrane. Plant Physiol 89:250–259

Depta H, Robinson DG (1986) The isolation and enrichment of coated vesicles from suspension-cultured carrot cells. Protoplasma 130:162–170

Depta H, Andreae M, Blaschek W, Robinson DG (1987a) Glucan synthase II activity in a coated vesicle fraction from zucchini hypocotyls. Eur J Cell Biol 45:219–223

Depta H, Freundt H, Hartmann D, Robinson DG (1987b) Preparation of a homogeneous coated vesicle fraction from bean leaves. Protoplasma 136:154–160

Dohrmann U, Hertel R, Pesci P, Cocucci S, Marrè G, Randazzo G, Ballio A (1977) Localization of in vitro binding of the fungal toxin fusicoccin to a plasma membrane-rich fraction from corn coleoptiles. Plant Sci Lett 9:291–299

Doohan ME, Palevitz BA (1980) Microtubules and coated vesicles in guard cell protoplasts of *Allium cepa* L. Planta 149:389–401

Emons AMC, Traas JA (1986) Coated pits and coated vesicles on the plasma membrane of plant cells. Eur J Cell Biol 41:57–64

Feyerabend M, Weiler EW (1987) Monoclonal antibodies against fusicoccin with binding characteristics similar to the putative fusicoccin receptor of higher plants. Plant Physiol 85:835–840

Feyerabend M, Weiler EW (1988) Characterization and localization of fusicoccin binding sites in leaf tissues of *Vicia faba* L. probed with a novel radioligand. Planta 174:115–122

Fichmann J, Taiz L, Gallagher S, Leonard RT, Depta H, Robinson DG (1980) Immunochemical comparison of the coated vesicle H^+-ATPases of plants and animals. Protoplasma, in press

Fink J, Jeblick W, Blaschek W, Kauss H (1987) Ca^{2+} and polyamines activate the plasma membrane-located 1,3-β-glucan synthase. Planta 171:130–135

Foissner I (1988) The relationship of echinate inclusions and coated vesicles on wound healing in *Nitella flexis* (Characeae). Protoplasma 142:164–175

Fowke LC, Rennie PJ, Constabel F (1983) Organelles associated with the plasma membrane of *Nicotiana tabacum* leaf protoplasts. Plant Cell Rep 2:292–295

Franke WW, Eckert WA, Krien S (1971) Cytomembrane differentiation in a ciliate *Tetrahymena pyriformis*. I Endoplasmic reticulum and dictyosomal equivalents. Z Zellforsch Mikrosk Anat 119:577–604

Gogarten JP (1988) Physical properties of the cell wall of photoautotrophic suspension cells from *Chenopodium rubrum* L. Planta 174:333–339

Goldstein JL, Brown MS, Anderson RGW, Russell DW, Schneider WT (1985) Receptor-mediated endocytosis: concepts emerging from the LDL receptor system. Annu Rev Cell Biol 1:1–39

Gradmann D, Robinson DG (1989) Does turgor prevent endocytosis in plant cells? Plant Cell Environ 12:151–154

Griffiths G, Warren I, Stuhlfauth I, Jockusch BM (1981) The role of clathrin-coated vesicles in acrosome formation. Eur J Cell Biol 26:52–60

Grotha R (1986) Tetracyclines, verapamil and nifedipine induce callose deposition at specific cell sites in *Riella hylicophylla*. Planta 169:546–554

Gubler F, Ashford AE, Jacobsen JV (1987) The release of α-amylase through gibberellin-treated barley aleurone cell walls: an immunocytochemical study with Lowicryl K4M. Planta 172:155–161

Gunning BES, Steer MW (1975) Ultrastructure and the biology of plant cells. Edward Arnold, London

Hall MA, Howarth CJ, Robertson D, Sanders IO, Smith PG, Starling RJ, Tang Z-D, Thomas CJR, Williams RAN (1987) Ethylene binding proteins. In: Fox JE, Jacobs M (eds) Molecular biology of plant growth control. UCLA Symposia on molecular biology 44, Alan R Liss, New York, pp 335–344

Hawes CR (1985) Conventional and high voltage electron microscopy of the cytoskeleton and cytoplasmic matrix of carrot (*Daucus carota* L.) cells grown in suspension culture. Eur J Cell Biol 38:201–210

Hawes CR, Martin B (1986) Deep etching of plant cells: cytoskeleton and coated pits. Cell Biol Int Rep 10:985–991

Hertel R (1987) Auxin transport: binding of auxins and phytotropins to the carriers. Accumulation into and efflux from membrane vesicles. In: Klämbt D (ed) Plant hormone receptors. NATO ASI Series, Vol H10, pp 81–92

Heuser JE (1980) Three-dimensional visualization of coated vesicle formation in fibroblasts. J Cell Biol 84:550–583

Heuser JE, Kirchhausen T (1985) Deep-etch views of clathrin assemblies. J Ultrastruct Res 92:1–27

Heuser JE, Reese TS (1981) Structural changes after transmitter release at the frog neuromuscular junction. J Cell Biol 88:564–580

Heuser JE, Reese TS, Dennis MJ, Jan Y, Jan L, Evans L (1979) Synaptic vesicle exocytosis captured by quick-freezing and correlated with quantal transmitter release. J Cell Biol 81: 275–300

Heuser JE, Keen JH, Amende LM, Lippoldt RE, Prasad K (1987) Deep-etch visualization of 27S clathrin: a tetrahedral tetramer. J Cell Biol 105:1999–2009

Hillmer S, Depta H, Robinson DG (1986) Confirmation of endocytosis in higher plant protoplasts using lectin-gold conjugates. Eur J Cell Biol 41:142–149

Hillmer S, Freundt H, Robinson DG (1988) The partially coated reticulum and its relationship to the Golgi apparatus in higher plant cells. Eur J Cell Biol 47:206–212

Hillmer S, Quader H, Robert-Nicoud M, Robinson DG (1989) Lucifer yellow uptake in cells and protoplasts of *Daucus carota* visualized by confocal laser scanning microscopy. J Exp Bot 40:417–423

Hornberg C, Weiler E (1984) High-affinity binding sites for abscisic acid on the plasmalemma of *Vicia faba* guard cells. Nature 310:321–324

Hübner R, Depta H, Robinson DG (1985) Endocytosis in maize root cap cells. Evidence obtained using heavy metal salt solutions. Protoplasma 129:214–222

Jackson AP, Seow H-F, Holmes N, Drickamer K, Parham P (1987) Clathrin light chains contain brain-specific insertion sequences and a region of homology with intermediate filaments. Nature 326:154–157

Joachim S, Robinson DG (1984) Endocytosis of cationic ferritin by bean leaf protoplasts. Eur J Cell Biol 34:212–216

Kanaseki T, Kadota K (1969) The "vesicle in a basket": a morphological study of the coated vesicle isolation from the nerve endings of the guinea pig brain with special reference to the mechanism of membrane movements. J Cell Biol 42:202–220

Kauss H, Jeblick W (1986) Synergistic activation of an 1,3-β-D-glucan synthase by Ca^{2+} and polyamines. Plant Sci 43:103–107

Keen JH (1987) Clathrin assembly proteins: affinity purification and a model for coat assembly. J Cell Biol 105:1989–1998

Kogel KH, Ehrlich-Rogozinski S, Reisener HJ, Sharon N (1984) Surface galactolipids of wheat protoplasts as receptors for soybean agglutinin and their possible relevance to host-parasite interaction. Plant Physiol 76:924–928

Kogel KH, Heck B, Kogel G, Moerschbacher B, Reisener HJ (1985) A fungal elicitor of resistance response in wheat. Z Naturforsch 40c:743–744

Köhle H, Jeblick W, Poten F, Blaschek W, Kauss H (1985) Chitosan-elicited callose synthesis in soybean cells as a Ca^{2+}-dependent process. Plant Physiol 77:544–551

Kristen U, Lockhausen J (1983) Estimation of Golgi membrane flow rates in ovary glands of *Aptenia cordifolia* using cytocalasin B. Eur J Cell Biol 29:262–267

Lerchl D, Hillmer S, Grotha R, Robinson DG (1989) Ultrastructural observations on CTC-induced callose formation in *Riella helicophylla*. Bot Acta 102:62–70

Libbenga KR, van Telgen HJ, Mennes AM, van der Linde PCG, van der Zaal EJ (1987) Characterization and function analysis of a high-affinity cytoplasmic auxin-binding protein. In: Fox JE, Jacobs M (eds) Molecular biology of plant growth control. UCLA Symposia on molecular biology 44. Alan R. Liss, New York, pp 229–244

Lisanti MP, Shapiro LS, Moskowitz N, Hua EL, Puszkin S, Schook W (1982) Isolation and preliminary characterization of clathrin-associated proteins. Eur J Biochem 125:463–470

Liu ZH, Srivastava LM (1987) In vitro binding of gibberellin A4 in epicotyls of dwarf pea and tall pea. In: Fox JE, Jacobs M (eds) Molecular biology of plant growth control. UCLA Symposia on molecular biology 44, Alan R. Liss, New York, pp 315–322

Löbler M, Klämbt HD (1985a) Auxin-binding protein from coleoptile membranes of corn (*Zea mays* L.). I Purification by immunological methods and characterization. J Biol Chem 260:9848–9853

Löbler M, Klämbt HD (1985b) Auxin-binding protein from coleoptile membranes of corn (*Zea mays* L.). II Localization of a putative auxin receptor. J Biol Chem 260:9854–9859

Low PS, Heinstein PF (1986) Elicitor stimulation of the defense response in cultured plant cells monitored by fluorescent dyes. Arch Biochem Biophys 249:472–479

Makarow M (1985) Endocytosis in *Saccharomyces cerevisiae*: internalization of α-amylase and fluorescent dextran into cells. EMBO J 4:1861–1866

Marchant HJ, Hines ER (1979) The role of microtubules and cell wall deposition in elongation of regenerating protoplasts of *Mougeotia*. Planta 146:41–48

Marrè E (1979) Fusicoccin: a tool in plant physiology. Annu Rev Plant Physiol 30:273–288

Merisko EM, Farquhar MG, Palade GE (1982) Coated vesicle isolation by immunoadsorption on *Staphylococcus aureus* cells. J Cell Biol 92:846–857

Mersey BG, Fowke LC, Constabel F, Newcomb EH (1982) Preparation of a coated vesicle-enriched fraction from plant cells. Exp Cell Res 141:459–463

Mersey BG, Griffing LR, Rennie PJ, Fowke LC (1985) The isolation of coated vesicles from protoplasts of soybean. Planta 163:317-327
Mollenhauer HH, Hass BS, Morré DJ (1976) Membrane transformations of Golgi apparatus of rat spermatids: a role for thick cisternae and two classes of coated vesicles in acrosome formation. J Microsc Biol Cell 27:33-36
Mooibroek MJ, Michiel DF, Wang JH (1987) Clathrin light chains are calcium-binding proteins. J Biol Chem 262:25-28
Morrow DL, Jones RL (1986) Localization and partial characterization of the extracellular proteins centrifuged from pea internodes. Physiol Plant 67:397-407
Mueller SC, Branton D (1984) Identification of coated vesicles in *Saccharomyces cerevisiae*. J Cell Biol 98:341-346
Ockleford CD, Whyte A (eds) (1980) Coated vesicles. Cambridge University Press, Cambridge
O'Neil RM, La Claire JW (1984) Mechanical wounding induces the formation of extensive coated membranes in giant algal cells. Science 225:331-333
Orci L, Ravazzola M, Amherdt M, Louvard D, Perrelet A (1985) Clathrin-immunoreactive sites in the Golgi apparatus are concentrated at the trans pole in polypeptide hormone-secreting cells. Proc Natl Acad Sci USA 42:5385-5389
Orci L, Glick BS, Rothman JE (1986) A new type of coated vesicular carrier that appears not to contain clathrin: its possible role in protein transport within the Golgi stack. Cell 46:171-184
Palevitz BA, Hepler PK (1985) Changes in dye coupling of stomatal cells of *Allium* and *Commelina* demonstrated by microinjection of lucifer yellow. Planta 164:473-479
Paull RE, Jones RL (1976) Studies on the secretion of maize root cap slime. V The cell wall as a barrier to secretion. Z Pflanzenphysiol 79:154-164
Pesacreta TC, Lucas WJ (1985) Presence of a partially-coated reticulum and a plasma membrane coat in angiosperms. Protoplasma 125:173-184
Pesci P, Tognoli N, Beffagna N, Marrè E (1979) Solubilization and partial purification of a fusicoccin receptor complex from maize microsomes. Plant Sci Lett 15:313-322
Phillips GD, Preshaw C, Steer MW (1988) Dictyosome vesicle production and plasma membrane turnover in auxin-stimulated outer epidermal cells of coleoptile segments from *Avena sativa* (L). Protoplasma 145:59-65
Quader H, Deichgräber G, Schnepf E (1986) The cytoskeleton of *Cobaea* seed hairs: patterning during cell-wall differentiation. Planta 168:1-10
Quader H, Herth W, Ryser U, Schnepf E (1987) Cytoskeletal elements in cotton seed hair development in vitro: their possible regulatory role in cell wall organization. Protoplasma 137:56-62
Quail PH (1979) Plant cell fractionation. Annu Rev Plant Physiol 30:425-484
Raven JA (1987) The role of vacuoles. New Phytol 196:357-422
Riezman H (1985) Endocytosis in yeast: several of the yeast secretory mutants are defective in endocytosis. Cell 40:1001-1009
Robertson JG, Lyttleton P (1982) Coated and smooth vesicles in the biogenesis of cell walls, plasma membranes, infection threads and peribacteroid membranes in root hairs and nodules of white clover. J Cell Sci 58:63-78
Robinson DG (1984) Membranes and secretion in higher plants. In: Boudet A (ed) Membranes and compartmentation in the regulation of plant function. Annu Rev Proc Phytochem Soc Eur 24, pp 147-161
Robinson DG (1985) Plant membranes: endo- and plasma membranes of plant cells. John Wiley, New York
Robinson DG, Depta H (1988) Coated vesicles. Annu Rev Plant Physiol Plant Mol Biol 39:53-99
Robinson DG, Andreae M, Depta H, Hartmann D, Hillmer S (1987) Coated vesicles in plants: characterization and function. In: Leaver C, Sze H (eds) Plant membranes: structure, function, biogenesis. UCLA Symposia on Molecular and Cellular Biology, New Series, Vol 63 Alan R Liss, New York, pp 341-358
Rothstein SJ, Lazarus CM, Smith WE, Baulcombe DC, Gatenby AA (1984) Secretion of a wheat α-amylase expressed in yeast. Nature 308:662-665
Rubinstein JLR, Fine RE, Luskey BD, Rothman JE (1981) Purification of coated vesicles by agarose gel electrophoresis. J Cell Biol 89:357-361

Ryser U (1979) Cotton fibre differentiation: occurrence and distribution of coated and smooth vesicles during primary and secondary wall formation. Protoplasma 98:223–239
Saxton MJ, Breidenbach RW (1988) Receptor-mediated endocytosis in plants is energetically possible. Plant Physiol 86:993–995
Scherrer R, Londen L, Gerhardt P (1974) Porosity of the yeast cell wall and membrane. J Bacteriol 118:534–540
Serva (1988) Serva VESIPOR: Vesikuläres Trennmaterial für Biopolymere. Serva Merkblatt 320. Serva Feinbiochemica, Heidelberg, FRG
Shannon TM, Steer MW (1984) The root cap as a test system for the evaluation of Golgi inhibitors. I Structure and dynamics of the secretory system and response to solvents. J Exp Bot 35:697–707
Steer MW (1988) Plasma membrane turnover in plant cells. J Exp Bot 39:987–996
Steinbiss HH, Stabel P (1983) Protoplasts derived from tobacco cells can survive capillary microinjection of the fluorescent dye lucifer yellow. Protoplasma 116:223–227
Steinman RM, Mellman IS, Mueller WA, Cohn ZA (1983) Endocytosis and the recycling of plasma membrane. J Cell Biol 96:1–27
Stout RG, Cleland RE (1980) Partial characterization of fusicoccin binding to receptor sites on oat root membranes. Plant Physiol 66:353–359
Tanchak MA, Fowke LC (1987) The morphology of multivesicular bodies in soybean protoplasts and their role in endocytosis. Protoplasma 138:173–182
Tanchak MA, Griffing LR, Mersey BG, Fowke LC (1984) Endocytosis of cationized ferritin by coated vesicles of soybean protoplasts. Planta 162:481–486
Tanchak MA, Rennie PJ, Fowke LC (1988) Ultrastructure of the partially coated reticulum and dictyosomes during endocytosis by soybean protoplasts. Planta 175:433–441
Tepfer M, Taylor IEP (1981) The permeability of plant cell walls as measured by gel filtration chromatography. Science 213:261–263
Traas JA (1984) Visualization of the membrane bound cytoskeleton and coated pits of plant cells by means of dry cleaving. Protoplasma 119:212–218
Van der Valk P, Fowke LC (1981) Ultrastructural aspects of coated vesicles in tobacco protoplasts. Can J Bot 59:1307–1313
Van der Valk P, Rennie PJ, Connolly JA, Fowke LC (1980) Distribution of cortical microtubules in tobacco protoplasts: an immunofluorescence microscopic and ultrastructural study. Protoplasma 105:27–43
Venis M (1987) Auxin-binding proteins in maize: purification and receptor function. In: Fox JE, Jacobs M (eds) Molecular biology of plant growth control. UCLA Symposia on molecular biology 44 Alan R Liss, New York, pp 219–228
Vigers GPA, Crowther RA, Pearse BMF (1986) Location of the 100 kD–50 kD accessory proteins in clathrin coats. EMBO J 5:2079–2085
Wehland J, Willingham MC, Gallo MG, Rutherford AV, Rudick J, Dickson RB, Pastan I (1982) Microinjection of anticlathrin antibodies into cultured fibroblasts: clathrin-coated structures in receptor-mediated endocytosis and in exocytosis. Cold Spring Harbor Symp 46:743–753
Wiedenhoeft RE, Schmidt GW, Palevitz BA (1988) Dissociation and reassembly of soybean clathrin. Plant Physiol 86:412–416
Willingham MC, Pastan I (1984) Endocytosis and exocytosis: current concepts of vesicle traffic in animal cells. Int Rev Cytol 92:51–92
Wink M (1984) Evidence for an extracellular lytic compartment of plant cell suspension cultures: the cell culture medium. Naturwissenschaften 71:635–637
Woodward MP, Roth TF (1978) Coated vesicles: characterization, selective dissociation and reassembly. Proc Natl Acad Sci USA 75:4394–4398
Young DH, Kauss H (1983) Release of calcium from suspension-cultured *Glycine max* cells by chitosan, other polycations and polyamines in relation to effects on membrane permeability. Plant Physiol 73:698–702
Zaremba S, Keen JH (1983) Assembly polypeptides from coated vesicles mediate reassembly of unique clathrin coats. J Cell Biol 97:1339–1347
Zhang Y-H, Robinson DG (1986) The endomembranes of *Chlamydomonas reinhardii:* a comparison of the wild-type with the wall mutants CW2 and CW15. Protoplasma 133:186–194

Chapter 11 Role of the Plasma Membrane in Cellulose Synthesis

D.P. DELMER[1]

1 Background

In one sense it may seem presumptuous to write an article about the role of the plasma membrane in cellulose synthesis when the synthetic reaction itself has never been directly demonstrated using plasma membranes isolated from higher plants or algae. This lack of success in finding a cellulose synthase activity in vitro has been well-documented in several recent reviews on cell wall biosynthesis (Delmer 1987; Bolwell 1988; Delmer and Stone 1988), but for the purposes of this discussion, I shall summarize again briefly our state of knowledge about the enzymology of cellulose biosynthesis. In spite of the paucity of information about the enzyme itself, there is good evidence that synthetic complexes reside and function within the plasma membrane, and that they are most likely subject to unique forms of regulation. Thus, the remainder of this review, in keeping with the focus of this book, will concentrate on the role of the plasma membrane in the process.

Only one cellulose synthase activity has been demonstrated in vitro with certainty; this is the enzyme of the bacterium *Acetobacter xylinum.* First

[1]Department of Botany, Institute of Life Sciences, The Hebrew University, Jerusalem 91904, Israel

Abbreviations: DCB, 2,6-dichlorobenzonitrile; TC's, terminal complexes; TG's, terminal globules.

C. Larsson, I.M. Møller (Eds):
The Plant Plasma Membrane

demonstrated by Glaser (1958), a membrane-localized UDP-glucose: (1→4)-β-D-glucan glucosyltransferase (EC 2.4.1.12, hereafter referred to as cellulose synthase) was detected, but activity was much lower in vitro than that observed in vivo. Many years later, work from the laboratory of Benziman in Israel clarified the effector requirements for the enzyme; the finding that the enzyme is activated by a unique compound, bis-(3′,5′)-cyclic diguanylic acid, allowed these workers to describe conditions which led to high rates of synthesis in vitro (Aloni et al. 1982; Ross et al. 1987). The synthase also requires Mg^{2+} and can be solubilized with digitonin under conditions where it still responds to the activator (Aloni et al. 1983); thus, one has reason to expect it will soon be purified to homogeneity and characterized in detail (see Lin and Brown 1989).

No such progress exists with respect to cellulose synthases from photosynthetic, eukaryotic organisms (reviewed extensively in Delmer 1987). Although good indirect evidence exists that UDP-glucose is also the precursor in these organisms (Carpita and Delmer 1981; Inouhe et al. 1986), no one has really succeeded in reproducibly demonstrating significant in vitro activity for a plasma membrane-localized cellulose synthase. Discovery of the unique activator in *A. xylinum* offered hope that this might be the missing requirement, but, to date, there is no evidence that this compound exists in plants, and addition of the cyclic dinucleotide from *A. xylinum* has not caused stimulation of any such activity using plant extracts (M. Benziman and D.P. Delmer, unpublished results). Thus, the requirements for demonstrating such activity in these organisms remain elusive. I shall return to this point later when aspects of regulation of cellulose synthesis are discussed.

2 Evidence for Plasma Membrane Localization

2.1 Bacteria

Bacteria such as *Acetobacter xylinum* and, to a lesser extent, species of *Rhizobium* and *Agrobacterium*, synthesize and secrete cellulose, not as a cell wall constituent, but as an extracellular pellicle. The presumed function of such material is to aid in the flotation of such nonmotile obligate aerobes as *A. xylinum*, or to aid in attachment to other organisms, as represented by the symbiotic *Rhizobium* or the pathogenic *Agrobacterium* species which interact with tissues of higher plants. These bacteria are of the gram-negative type and possess two surface membranes, the so-called inner and outer membranes. The inner membrane is most analogous to the plasma membrane of eukaryotes, and is the site of most transport systems and electrogenic pumps; the outer membrane is of lipopolysaccharide and is permeable to most small molecules but restricts the passage of macromolecules. Until recently, all the enzymological studies with *A. xylinum* had utilized membrane preparations which contained both membranes, and it was unclear where the synthase resided. Cytological studies by Zaar (1979) and by Brown and co-workers (Brown et al. 1976; Brown 1985;

review Haigler 1985) indicated that a row of 12 nm particles, thought to be the synthases, can be detected by freeze-fracture of the *A. xylinum* membranes, and that cellulose microfibrils synthesized by these complexes are extruded through pores in the outer membrane. Based on interpretations of the freeze-fracture images, Haigler and Benziman (1982) stated that it appeared most likely that the synthetic complexes resided within the outer membrane. However, Bureau and Brown (1987) recently achieved in vitro separation of the inner and outer membranes of *A. xylinum,* and measurement of the in vitro activity for the synthase indicated clearly that the enzyme resides on the inner membrane. Since this membrane is apparently not permeable to the substrate UDP-glucose, this localization seems most logical, assuming that the enzyme is a transmembrane complex which could accept UDP-glucose from the cytoplasm and extrude the glucan product from the outer face.

2.2 *Algae*

Many, but certainly not all, algae secrete cellulose as a cell wall component. As for higher plants, no cellulose synthase activity has been reported for algae, so localization studies have relied almost exclusively on cytological observations. One of the earliest cytological studies related to the localization of cellulose synthesis in algae was performed using *Pleurochrysis*, a genus which possesses a unique cell wall consisting of scales composed largely of cellulose. Brown et al. (1969) documented with elegant electron micrographs that such algae possess one large Golgi apparatus and that the cellulosic scales are synthesized within this organelle and then secreted to the wall. However, such algae appear to be the exception rather than the rule, and most other families of cellulosic algae apparently synthesize cellulose via organized synthase complexes embedded within the plasma membrane. The microfibrils synthesized by such complexes can be quite large (compare the diameter of about 30 nm for the microfibrils of the alga *Valonia* to a diameter of about 3.5 nm for that found in the primary wall of higher plants; see Delmer 1987); furthermore, the microfibrils are often highly crystalline and oriented in precise patterns within the algal walls.

The elegant freeze-fracture images of the plasma membrane of the cellulosic alga *Oocystis solitaria* by Brown and Montezinos (1976) provided the first images of apparent cellulose-synthesizing complexes in eukaryotes. Called "terminal complexes" or "TC's" by these authors, the apparent synthases are seen as organized linear arrays of protein subunits which fracture on the outer (EF) face of the plasma membrane; impressions of cell wall microfibrils can be visualized at the ends of the complexes indicating that the microfibrils are synthesized from the TC's (see Fig. 1 for examples of these linear complexes). Since that first report, a number of other reports have appeared documenting the occurrence of similar linear TC's in a variety of algae (see Quader and Robinson 1981; Brown 1985; Haigler 1985; and Delmer 1987 for reviews which include extensive discussion of TC's). As a rough estimate, it generally appears that the

Fig. 1. Freeze-fracture image of the plasma membrane E-face of *Oocystis solitaria* showing linear terminal complexes (TC's) at the ends of imprints of microfibrils. *Bar* = 0.2 μm (Photograph courtesy of H. Quader)

number of subunits found in the TC is similar to the number of glucan chains in the associated microfibril (Brown and Montezinos 1976; Itoh et al. 1984), a finding which agrees very nicely with Preston's insightful "ordered granule hypothesis" (Preston 1964) which predicted that microfibrils would be synthesized by organized protein complexes, each subunit of which would catalyze the synthesis of a single glucan chain. In at least one case (*Valonia;* Itoh and Brown 1984), the TC's may fracture to either face of the plasma membrane, supporting the notion that such TC's are transmembrane complexes which accept substrate from the cytoplasm and extrude the glucan product to the wall.

Not all cellulosic algae possess the linear type of TC; as first demonstrated by Giddings et al. (1980), other algae such as *Micrasterias denticulata* possess apparent synthase complexes of quite a different structure. In this alga, organized hexagonal arrays of rosette-shaped structures can be observed on the P-fracture face; in similar locations on the E-face, Giddings et al. (1980) sometimes observed arrays of particles that appeared to be complementary to the holes in the center of the rosettes. All of these organized arrays appeared at the ends of microfibrils and were suggested to be components of the cellulose synthases. This conclusion was further supported by the observation that the center-to-center spacing between rosette-particle arrays matches the average center-to-center spacing between microfibrils (see also Chap. 12). Brown (1985) gives a listing of all the algae studied to date, comparing those which possess linear TC's with those which possess the rosette-particle structure.

2.3 Lower and Higher Plants

Although direct demonstration of a cellulose synthase activity is lacking, there is some biochemical, in addition to cytological, evidence to support the notion that plants synthesize cellulose at the plasma membrane. In vivo labeling studies using supplied substrates such as [^{14}C]-glucose indicate that many high-molecular-weight polysaccharides can be detected within Golgi and Golgi-derived membrane vesicles (see Delmer and Stone 1988 for discussions of secretory processes involved in cell wall deposition). Kinetic studies, including pulse-chase experiments and use of inhibitors to disrupt Golgi function, support the notion that these polysaccharides are polymerized within the Golgi apparatus and move in Golgi-derived vesicles to fuse with the plasma membrane and thus deposit the polymers to the cell wall space. The limited analyses of the structure of these polysaccharides indicate that they resemble the matrix polysaccharides, and there is no report documenting, either by biochemical or cytological techniques, the existence of intracellular microfibrils of (1→4)-β-glucan. Furthermore, the existence of (usually isolated) rosettes which fracture on the P-face and particles (called terminal globules or TG's) on the E-face of the plasma membrane (see Fig. 2 for examples), support the notion that lower and higher plants possess TC's similar to those found in some of the algae (for reviews, see Brown 1985; Haigler 1985; Delmer 1987). Evidence that these structures are involved in cellulose synthesis in plants comes from several observations: (1) these complexes, like the activity of the enzyme, appear to be extremely labile and require special care to preserve their structure during preparation of samples for freeze-fracture (Herth and Weber 1984); (2) the abundance and localization of rosettes and globules are directly correlated with the rate and pattern of cellulose synthesis during xylogenesis, a case wherein highly ordered and developmentally regulated synthesis occurs (Schneider and Herth 1986); and (3) dichlorobenzonitrile (DCB), considered to be a very specific inhibitor of cellulose synthesis in plants (Delmer et al. 1987), causes a

Fig. 2A Freeze-fracture image of the plasma membrane P-face of a wheat root xylem element showing rosette structures (some of which are indicated by *arrows*) thought to be components of the cellulose synthase. **B** Freeze-fracture image of the plasma membrane E-face of a wheat root cortex cell showing terminal globules (*TG's*, two of which are indicated by *arrows*) which may also be components of the cellulose synthase complex. *Bars* for **A** and **B** = 0.1 μm (Photographs courtesy of W. Herth)

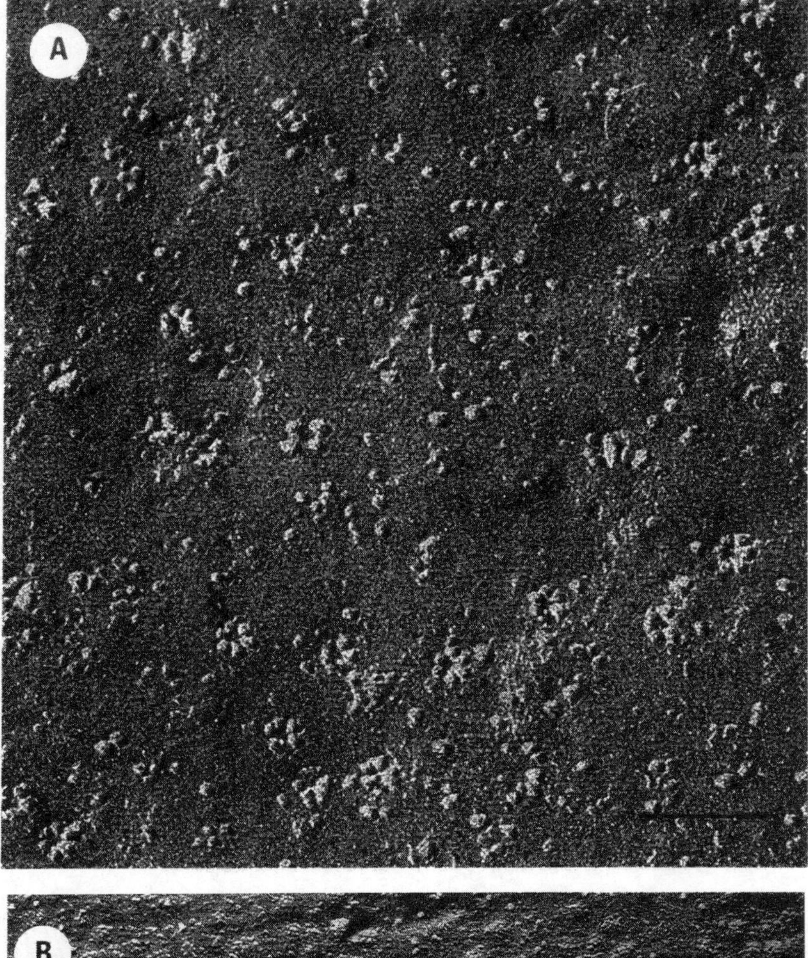
A

B

transient accumulation of rosettes followed by their gradual disappearance from the membrane (Herth 1989).

In summary, it would appear that all organisms which synthesize cellulose, with the exception of the scaled algae such as *Pleurochrysis,* do so via the catalytic activity of organized protein complexes, the cellulose synthases, embedded as transmembrane structures, within the plasma membrane. It must be noted that direct biochemical evidence that the TC's observed by freeze-fracture are, in fact, cellulose synthases is lacking. However, the indirect evidence that they are the synthases is rather compelling at present.

3 Other Glycosyltransferases on the Plasma Membrane of Higher Plants

As indicated earlier, cellulose is unique among cell wall polysaccharides in being synthesized at the plasma membrane. However, as is well known to all workers on plant plasma membranes, there is one additional glycosyltransferase known to exist within this membrane; this enzyme, originally called glucan synthase II by Ray (1979), is more clearly defined as UDP-glucose: (1→3)-β-D-glucan glucosyltransferase (EC 2.4.1.34). Since the product in plants is also called callose, the enzyme is also referred to as a callose synthase, and has become a classic marker enzyme for the plasma membrane of plants (Chap. 2). Although there are some exceptions (see Delmer 1987), callose is not a normal wall constituent, but rather is synthesized rapidly in response to mechanical wounding, stress, and/or pathogen invasion (Chap. 14). The enzyme, like cellulose synthase, uses UDP-glucose as substrate and appears to be a high-molecular-weight complex; however, it is unique in that it is activated by micromolar concentrations of Ca^{2+}, a property which may explain its latent nature in normal cells and its rapid activation upon cell perturbation (Chap. 14). Delmer (1977) first suggested that this enzyme might be an altered form of the cellulose synthase since the two enzymes share a common localization and substrate and seem to be active under opposite conditions, i.e., an intact cell synthesizes cellulose and a perturbed cell switches to synthesis of callose. This hypothesis has been further elaborated more recently (Jacobs and Northcote 1985; Delmer 1987). The enzyme is easily detected and exhibits high rates of synthesis in vitro when the appropriate substrate and effectors are supplied. Since progress is being made on partial purification of the enzyme and identification of the catalytic subunit (Delmer 1989; Wasserman and Sloan 1989), it is expected that antibodies or other probes for this enzyme may soon become available; when used for localization of the enzyme and its relationship to TC's, and/or for cloning and expression studies of the gene(s) for the enzyme, one can hope that its relationship to cellulose synthesis may be more directly tested in the near future.

4 The Life Cycle of a Cellulose Synthase

Indirect evidence is accumulating which suggests that the timing of synthesis, secretion, and turnover of cellulose synthase complexes may be an important factor in the regulation of the process in algae and higher plants. This may not be true for the bacterial synthase; in *A. xylinum* synthesis of cellulose can proceed for many hours in "resting cells", i.e., cells incubated in buffer supplied with glucose but lacking other nutrients including a nitrogen source to support continued protein synthesis. Also, no data exists which indicate that activity varies substantially throughout the cell division cycle or growth cycle in culture. Nevertheless the interesting activation in vitro by cyclic diguanylic acid does suggest that some form of precise regulation occurs in vivo, but does not occur primarily at the level of transcription, translation, posttranslational processing, or turnover of the synthase proteins.

In algae and higher plants, it is clear that large changes in the rate of cellulose synthesis can occur, primarily in response to developmental signals which program the onset of secondary wall synthesis. For example, Meinert and Delmer (1977) have calculated that the rate of cellulose synthesis increases $>$ 100-fold at the onset on secondary wall synthesis in developing cotton fibers. In vitro translation of RNA isolated before, during, and after this transition indicates that several abundantly translatable mRNA's increase substantially during the transition (Delmer et al. 1985), but the polypeptides coded for by these mRNA's have not been identified. One candidate may be an apparent protein receptor for the herbicide, DCB, a compound that specifically inhibits cellulose synthesis; the ability to detect the receptor increases substantially at this time. However, the function of this protein with respect to cellulose synthesis has not yet been clarified (Delmer et al. 1987), although Delmer has speculated that it may be a regulatory protein which modulates the rate of cellulose and callose synthesis in vivo. Thus we have hints, but little specific proof, that cellulose synthesis may, at least in part, be regulated during development by control of gene expression. It can be hoped, and perhaps reasonably anticipated, that the rapid advances in plant molecular biology will shed further light on this question in the near future (Chap. 16).

Since the synthase is a membrane complex, it is reasonable to assume that its polypeptides are synthesized on ER-associated polysomes. A recent report by Rudolph (1987) showing rosette structures within the ER membrane of the alga *Funaria* suggests that the component polypeptides may even be assembled at this early stage. Certainly rosettes have been clearly seen by freeze-fracture of membranes of dictyosome cisternae and secretory vesicles (Giddings et al. 1980; Haigler and Brown 1986). However, one note of caution is necessary; structures resembling rosettes have also been observed in membranes from organisms that do not synthesize cellulose (e.g., Henry et al. 1985), a fact that emphasizes how important it would be to have biochemical probes for the synthase.

Intriguing data exist which indicate that cellulose synthases in plants and algae undergo relatively rapid turnover. Quader and Robinson (1981) concluded from studies using inhibitors that cellulose synthesis was dependent upon continued protein synthesis, and indicated that the TC's have a half-life of less than 4 h (Robinson and Quader 1981). The groups of Herth and Schnepf in Heidelberg have published a number of papers which discuss the half-life of rosettes in algae and higher plants, and they have concluded that the half-life can be as short as 11 min (Reiss et al. 1984; Schnepf et al. 1985; Schneider and Herth 1986). Recently, Rudolph and Schnepf (1988) have shown that monensin, which blocks secretion from the Golgi, causes a rapid reduction in the number of rosettes observed in the plasma membrane of *Funaria*. From these results, they calculate that the lifetime of a single rosette is 21 min. Such estimates of the very rapid turnover of rosettes are interesting from several standpoints. One is that Reiss et al. (1984) estimate that this is very close to the time required for synthesis of a single microfibril, thus raising the intriguing possibility that the lifetime of a rosette is dictated by the time required for synthesis of a single microfibril. However, another possibility must also be considered. Steer (1988) has elegantly discussed the topic of plasma membrane turnover in plant cells, and has shown that the time required for complete turnover of the plant plasma membrane can be as short as 10 min for a nonexpanding secretory cell to up to 3 h for an elongating epidermal cell. Thus, the half-life of TC's may be dictated by the turnover rate of the plasma membrane; one might even go further and hypothesize that the length of a single microfibril might also be limited by this parameter. The interesting observation that DCB causes a transient accumulation of rosettes (Herth 1989) may indicate that the mode of action of this herbicide is somehow related to the turnover of the complexes. In this regard, Delmer (1989) recently showed preliminary results consistent with the conclusion that the DCB receptor of cotton fibers may form complexes with ubiquitin, a protein known to be involved in the turnover of eukaryotic proteins with short half-lives (Rechsteiner 1988); however, direct association with ubiquitin has not yet been proven.

Assuming that turnover does occur, we have no indication as to whether the subunits can be recycled; at least in *Oocystis*, it would appear that new subunits must be continually synthesized; no similar information is available for higher plants. Whatever their ultimate fate, based on recent studies with coated vesicles in higher plants, it is reasonable to speculate that the synthases would be internalized via such vesicles (see Chap. 10 for more discussion of this topic).

5 Why the Plasma Membrane?

5.1 Role in Microfibril Orientation

Cellulosic microfibrils are long, insoluble, often crystalline, and, in many cases, are deposited in the cell wall in a highly oriented manner which can circumscribe the entire cell. For all of these reasons, it is difficult to imagine how this could be achieved if synthesis occurred in the Golgi as opposed to the plasma membrane.

In fact, the only known case of Golgi-localized cellulose synthesis is for the scaled cellulosic algae where an entire cellulosic scale is synthesized and then secreted.

In all other algae and lower and higher plants, cellulose synthesis occurs at the plasma membrane and, in most, if not all, cases, evidence indicates that the mechanism of orientation is somehow directed by cortical microtubules which underlie this membrane (for recent reviews, see references in Giddings and Staehelin 1988; the interaction of the plasma membrane with the cytoskeleton is also treated in Chap 11). Current theories, discussed in the cited reviews, propose that the cellulose synthase complexes move by the force generated from polymerization of the rigid microfibrils, and that underlying microtubules, which are almost always found aligned in the direction of newly synthesized microfibrils, guide the direction of this movement. It is believed that there is not a direct connection between the synthases and the microtubules, but rather that the latter set up channels (or tracks) in the plane of the membrane which roughly guide synthase movement. However, occasionally some connections between the microtubules and the plasma membrane are apparent (Giddings and Staehelin 1988). Although this is believed to be the general case, there may be some cases, for example, during secondary wall synthesis and in synthesis of root hair walls, in which microtubules play no role in orientation (Giddings and Staehelin 1988).

5.2 *Role in Regulation*

It is clear from many of the chapters in this book that the plasma membrane plays a major role in perception and signaling in plants. It is also clear from what is described in this Chapter that the process of cellulose synthesis may be regulated at a variety of levels. The very rapid, apparent "inactivation" of synthase activity upon cell perturbation or breakage is an intriguing event; this author suspects that this phenomenon is in some way related to some form of rapid post-translational regulation of activity in vivo which somehow also relates to the classic "cellulose-to-callose switch" which occurs upon perturbation.

If one asks which factors change upon perturbation and how these may relate to this "switch", the first obvious change is that the level of intracellular Ca^{2+} rises and is one factor in the induction of callose synthesis (see Chap. 14). Evidence that such a rise directly inhibits cellulose synthesis is sparse; however, Quader and Robinson (1979) have shown that Ca^{2+} ionophores can interfere with cellulose synthesis in *Oocystis*. Nevertheless, it is clear that changes in intracellular concentration of free Ca^{2+} can signal a variety of secondary responses, such as protein phosphorylation, changes in the status of ATPases, ion channels and membrane potential, cytoskeletal organization, enzyme activities, and secretory activity (Allan and Trewavas 1987; Chaps. 5–9, 12 and 14). In this regard, reports of effects of transmembrane electrical potential on β-glucan synthesis (Bacic and Delmer 1981; Delmer et al. 1982), and of protein kinase on the activity of callose synthase (Paliyath and Poovaiah 1988), may offer some initial clues. Thus, one might envision any number of mechanisms whereby signal sensing at the plasma membrane may directly or indirectly affect cellulose synthase activity. Understanding of such signaling and its relationship to cellulose synthesis is an exciting challenge for the future.

Acknowledgments. The author is grateful to Drs. Werner Herth and Hartmut Quader of the University of Heidelberg for contributing the photographs of terminal complexes shown in Figs. 1 and 2.

References

Allan EF, Trewavas AJ (1987) The role of calcium in metabolic control. In: Stumpf PK, Conn EE (eds) The biochemistry of plants, Vol 12, Academic Press, San Diego, pp 117–149

Aloni Y, Delmer DP, Benziman M (1982) Achievement of high rates of in vitro synthesis of 1,4-β-glucan: activation by cooperative interaction of the *Acetobacter xylinum* enzyme system with GTP, polyethylene glycol, and a protein factor. Proc Natl Acad Sci USA 79:6448–6452

Aloni Y, Cohen R, Benziman M, Delmer DP (1983) Solubilization of the UDP-glucose: 1,4-β-D-glucosyltransferase (cellulose synthase) from *Acetobacter xylinum*. J Biol Chem 258: 4419–4423

Bacic A, Delmer DP (1981) Stimulation of membrane-associated polysaccharide synthetases by a membrane potential in developing cotton fibers. Planta 152:346–351

Bolwell GP (1988) Synthesis of cell wall components: aspects of control. Phytochemistry 27:1235–1253

Brown RM Jr (1985) Cellulose microfibril assembly and orientation: recent developments. J Cell Sci Suppl 2:13–32

Brown RM Jr, Montezinos D (1976) Cellulose microfibrils: visualization of biosynthetic and orienting complexes in association with the plasma membrane. Proc Natl Acad Sci USA 73:143–147

Brown RM Jr, Franke WW, Kleinig H, Falk H, Sitte P (1969) Cellulosic wall component produced by the Golgi apparatus of *Pleurochrysis scherffelii*. Science 166:894–897

Brown RM Jr, Willison JHM, Richardson CL (1976) Cellulose biosynthesis in *Acetobacter xylinum*: visualization of the site of synthesis and direct measurement of the in vivo process. Proc Natl Acad Sci USA 73:4565–4569

Bureau TE, Brown RM Jr (1987) In vitro synthesis of cellulose II from a cytoplasmic membrane fraction of *Acetobacter xylinum*. Proc Natl Acad Sci USA 84:6985–6989

Carpita NC, Delmer DP (1981) Concentration and metabolic turnover of UDP-glucose in developing cotton fibers. J Biol Chem 256:308–315

Delmer DP (1977) Biosynthesis of cellulose and other plant cell wall polysaccharides. Recent Adv Phytochem 11:45–77

Delmer DP (1987) Cellulose biosynthesis. Annu Rev Plant Physiol 38:259–290

Delmer DP (1989) The relationship between the synthesis of cellulose and callose in higher plants. Appl Polymer J, in press

Delmer DP, Stone BA (1988) Biosynthesis of plant cell walls. In: Stumpf PK, Conn EE (eds) The biochemistry of plants, Vol 14, Academic Press, San Diego pp 373–420

Delmer DP, Benziman M, Padan E (1982) Requirement for a membrane potential for cellulose synthesis in intact cell of *Acetobacter xylinum*. Proc Natl Acad Sci USA 79:5282–5286

Delmer DP, Cooper G, Alexander D, Cooper J, Hayashi T, Nitsche C, Thelen M (1985) New approaches to the study of cellulose biosynthesis. J Cell Sci Suppl 2:33–50

Delmer DP, Cooper G, Read SR (1987) Identification of a receptor protein in cotton fibers for the herbicide 2,6-dichlorobenzonitrile. Plant Physiol 84:415–420

Giddings TH, Staehelin LA (1988) Spatial relationship between microtubules and plasma-membrane rosettes during the deposition of primary wall microfibrils in *Closterium* sp. Planta 173:22–30
Giddings TH, Brower DL, Staehelin LA (1980) Visualization of particle complexes in the plasma membrane of *Micrasterias denticulata* associated with the formation of cellulose microfibrils in primary and secondary walls. J Cell Biol 84:327–339
Glaser L (1958) The synthesis of cellulose in cell-free extracts of *Acetobacter xylinum*. J Biol Chem 232:627–636
Haigler CH (1985) The functions and biogenesis of native cellulose. In: Nevell RP, Zeronian SH (eds) Cellulose chemistry and its applications. Horwood, Chichester, pp 30–83
Haigler CH, Benzimann RM Jr (1982) Biogenesis of cellulose I microfibrils occurs by cell-directed self-assembly in *Acetobacter xylinum*. In: Brown RM Jr (ed) Cellulose and other natural polymer systems. Plenum, New York, pp 273–297
Haigler CH, Brown RM Jr (1986) Transport of rosettes from the Golgi apparatus to the plasma membrane in isolated mesophyll cells of *Zinnia elegans* during differentiation to tracheary elements in suspension cultures. Protoplasma 134:111–120
Henry Y, Pouphile M, Gulik-Krzywicki T, Wiessner W, Lefort-Tran M (1985) Freeze-fracture study of ordered arrays of particles in the plasma membrane of *Chlamydobotrys stillata* Korsch. Protoplasma 126:100–113
Herth W (1989) Inhibitor effects on putative cellulose synthetase complexes of vascular plants. Appl Polymer J, in press
Herth W, Weber G (1984) Occurrence of the putative cellulose-synthesizing "rosettes" in the plasma membrane of *Glycine max* suspension culture cells. Naturwissenschaften 71:153–154
Inouhe M, Yamamoto R, Masuda Y (1986) Inhibition of IAA-induced cell elongation in *Avena* coleoptile segments by galactose: its effect on UDP-glucose formation. Physiol Plant 66:370–376
Itoh T, Brown RM Jr (1984) The assembly of cellulose microfibrils in *Valonia macrophysa* Kutz. Planta 160:373–381
Itoh T, O'Neil RM, Brown RM Jr (1984) Interference of cell wall regeneration of *Boergesenia forbesii* protoplasts by Tinopal LPW, a fluorescent brightening agent. Protoplasma 123:174–183
Jacobs SR, Northcote D (1985) In vitro glucan synthesis by membranes of celery petioles: the role of the membrane in determining the linkage formed. J Cell Sci Suppl 2:1–11
Lin FC, Brown RM Jr (1989) Purification of cellulose synthase from *Acetobacter xylinum*. Appl Polymer J, in press
Meinert M, Delmer DP (1977) Changes in biochemical composition of the cell wall of the cotton fiber during development. Plant Physiol 59:1088–1097
Paliyath G, Poovaiah BW (1988) Promotion of β-glucan synthase activity in corn microsomal membranes by calcium and protein phosphorylation. Plant Cell Physiol 29:67–73
Preston RD (1964) Structural and mechanical aspects of plant cell walls with particular reference to synthesis and growth. In: Zimmermann MH (ed) The formation of wood in forest trees. Academic Press, New York, pp 169–201
Quader H, Robinson DG (1979) Structure, synthesis, and orientation of microfibrils. VI. The role of ions in microfibril deposition in *Oocystis solitaria*. Eur J Cell Biol 20:51–56
Quader H, Robinson DG (1981) *Oocystis solitaria*: a model organism for understanding the organization of cellulose synthesis. Ber Dtsch Bot Ges 94:75–84
Ray PM (1979) Separation of maize coleoptile cellular membranes that bear different types of glucan synthetase activity. In: Reid E (ed) Plant organelles. Horwood, Chichester, pp 135–146
Rechsteiner M (1988) Ubiquitin. Plenum, New York
Reiss HD, Schnepf E, Herth W (1984) The plasma membrane of the *Funaria* caulonema tip cell: morphology and distribution of particle rosettes, and the kinetics of cellulose synthesis. Planta 160:428–435
Robinson DG, Quader H (1981) Structure, synthesis, and orientation of microfibrils. IX. A freeze-fracture investigation of the *Oocystic* plasma membrane after inhibitor treatment. Eur J Cell Biol 25:278–288
Ross P, Aloni Y, Weinhouse C, Michaeli D, Weinberger-Ohana P, Meyer R, Benziman M (1987) Regulation of cellulose synthesis in *Acetobacter xylinum* by cyclic diguanylic acid. Nature 325:279–281

Rudolph U (1987) Occurrence of rosettes in the ER membrane of young *Funaria hygrometrica* protonemata. Naturwissenschaften 74:439

Rudolph U, Schnepf E (1988) Investigations of the turnover of the putative cellulose-synthesizing particle "rosettes" within the plasma membrane of *Funaria hygrometrica* protonema cells. I. Effects of monensin and cytochalasin B. Protoplasma 143:63–73

Schnepf E, Witte O, Rudolph U, Deichgraber G, Reiss HD (1985) Tip cell growth and the frequency and distribution of particle rosettes in the plasmalemma: experimental studies in *Funaria* protonema cells. Protoplasma 127:222–229

Schneider B, Herth W (1986) Distribution of plasma membrane rosettes and kinetics of cellulose formation in xylem development of higher plants. Protoplasma 131:142–152

Steer MW (1988) Plasma membrane turnover in plant cells. J Exp Bot 39:987–996

Wasserman BP, Sloan ME (1989) Molecular approaches for probing the structure and function of callose and cellulose synthases. In: Haigler C, Weimer P (eds) Marcel Dekker, New York, in press

Zaar K (1979) Visualization of pores (export sites) correlated with cellulose production in the envelope of the gram-negative bacterium *Acetobacter xylinum*. J Cell Biol 80:773–777

Chapter 12 The Plasma Membrane-Associated Cytoskeleton

J.A. Traas[1]

[1]I.N.R.A., Physiopathologie, B.V. 1540, F-21034 Dijon Cédex, France

Abbreviation: PPB, preprophase band.

C. Larsson, I.M. Møller (Eds):
The Plant Plasma Membrane

1 Introduction

The cytoplasm of all eukaryotic cells contains a three-dimensional network of protein fibers which has been termed the cytoskeleton. This extremely complex structure interconnects all cytoplasmic elements and joins the nuclear envelope with the plasma membrane. The cytoskeleton is composed of different types of fibers: actin filaments, microtubules, and intermediate filaments. In addition, a number of associated proteins have been identified which influence the structure, function, and distribution of the backbone polymers. Different domains can be distinguished in this network: the cytoskeletal elements surrounding the nucleus, the filamentous network in the subcortical cytoplasm, and the cortical or membrane-associated cytoskeleton which is the topic of the present chapter.

The cortical cytoskeleton of plant cells has been the subject of numerous studies and over recent years our understanding of its structure and function has undergone rapid progress. This cortical network determines many dynamic processes in the plant cell including morphogenesis, redistribution of surface components, endo-exocytosis, and the positioning of cytoplasmic elements. Moreover, by forming a coherent network the cytoskeleton mediates interactions between the plasma membrane and all parts of the cell, playing a key role in the transmission and processing of environmental stimuli.

An extensive summary of such a broad field would be beyond the scope of this chapter, which therefore restricts itself mainly to the cytoskeleton in higher plants. However, some of the data obtained on lower "green" plants must be presented to fill gaps in our knowledge of higher plant cells. For additional information the reader will be referred to specific publications and review articles. Two aspects of cytoskeleton-membrane interactions, i.e., endo- and exocytosis via coated pits and cell wall synthesis, are discussed in Chapters 10 and 11, respectively, and will only be treated briefly here.

2 The Cytoskeleton of Plant Cells

Although the (higher) plant cytoskeleton undoubtedly contains many proteins, only the microtubular and F-actin system have been well characterized (reviews, Gunning and Hardham 1982; Lloyd 1987; Staiger and Schliwa 1987). Recently, the presence of other elements was demonstrated such as the fibrillar bundle proteins and the cytoskeleton-associated proteins myosin and troponin-T (Dawson et al. 1985; Lim et al. 1986; Parke et al. 1986). Some relevant characteristics of these elements will be presented in this chapter. More details can be found in a number of reviews on plant microtubules and microfilaments (Gunning and Hardham 1982; Jackson 1982; Lloyd 1982, 1984, 1987; Dawson and Lloyd 1987; Staiger and Schliwa 1987). More general aspects are also covered in different reviews on the cytoskeleton in animals and lower plants (Geiger 1983; Dustin 1985; Pollard and Cooper 1986; Schliwa 1986; Huffaker et al. 1987; Lazarides 1987).

2.1 Microtubules

Microtubules are hollow bundles of protofilaments of about 20 nm in diameter. They occur throughout the cytoplasm but in higher plants they are almost exclusively associated with the plasma membrane during interphase. Depending on the stage of cell development they are able to de- and repolymerize into functionally distinct arrays. Biochemical analyses have shown that plant microtubules, like their animal counterparts, are composed of tubulin, a heterodimer containing one alpha and one beta subunit with molecular weights of 50–55 kD. A number of isotypes have been identified for each subunit (e.g., Dawson and Lloyd 1987; Hussey et al. 1987). Although there is no evidence that they are used for different types of microtubules, the presence of these isotypes apparently varies during plant cell development (Hussey et al. 1988). Like most cytoskeletal proteins, tubulins have been highly conserved throughout evolution. Yet there are important differences between tubulins from plant and animal sources as far as their migration behavior in gels and their response to antimicrotubular drugs are concerned (review, Dawson and Lloyd 1987).

Tubulin genes of different plant species have been identified and sequenced (e.g., Guiltinan et al. 1987; Marks et al. 1987; Ludwig et al. 1988). They show a high degree of homology with tubulin genes from distantly related species. The derived amino acid sequence of β-tubulins of soybean shows 79–83% homology to β-tubulins of vertebrate species. Both α- and β-tubulins are encoded by multigene families of variable size.

2.2 Microfilaments

Microfilaments form bundles of different sizes or cables which are organized into a network, filling the cytoplasm. Their major constituent is actin, a 42 kD globular protein (G-actin) which can be polymerized in vitro into 5–7 nm filaments (F-actin) in the presence of ATP and high ionic strength. In vivo, plant actin forms microfilaments in association with other proteins, probably including myosin and troponin-T. Due to the instability of plant actin in vitro and the high proteolytic activity in extracts, it has been extremely difficult to characterize the protein (reviews: Jackson 1982; Staiger and Schliwa 1987). However, actin genes have been identified and sequenced (e.g., Shah et al. 1982). Plant cells of different species contain multigene families which show a high degree of homology with animal actin genes. Despite this, important heterogeneity has been observed between genes coding for plant and animal actin and even between plant genes within one species (e.g., Hightower and Meagher 1985; review, Staiger and Schliwa 1987).

2.3 Fibrillar Bundles

In addition to microtubules and F-actin, a class of 7–10 nm filaments has been observed which in carrot protoplasts form large bundles with a diameter of 50–100 nm (Powell et al. 1982). These bundles contain a number of different proteins some of which show immunological homology with animal intermediate filaments (Dawson et al. 1985). In meristematic cells of onions a 50 kD protein was also identified which cross-reacted with antibodies against a conserved epitope of intermediate filament proteins from animals. This protein seems to associate with the microtubular arrays (Dawson et al. 1985; see also Parke et al. 1987). Whether these proteins form the equivalent of intermediate filaments in plant cells remains to be established. If they do occur – as the immunological data indicates – they would form a fine network codistributing with microtubules (see also Hargreaves et al. 1989).

2.4 Cytoskeleton-Associated Proteins

The only cytoskeleton-associated protein which has been identified so far is myosin. It has been purified from *Nitella* and myosin-like polypeptides have also been found in different higher plant cells (Kato and Tonomura 1977; Vahey et al. 1982; Parke et al. 1986). Plant myosin apparently resembles its animal equivalent which is a hexamer of two heavy chains (200 kD) and four light chains (20 kD).

The presence of a troponin T-like protein has also been reported (Lim et al. 1986). Troponin-T is a Ca^{2+}-binding polypeptide which interacts with actin and probably also with tubulin.

2.5 Coated Pits and Coated Vesicles

Coated pits and vesicles are membrane areas with polygonal assemblies of clathrin, a hexamer composed of three 180 kD side chains associated with three light chains (e.g., Coleman et al. 1987; Robinson and Depta 1988). Although coated pits are usually considered separately, they form part of the membrane-associated network and can therefore be regarded as cytoskeletal proteins (Geiger 1983; for plant cells: Traas 1984; Emons and Traas 1986). In plant cells coated pits are involved in endocytosis and regulation of membrane traffic (Joachim and Robinson 1984; Tanchak et al. 1984; Emons and Traas 1986). A role in the uptake of specific molecules – as was shown for animal cells – has not been demonstrated for plants.

Coated pits are discussed in detail in Chapter 11.

3 The Plasma Membrane-Associated Cytoskeleton

It is difficult to define exactly the boundaries of the cortical cytoskeleton. In a strict sense it includes only those proteins that are directly bound to integral membrane components. However, the molecular basis of membrane-cytoskeleton associations in higher plant cells is poorly understood and to date none of the proteins responsible for the interactions has been identified. In the absence of biochemical data we have to rely on ultrastructural criteria such as the proximity to the membrane and the presence of visible links. As our view on the membrane-cytoskeleton complex largely depends on the possibilities and limitations of the techniques used, the latter will be briefly discussed in the next section.

3.1 Methods to Study the Cortical Cytoskeleton

Until about 10 years ago, the only way of studying the membrane-associated cytoskeleton was to use thin sectioning in combination with chemical fixation and electron microscopy. This technique has given important evidence for the existence of links between microtubules and the plasma membrane and serial sectioning has been very useful in characterizing the structure of cortical microtubule arrays (e.g., Gunning et al. 1978; Hardham and Gunning 1978; Seagull and Heath 1980a). More recently, this method was greatly improved with the introduction of freeze substitution as a means of stabilizing rapidly the cytoskeleton which can otherwise be highly sensitive to fixation (Tiwari et al. 1984; Hepler 1985; Lancelle et al. 1986, 1987). The technique has also some restrictions, since it only gives two-dimensional information on a limited number of cells and even serial sectioning does not allow a faithful three-dimensional reconstruction of the integral network.

An important complement to electron microscope techniques is the immunofluorescence technique introduced by Lloyd et al. (1979) for the labeling of cortical microtubular arrays in suspension culture cells. Immunofluorescence has been very useful in characterizing the dynamics of the cortical cytoskeleton, as it allows the observation of the entire microtubule or F-actin network in a large number of cells (review, Lloyd 1987). In addition to antibodies, fluorescent phallotoxins are now widely used to visualize F-actin (for a review on plant cells, see Staiger and Schliwa 1987). Although fluorescence microscopy offers enormous advantages, it still needs to be correlated with ultrastructural research, especially where the detailed organization of the cytoskeleton is concerned. In view of the limitations of thin sectioning, electron microscope techniques have been developed to study the cortical cytoskeleton in nonsectioned material. Thus, preparations of osmotically burst, negatively stained protoplasts have been used (e.g., Van der Valk et al. 1980); and for more general applications on tissues the dry cleaving technique (Mesland et al. 1981) has been adapted for plant cells (Traas 1984; Traas et al. 1984, 1985; Quader et al. 1986). This method allows the visualization of the plasma membrane and its adherent structures: fixed cells are attached to grids and ruptured after staining, dehydration, and critical point drying. After this procedure large areas of membrane and the adherent cytoplasm can be studied directly in the electron microscope. The technique has also been adapted for immunogold labeling making it a powerful tool for the characterization of the cortical cytoskeleton (Traas and Kengen 1986; Traas 1989).

Each of these techniques gives a different view of the membrane-associated cytoskeleton (cf. Fig. 1A-G). Looking at sections, one tends to consider only those elements that are tightly appressed to the plasma membrane, whereas immunofluorescence and dry cleaving show us the cytoskeleton as a coherent network which has in a broad sense all its components linked to the membrane.

Clearly our conclusions on membrane-filament interactions should always be based on different techniques and as new microscope techniques and biochemical methods are developed our view on those interactions will undoubtedly change.

3.2 Composition of the Membrane-Bound Cytoskeleton

The use of electron microscopy in combination with techniques such as dry cleaving, freeze etching, and freeze substitution reveal the presence of a network of filaments in the cell cortex (Traas 1984; Traas et al. 1984, 1985; Hawes 1985;

Fig. 1A-G. Different microscopic views of the cortical cytoskeleton. **A,B** The cortical cytoplasm in freeze-"slammed", deep-etched carrot protoplasts (Hawes and Martin 1986). Microtubules (*arrows*) can be observed close to the plasma membrane (*PM*), linked to each other by cross-bridges. In **B** also a coated pit (*CP*) in visible; *bars* = 200 nm (Courtesy of C. Hawes). **C,D** Freeze-substituted cells of *Tradescantia* staminal hairs. Cells were first rapidly frozen and subsequently fixed at low temperature in a mixture of acetone and osmic acid. Sections show that the plasma membrane and the cortical cytoplasm are well preserved. Microtubules (*arrows*) can be observed in tangential and transverse sections. In **C** also filaments parallel to the microtubules are present (*arrowheads*). *CP* coated pit; *bars* = 150 nm (Lancelle et al. 1986; courtesy of P. Hepler). **E,F** Dry-cleaved protoplast (**E**) and meristematic root cells (**F**). Fixed cells are attached to grids and disrupted after critical point drying. This technique allows the visualization of the integral membrane-associated cytoskeleton in detail (**E**) and at low magnification (**F**). In **E** microtubules (*arrows*), coated pits (*CP*), and a random network of filaments can be observed; *bar* in **E** 200 nm; in **F** 1 μm. **F** reprinted with permission from Traas et al. (1984), Eur J Cell Biol 34:229–238. **G** Immunofluorescence preparation of an onion root cell showing the cortical microtubular helix. This technique is very useful in studying the overall organization of the cytoskeleton; *bar* = 5 μm

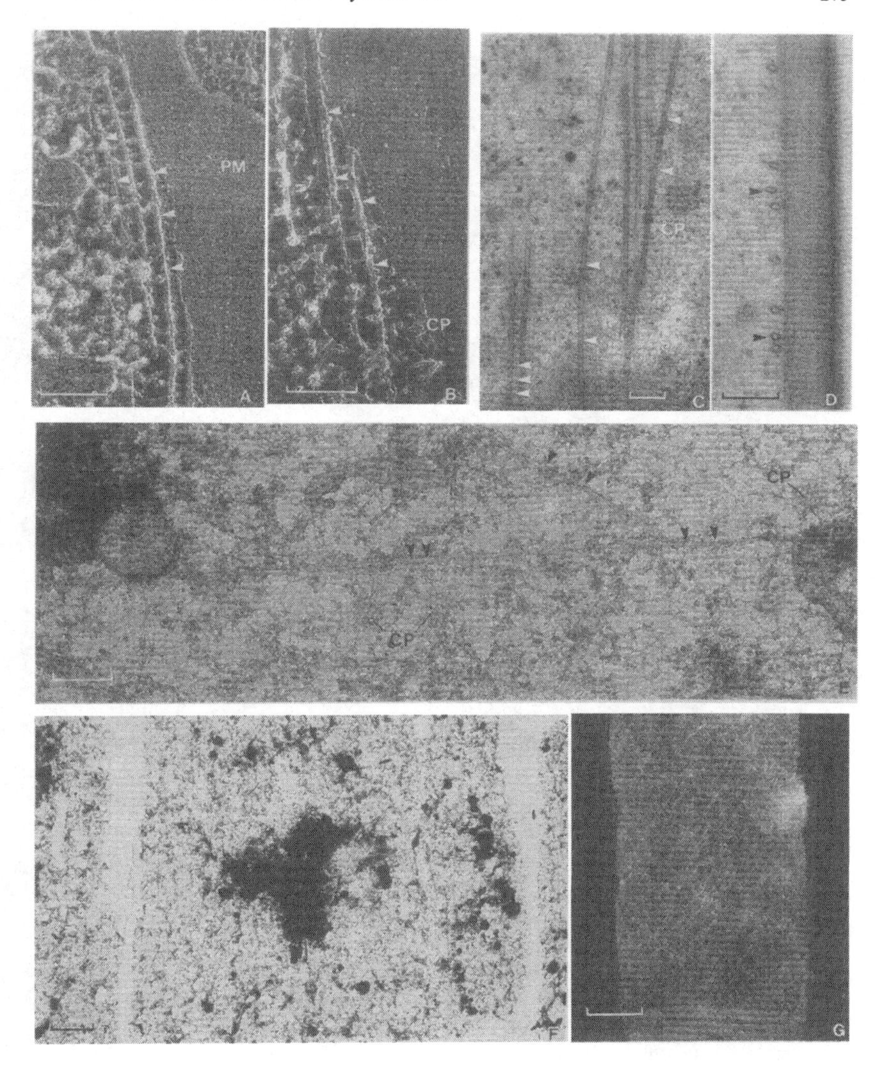
PM
CP
A
B
CP
C
D
CP
CP
E
F
G

Hawes and Martin 1986; Lancelle et al. 1986, 1987). Only microtubules and coated pits have been identified unequivocally as components of this network, both systems being tightly linked to the membrane. Correlating these results with immunofluorescence has shown that F-actin (Traas et al. 1987), fibrillar bundle proteins (Dawson et al. 1985; Hargreaves et al. 1989), and troponin T-like polypeptides (Lim et al. 1986) all codistribute with the cortical microtubules. Moreover, using rhodamine phalloidin as a probe it was demonstrated that F-actin also forms an extensive network near the plasma membrane (e.g., Parthasarathy et al. 1985; Derksen et al. 1986; Kakimoto and Shibaoka 1987; Seagull et al. 1987; Traas et al. 1987; Lloyd and Traas 1988).

4 Organization and Dynamics of the Plasma Membrane-Associated Cytoskeleton

4.1 The Cortical Microtubule Array

4.1.1 Interphase

Electron microscope data initially suggested that membrane-associated microtubules are relatively short overlapping elements that form hoops around the cell (e.g., Hardham and Gunning 1978; Gunning and Hardham 1982). This concept was radically modified with the introduction of immunofluorescence showing that in many plant cells microtubules are organized in helical arrays of variable pitch (Lloyd 1983, 1984; Traas et al. 1984; Fig. 1G). A model has been proposed in which populations of long, stable microtubules form helices associated with the membrane which can wind or unwind, thus reorientating as integral units (Lloyd 1984; Lloyd and Seagull 1985). This means that rearrangement is achieved via a gradual change in the helix pitch and not via a complete de- and repolymerization of the array. However, it was proposed that some de- and repolymerization and intermicrotubule sliding do take place to compensate for the changes in length (or width) of the helix that would otherwise occur. The model requires that the stability and dynamics of the helical array depend equally on interactions between individual microtubules and on associations with the plasma membrane via cross-bridges.

Several observations support the above model. Roberts et al. (1985), for instance, used ethylene to induce a 90° shift in the orientation of microtubules of mung bean and pea cortical and epidermal cells. In the presence of the hormone the proportion of transverse helices gradually decreased, whereas the proportion of more steeply pitched helices increased. Comparable gradual transitions have been observed during the differentiation of a number of plant cell types such as root cells and cotton seed hairs (Traas et al. 1984; Seagull 1986; Quader et al. 1987).

This highly dynamic concept was further extended by Traas et al. (1984, 1985) who quantified length, density, and orientation of cortical microtubules in dry cleaved root cells. These analyses confirmed that helical arrays are a common feature of most plant cells. It was also shown that the long elements which form the helix are in fact composed of shorter, overlapping, interconnected microtubules as was suggested earlier using thin sectioning. Although most microtubules are relatively short in these root cells, their length is highly variable (0.5–45 μm). The length appears to be correlated with the stage of cell development and increases three- to fourfold when growth ceases.

A number of observations demonstrate that the helical model does not apply to all cells and that there are different ways of rearrangement. Under certain conditions the helix can be disassembled into a random array. This happens, for instance, when protoplasts are prepared from plasmolyzed plant cells, but a "randomization" of microtubules has also been observed in tissues and seems to be related to de- and redifferentiation processes (Wilms and Derksen 1988) occurring after wounding or during artificially induced organogenesis/embryogenesis in vitro. This randomization is not necessarily in conflict with the helical model the main feature of which is that the array is an integral one, not composed of disconnected units. In some cells, density and length of the microtubules could be insufficient to allow a helical arrangement and under certain conditions there might not be enough cross-linking proteins available to stabilize the cortical array. For instance in *Spirogyra* cells, microtubules depolymerize in the presence of amiprophosmethyl (APM). Upon removal of the drug the microtubules reform, first in random orientation, but as polymerization progresses and microtubule density (and length?) increases, they rearrange to form a transverse or flat helical array (Hogetsu 1987). Similar results were obtained after the depolymerization of microtubules in the alga *Closterium* at low temperature (Hogetsu 1986). These data demonstrate that the helical arrangement is not imposed on the plasma membrane by a preexisting template and that microtubular orientation is largely determined by the properties of the array itself.

4.1.2 Cell Division

During cell division the distribution of cytoskeletal elements changes following a well-defined program. In most cells from organized tissues of higher plants, the first of these rearrangements is the formation of a membrane-bound ring of densely packed microtubules around the nucleus which accurately indicates the future plane of division (reviews, Gunning and Hardham 1982; Lloyd 1987). This structure usually appears just before prophase, hence its name preprophase band (PPB). Immunofluorescence studies indicate that the PPB arises directly from the cortical array by a "bunching up" of membrane-associated microtubules in the plane of division (Fig. 2A-E). Evidence comes from the observation of very broad as well as double bands at early stages of PPB formation, suggesting that the cell gradually determines its site of division with increasing precision (Wick and Duniec 1983; Doonan et al. 1987). Those microtubules that do not participate in the formation of the PPB are depolymerized. As the first

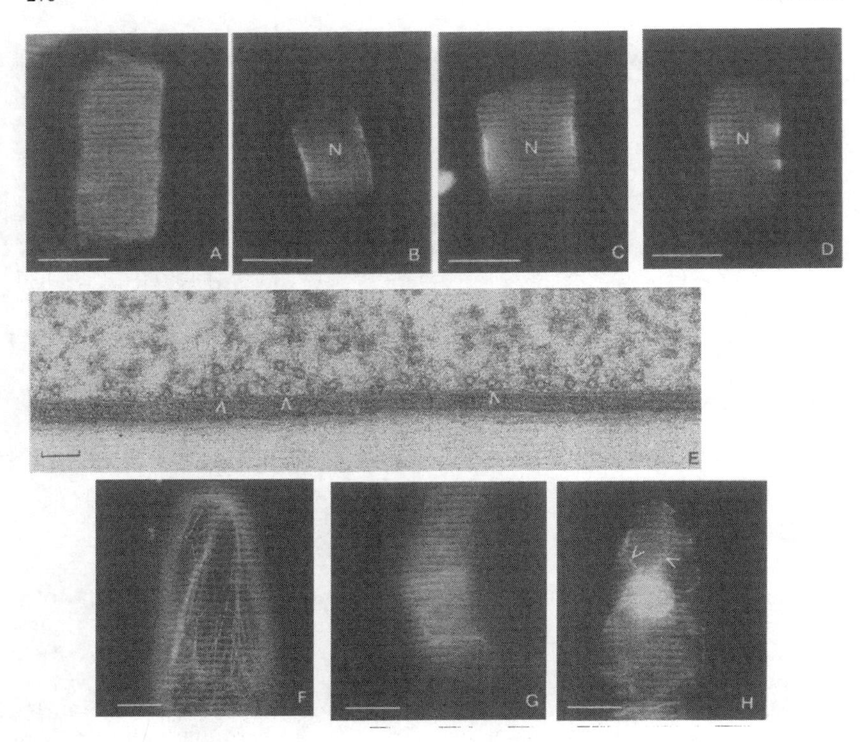

Fig. 2A-H. The cortical cytoskeleton during cell division (**A-D** and **F-H**, immunofluorescence micrographs; **E**, electron micrograph). **A** The interphase array of meristematic onion root cells, showing transverse arrays of cortical microtubules. **B-D** Different stages of preprophase band formation. Initially, a broad band or even multiple bands are formed (**B**), which then tighten and concentrate at the future division site. In **D** this process is almost completed. Besides the main band surrounding the nucleus, another band is still present. *N* Nucleus; *bars* = 10 μm. **E** Preprophase band in a freeze-substituted cell from a *Gibasis* staminal hair, showing that microtubules are linked to each other and the plasma membrane (links with the plasma membrane indicated by *arrowheads*); *bar* = 100 nm (Lancelle et al. 1986; courtesy of P. Hepler). **F** F-actin array in stamen hair cell of *Tradescantia*, showing a fine system of transverse cortical bundles and axially oriented cables deeper in the cytoplasm; *bar* = 20 μm. **G-H** F-actin in dividing carrot suspension culture cells. Actin is associated with the preprophase band microtubules (**G**) and with the spindle (**H**). A network of filament bundles is present throughout division and links the spindle to the cortex (*arrowheads*); *bars* = 10 μm. **F-H** from Traas et al. (1987) with permission of the Rockefeller University Press

nuclear-associated spindle microtubules appear, the PPB microtubules dissociate and at metaphase no microtubules can be found outside the mitotic apparatus. The structure and development of the spindle and the phragmoplast are extremely complex and will not be discussed here in detail (reviews, Gunning and Hardham 1982; Hepler 1985; Lloyd 1987; see also Wick and Duniec 1984; Schmit et al. 1985). At late anaphase the new membrane and cell wall are formed in the phragmoplast which contains numerous short microtubules perpendicular to the plane of division. The phragmoplast grows out and fuses with the plasma membrane exactly bisecting the former PPB site (Gunning et al. 1978). Already during telophase the first cytoplasmic microtubules reappear, apparently from microtubule nucleation sites at the surface of the nuclear envelope. They first radiate from the latter to the plasma membrane where they subsequently reform a flat, helical array. This was seen in immunofluorescence studies (Lloyd et al. 1985; Wick 1985). Later, studies using antisera to detect amorphous microtubule nucleation sites confirmed that these sites are clustered around the nucleus (Clayton et al. 1985).

4.2 The Cortical Actin Array

4.2.1 Interphase

The use of fluorescent phallotoxins and, to a lesser extent, antibodies have provided us with extensive information about the actin network in a range of plant cells (e.g., Pesacreta et al. 1982; Parthasarathy et al. 1985). From these studies it can be concluded that plant cells possess a dense network of actin-containing bundles which can be divided into different interconnected arrays: a perinuclear basket, a system of thick cables deep in the cytoplasm, and a fine network close to the plasma membrane (Derksen et al. 1986; Seagull et al. 1987; Traas et al. 1987). In addition, correlating fluorescence microscopy with electron microscopy, it became evident that there is a fourth array of microfilaments which parallels the cortical microtubules and associates with the plasma membrane (Fig. 2F). This has now been observed for a number of higher (Kakimoto and Shibaoka 1987; Traas et al. 1987) and lower (Menzel and Schliwa 1986) plant cells. The stabilization and organization of cortical F-actin is at least partially determined by interactions with microtubules. It was demonstrated that specific antimicrotubule drugs such as cremart can severely perturb the arrangement of actin filaments (Traas et al. 1987; Kobayashi et al. 1988). On the other hand, disruption of actin filaments by cytochalasin B inhibits the rearrangement of microtubules during the differentiation of *Zinnia* tracheary elements. From the data discussed above, it is evident that the organization of the cortical cytoskeleton depends on two processes: first, the interaction between the individual filaments and second, as only cortical components are found in highly ordered arrays, on the interaction with the plasma membrane.

4.2.2 *Cell Divison*

In contrast to the interphase cytoplasmic and cortical microtubules, the actin network does not disappear during cell division (Schmit and Lambert 1987; Traas et al. 1987). Instead, a network of fine bundles remains throughout the cytoplasm (Fig. 2G, H). This network is concentrated in the plane of division, linking first the dividing nucleus and later the phragmoplast to the PPB site throughout mitosis. In addition, F-actin remains associated with the microtubular arrays in the PPB, in the mitotic spindle, and in the phragmoplast for as long as those microtubular arrays persist (see also Clayton and Lloyd 1985; Palevitz 1987; Seagull et al. 1987). It was proposed that cortical microtubules and therefore probably also interactions with the plasma membrane play a major role in the concentration of F-actin in the division plane (Lloyd and Traas 1988; see also Sect. 5.1).

5 Role of the Plasma Membrane-Cytoskeleton Complex

The plasma membrane-associated cytoskeleton is involved in a wide range of physicochemical reactions in the cell, ranging from endocytosis to the movements of organelles and the positioning of the nucleus. In this section a number of processes, all involving cytoskeleton-plasma membrane interactions, will be discussed. Arbitrarily, these processes fall into two categories: those related strictly to morphogenesis and those more generally involved in cell organization.

5.1 Morphogenesis

In contrast to animal cells, which can change their form and even their position during morphogenesis, plant cells are surrounded by a rigid cell wall which strongly constrains their ability to move. As the possibilities for "correction" are limited, both cell division and cell growth must be controlled with great precision. In the establishment of the plane of division and the control of cell elongation the membrane-associated cytoskeleton plays an important role.

5.1.1 *Determination of the Plane of Division*

As was discussed in Section 4.1.2, the plane of division in cells of highly organized tissues is predicted exactly by the PPB. Still, the precise function of this cortical band of microtubules remains a major unsolved problem. It is not clear whether microtubules are actively involved in determining the division site or whether they merely reflect changes occurring in the plasma membrane or the cortical cytoplasm. An active role for the PPB microtubules has been proposed in positioning the nucleus, and the initiation of spindle formation (reviews,

Gunning and Hardham 1982; Hepler 1985). Yet, it is evident that the PPB site itself also has unique properties and that it must play a key role in memorizing the programmed division plane when the cortical microtubules have disappeared. For instance, displacing the nucleus by micromanipulation does not alter the plane of division. Alternatively, mechanical disruption of the plasma membrane at the PPB site does prevent fusion of the phragmoplast (Gunning and Wick 1985; Hepler 1985). Galatis et al. (1984) used centrifugation to pull the nucleus away from the normal division plane in mitotic cells from *Triticum*. Although cell division was disturbed, at least one edge of the phragmoplast usually extended to the PPB site. It can be concluded that the membrane-bound site somehow attracts the outgrowing phragmoplast and that it is even physically linked to it. Evidence for the presence of such a link between the cytokinetic apparatus and the cortical division site was provided by observations on highly vacuolated, mitotic cells. At prophase a system of cytoplasmic strands, the phragmosome (Sinnot and Bloch 1940), connects the nucleus to the PPB site (Venverloo et al. 1980). The observation of 8 nm filaments in the phragmosome suggests that they contain actin (Goossen-de Roo et al. 1984). A role for F-actin in determining the plane of division was also proposed by Palevitz (1980), who observed that cytochalasin B (an agent disrupting F-actin) can severely affect the realignment of the phragmoplast in *Allium* guard mother cells. The first direct evidence was provided by Traas et al. (1987) who showed the presence of F-actin in the PPB site as part of a network which remains in the plane of division. Based on these observations, Lloyd and Traas (1988) proposed a model which assumes that the cortical F-actin is drawn into the division plane together with the cortical microtubules during PPB formation. This leads to a redistribution of the cytoplasmic actin into a transvacuolar disk, which provides a memory for the predetermined division plane. A system of cytoplasmic strands at right angles to this transverse phragmosome links the spindle poles to the plasma membrane, thus stabilizing the position of the spindle. Whether cortical microtubules play an active role in redistributing actin filaments remains unclear; in some organisms cell division is strictly controlled without the presence of a PPB (e.g., Gunning and Hardham 1982; Doonan et al. 1987). This suggests that other elements present in the cell cortex are responsible for this reorganization and that cortical microtubules passively follow. On the other hand, regeneration experiments using protoplasts have shown that a cortical microtubule network is a prerequisite for cell division (Hahne and Hoffmann 1984b, 1985).

It seems difficult to present a unifying hypothesis for all plant cells and one should keep in mind that different mechanisms for cell division have evolved during evolution. This is most clearly so for cells from "unorganized" tissues, such as endosperm or in pollen mother cells at meiosis. Such cells lack an ordered cortical microtubular array, and no PPB is observed in these cells. Yet, cortical actin is clearly present in the division plane and forms part of a dense network preceding the phragmoplast (Schmit et al. 1985; Schmit and Lambert 1987).

5.1.2 Cell Growth and Differentiation

Both ultrastructural and cytophysiological evidence supports the view that in plants the cytoskeleton does not directly determine cell shape. Instead, it is generally assumed that the cortical microtubules play an indirect role in morphogenesis by controlling the orientation of the microfibrils in the cell wall. In this way they establish the texture and the physical properties of the cell wall, i.e., the way in which the cell can extend. This concept is based on the observation that in many plant cells microtubules are aligned parallel to the microfibrils and that inhibitors of the microtubular system equally affect the orientation of deposition of cellulose microfibrils (for reviews, Heath and Seagull 1982; Robinson and Quader 1982; Lloyd 1984, 1987; Hepler 1985; Chap. 11).

Different models have been proposed to explain the mechanism behind microtubule-mediated control of microfibril orientation (e.g., Heath and Seagull 1982). More and more evidence supports the hypothesis originally put forward by Herth (1980): that microtubules create channels (or tracks) in the membrane via cross-bridges (Herth 1985; Schneider and Herth 1986; Giddings and Staehelin 1988). The cellulose synthase (EC 2.4.1.12) complexes in the plasma membrane move within the limits of these channels, more or less parallel to the microtubules. The driving force for the movement of these synthases along the plasma membrane is probably generated by the crystallization force of the cellulose fibril itself (Herth 1985; Chap. 11). However, also cytoskeleton-membrane interactions could be involved in directing the membrane flow. Although this model for the role of microtubules in cell morphogenesis is generally accepted, there are a number of observations which are seemingly inconsistent with the model. In some cells microtubules and microfibrils are clearly not aligned (Emons 1982, 1986; Traas et al. 1985; Mizuta and Okuda 1987), whereas colchicine depolymerization of microtubules does not necessarily alter microfibril orientation (Robinson and Quader 1982; Emons 1986). Indeed, depolymerization of microtubules in *Oocystis* only inhibited the regular shifts in microfibril orientation. These observations do not necessarily exclude a role for microtubules in microfibril orientation. They demonstrate, however, that other factors may also play a role. For instance, in *Oocystis* the shape of the microfibril-synthesizing complex (a polarized bar) could help to maintain the orientation of microfibril deposition. Clearly, the importance of each factor, including the microtubules, may vary from cell type to cell type.

A more direct role for the cytoskeleton in determining cell shape has also been proposed (Gunning and Hardham 1982; Traas et al. 1984). Lloyd et al. (1980) observed that the presence of cortical microtubules was necessary for maintaining cell shape in carrot cells. Furthermore, Hahne and Hoffman (1984a) also reported that protoplasts with an irregular shape will become spherical after cytochalasin B treatment which causes a depolymerization of F-actin. For tip-growing cells it was observed that inhibitors of the cytoskeleton can cause changes in morphology. For example, root hairs treated with cytochalasin B swell at their tip (Seagull and Heath 1980b). The cell wall in this tip region is still very thin and apparently incapable of maintaining cell shape without an intact cytoskeleton. Such observations indicate that microtubules and microfilaments initially determine the form of such cells. This function is later taken over by the cell wall as soon as it is strong enough.

Another important function of the cytoskeleton in cell differentiation is the establishment and maintenance of cell polarity (review, Schnept 1986). In tip-growing cells F-actin and usually also the cortical microtubules are found in axial orientations (e.g., Seagull and Heath 1980a; Derksen et al. 1985; Lloyd and Wells 1985; Traas et al. 1985; Pierson et al. 1986). Codistribution of organelles with microtubules indicates that the latter play a role in the transport of cytoplasm and both cytochalasin B and antimicrotubule agents can severely disrupt polarity in tip-growing cells (Seagull and Heath 1980b; Mizukami and Wada 1983; Traas et al. 1985; Doonan et al. 1987). Likewise, Hensel (1984, 1986, 1988) proposed a role of membrane-associated microtubules and microfilaments in the polar development of statocytes from cress roots.

5.2 The Role of the Cortical Cytoskeleton in Cytoplasmic Organization

Most of the evidence for the participation of the cytoskeleton in intracellular movements comes from experiments using antimicrotubule or anti-F-actin drugs, in combination with microscope analysis (Gunning and Hardham 1982; Williamson 1986; Lloyd 1987; Staiger and Schliwa 1987). Thus, it has been shown that both cortical microtubules and microfilaments are responsible for organelle movements in both higher and lower plant cells.

To date, the exact function of the cortical cytoskeleton of higher plants in particle movement is poorly understood and much of what we know is based on work using giant algae (reviews, Williamson 1986; Staiger and Schliwa 1987). The (sub)cortical cytoplasm of these cells has numerous bundles of both microtubules and microfilaments which reflect the patterns of endoplasmic streaming. Myosin-coated, plastic beads can move along the cortical actin bundles of Characean algae. This process is Ca^{2+}-dependent when tropomyosin is incorporated into *Chara* actin bundles. Tropomyosin is one of the components which mediate the Ca^{2+} regulation of the actin-myosin interaction. Therefore, it was proposed that cytoplasmic streaming is under the control of regulated Ca^{2+} influx across the plasma membrane (reviews, Shimmen and Yano 1986; Tazawa et al. 1987).

5.3 The Role of the Cytoskeleton in the Distribution of Membrane Components

The major filamentous systems, microtubules, and the F-actin network are closely associated with the plasma membrane in higher plant cells. Therefore, they must also influence the distribution of membrane components. This area is still poorly defined, yet a number of studies have demonstrated that the mobility of specific membrane elements depends on interactions with the cortical cytoskeleton. In developing xylem elements, Schneider and Herth (1986) observed that membrane particles (rosettes) reflect the underlying microtubular pattern and similar observations were also reported for the alga *Closterium* (Giddings and Staehelin 1988). Evidence comes also from photobleaching experiments using monoclonal antibodies to monitor the lateral diffusion of a specific membrane component of soybean protoplasts (Metcalf et al. 1986). Soybean agglutinin substantially reduces the mobility of the labeled membrane component. Colchicine and cytochalasin B completely reverse this effect. It was concluded from these results that the effect of the agglutinin on the labeled membrane receptor was mediated by the cytoskeleton (see also Dugas et al. 1989).

6 Regulation of Cytoskeletal Organization and Function

Cytoskeletal function, organization, and its associations with the plasma membrane are apparently regulated by a number of factors such as Mg^{2+}, Ca^{2+}, and different hormones. For plant cells the exact role of these elements remains an open question and the literature is still fragmentary or even contradictory. But, in view of their importance in cytoskeleton-membrane mediated processes, they will be briefly mentioned in this section.

6.1 Calcium Ions and Calmodulin

Ca^{2+} influences the stability of cytoskeletal components and regulates their interaction both in vitro and in vivo. Ca^{2+} causes the depolymerization of microtubules in vitro, whereas it can induce polymerization of G-actin (Dustin 1985; Kakimoto and Shibaoka 1986; Pollard and Cooper 1986). In vivo it seems to play an important role in the control of cytoskeleton-mediated processes, such as cytoplasmic streaming, mitosis, and morphogenesis (Hepler and Wayne 1985). For instance, fluctuations in the cytoplasmic concentration of free Ca^{2+} seem to coincide with the breakdown and repolymerization of microtubules during mitosis of *Tradescantia* stamen hair cells (e.g., Chen and Wolniak 1987; Hepler and Callaham 1987) and growth, morphology, and polarity of pollen tubes depend on intracellular Ca^{2+} (e.g., Picton and Steer 1985; Reiss and Herth 1985). As already mentioned the cytoplasmic concentration of free Ca^{2+} is under the control of the plasma membrane. Reiss and Herth (1985) have demonstrated the existence of Ca^{2+} channels in the plasma membrane of growing pollen tubes. These channels participate in the maintenance of a polar gradient of intracellular Ca^{2+}. It was proposed that the opening of these channels depends on voltage-dependent gating much as in animal cells (Chap. 8). Several lines of evidence suggest that the effects of Ca^{2+} are at least in part mediated by calmodulin, a highly conserved Ca^{2+}-binding protein. The protein has been detected immunologically in the mitotic apparatus of higher plant cells (Vantard et al. 1985; Wick et al. 1985), where it could play a role in the regulation of microtubule dynamics. In pollen tubes it is concentrated at the tip, where it might be in part associated with the microfilament network, regulating vesicle transport and cytoplasmic streaming (Hausser et al. 1984). For more details on the effect of Ca^{2+} the reader is referred to a number of recent reviews (e.g., Dustin 1985; Hepler and Wayne 1985; Pollard and Cooper 1986; Chaps. 9 and 14).

6.2 The Effect of Hormones

Gibberellins, cytokinins, and ethylene have all been reported to induce changes in the organization of the plasma membrane-bound cytoskeleton, whereas microtubule inhibitors can reverse their effect.

Gibberellic acid treatment results in a rearrangement of the cortical microtubules in epidermal cells of pea internodes from longitudinal to transverse (Akashi and Shibaoka 1987). In onion leaf sheath cells the hormone increases the number of transverse microtubules, whereas it prevents the depolymerization of microtubules by colchicine, cremart, and low temperature (Mita and Shibaoka 1984). Likewise, treatment with gibberellic acid seems to stabilize the transverse cortical array and to prevent reorientation in sections of excised lettuce hypocotyls (Durnam and Jones 1982). In some cases, however, gibberellin causes a destabilization of the cortical microtubules (Akashi and Shibaoka 1987). This, in combination with the long incubation times needed, strongly suggests that the hormone only acts indirectly on the cytoskeleton by changing its cytoplasmic environment.

Ethylene can induce changes in cell expansion which coincide with changes in the orientation of the cortical microtubules and the orientation of microfibril deposition (e.g., Steen and Chadwick 1981; Roberts et al. 1985). This process is reversed by colchicine treatment, and the microtubule-stabilizing agent D_2O can mimic the ethylene effect. It was proposed that the hormone binds to cytoskeletal components, affecting microtubule-microtubule or microtubule-plasma membrane interactions. Changes in the cortical system can be observed as early as 20 min after incubation with ethylene, which suggests a short, or at least rapid, chain between target and effector (Roberts et al. 1985).

The effect of cytokinins in relation to cytoskeletal function has been mainly characterized in mosses such as *Funaria* and *Physcomitrella*, where it induces the formation of buds that lead to leafy shoots (Conrad et al. 1986; Doonan et al. 1987). Doonan et al. (1987) observed that high concentrations of benzylaminopurine (100 μM) cause the depolymerization of microtubules in tip-growing cells of *Physcomitrella*. The effect of cytokinins are probably mediated by Ca^{2+} which in turn influences cytoskeletal functioning. Cytokinins modulate the intracellular Ca^{2+} concentration and Saunders (1986) has suggested that these hormones activate ion channels in the apical part of the plasma membrane of target cells. The polarity of these cells can be changed with cytochalasin B, which could mean that microfilaments are involved in the distribution of the ion channels in the membrane.

7 Concluding Remarks

We now have detailed information about the dynamics of the cortical microtubule and F-actin arrays and the results discussed in the present chapter demonstrate the importance of cytoskeleton-plasma membrane interactions in many processes. However, surprisingly little is known about the molecular basis of these interactions; to date, there is only morphological evidence for the presence of links between the plasma membrane and microtubules. Clearly, in the immediate future we need to know more about the composition of the cytoskeleton of higher plant cells and identify the proteins that are involved in its association with the plasma membrane. What mechanisms for plasma membrane-cytoskeleton interactions can we expect to find in higher plants? An answer to this question may be partly derived from work on animals and some lower plants about which much more is known. Since many cytoskeletal components have been highly conserved throughout evolution, the available information on animals or slime molds might form an important basis for future research on higher plant cells. In the past, for example, antibodies against animal cytoskeletal components have helped to identify their counterparts in plants. Cytoskeleton-plasma membrane interactions have been studied extensively in slime molds such as *Dictyostelium* or *Acanthamoeba*. Purified plasma membranes from these organisms are able to bind actin and even seem to organize actin into filaments. A trans-plasma membrane protein, ponticulin, has been identified which is important for connecting the microfilament network in *Dictyostelium* (Wuesthube and Luna 1987). Both *Acanthamoeba* and *Dictyostelium* possess proteins which are closely related to spectrin, a tetrameric protein responsible for actin-plasma membrane interactions in the red blood cell (Pollard 1984; Pollard and Cooper 1986) and we can therefore expect to find similar polypeptides in higher plants. Still, important differences do exist between animal and plant cytoskeletons and generalizing conclusions obtained on one system should be made with great caution.

Work on animal cells and lower plants might also provide a framework for the experimental approach which has to be followed in order to characterize the plasma membrane-cytoskeleton interactions in higher plants. This has been the case up to now for microscope studies; techniques originally developed for animal cells such as immunofluorescence, dry cleaving, and freeze substitution have been successfully adapted to plants. Clearly, these techniques will continue to be essential in the future. Several biochemical approaches have been used to study the association of cytoskeletal proteins with integral plasma membrane components in animals and lower plants (for review, see Geiger 1983; Pollard and Cooper 1986). Cytoskeletal components can be copurified with plasma membranes to demonstrate associations between specific components (e.g., Ramaekers et al. 1982). Another example comes from studies on *Dictyostelium* in which purified actin has been used in affinity chromatography to identify actin-binding proteins in purified plasma membranes (Wuesthube and Luna 1987; see also Schwartz and Luna 1986). It seems, however, that techniques which are routinely used for the purification and characterization of cytoskeletons from animals and lower plants cannot easily be applied to higher plant cells. This is due to different properties of plant cells such as the presence of cell walls and vacuoles. Plant actin, for instance, is difficult to repolymerize in vitro under conditions where animal actin repolymerizes (Vahey et al. 1982; J.A. Traas and A. Hargreaves, unpublished results). Therefore, much of our attention must be focused on the search for new techniques adapted to the specific properties of plant cells. Recently, Hussey et al. (1987) developed a method which allows the isolation of the integral detergent-resistent cytoskeleton of protoplasts. Such techniques will prove very useful in future attempts to analyze the composition of the network as well as its interactions with the plasma membrane.

Acknowledgments. I would like to thank Dr. C.W. Lloyd for helpful comments and Drs. C. Hawes and P. Hepler for sending me some of their micrographs.

References

Akashi T, Shibaoka H (1987) Effects of gibberellin on the arrangement and the cold stability of cortical microtubules in epidermal cells of pea internodes. Plant Cell Physiol 28:339–348
Chen TL, Wolniak SM (1987) Mitotic progression in stamen hair cells of *Tradescantia* is accelerated by treatment with ruthenium red and Bay K-8644. Eur J Cell Biol 45:16–22
Clayton L, Lloyd CW (1985) Actin organization during the cell cycle in meristematic plant cells. Actin is present in the cytokinetic phragmoplast. Exp Cell Res 156:231–238
Clayton L, Black CM, Lloyd CW (1985) Microtubule nucleating sites in higher plant cells identified by an auto-antibody against pericentriolar material. J Cell Biol 101:319–324
Coleman J, Evans D, Hawes C, Horsley D, Cole L (1987) Structure and molecular organization of higher plant coated vesicles. J Cell Sci 88:35–45
Conrad PA, Steucec GL, Hepler PK (1986) Bud formation in *Funaria*: organelle redistribution following cytokinin treatment. Protoplasma 131:211–223
Dawson PJ, Lloyd CW (1987) A comparison of plant and animal tubulins. In: Davies DD (ed). The biochemistry of plants: a comprehensive treatise, vol 12. Academic Press, London, pp 3–47
Dawson PJ, Hulme JH, Lloyd CW (1985) Monoclonal antibody to intermediate filament antigen cross-reacts with higher plant cells. J Cell Biol 100:1793–1798
Derksen JWM, Pierson ES, Traas JA (1985) Microtubules in vegetative and generative cells of pollen tubes. Eur J Cell Biol 38:142–148
Derksen JWM, Traas JA, Oostendorp T (1986) Distribution of actin filaments in differentiating cells of *Equisetum hyemale* root tips. Plant Sci 43:77–81
Doonan JH, Cove DJ, Corke FMK, Lloyd CW (1987) Pre-prophase band of microtubules, absent from tip-growing moss filaments, arises in leafy shoots during transition to intercalary growth. Cell Motil Cytoskel 7:138–153
Dugas CM, Li Q, Khan IA, Nothnagel EA (1989) Lateral diffusion in the plasma membrane of maize protoplasts with implications for cell culture. Planta 179:387–396
Durnam DJ, Jones RL (1982) The effects of colchicine and gibberellic acid on growth and microtubules in excised lettuce hypocotyls. Planta 154:204–211

Dustin P (1985) Microtubules. Springer, Berlin Heidelberg New York Tokyo
Emons AMC (1982) Microtubules do not control microfibril orientation in a helicoidal cell wall. Protoplasma 113:85–87
Emons AMC (1986) Cell wall formation in root hairs. Thesis, University of Nijmegen, Nijmegen
Emons AMC, Traas JA (1986) Coated pits and coated vesicles on the plasma membrane of plant cells. Eur J Cell Biol 41:57–64
Galatis B, Apostolakos P, Katsaros C (1984) Experimental studies on the function of the cortical cytoplasmic zone of the preprophase microtubule band. Protoplasma 122:11–26
Geiger B (1983) Membrane-cytoskeleton interactions. Biochim Biophys Acta 737:305–341
Giddings TH, Staehelin LA (1988) Spatial relationship between microtubules and plasma-membrane rosettes during the deposition of primary wall microfibrils in *Closterium* sp. Planta 173:22–30
Goossen-de Roo L, Bakhuizen R, Van Spronsen P, Libbenga KR (1984) The presence of extended phragmosomes containing cytoskeletal elements in fusiform cambial cells of *Fraxinus excelsior* L. Protoplasma 122:145–152
Guiltinan MJ, Ma DP, Barker RF, Bustos MM, Cyr RJ, Yadegari R, Fosket DE (1987) The isolation, characterization and sequence of two divergent beta-tubulin genes from soybean (*Glycine max* L). Plant Mol Biol 10:171–184
Gunning BES, Hardham AR (1982) Microtubules. Annu Rev Plant Physiol 33:651–698
Gunning BES, Wick SM (1985) Preprophase bands, phragmoplasts, and spatial control of cytokinesis. J Cell Sci Suppl 2:157–179
Gunning BES, Hardham AR, Hughes JE (1978) Pre-prophase bands of microtubules in all categories of formative and proliferative cell division in *Azolla* roots. Planta 143:145–160
Hahne G, Hoffmann F (1984a) The effect of laser microsurgery on cytoplasmic strands and cytoplasmic streaming in isolated plant protoplasts. Eur J Cell Biol 33:175–179
Hahne G, Hoffmann F (1984b) Dimethyl sulfoxide can initiate cell divisions of arrested callus protoplasts by promoting cortical microtubule assembly. Proc Natl Acad Sci USA 81:5449–5453
Hahne G, Hoffmann F (1985) Cortical microtubular lattices: absent from mature mesophyll and necessary for cell division? Planta 166:309–313
Hardham AR, Gunning BES (1978) Structure of cortical microtubule arrays in plant cells. J Cell Biol 77:14–34
Hargreaves AJ, Dawson PJ, Butcher GW, Larkins A, Goodbody KC, Lloyd CW (1989) A monoclonal antibody against cytoplasmic fibrillar bundles from carrot cells, and its cross-reaction with animal intermediate filaments. J Cell Sci 92:371–378
Hausser I, Herth W, Reiss HD (1984) Calmodulin in tip growing plant cells, visualized by fluorescing calmodulin-binding phenothiazines. Planta 162:33–39
Hawes CR (1985) Conventional and high voltage electron microscopy of the cytoskeleton and cytoplasmic matrix of carrot (*Daucus carota*). Eur J Cell Biol 38:201–210
Hawes CR, Martin B (1986) Deep etching of plant cells: cytoskeleton and coated pits. Cell Biol Int Rep 10:985–992
Heath IB, Seagull RW (1982) Oriented cellulose microfibrils and the cytoskeleton: a critical comparison of models. In: Lloyd CW (ed) The cytoskeleton in plant cell growth and development. Academic Press, London, pp 163–187
Hensel W (1984) Microtubules in statocytes from roots of cress. Protoplasma 119:121–134
Hensel W (1986) Cytodifferentiation of polar plant cells; use of antimicrotubular agents during the differentiation of statocytes from cress roots (*Lepidium sativum* L.) Planta 169:293–303
Hensel W (1988) Demonstration by heavy-meromyosin of actin microfilaments in extracted cress (*Lepidium sativum* L.) root statocytes. Planta 173:142–143
Hepler PK (1985) The plant cytoskeleton. In: Robards AW (ed) Botanical microscopy 1985. Oxford University Press, Oxford, pp 233–262
Hepler PK, Callaham DA (1987) Free calcium increases during anaphase in stamen hair cells of *Tradescantia*. J Cell Biol 105:2137–2143
Hepler PK, Wayne RO (1985) Calcium and plant development. Annu Rev Plant Physiol 36:397–439
Herth W (1980) Calcofluor white and congo red inhibit chitin microfibril assembly of *Poterioochromonas:* evidence for a gap between polymerization and microfibril formation. J Cell Biol 87:442–450

Herth W (1985) Plant cell wall formation. In: Robards AW (ed) Botanical microscopy 1985. Oxford University Press, Oxford, pp 285–310
Hightower RC, Meagher RB (1985) Divergence and differential expression of soybean and actin genes. EMBO J 4:1–8
Hogetsu T (1986) Re-formation of microtubules in *Closterium ehrenbergii Meneghini* after cold induced depolymerization. Planta 167:437–443
Hogetsu T (1987) Re-formation and ordering of wall microtubules in *Spirogyra* cells. Plant Cell Physiol 28:875–883
Huffaker TC, Hoyt MA, Botstain D (1987) Genetic analysis of the yeast cytoskeleton. Annu Rev Genet 21:259–284
Hussey PJ, Traas JA, Gull K, Lloyd CW (1987) Isolation of cytoskeletons from synchronized plant cells: the interphase microtubular array utilizes multiple tubulin isotypes. J Cell Sci 88:225–230
Hussey PJ, Lloyd CW, Gull K (1988) Differential and developmental expression of beta-tubulins in a higher plant. J Biol Chem 263:5474–5479
Jackson WT (1982) Actomyosin. In: Lloyd CW (ed) The cytoskeleton in plant cell growth and development. Academic Press, London, pp 3–29
Joachim S, Robinson DG (1984) Endocytosis of cationic ferritin by bean leaf protoplasts. Eur J Cell Biol 34:212–216
Kakimoto T, Shibaoka H (1986) Calcium-sensitivity of cortical microtubules in the green alga *Mougeotia*. Plant Cell Physiol 27:91–101
Kakimoto T, Shibaoka H (1987) Actin filaments and microtubules in the preprophase band and phragmoplast of tobacco cells. Protoplasma 140:151–156
Kato T, Tonomura Y (1977) Identification of myosin in *Nitella flexilis*. J Biochem 82:777–782
Kobayashi H, Fukuda H, Shibaoka H (1988) Interrelation between the spacial disposition of actin filaments and microtubules during the differentiation of tracheary elements in cultured *Zinnia* cells. Protoplasma 143:29–36
Lancelle SA, Callaham DA, Hepler PK (1986) A method for rapid freeze fixation of plant cells. Protoplasma 131:153–165
Lancelle SA, Cresti M, Hepler PK (1987) Ultrastructure of the cytoskeleton in freeze-substituted pollen tubes of *Nicotiana alata*. Protoplasma 140:141–150
Lazarides E (1987) From genes to structural morphogenesis: the genesis of a red blood cell. Cell 51:345–356
Lim SS, Hering GE, Borisy GG (1986) Widespread occurrence of anti-troponin T crossreactive components in non-muscle cells. J Cell Sci 85:1–19
Lloyd CW (ed) (1982) The cytoskeleton in plant growth and development. Academic Press, London
Lloyd CW (1983) Helical microtubular arrays in onion root hairs. Nature 305:311–315
Lloyd CW (1984) Towards a dynamic helical model for the influence of microtubules on wall patterns in plants. Int Rev Cytol 86:1–51
Lloyd CW (1987) The plant cytoskeleton: the impact of fluorescence microscopy. Annu Rev Plant Physiol 38:119–139
Lloyd CW, Seagull RW (1985) A new spring for plant cell biology: microtubules as dynamic helices. Trends Biochem Sci 10:476–478
Lloyd CW, Traas JA (1988) The role of F-actin in determining the division plane of carrot suspension cells. Drug studies. Development 102:211–221
Lloyd CW, Wells B (1985) Microtubules are at the tips of root hairs and form helical patterns corresponding to inner wall fibrils. J Cell Sci 75:225–238
Lloyd CW, Slabas AR, Powell AJ, Macdonald G, Badley RA (1979) Cytoplasmic microtubules of higher plant cells visualized with antitubulin antibodies. Nature 279:239–241
Lloyd CW, Slabas AR, Powell AJ, Lowe SB (1980) Microtubules, protoplasts and plant cell shape. An immunofluorescence study. Planta 147:500–506
Lloyd CW, Clayton L, Dawson PJ, Doonan J, Hulme JS, Roberts IN, Wells B (1985) The cytoskeleton underlying crosswalls and side walls in plants; molecules and macromolecular assemblies. J Cell Sci Suppl 2:143–145
Ludwig SR, Oppenheimer DG, Silflow CD, Snustad DP (1988) The alpha 1-tubulin gene of *Arabidopsis thaliana*: primary structure and preferential expression in flowers. Plant Mol Biol 10:311–321

Marks DM, West J, Weeks DP (1987) The relatively large beta-tubulin gene family of *Arabidopsis* contains a member with an unusual transcribed 5′ noncoding sequence. Plant Mol Biol 10:91–104

Menzel D, Schliwa M (1986) Motility in the siphonous green alga *Bryopsis* I. Eur J Cell Biol 40:275–285

Mesland DAM, Spiele H, Roos E (1981) Membrane associated cytoskeleton and coated vesicles in cultured hepatocytes visualized by dry cleaving. Exp Cell Res 132:169–184

Metcalf TN III, Villanueva MA, Schindler M, Wang JL (1986) Monoclonal antibodies directed against protoplasts of soybean cells: analysis of the lateral mobility of plasma membrane-bound antibody MVS-1. J Cell Biol 102:1350–1357

Mita T, Shibaoka H (1984) Gibberellin stabilizes microtubules in onion leaf sheath cells. Protoplasma 119:100–109

Mizukami M, Wada S (1983) Morphological anomalies induced by antimicrotubule agents in *Bryopsis plumosa*. Protoplasma 114:151–162

Mizuta S, Okuda K (1987) *Bloodlea* cell wall microfibril orientation unrelated to cortical microtubule arrangement. Bot Gaz 148:297–307

Palevitz BA (1980) Comparative effects of cytochalasin B and phalloidin on motility and morphogenesis in *Allium*. Can J Bot 58:773–785

Palevitz BA (1987) Actin in the preprohase band of *Allium cepa*. J Cell Biol 104:1515–1519

Parke J, Miller C, Anderton BH (1986) Higher plant myosin heavy chain identified using a monoclonal antibody. Eur J Cell Biol 41:9–13

Parke J, Miller CCJ, Cowell I, Dodson A, Dowding A, Downes M, Duckett JG, Anderton BJ (1987) Monoclonal antibodies against plant proteins recognise animal intermediate filaments. Cell Motil Cytoskel 8:312–323

Parthasarathy MV, Perdue TD, Witzmun A, Alvernaz J (1985) Actin network as a normal component of the cytoskeleton in many higher vascular plant cells. Am J Bot 72:1318–1323

Pesacreta TC, Carley WW, Webb WW, Parthasarathy MV (1982) F-actin in conifer roots. Proc Natl Acad Sci USA 79:2898–2901

Picton JM, Steer M (1985) The effects of ruthenium red, lanthanum, fluorescein isothiocyanate and trifluoperazine on vesicle transport, vesicle fusion and tip extension in pollen tubes. Planta 163:20–26

Pierson E, Derksen J, Traas J (1986) Organization of microfilaments and microtubules in pollen tubes grown in vitro or in vivo in various angiosperms. Eur J Cell Biol 41:14–18

Pollard TD (1984) Purification of a high molecular weight actin filament gelation protein from *Acanthamoeba castellanii* that shares antigenic determinants with vertebrate spectrins. J Cell Biol 99:1970–1980

Pollard TD, Cooper JA (1986) Actin and actin binding proteins. A critical evaluation of mechanisms and functions. Annu Rev Biochem 55:987–1035

Powell AJ, Peace GW, Slabas AR, Lloyd CW (1982) The detergent resistant cytoskeleton of higher plant protoplasts contains nucleus associated fibrillar bundles in addition to microtubules. J Cell Sci 56:319–335

Quader H, Deichgraber G, Schnepf E (1986) The cytoskeleton in *Cobaea* seed hairs: patterning during cell wall differentiation. Planta 168:1–10

Quader H, Herth W, Ryser U, Schnepf E (1987) Cytoskeletal elements in cotton seed hair development in vitro – their possible role in cell wall organization. Protoplasma 137:56–62

Ramaekers FCS, Dunia I, Dodemont HJ, Benedetti EL, Bloemendal H (1982) Lenticular intermediate-sized filaments: biosynthesis and interaction with plasma membrane. Proc Natl Acad Sci USA 79:3208–3212

Reiss HD, Herth W (1985) Nifedipine-sensitive calcium channels are involved in polar growth of lily pollen tubes. J Cell Sci 76:247–254

Roberts IN, Lloyd CW, Roberts K (1985) Ethylene-induced microtubule reorientations: mediation by helical arrays. Planta 164:439–447

Robinson DG, Quader H (1982) The microtubule-microfibril syndrome. In: Lloyd CW (ed) The cytoskeleton in plant cell growth and development. Academic Press, London, pp 109–126

Robinson DG, Depta H (1988) Coated vesicles. Annu Rev Plant Physiol Plant Mol Biol 39:53-99
Saunders MJ (1986) Cytokinin activation and redistribution of plasma membrane ion channels in *Funaria* A vibrating-microelectrode and cytoskeleton-inhibitor study. Planta 167:402-409
Schliwa M (1986) The cytoskeleton, an introductory survey. Springer, Vienna New York
Schmit AC, Lambert AM (1987) Characterization and dynamics of cytoplasmic F-actin in higher plant endosperm cells during interphase, mitosis, and cytokinesis. J Cell Biol 105:2157-2166
Schmit AC, Vantard M, Lambert AM (1985) Microtubule and F-actin rearrangement during the initiation of mitosis in acentriolar higher plant cells. In: Ishikawa H, Hatano S, Sato H (eds) Cell motility: mechanism and regulation. University of Tokyo Press, pp 415-433
Schneider B, Herth W (1986) Distribution of plasma membrane rosettes and kinetics of cellulose formation in xylem development of higher plants. Protoplasma 131:142-152
Schnepf E (1986) Cellular polarity. Annu Rev Plant Physiol 37:23-47
Schwarz MA, Luna EJ (1986) Binding and assembly of actin filaments by plasma membranes from *Dictyostelium discoideum*. J Cell Biol 102:2067-2075
Seagull RW (1986) Changes in microtubule organization and wall microfibril orientation during in vitro cotton fiber development. Can J Bot 64:1373-1381
Seagull RW, Heath IB (1980a) The organization of the cortical microtubule array in the radish root hair. Protoplasma 103:205-229
Seagull RW, Heath IB (1980b) The differential effects of cytochalasin B on microfilament populations and cytoplasmic streaming. Protoplasma 103:231-240
Seagull RW, Falconer MM, Weerdenburg CA (1987) Microfilaments: dynamic arrays in higher plant cells. J Cell Biol 104:995-1004
Shah DM, Hightower RC, Maegher RB (1982) Complete nucleotide sequence of a soybean actin gene. Proc Natl Acad Sci USA 79:1022-1026
Shimmen T, Yano M (1986) Regulation of myosin sliding along *Chara* actin bundles by native skeletal muscle tropomyosin. Protoplasma 132:129-136
Sinnot EW, Bloch R (1940) Cytoplasmic behaviour during division of vacuolate plant cells. Proc Natl Acad Sci USA 26:223-227
Staiger CJ, Schliwa M (1987) Actin localization and function in higher plants. Protoplasma 141:1-12
Steen DA, Chadwick AV (1981) Ethylene effects in pea stem tissue. Plant Physiol 67:460-466
Tanchak M, Griffing L, Mersey B, Fowke L (1984) Endocytosis of cationised ferritin by coated vesicles of soybean protoplasts. Planta 162:481-486
Tazawa M, Shimmen T, Mimura T (1987) Membrane control in the Characeae. Annu Rev Plant Physiol 38:95-117
Tiwari SC, Wick SM, Williamson RE, Gunning BES (1984) Cytoskeleton and integration of cellular function of cells of higher plants. J Cell Biol 99:63s-69s
Traas JA (1984) Visualization of the membrane bound cytoskeleton and coated pits of plant cells by means of dry cleaving. Protoplasma 119:212-218
Traas JA (1989) Gold labeling of cortical microtubules in cleaved whole mounts of cells. In: Hayat MA (ed) Colloidal gold: methods and applications. Academic Press, Orlando-London, in press
Traas JA, Kengen HMP (1986) Gold labeling of microtubules in cleaved whole mounts of cortical roots cells. J Histochem Cytochem 34:1501-1504
Traas JA, Braat P, Derksen JWM (1984) Changes in microtubule arrays during the differentiation of cortical root cells of *Raphanus sativus*. Eur J Cell Biol 34:229-238
Traas JA, Braat P, Emons AMC, Meekes HJTM, Derksen JWM (1985) Microtubules in root hairs. J Cell Sci 76:303-320
Traas JA, Doonan JH, Rawlins DJ, Shaw PJ, Watts J, Lloyd CW (1987) An actin network is present in the cytoplasm throughout the cell cycle of carrot cells and associates with the dividing nucleus. J Cell Biol 105:387-395
Vahey M, Titus M, Trautwein R, Scordilis S (1982) Tomato actin and myosin: contractile proteins from a higher land plant. Cell Motil 2:131-147
Van der Valk P, Rennie P, Conolly JA, Fowke LC (1980) Distribution of cortical microtubules in tobacco protoplasts. An immunofluorescence microscopic and ultrastructural study. Protoplasma 105:27-43

Vantard M, Lambert AM, De Mey J, Picquot P, Van Eldik LJ (1985) Characterization and immunocytochemical distribution of calmodulin in higher plant endosperm cells: localization in the mitotic apparatus. J Cell Biol 101:488–499

Venverloo CJ, Hovenkamp PH, Weeda AJ, Libbenga KR (1980) Cell division in *Nautilocalyx* explants. I. Phragmosome, preprophase band and plane of division. Z Pflanzenphysiol 150:161–174

Wick SM (1985) Immunofluorescence microscopy of tubulin and microtubule arrays in plant cells. II. Transition between cytokinetic and interphase microtubule arrays. Cell Biol Int Rep 9:357–371

Wick SM, Duniec J (1983) Immunofluorescence microscopy of tubulin and microtubular arrays in plant cells I. Preprophase band development and concomitant appearance of nuclear envelope-associated tubulin. J Cell Biol 97:235–243

Wick SM, Duniec J (1984) Immunofluorescence microscopy of tubulin and microtubule arrays II. Transition between the preprophase band and the mitotic spindle. Protoplasma 122:45–55

Wick SM, Muto S, Duniec J (1985) Double immunofluorescence labeling of calmodulin and tubulin in dividing plant cells. Protoplasma 126:198–206

Williamson RE (1986) Organelle movements along actin filaments and microtubules. Plant Physiol 82:631–634

Wilms FHA, Derksen JWM (1988) Reorganization of cortical microtubules during cell differentiation in tobacco explants. Protoplasma 146:127–132

Wuesthube LJ, Luna EJ (1987) F actin binds to the cytoplasmic surface of ponticulin a 17 kD integral glycoprotein from *Dictyostelium discoideum* plasma membranes. J Cell Biol 105: 1741–1752

Chapter 13 Responses of the Plasma Membrane to Cold Acclimation and Freezing Stress

S. YOSHIDA[1] and M. UEMURA[2]

1 Introduction

Among various environmental stresses, low temperature is one of the most important factors limiting the productivity and distribution of plants. An untimely frost or extremely cold winter in a major production area causes significant losses in productivity of many crop plants. An understanding of the molecular mechanisms of freezing injury and cold acclimation of plants may provide a basis for developing new crops and better production systems. As early

[1]Institute of Low Temperature Science, Hokkaido University, Sapporo, Japan
[2]Department of Biology, Faculty of Science, Koube University, Koube, Japan

Abbreviations: DCCD, *N,N'*-dicyclohexylcarbodiimide; DPH, 1,6-diphenyl-1,3,5-hexatriene; DSC, differential scanning calorimetry; ESR, electron spin resonance; IMP, intramembranous protein particle(s): TSAI, tolerable surface area increment.

C. Larsson, I.M. Møller (Eds):
The Plant Plasma Membrane

as 1912, it was suggested that freezing injury of cells is due to a freeze-induced removal of water from the protoplasmic surface, resulting in coagulation or stiffening of the protoplasmic surface (Maximov 1912). This hypothesis was further developed by Levitt and Scarth (1936a,b), Siminovitch and Scarth (1938), Levitt and Siminovitch (1940), Scarth et al. (1940), Scarth (1941, 1944), Siminovitch and Levitt (1941), and Levitt (1972). Using light microscopy, they observed that the properties of the protoplasmic surface changed in parallel with the development of freezing tolerance. The focus was then shifted to the relationship between biochemical changes in cellular membranes, especially changes in proteins and lipids, and freezing tolerance in a wide variety of plants. However, these studies did not lead to any clear conclusion concerning the role of the plasma membrane in the development of freezing tolerance, because the results were mostly concerned with total cellular membranes, and not specifically with the plasma membrane. A major problem has been the difficulty in isolating plasma membranes of high purity needed for detailed studies. Recently, major advances in plasma membrane isolation have been achieved in many laboratories (Chap. 3). Aqueous polymer two-phase partitioning is particularly useful, being applicable to the isolation of plasma membrane from a wide variety of plants. This encouraged us to start our investigations on the biochemical alterations of plasma membranes related to cold acclimation and freezing injury. In this chapter, we will discuss how plasma membranes are altered biochemically and biophysically during a lethal freezing, and during cold acclimation.

2 Process of Cell Freezing

2.1 Mode of Ice Formation

Upon cooling of plant cells to subzero temperatures, ice is first nucleated outside the cell walls and ice crystals are initiated, resulting in a disequilibrium between the chemical potential of water in the intracellular solution and that of extracellularly frozen water. The disequilibrium is thermodynamically relieved by cellular dehydration and growing of ice crystals on the external cell surface or by intracellular ice formation. The manner of ice formation, i.e., extracellular or intracellular freezing, depends on the speed of cooling and the speed of withdrawal of water molecules from the cells. Under natural conditions the speed of cooling is sufficiently slow to allow extracellular ice formation, therefore, intracellular freezing rarely occurs. Intracellular freezing, as a rule, causes lethal damage to cells due to mechanical destruction of cytoplasmic structures by the ice crystals. The ability of cells to withstand extracellular freezing varies significantly depending on plant species and season.

2.2 *Cryomicroscopy*

Extensive studies on the morphological changes in cells during a freeze-thaw cycle have been carried out by light microscopy (Molish 1897; Siminovitch and Scarth 1938; Asahina 1956). Recent advances in optics, video-image analysis and related techniques have enabled us to use more sophisticated cryomicroscopy and to follow visually the detailed changes in cell structures during a freeze-thaw cycle.

Steponkus and co-workers (Steponkus et al. 1982, 1983, 1984; Dowgert and Steponkus 1984) observed the behavior of winter rye protoplasts during osmotic manipulation and a freeze-thaw cycle, particularly with regard to structural alterations of the plasma membrane, using a high resolution cryomicroscope. Upon freezing, protoplasts were dehydrated by extracellular freezing and the cell volume decreased with decreasing temperature. During freeze-induced or hyperosmotic dehydration, endocytotic vesiculation occurred in nonacclimated protoplasts, whereas exocytotic extrusions were formed on the surface of cold-acclimated protoplasts (Steponkus et al. 1983, 1984). During osmotic expansion caused by thawing, the endocytotic vesicles remained in the cytoplasm of nonacclimated protoplasts and the protoplasts lysed. In contrast, the exocytotic extrusions were resorbed into the surface of cold-acclimated protoplasts as the protoplasts regained their original volume and surface area. Thus, cells behave differently upon freeze-dehydration depending on their freezing tolerance. From these results, it was postulated that injury of nonacclimated protoplasts is due to expansion-induced lysis of the plasma membrane during thawing (Steponkus 1984). On the other hand, injury of cold-acclimated protoplasts is due to a loss of osmotic responsiveness during freezing to a lethal temperature, suggesting a different mechanism (Dowgert and Steponkus 1984).

2.3 *Electron Microscopy*

Electron microscopy will undoubtedly provide more detailed information concerning alterations in the molecular structure of cellular membranes. Extensive studies employing electron microscopy have been published and these were recently reviewed by Fujikawa (1989a,b).

Using thin-section electron microscopy, formation of osmiophilic globules, presumably lipid droplets, has been observed on the plasma membrane during dehydration induced by a slow freezing to a lethal temperature (Singh 1979; Pearce 1982), indicating an alteration of the molecular structure of the plasma membrane. Nearly the same structural changes are observed during a hyperosmotic treatment without freezing (Giles et al. 1974; 1976). However, the relation between the formation of osmiophilic globules on the plasma membrane and freeze-induced or hyperosmotic dehydration may not be universal. In Brome grass culture cells, the formation of osmiophilic material on the plasma membrane during a slow freezing may be an artifact due to an inadequate long-term freeze-substitution of freeze-dehydrated cells prior to electron microscope observation (T. Niki, personal communication).

Recent advances in freeze-fracture electron microscopy, including ultrarapid cryo-fixation of cells, enabled us to get more insight into the real image of cell structures, especially the molecular architecture of cell membranes. Upon freezing of plant cells to a lethal temperature by slow cooling prior to the ultrarapid cryo-fixation, distinct changes are observed in the distribution of intramembranous protein particles (IMP) of plasma membranes (Pearce and Willison 1985; Fujikawa and Miura 1986; Fujikawa 1987). The formation of IMP-free patches and IMP aggregation are closely related to cell injury. The size of the IMP-free patches once formed during freezing are further extended upon thawing, suggesting an irreversible process. Since these morphological changes in the IMP on plasma membranes could not be induced by supercooling by itself, the changes are considered to depend solely on cell dehydration and the resulting severe contraction of cells. The formation of IMP-free patches probably involves a redistribution of membrane proteins. Temperature-dependent phase separation of membrane lipids involves segregation of membrane proteins from solidified lipid domains into still fluid domains of the lipid bilayer, forming IMP-free patches and IMP aggregation (Ono and Murata 1982). Since the formation of IMP-free patches during freezing is not produced by low temperature per se, but is produced by cell dehydration, the cause cannot simply be ascribed to temperature-induced changes in the physical state of membrane lipids from a liquid crystalline to a gel phase. According to Fujikawa (1987) and Fujikawa and Miura (1986), the occurrence of IMP aggregation in plasma membranes of tertiary hyphae of several basidiomycete species during lethal freezing is closely related to a direct contact between cellular membranes and membrane-membrane fusion. These, in turn, are the result of severe cell dehydration and mechanical cell deformation caused by the formation of extracellular ice crystals (Fig. 1). The occurrence of IMP aggregation is demonstrated not only in plasma membranes but also in endomembranes such as tonoplasts, plastid membranes, and nuclear envelopes upon freezing of both higher plant cells and tertiary hyphae of basidiomycetes (Fujikawa 1989a).

Gordon-Kamm and Steponkus (1984) showed that hyperosmotic dehydration of nonacclimated rye leaf cells causes a lamellar to hexagonal$_{II}$ phase transition of lipids in the plasma membranes. However, this phase transition is not found in all plants (wheat leaf cells, Pearce and Willison 1985; Brome grass culture cells, T. Niki, personal communication). The formation of the hexagonal$_{II}$ phase was not observed in plasma membranes of tertiary hyphae from several species of basidiomycetes during lethal freezing, instead, it occurred in the tonoplast (Fujikawa 1989a). Thus, the lamellar to hexagonal$_{II}$ phase transition observed in plasma membranes upon a lethal freezing may not be a widespread phenomenon in plants, and may differ between different cell membranes and different types of plant cells.

As mentioned in the previous section, endovesiculation of plasma membranes is observed in nonacclimated winter rye protoplasts upon osmotic

Fig. 1. Freeze-etch electron micrograph showing ultrastructural changes in the plasma membrane caused by slow freezing in tertiary hyphae of *Corpinus micaceus*. Intramembranous protein particle (IMP) aggregation took place in the membrane regions where the plasma membranes were in direct contact (*upper half* of photograph), but not in the membrane regions where a layer of cytoplasm existed between the plasma membranes (*lower half* of photograph) (S. Fujikawa 1988, Application Photos II, JEOL Ltd., Tokyo, p. 29)

dehydration and freeze-induced dehydration (Steponkus 1984). In Brome grass cultures, however, no endovesiculation or exocytotic extrusion of the plasma membrane was observed in thin-section electron micrographs of freeze-dehydrated cells prepared by an improved freeze-substitution technique using a solvent-fixative mixture (T. Niki, personal communication). Even when cells were frozen to a lethal temperature (-10°C), plasma membranes showed a smooth texture; instead, small vesicles originating from the endoplasmic reticulum were conspicuously located adjacent to the plasma membranes. Fujikawa (1986) detected endovesiculation of plasma membranes in only one of more than ten basidiomycete species (Fig. 2).

3 Modern Theories of Freezing Injury

Several hypotheses have been proposed to explain the mechanism of freezing injury of plant cells as reviewed by several authors (Mazur 1969; Levitt 1972; Meryman 1974; Steponkus 1984; Sakai and Larcher 1987). The hypotheses

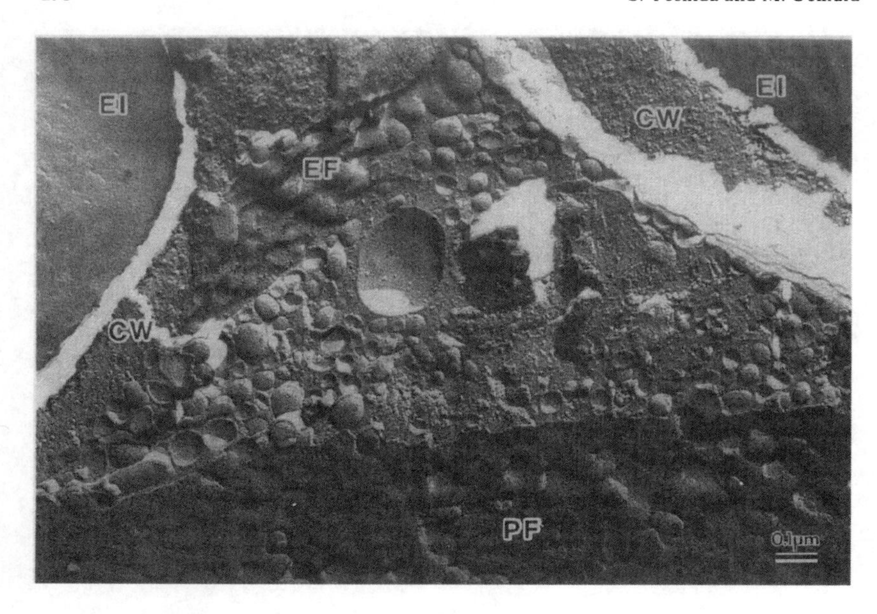

Fig. 2. Freeze-fracture electron micrograph showing endoplasmic vesiculation of plasma membranes in a tertiary hyphae of *Naematoloma sublateritium*. *CW* Cell wall; *EI* extracellular ice crystals; *EF* exoplasmic face of plasma membrane; *PF* protoplasmic face of plasma membrane (Fujikawa 1989b)

recently proposed by Steponkus (1984) and by Meryman and Williams (1985) are mainly based on the volumetric behavior of cells upon hyperosmotic or freeze-induced dehydration. The stress-strain relationship of isolated protoplasts has been corroborated by Wolfe and Steponkus (1983) using a micropipette aspiration method. They showed that elastic surface area contraction and expansion of the plasma membrane is only 2%. The surface area deformations which occur during osmotic manipulation or during a freeze-thaw cycle greatly exceed this value. Therefore, they have proposed the existence of a membrane reservoir into which plasma membrane material is deposited during contraction and from which material is recaptured during reexpansion. This is the theoretical background for their TSAI (tolerable surface area increment) concept to explain the expansion-induced lysis of nonacclimated rye protoplasts (Steponkus 1984). Following cold acclimation, the $TSAI_{50}$, the value which causes 50% cell lysis, increased from 1000 to 3000 μm^2, which well exceeds the extent of contraction and expansion that would be incurred if the protoplasts were to remain spherical when completely dehydrated and subsequently returned to isotonic conditions. Direct cryomicroscope observations confirmed that the expansion-induced lysis did not occur in the cold-acclimated protoplasts, suggesting a different mechanism of injury.

Meryman (1968) proposed a minimum volume theory based on the postulate that for every cell there is a critical volume beyond which it cannot be reduced without injury. Meryman and Williams (1985) recently modified this theory in terms of a critical surface area of the cell beyond which it cannot be reduced without extrusion of lipids from the plasma membrane. The capability of recapturing the extruded lipids upon deplasmolysis or thawing would depend on the nature of the phospholipids and carrier molecules which mediate lipid transport between the lipid deposit and the plasma membrane.

Freezing injury has long been attributed to the increased concentration of solutes, especially electrolytes (Maximov 1912; Lovelock 1953). Dysfunction of isolated thylakoid membranes by in vitro freezing has been corroborated by Heber and Santarius (1973) and Santarius (1982). The cause of freeze-induced damage of thylakoid membranes in vitro is attributed to the increased concentration of electrolytes and loss of specific proteins (Mollenhauer et al. 1983). The membrane is well protected against damage by sugars and related compounds.

4 Biochemical and Biophysical Changes in the Plasma Membrane During Freezing

4.1 Phospholipid Degradation in Whole Tissues and Isolated Membranes

Theories on mechanisms of freezing injury have so far been more or less based on the behavior of intact cells or protoplasts upon in vivo freezing or hyperosmotic manipulation. By optical microscopy, we cannot directly determine the molecular background, especially the biochemical and biophysical background, of the effects of freeze-dehydration or hyperosmotic stress on cellular membranes. Degradation of phospholipids by phospholipase D (EC 3.1.4.4) is initiated in the cortical cells of poplar and black locust trees during freezing at lethal temperatures (Yoshida and Sakai 1974). The freeze-induced breakdown of membrane phospholipids was confirmed with other herbaceous plants (Wilson and Rinne 1976; Sikorska and Kacperska-Palecz 1980; Clark et al. 1982; Borochov et al. 1987). Since phospholipase D is bound to several membranes including the plasma membrane (Yoshida 1979a), it was thought that freeze-dehydration might activate the degradative enzyme through an alteration of the membrane structure (Yoshida 1979c). Ca^{2+} is required for the activity of the enzyme (Kate and Sastry 1969). Through kinetic studies using microsomal membranes isolated from living bark tissues of black locust trees (*Robinia pseudoacacia* L.), the activity of the membrane-bound phospholipase D was found to be regulated by a competitive binding of divalent cations (Yoshida 1979b). Binding of Ca^{2+} at high concentrations modified the pH activity profile, shifting the optimum pH toward neutral thereby increasing the activity at neutral pH. Mg^{2+}, on the other hand, inhibited competitively these effects of Ca^{2+}. The membrane-bound phospholipase D became more sensitive to Ca^{2+}

and less sensitive to Mg^{2+} as the cold hardiness decreased. Freeze-thawing of the microsomal membranes isolated from living bark tissues of *Robinia pseudoacacia* L., especially from less hardy tissues, caused a drastic increase in the Ca^{2+} sensitivity and a complete loss of the regulatory action of Mg^{2+}. These facts may indicate that some qualitative changes in membranes and/or the membrane-bound enzyme are involved in the hardiness changes and also in the susceptibility of membrane phospholipids to degradation by the enzyme.

4.2 Changes in Lipid Composition

Uemura and Yoshida (1986) studied the compositional alterations of plasma membranes during freezing of tissues in vivo. Jerusalem artichoke tubers collected in the field in mid-winter and stored at 2 °C were slowly frozen to a lethal temperature of -10 °C and kept there for 2 h sufficient for temperature equilibration, then the plasma membranes were isolated from the frozen tissues. Nonlethal freezing at -5 °C did not cause any detectable change in lipid composition (Table 1). However, freezing to -10 °C, lethal to the tissues, resulted in a large decrease in the total amount of sterols. Since the plasma membranes of Jerusalem artichoke tubers contain a high proportion of free sterols, 45 mol% of total lipids, it was considered that free sterols were specifically lost from the plasma membranes during the lethal freezing. Compositional changes were also observed in the phospholipids during freezing at -10 or -15 °C (Table 2). The relative content of phosphatidylethanolamine decreased during freezing below -5 °C. Based on these results, it was suggested that sterols and phosphatidylethanolamine are sorted out or segregated into specific domains in the plasma membrane as a result of severe freeze-dehydration of cells at sublethal temperatures and that the specific membrane domains enriched in these lipids may be lost specifically during freezing or during cell disruption of the freezing

Table 1. Sterol and phospholipid contents [μmol (mg protein)$^{-1}$] of plasma membranes isolated from Jerusalem artichoke tubers frozen to various temperatures in vivo (Uemura and Yoshida 1986)[a]

	Unfrozen	Frozen to		
		-5 °C	-10 °C	-15 °C
Sterols	1.65	1.67	1.42	1.18
Phospholipids	1.52	1.47	1.47	1.47
Sterols/phospholipid	1.09	1.14	0.94	0.80

[a]Tissues of Jerusalem artichoke tubers collected in mid-December in the field were slowly (5 °C h^{-1}) frozen to various temperatures as indicated and kept there for 2 h. The frozen tissues were immediately homogenized and plasma membranes were separated by aqueous polymer two-phase partitioning. Lipids were extracted and separated by silicic acid column chromatography. The tissues were killed by freezing below -8 °C.

Table 2. Changes in composition (mol%) of plasma membrane phospholipids in Jerusalem artichoke tubers during freezing in vivo at various temperatures (Uemura and Yoshida 1986)[a]

		Frozen to		
Phospholipids	Unfrozen	-5°C	-10°C	-15°C
PC[b]	39.3	39.0	39.0	40.1
PS + PI	10.6	13.6	14.2	17.9
PE	43.6	41.0	36.1	31.8
PG	2.6	3.0	3.5	4.5
PA	3.9	3.4	5.2	5.7

[a] Lipids were extracted from plasma membranes isolated from frozen tissues as described in Table 1.
[b] PC, phosphatidylcholine; PS, phosphatidylserine; PI, phosphatidylinositol; PE, phosphatidylethanolamine; PG, phosphatidylglycerol; PA, phosphatidic acid.

cells and subsequent membrane isolation. The moderate increase in phosphatidic acid content during freezing (Table 2) clearly shows that phospholipase D-mediated phospholipid degradation is also induced. Biochemical and ultrastructural investigations show that slow freezing of erythrocytes causes a segregation of cholesterol and results in the formation of cholesterol-enriched, protein-free vesicles (Araki 1979).

4.3 Changes in Polypeptide Composition

As indicated in Fig. 3, SDS-polyacrylamide gel electrophoresis of plasma membranes isolated from frozen artichoke tubers revealed a specific loss of some polypeptides (spots 1 to 8) and an increase in others (spots a to f) during freezing at a lethal temperature of -15°C (Uemura and Yoshida 1986). Only minor changes were detected in the polypeptide composition during freezing at -5°C which is innocuous for the tissue. These changes in the polypeptide composition may suggest that proteins are decomposed during lethal freezing by a specific protease(s) and as a result spots like a to f may be newly formed. Nearly the same result was obtained with orchard grass seedlings (Uemura and Yoshida 1986). At the moment it is difficult to explain precisely how these lipid and protein changes are brought about in the plasma membranes and how these changes are interrelated. However, it is possible that plasma membrane proteins first undergo conformational changes in an irreversible manner as a consequence of the lowered temperature and/or severe freeze-dehydration. This, in turn, may cause specific molecular interactions between lipids and proteins to be significantly disturbed and may result in either/both denaturation of proteins to make them susceptible to a specific protease(s) or/and activation of the protease(s). The lamellar to hexagonal$_{II}$ phase transition in plasma membrane lipid bilayers could be a related phenomenon (Sect. 2.3).

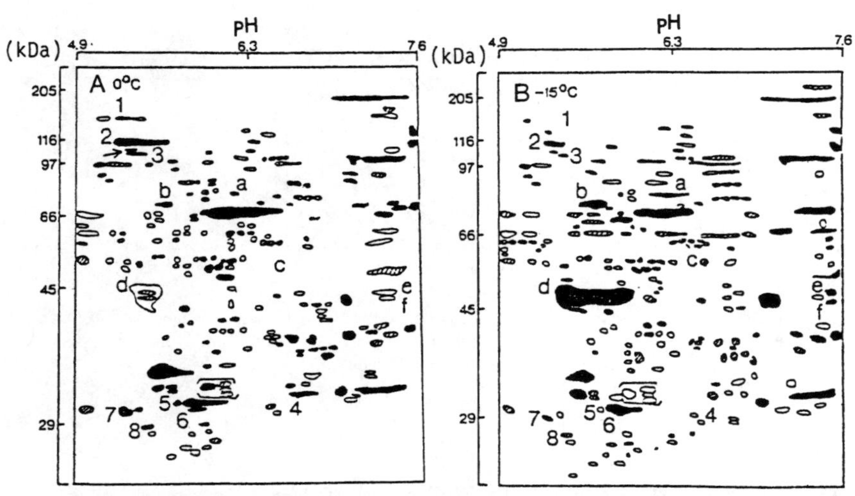

Fig. 3A,B. Changes in plasma membrane proteins of Jerusalem artichoke tubers during freezing at a sublethal temperature. Plasma membranes were isolated from unfrozen (**A**) and frozen (−15 °C) (**B**) tissues and the membrane proteins were separated with two-dimensional SDS-polyacrylamide gel electrophoresis. The gels were silver stained. Figures are tracings of the electrophoretograms. Major spots changing in staining intensity upon freezing are marked by *numbers* (decreased) or *letters* (increased) (Uemura and Yoshida 1986)

4.4 Temperature-Dependent Biophysical Changes

When intact cells are frozen, the plasma membranes experience complex stresses and, therefore, it might be difficult to distinguish between the factors responsible for membrane changes. These factors may involve a physical effect of the temperature reduction per se (Pringle and Chapman 1981), a freeze-induced condensation of solutes both in the cell interior and exterior (Franks 1981), a freeze-induced dehydration of membrane molecules (Franks 1981), a freeze-induced cell shrinkage and molecular packing of membrane constituents (Pringle and Chapman 1981), or a combination of these. From this point of view, isolated, sealed plasma membrane vesicles could be a useful model system for investigating the molecular mechanisms of freezing injury of plant cells. Steady-state fluorescent polarization of an embedded fluorophore, 1,6-diphenyl-1,3,5-hexatriene (DPH) has been widely used for detecting thermotropic phase transitions in lipid bilayers of biological membranes and liposomes made from lipids extracted from them (Shinitzky and Barenholz 1978; Livingstone and Schachter 1980), and also for the evaluation of the relative motional freedom of lipid molecules. Plasma membrane vesicles isolated from nonacclimated and cold-acclimated orchard grass seedlings were labeled with DPH and the anisotropy parameter values were measured at different subfreezing temperatures (Yoshida 1984a). To avoid freezing even at −20 °C, the membrane

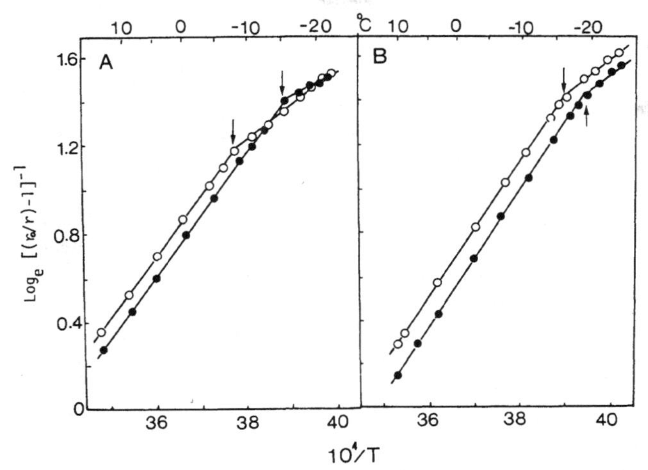

Fig. 4. Arrhenius plots of fluorescence anisotropy parameters, $[(r_0/r)-1]^{-1}$, of DPH embedded into intact plasma membranes (**A**) and liposomes (**B**) prepared from total lipid extracts therefrom. Plasma membranes were isolated from orchard grass seedlings in different seasons. ○, Membrane sample from early October (frost hardiness: −7.5°C); ●, membrane sample from early December (frost hardiness: −15°C). *Arrows* indicate inflection points (Yoshida and Uemura 1984)

suspension contained 35% (v/v) ethylene glycol. Distinct inflections could be observed in the Arrhenius plots of the anisotropy parameters around -8 and -16°C, respectively, for the membrane samples isolated from nonacclimated and cold-acclimated plants (Fig. 4A). These inflection temperatures appeared to coincide with those for the frost-killing of the tissues. Fluorescent polarization measurements were also performed with liposomes prepared from the total lipids extracted from plasma membranes. No inflection was observed in the Arrhenius plots down to -17°C or below in liposomes prepared from either membrane (Fig. 4B). Thus, there was a marked difference in the thermotropic properties between intact membrane vesicles and liposomes made from the extracted lipids. By treatment of the plasma membrane vesicles with pronase (mixture of proteinases prepared from *Streptomyces griseus*), the inflection temperatures on the Arrhenius plots of the anisotropy parameter was shifted downward by 10°C (Yoshida 1984a). Thus, the effect of membrane proteins on the thermotropic properties of lipid bilayers must be considered when looking at low temperature-induced phase changes in plasma membrane lipids.

4.5 Changes in ATPase Activity

According to Palta et al. (1982), electrolyte leakage from freeze-thawed onion epidermal cells is reversible unless the cells are frozen beyond a certain temperature range. They postulated that the properties of membrane transport proteins are altered in the initial stage of injury following a freeze-thaw cycle. Arora and Palta (1988) have recently shown that the enhanced efflux of K^+ from freeze-thawed onion bulb cells is due to the loss of plasma membrane-associated Ca^{2+}. Hellergren et al. (1987) reported that a lethal freezing of pine needles resulted in a marked decrease of plasma membrane ATPase activity. They prepared the plasma membrane samples from freeze-thawed tissues which were kept on ice overnight after freezing to various temperatures. Hence, it is difficult to decide whether the changes occurred during freezing or after thawing of the tissues as a result of injury. To avoid this problem, the plasma membranes should be prepared directly from the frozen tissues. In Jerusalem artichoke tubers (Uemura and Yoshida 1986), in vivo freezing of tissues to lethal temperatures resulted in a partial loss of plasma membrane ATPase activity, and the N,N′-dicyclohexylcarbodiimide (DCCD) sensitivity of the enzyme was markedly affected, i.e., the inhibition by 100 μM DCCD increased from 18 to 45%. From this result, it was concluded that either the ATPase was partially modified, especially at the DCCD binding site or DCCD had better access to the enzyme as a result of alterations in membrane lipids and/or proteins.

5 Alterations of Plasma Membranes Related to Cold Acclimation

5.1 Seasonal Changes in Freezing Tolerance

Perennial and biennial plants growing in cold climates show remarkable seasonal changes in freezing tolerance (Sakai and Larcher 1987). During the growing season most of them are very susceptible to freezing and are easily injured even by one night with light frost. However, their freezing tolerance increases gradually through autumn to early winter as the environmental temperature decreases. The ability to cold-acclimate and the magnitude of the freezing tolerance to be achieved in plants vary with species and are genetically controlled. As discussed above, there is a consensus that plasma membranes play a central role in freezing injury of plant cells. Therefore, it is quite natural to assume that plasma membranes undergo molecular changes during cold acclimation so as to adjust their properties against freezing stress. Levitt and Scarth (1936b) found that increases in the permeability of cells to polar substances such as water, urea, and KNO_3 were correlated with development of freezing tolerance during cold acclimation. This was corroborated by Fennell and Li (1986) using leaf petiole cells of *Solanum* species. The permeability change is considered to be a reflection of molecular alterations in the plasma membrane. No direct evidence, however, has been presented until recently as to what molecular changes are produced in plasma membranes and how these changes contribute to the tolerance against freezing damage.

5.2 Changes in Lipid Composition

Many studies have been carried out to find a causal relationship between lipid composition and freezing tolerance in plants (reviewed by Willemot 1979). The results reported so far are, however, variable. A problem is that lipids were extracted from whole cells or, at best, from isolated crude membrane fractions such as total microsomal fractions which may include only 10% plasma membranes as well as a variety of other cellular membranes. The lipid composition may vary between different cellular membranes. Therefore, the obtained results cannot be ascribed to the plasma membrane per se. In etiolated mung bean seedlings (*Vigna radiata* L.), the lipid composition of plasma membrane and tonoplast are quite different (Table 3; Yoshida and Uemura 1986). The plasma membrane is characterized by a high content of sterols, especially free sterols. During cold acclimation, the content of free sterols in the plasma membrane generally increased in most herbaceous plants (Uemura and Yoshida 1984; Yoshida and Uemura 1984; Ishikawa and Yoshida 1985; Lynch and Steponkus 1987).

In orchard grass seedlings (*Dactylis glomerata* L.), the phospholipids of the plasma membrane are less unsaturated than those of mitochondrial membranes and the fluidity of the plasma membranes as assessed by fluorescent polarization of embedded fluorophore, DPH, is therefore usually lower than in endomembranes such as ER, tonoplasts, and mitochondrial membranes, suggesting that the plasma membrane is more rigid (Yoshida 1984a). Plasma membrane phospholipids in orchard grass seedlings did not show any appreciable change in the fatty acid composition following cold acclimation, although the proportion of unsaturated fatty acids increased to some extent in phospholipids in other cellular membranes (Table 4; Yoshida and Uemura 1984). The situation was nearly the same in winter rye seedings (Uemura and Yoshida 1984) and Jerusalem artichoke tubers (Ishikawa and Yoshida 1985). In winter rye seedlings, a moderate increase in linolenic acid (18:3) content with a

Table 3. Lipid composition of plasma membranes and tonoplasts isolated from etiolated mung bean hypocotyls (*Vigna radiata* L.) (Yoshida and Uemura 1986)[a]

Lipids	Composition (mol%)	
	Plasma membrane	Tonoplast
Phospholipids	48.9	51.0
Free sterols	39.5	18.2
Acylated sterolglycosides	1.5	7.4
Sterolglycosides	2.3	2.3
Glucocerebroside	6.8	16.6

[a] Plasma membranes and tonoplasts were isolated from hypocotyls excised from 3.5-day-old etiolated seedlings.

Table 4. Changes in fatty acid composition of phospholipids in different cellular membranes following cold acclimation of orchard grass (Yoshida and Uemura 1984)[a]

	Membrane fraction			
	Plasma membranes (Upper phase)		Endomembranes (Lower phase)	
Fatty acid	2 Oct.	8 Dec.	2 Oct.	8 Dec.
16:0	26.1	26.1	23.9	21.0
18:0	0.7	0.6	0.8	0.4
18:1	7.5	6.1	9.6	6.1
18:2	40.6	41.5	34.4	38.6
18:3	23.1	23.8	30.2	32.4
20:0	1.2	1.0	0.7	0.7
Unsaturated/saturated	2.54	2.57	2.92	3.49

[a] Microsomal membrane fractions were subjected to aqueous polymer two-phase partitioning. The upper phase (plasma membranes) and the lower phase (mixture of endomembranes) were used for lipid analyses.

concomitant decrease in linoleic acid (18:2) content was observed during early stages of cold acclimation (10 days) where development of freezing tolerance was slight (Fig. 5). However, no detectable change was observed thereafter up to the end of the cold acclimation period (30 days) at which maximum freezing tolerance was achieved. Thus, there seems to be no direct correlation between lipid unsaturation and development of freezing tolerance in these plants. In winter rye seedlings (Uemura and Yoshida 1984), phospholipids in endomembranes, which were partitioned into the lower phase after phase partitioning of crude microsomal membranes, were highly unsaturated and the content of linolenic acid accounted for over 50 mol% of total fatty acids. In none of these herbaceous plants did the relative proportions of the individual phospholipid classes change significantly. The lipid changes in plasma membranes, however, seem to depend on the magnitude of freezing tolerance in plants. Most herbaceous plants, such as orchard grass, winter rye, and Jerusalem artichoke tubers are all intermediately hardy plants, tolerant to freezing to −10 to −18 °C. In extremely cold-hardy woody species, such as mulberry trees, the cortical parenchyma cells can withstand freezing below −40 °C or even in liquid nitrogen (−196 °C) in mid-winter, although they are easily injured even by a freeze-thaw cycle above −3 °C in mid-summer. In mulberry bark tissues, plasma membrane phospholipids became highly unsaturated during cold acclimation with a concomitant increase in freezing tolerance under natural conditions (Fig. 6; Yoshida 1984b). Therefore, the unsaturation of phospholipids in plasma membranes seems to be largely dependent on the magnitude of freezing tolerance in plants.

Glucocerebroside is one of the major lipids in plasma membranes and tonoplasts (Yoshida and Uemura 1986). In mung bean seedlings, the contents

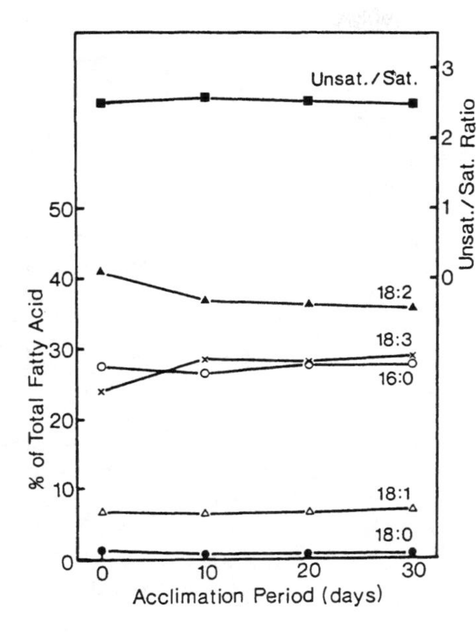

Fig. 5. Changes in the fatty acid composition of plasma membrane phospholipids from winter rye seedlings (*Secale cereale* L. cv. Puma) during cold acclimation. ○, Palmitic acid; ●, stearic acid; △, oleic acid; ▲, linoleic acid; ×, linolenic acid; ■, unsaturated/saturated ratio [(oleic + linoleic + linolenic)/(palmitic + stearic)] (Uemura and Yoshida 1984)

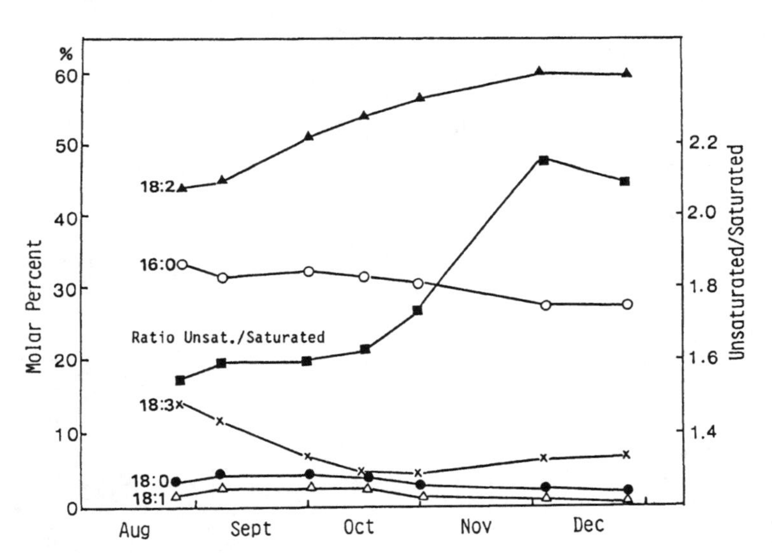

Fig. 6. Seasonal changes in fatty acid composition of plasma membrane phospholipids in living bark cells of mulberry trees. ○, Palmitic acid; ●, stearic acid; △, oleic acid, ▲, linoleic acid; ×, linolenic acid; ■, unsaturated/saturated ratio [(oleic + linoleic + linolenic)/palmitic + stearic)] (Yoshida 1984b)

are 6.5 and 17.0 mol% of total lipids, respectively (Table 3). According to Lynch and Steponkus (1987), the content of glucocerebroside decreased from 16.2 to 6.8 mol% in plasma membranes from winter rye following cold acclimation. Furthermore, the amide-linked fatty acid composition of the glycolipid was significantly altered following cold acclimation. Although the fatty acids were mostly hydroxylated at the 2-position, the relative proportion of unsaturated hydroxy fatty acid (24h:1) increased substantially, whereas that of the other hydroxy fatty acids such as 22h:0 and 24h:0 decreased. This is a very interesting finding in light of the thermotropic properties. In chilling-sensitive mung bean seedlings, glucocerebroside in the tonoplast shows a high temperature of initiation (35°C) for phase transition (Yoshida et al. 1988). Detailed analysis by differential scanning calorimetry (DSC) of liposomes prepared from extracted total lipids and intact tonoplast vesicles revealed that the glucocerebroside species with a high melting point undergoes a phase separation around 8°C (Yoshida et al. 1988) below which seedlings suffered chilling injury. In contrast to mung bean seedlings, the phase transition temperature of glucocerebroside in tonoplasts isolated from chilling-resistant pea seedlings was 10°C lower than that of mung been seedlings (K. Washio and S. Yoshida, unpublished results). Furthermore, no phase transition was observed in liposomes prepared from the total lipids extracted from these membranes. Recently S. Ito and M. Ohnishi (personal communication) have examined the distribution of the molecular species with a high melting point in a wide variety of plants and found a close correlation between the occurrence of this molecular species and the chilling sensitivity of plants. Further studies on the glucocerebrosides in plasma-membranes should be performed on a wide variety of plants in view of a possible correlation with the thermotropic properties and membrane instability to freeze-dehydration.

As mentioned above, the compositional changes in plasma membrane lipids, especially the degree of fatty acid unsaturation in the phospholipids during cold acclimation, are rather variable with plant species. However, there seems to be a general pattern in the lipid changes in plasma membranes. As shown in Fig. 7, the relative content of phospholipids per mg protein increased significantly in mulberry plasma membranes and this increase was associated with an increase in freezing tolerance (Yoshida 1984b). Upon an artificial cold acclimation of the excised twigs at 0°C for 2 weeks, the relative content of phospholipids showed a significant increase. This was also true for herbaceous plant species (Uemura and Yoshida 1984; Yoshida and Uemura 1984; Ishikawa and Yoshida 1985). After sucrose density gradient centrifugation of microsomal membranes isolated from cortical cells of wintering hardy and deacclimated mulberry twigs, significant shifts of buoyant densities of cellular membranes were observed upon deacclimation. Especially, ER, tonoplasts, and plasma membranes increased their buoyant density following deacclimation, indicating that the relative content of lipids, in particular phospholipids, in these membranes vary depending on the degree of freezing tolerance (S. Yoshida, unpublished results).

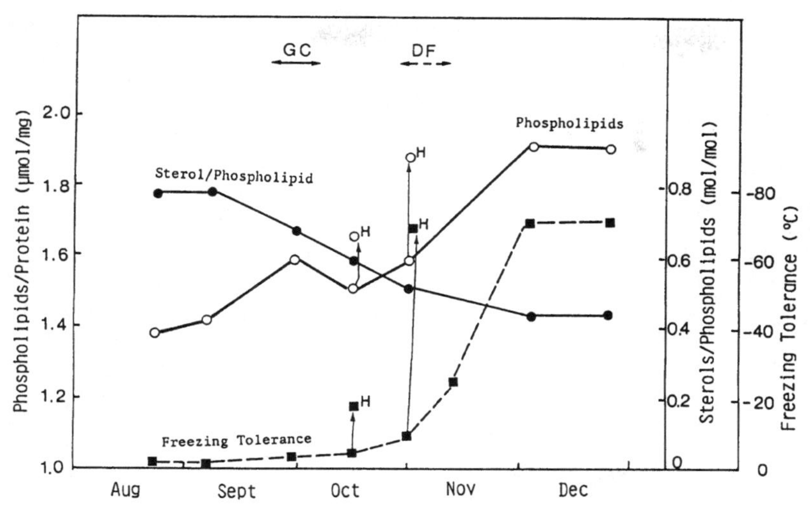

Fig. 7. Seasonal changes in phospholipids and sterols in plasma membranes isolated from living bark cells of mulberry tree. ○, Phospolipid content (μmol/mg protein); ●, molar ratio of total sterols to phospholipids; ■, freezing tolerance. *H* corresponds to the value after an artificial cold acclimation of excised twigs of 0 °C for 3 weeks. *GC* Growth cessation; *DF* defoliation period (Yoshida 1984b)

The capacity of phospholipid synthesis in plants is reported to increase following growth at low temperature. In wheat seedlings the incorporation of [^{14}C]acetate (Grenier et al. 1975) and [^{32}P] (Willemot 1975) into phospholipids increased during cold acclimation, but only in cold-hardy varieties. Recently, it was reported that the enzyme systems for phosphatidylcholine biosynthesis in rye roots are activated by low temperature (Kinney et al. 1987). The incorporation of [^{14}C]choline and [^{14}C]glycerol into root segments was significantly stimulated by pregrowth at 5 °C. Measurements in vitro of the enzymes of the nucleotide pathway showed that activities of choline kinase (EC 2.7.1.32), choline-phosphate cytidyltransferase (EC 2.7.7.15), and choline phosphotransferase (EC 2.7.8.2) were significantly higher in roots grown at 5 °C than in roots grown at 20 °C. According to Hatano et al. (1981), the activity of the fatty acid synthetase system in *Chlorella* is markedly enhanced by cold acclimation. Thus, a high capacity for lipid synthesis seems to be a prerequisite for the development of cold hardiness in plants.

5.3 Ultrastructural Changes

The phospholipid enrichment of plasma membranes following cold acclimation are consistent with the observed alteration in the population of intramembranous protein particles (IMP) in plasma membranes during cold acclimation and deacclimation. Sugawara and Sakai (1978) showed by freeze-fracture electron microscopy that the population of the IMP on the cytoplasmic fracture face on the plasma membrane was significantly reduced following cold acclimation of Jerusalem artichoke tuber callus. The population of IMP in plasma membranes of cambial-zone cells of *Salix* increased significantly from winter to spring (Parish 1974). Nearly the same results are obtained in cortical parenchyma cells of mulberry trees (S. Fujikawa, personal communication). At the moment, the effect of the phospholipid enrichment or depletion on the physical properties of plasma membranes is unclear. However, it is assumed that the changes in the relative content of phospholipids, in particular unsaturated phospholipids, in plasma membranes are an important part of the molecular rearrangements which occur during cold acclimation, including changes in membrane polypetides (Sect. 5.5).

5.4 Changes in Membrane Fluidity

Using electron spin resonance (ESR) spectroscopy, Vigh et al. (1979) found that plasma membranes of protoplasts isolated from winter wheat leaves became more fluid with cold acclimation. However, it is not clear whether the ESR probe, a fatty acid derivative, was specifically incorporated into the protoplast surface nor is it clear whether the incorporated probe molecule was kept intact during the measurement. According to our studies (Yoshida and Uemura 1984; Ishikawa and Yoshida 1985) using isolated plasma membrane vesicles and a fluorescent probe, DPH, the fluidity of plasma membranes in most herbaceous plants showed only a slight increase after cold acclimation. However, in extremely hardy woody plants like the mulberry tree, plasma membranes became more fluid with increasing frost tolerance (Yoshida 1984b). This was closely associated with an increase in phospholipid unsaturation and a decrease in the sterol to phospholipid ratio as a result of a relative increase in phospholipids. Thus, the changes in membrane fluidity during cold acclimation depend on whether the plant species is moderately or extremely cold hardy.

5.5 Changes in Polypeptide Composition

Plasma membranes have many important physiological functions such as cell wall synthesis (Chap. 11), active transport of ions and metabolites (Chap. 7), hormone action and responses (Chap. 9), disease resistance (Chap. 14), and stress responses. Therefore, the plasma membrane should contain several types of proteins involved in those functions, in addition to structural proteins. These membrane proteins are probably in a dynamic state, changing the structure and composition of the plasma membrane in response to physiological changes including growth, development, and environmental adaptation. The polypeptide composition of plasma membranes from cortical cells of mulberry tree (Yoshida 1984b) and winter rye seedlings (Uemura 1984) varied with growth, i.e., from active growth to cessation in the former plant and leaf development in the latter plant, independently of development of freezing tolerance. However, in these plants, specific changes were also determined in the polypeptide composition of plasma membranes during cold acclimation (Uemura and Yoshida 1984; Yoshida and Uemura 1984; Ishikawa and Yoshida 1985). After two-dimensional gel electrophoresis of denatured plasma membrane polypeptides of orchard grass and winter rye seedlings, specific proteins appeared to be degraded or newly synthesized during cold acclimation. Figure 8 shows the changes in plasma membrane polypeptides in winter rye seedlings following cold acclimation. These results indicate that plasma membranes undergo a dynamic alteration in the protein composition in response to a cold environment. However, it is uncertain whether the changes are specific for the development of cold hardiness or not. Experiments using different wheat cultivars, less hardy (cv. Chinese spring) and hardy (cv. Norstar) wheats, demonstrated that about 13 polypeptides increased specifically in the plasma membrane from the hardy cultivar (cv. Norstar) during cold acclimation (B. Zhou and S. Yoshida, unpublished results). Therefore, these polypeptides are considered to have a specific role in stabilization of plasma membranes against freezing stress.

To get a better insight into the structural features of proteins, we attempted to analyze the proteins under different conditions including a relatively mild solubilization with a neutral detergent, Triton X-114 (M. Uemura and S. Yoshida, unpublished results). The solubility of plasma membrane proteins of Jerusalem artichoke tubers into Triton X-114 changed during cold acclimation. As the frost tolerance of the tubers increased from -3°C (September) to -8.8°C (December), the plasma membrane proteins became more soluble in this neutral detergent (increase from 52 to 71%). This might indicate either that membrane proteins as a whole became more soluble in the detergent or that the detergent-insoluble membrane proteins were replaced by more detergent-soluble ones during cold acclimation. As indicated in Fig. 9, however, there was a distinct difference in the polypeptide composition between the detergent-soluble and -insoluble proteins, consistent with the latter possibility. Of the detergent-insoluble polypeptides, those at 52, 46, 30 and 25 kD decreased in staining intensity, whereas the bands F-3 (one of the frost-susceptible polypeptides), B,C,F,H,I,J, and L of the detergent-soluble proteins increased with development of frost tolerance during cold acclimation.

Triton X-114 is a polydisperse, nonionic detergent and can be homogeneously dispersed at 0°C, but reaches a clouding point above 20°C. After warming to 20°C, a diluted solution of the detergent (1%, w/v) separates into two phases, a detergent-rich and a detergent-poor phase. During this phase separation, the solubilized membrane proteins partition between the phases depending on their hydrophobicity, although the partition may also be influenced by the molecular shapes and the conformations (Bordier 1981). As

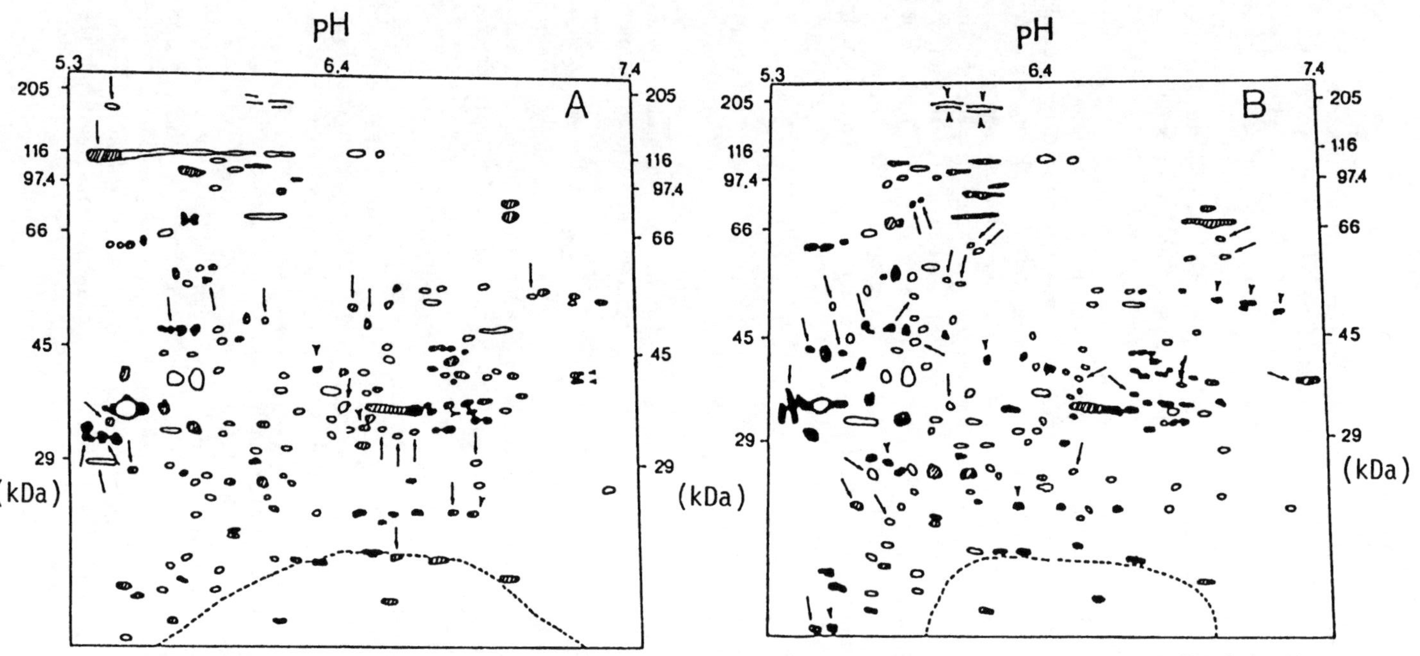

Fig. 8. Tracings of two-dimensional SDS-polyacrylamide gel electrophoretograms of plasma membrane proteins from winter rye seedlings before (**A**) and after (**B**) cold acclimation. Disappearance or appearance of polypeptides following cold acclimation are marked by *arrows*, and spots showing quantitative changes are marked by *arrowheads*. The areas *circled by dotted lines* are stained heavily by the presence of ampholytes. Heavily stained spots are filled in *solid*, and other spots are *hatched* or *outlined* according to the degree of staining (Uemura and Yoshida 1984)

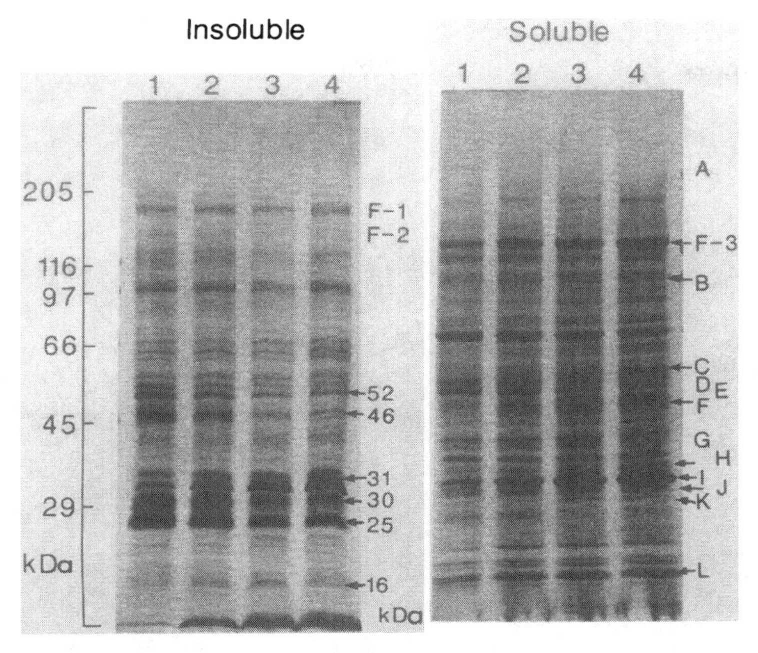

Fig. 9. SDS-polyacrylamide gel electrophoretograms of Triton X-114 insoluble and soluble plasma membrane proteins. Plasma membranes were isolated from Jerusalem artichoke tubers at different times of the year. The isolated plasma membranes were solubilized with 1% (w/v) Triton X-114 at 0°C. The insoluble and the soluble protein fractions were separated on a SDS-polyacrylamide gradient (5–15%) gel. After electrophoresis gels were stained with Coomassie brilliant blue. Lanes *1–4* correspond to sampling dates, 25 Sept., 22 Oct.,19 Nov., and 16 Dec., respectively. *F-1,2,* and *3* show the frost-susceptible proteins which are decomposed into smaller polypeptides following a lethal freezing in vivo (Uemura and Yoshida 1986). The polypeptide bands changing during cold acclimation are marked by *numbers* (kD) (insoluble fraction) and *letters* (soluble fraction) (M. Uemura and S. Yoshida, unpublished data)

indicated in Fig. 10, polypeptide bands I (31 kD) and J (30 kD) mainly partitioned into the detergent-rich phase, suggesting that they are the most hydrophobic. These polypeptides showed a marked increase during cold acclimation. The role of these polypeptides on membrane stability during freezing stress has still to be determined. However, there may be a relationship between the phospholipid enrichment of plasma membrane and the specific increase in those hydrophobic membrane proteins during cold acclimation.

6 Concluding Remarks

Electron microscopy of freezing plant cells has provided evidence that the molecular structure of plasma membranes is dramatically altered by a lethal freezing. At the moment it is still difficult to explain precisely how these

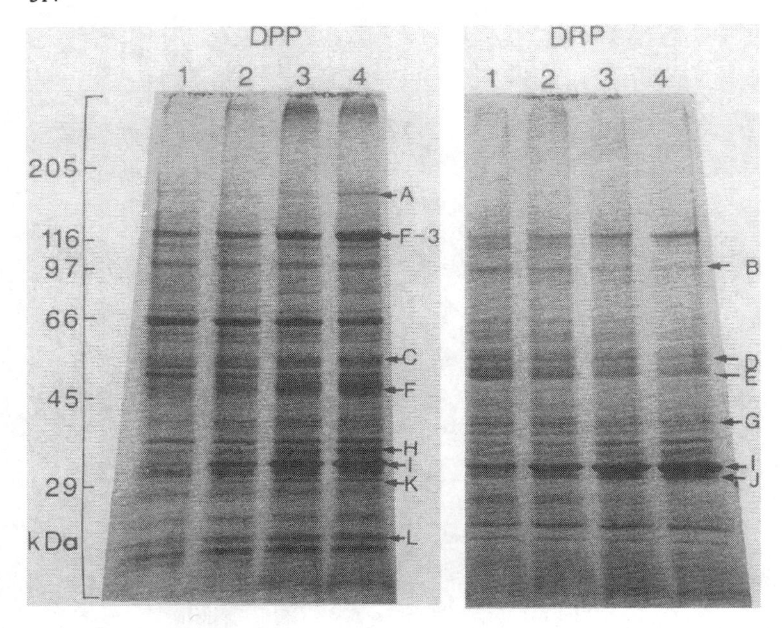

Fig. 10. SDS-polyacrylamide gel electrophoresis of Triton X-114 soluble protiens after separation into detergent-poor (DPP) and detergent-rich (DRP) phases. The Triton X-114 soluble fraction was warmed to 20°C and centrifuged, then the separated DPP and DRP fractions were subjected to SDS-polyacrylamide gradient (5–15%) gel electrophoresis. Gels were stained with Coomassie brilliant blue. Lanes *1–4* refer to the legend of Fig. 7. The bands changing in staining intensity during cold acclimation are indicated by *letters* on the side of the panels (M. Uemura and S. Yoshida, unpublished data)

structural changes are brought about by freezing and how cold acclimation can mitigate the deleterious changes. There is a great diversity in the ultrastructural changes, i.e., hexagonal$_{II}$ phase transition, formation of smooth patches, aggregation of IMP, formation of multiple lamellars, and osmiophilic globules, depending not only on plant species but also on the experimental conditions. If a hexagonal$_{II}$ phase transition in the plasma membrane is a general feature of freeze-injured cells, then the physical state of lipids and, therefore, the lipid composition per se might be a major factor in the mechanism of freezing injury. However, the universal occurrence of the hexagonal$_{II}$ phase transition in plant plasma membranes has not been confirmed. Due to the phenomenological similarity between hyperosmotic manipulation and freeze-induced cell dehydration, effects of low temperature per se are not considered as a major factor in freezing injury. However, one should remember that the physical state of lipid bilayers can be markedly influenced by subzero temperatures without freezing and that this influence is highly dependent on the membrane proteins, or rather on the interaction between lipids and proteins. Modern theories to explain freezing injury mostly consider freeze-induced reductions in cell volume and, therefore, the cell surface area, and removal of hydrating water molecules from polar head groups of membrane lipids. However, removal of hydrating water molecules from membrane proteins should not be neglected. The SH hypothesis of Levitt (1972), which is based on the assumption that intermolecular S-S bonds are formed between protein molecules as a result of dehydration-induced molecular packing of membranes, has been strongly criticized by many investigators (Steponkus 1984 and references therein). However, it is possible that the conformation of membrane protein is severely affected by removal of the hydrating water molecules by freezing. Indeed, in vivo freezing of plant tissues causes significant changes in the polypeptide composition of the plasma membranes. Therefore, the removal of water molecules from membrane proteins may influence protein-protein and protein-lipid interactions in the membranes. According to Sakai and Yoshida (1968), sugars and related compounds, which cannot penetrate to the cell interior, effectively protect plant cells against freezing injury, when intact tissues are frozen in these solutions. The protection site is considered to be proteins on the outer surface of the plasma membrane. The protecting effects of sugars are also dependent on the developmental stage of the plant, suggesting that the properties of the protection site(s) on the plasma membrane vary with season. It should be considered that these protective substances do not have any effect on the extent of reductions in cell volume and cell surface area upon cell freezing (S. Yoshida, unpublished results). Thus, the effects might be ascribed to the protection against freeze-induced removal of hydrating water molecules from membrane constituents.

During cold acclimation, qualitative changes in plasma membrane proteins concomitantly with an increase in detergent solubility and phospholipid enrichment are often observed, regardless of plant species and the magnitude of freezing tolerance, suggesting that these are fundamental factors. This indicates that the process of cold acclimation involves molecular rearrangements in the plasma membrane. The degree of phospholipid unsaturation, on the other hand, depends on plant species and presumably the magnitude of freezing tolerance. The importance of both quantitative and qualitative changes in sterols and glucocerebrosides should be further investigated in light of their possible relation to cold acclimation. It is conceivable that different biochemical traits have been sequentially introduced into plants in proportion to the degree of freezing tolerance in the long history of evolution. In this sense, other cytoplasmic factors such as increases in sugars (Sakai 1962) and soluble proteins (Siminovitch et al. 1968), and changes in cell structures including an augmentation of cytoplasm and reduction of vacuolar size (Pomeroy and Siminovitch 1971; Niki and Sakai 1981) should also be taken into consideration. These factors might contribute to a reduction in the mechanical stress including close contact of cellular membranes and membrane fusion caused by severe cell deformation during extreme freeze-dehydration (Fujikawa 1987; 1989a,b), mitigating the effect of dehydration of membrane molecules.

Finally, we should go back to the original concept proposed by Maximov (1912) that freezing injury is the result of the freeze-induced removal of water from the surface of cells, namely, the plasma membrane surface.

Acknowledgment. This work was supported in part by Grants-in-aid for Scientific Research from the Ministry of Education, Science and Culture of Japan. We wish to thank Dr. S. Fujikawa of Institute of Low Temperature Science, Hokkaido University, Sapporo for valuable suggestions and for providing original copies of electron micrographs. We should also like to express our thanks to those who have helped us by means of personal communications and by providing us with unpublished data.

References

Araki T (1979) Release of cholesterol-enriched microvesicles from human erythrocytes caused by hypertonic saline at low temperatures. FEBS Lett 97:237–240

Arora R, Palta JP (1988) In vivo perturbation of membrane associated calcium by freeze-thaw stress in onion bulb cells. Plant Physiol 87:622–628

Asahina E (1956) The freezing process of plant cells. Contrib Inst Low Temp Sci 10:83–126

Bordier C (1981) Phase separation of integral membrane proteins in Triton X-114 solution. J Biol Chem 256:1604–1607

Borochov A, Walker MA, Kendall EJ, Pauls KP, McKersie BD (1987) Effect of a freeze-thaw cycle on properties of microsomal membranes from wheat. Plant Physiol 84:131–134

Clark A, Coulson G, Morris J (1982) Relationship between phospholipid breakdown and freezing injury in a cell wall-less mutant of *Chlamydomonas reinhardii*. Plant Physiol 70:97–103

Dowgert MF, Steponkus PL (1984) Behavior of the plasma membrane of isolated protoplasts during a freeze-thaw cycle. Plant Physiol 75:1139–1151

Fennell A, Li PH (1986) Temperature response of plasma membranes in tuber-bearing *Solanum* species. Plant Physiol 80:470–472

Franks F (1981) Biophysics and biochemistry of low temperatures and freezing. In: Morris GJ, Clark A (eds) Effects of low temperatures on biological membranes. Academic Press, New York, pp 3–19

Fujikawa S (1986) Membrane ultrastructural changes caused by freezing as a possible source of artifacts with specimen preparation for freeze fracture. Proceedings of XIth International Congress on Electron Microscopy/1986, Kyoto, Vol III: pp 2209–2210

Fujikawa S (1987) Intramembrane particle aggregation caused by membrane to membrane direct contact during freezing. J Electron Microsc 36:224–227

Fujikawa S (1989a) Artificial biological membrane ultrastructural changes caused by freezing. Electron Microsc Res 1:113–140

Fujikawa S (1989b) Plasma membrane ultrastructural changes produced by freezing during the specimen preparation process for freeze fracture electron microscopy. In: Begna GH (ed) Water transport in biological membranes. CRC Press, FL, in press

Fujikawa S, Miura K (1986) Plasma membrane ultrastructural changes caused by mechanical stress in the formation of extracellular ice as a primary cause of slow freezing in fruit-bodies of Basidomycetes [*Lyophyllum ulmarium* (Fr.) Kuhner]. Cryobiology 23:371–382

Giles KL, Beardsell MF, Cohen D (1974) Cellular and ultrastructural changes in mesophyll and bundle sheath cells of maize in response to water stress. Plant Physiol 54:208–212

Giles KL, Cohen D, Beardsell MF (1976) Effects of water stress on the ultrastructure of leaf cells of *Sorghum bicolor*. Plant Physiol 57:11–14

Gordon-Kamm WJ, Steponkus PL (1984) Lamellar-to-hexagonal$_{II}$ phase transitions in the plasma membrane of isolated protoplasts after freeze-induced dehydration. Proc Natl Acad Sci USA 81:6373–6377

Grenier G, Hope H, Willemot C, Therrien HP (1975) Sodium-1,2-^{14}C acetate incorporation in roots of frost-hardy and less hardy alfalfa varieties under hardening conditions. Plant Physiol 55:906–912

Hatano S, Kabata K, Sadakane H (1981) Transition of lipid synthesis from chloroplasts to a cytoplasmic system during hardening in *Chlorella ellipsoidea*. Plant Physiol 67:216–220

Heber U, Santarius KA (1973) Cell death by cold and heat and resistance to extreme temperatures. Mechanism of hardening and dehardening. In: Precht H, Christophersen J, Hensel H, Larcher W (eds) Temperature and life. Springer, Berlin Heidelberg New York, pp 232–263

Hellergren J, Widell S, Lundborg T (1987) Freezing injury in purified plasma membranes from cold acclimated and non-acclimated needles of *Pinus silvestris*: Is the plasma membrane bound ion-stimulated ATPase the primary site of freezing injury? In: Li PH (ed) Plant cold hardiness, Alan R Liss Inc, New York, pp 211–220

Ishikawa M, Yoshida S (1985) Seasonal changes in plasma membranes and mitochondria isolated from Jerusalem artichoke tubers. Possible relationship to cold hardiness. Plant Cell Physiol 26:1331–1344

Kate M, Sastry PS (1969) Phospholipase D. Methods Enzymol 14:197–203

Kinney AJ, Clarkson DT, Loughman BC (1987) The regulation of phosphatidylcholine biosynthesis in rye (*Secale cereale*) roots. Stimulation of the nucleotide pathway by low temperature. Biochem J 242:755–759

Levitt J (1972) Responses of plants to environmental stresses. Academic Press, New York, pp 188–228

Levitt J, Scartch GW (1936a) Frost hardening studies with living cells. I. Osmotic and bound water changes in relation to frost resistance and the seasonal cycle. Can J Res Sect C 14:267–284

Levitt J, Scartch GW (1936b) Frost hardening studies with living cells. II. Permeability in relation to frost resistance and the seasonal cycle. Can J Res Sect C 14:285–305

Levitt J, Siminovitch D (1940) The relation between frost resistance and the physical state of protoplasm. I. Protoplasm as a whole. Can J Res Sect C 18:550–561

Livingstone CJ, Schachter D (1980) Lipid dynamics and lipid-protein interaction in rat hepatocyte plasma membranes. J Biol Chem 255:10902–10908

Lovelock J (1953) Haemolysis of human red blood cells by freezing and thawing. Biochim Biophys Acta 10:414–426

Lynch DV, Steponkus PL (1987) Plasma membrane lipid alterations associated with cold acclimation of winter rye (*Secale cereale* L. cv. Puma). Plant Physiol 83:761–767

Maximov NA (1912) Chemische Schutzmittel der Pflanzen gegen Erfrieren. Ber Dtsch Bot Ges 30:52–65, 293–305, 504–516

Mazur P (1969) Freezing injury in plants. Annu Rev Plant Physiol 20:419–448

Meryman HT (1968) Modified model for the mechanism of freezing injury in erythrocytes. Nature 218:333–336

Meryman HT (1974) Freezing injury and its prevention in living cells. Annu Rev Biophys Bioenerg 3:341–363

Meryman HT, Williams RJ (1985) Basic principles of freezing injury to plant cells: natural tolerance and approaches to cryopreservation. In: Kartha KK (ed) Cryopreservation of plant cells and organs, CRC Press, Orlando, pp 13–47

Mollenhauer A, Schmitt JM, Coughlan S, Heber U (1983) Loss of membrane proteins from thylakoids during freezing. Biochim Biophys Acta 728:331–338

Molish H (1897) Untersuchungen über das Erfrieren der Pflanzen. Reprinted in English 1982 in Cryo-Letters 3:332–390

Niki T, Sakai A (1981) Ultrastructural changes related to frost hardiness in the cortical parenchyma cells from mulberry twigs. Plant Cell Physiol 22:171–183

Ono T, Murata N (1982) Chilling-sensitivity of the blue-green alga *Anacystis nidulans*. III. Lipid phase of cytoplasmic membrane. Plant Physiol 69:125–129

Palta JP, Jensen KG, Li PH (1982) Cell membrane alterations following a slow freeze-thaw cycle: ion leakage, injury and recovery. In: Li PH, Sakai A (eds) Plant cold hardiness and freezing stress, vol 2. Academic Press, New York, pp 221–242

Parish GR (1974) Seasonal variation in the membrane structure of differentiating shoot cambial-zone cells demonstrated by freeze-etching. Cytobiology 9:131–143

Pearce RS (1982) Ultrastructure of tall fescue (*Fetuca arundinacea* Schreb. cv. s170) cells fixed while exposed to lethal or non-lethal extracellular freezing. New Phytol 92:259–272

Pearce RS, Willison JHM (1985) A freeze-etch study of the effects of extracellular freezing on cellular membranes of wheat. Planta 163:304–316

Pomeroy MK, Siminovitch D (1971) Seasonal cytological changes in secondary phloem parenchyma cells in *Robinia pseudoacacia* in relation to cold hardiness. Can J Bot 49:787–795

Pringle MJ, Chapman D (1981) Biomembrane structure and effects of temperature. In: Morris GJ, Clark A (eds) Effects of low temperatures on biological membranes. Academic Press, New York, pp 21–37

Sakai A (1962) Studies on the frost hardiness of woody plants. I. The causal relation between sugar content and frost hardiness. Contrib Inst Low Temp Sci B11:1–40

Sakai A, Larcher W (1987) Frost survival of plants: responses and adaptation to freezing stress. Springer, Berlin Heidelberg New York Tokyo, pp 39–47

Sakai A, Yoshida S (1968) The role of sugar and related compounds in variations of freezing resistance. Cryobiology 5:160–174

Santarius KA (1982) The mechanism of cryoprotection of biomembrane systems by carbohydrates. In: Li PH, Sakai A (eds) Plant cold hardiness and freezing stress, vol 2. Academic Press, New York, pp 475–486

Scarth GW (1941) Dehydration injury and resistance. Plant Physiol 16:171–179

Scarth GW (1944) Cell physiological studies of frost resistance. New Phytol 43:1–2

Scarth GW, Levitt J, Siminovitch D (1940) Plasma membrane structure in the light of frost-hardening changes. Cold Spring Harbor Symp Quant Biol 8:102–109

Shinitzky M, Barenholz Y (1978) Fluidity parameters of lipid regions determined by fluorescence polarization. Biochim Biophys Acta 515:367–399

Sikorska E, Kacperska-Palacz A (1980) Frost-induced phospholipid changes in cold-acclimated and non-acclimated rape leaves. Physiol Plant 48:201–206

Siminovitch D, Levitt J (1941) The relation between frost resistance and the physical state of protoplasm. II. The protoplasmic surface. Can J Res Ser C 19:9–20

Siminovitch D, Scarth GW (1938) A study of the mechanism of frost injury to plants. Can J Res Ser C 16:467–481

Siminovitch D, Rheaume B, Pomeroy K, Lepage M (1968) Phospholipid, protein, and nucleic acid increases in protoplasm and membrane structures associated with development of extreme freezing resistance in black locust tree cells. Cryobiology 5:202–225

Singh J (1979) Ultrastructural alterations in cells of hardened and nonhardened winter rye during hyperosmotic and extracellular freezing stress. Protoplasma 98:329–341

Steponkus PL (1984) Role of plasma membrane in freezing injury and cold acclimation. Annu Rev Plant Physiol 35:543–584

Steponkus PL, Evans RY, Singh J (1982) Cryomicroscopy of isolated rye mesophyll cells. Cryo-Letters 3:101–114

Steponkus PL, Dowgert MF, Gordon-Kamm WJ (1983) Destabilization of the plasma membrane of isolated plant protoplasts during a freeze-thaw cycle: the influence of cold acclimation. Cryobiology 20:448–465

Steponkus PL, Dowgert MF, Ferguson JR, Levin RL (1984) Cryomicroscopy of isolated plant protoplasts. Cryobiology 21:209–233

Sugawara Y, Sakai A (1978) Cold acclimation of callus cultures of Jerusalem artichoke. In: Li PH, Sakai A (eds) Plant cold hardiness and freezing stress. Vol 1. Academic Press, New York, pp 197–210

Uemura M (1984) An aspect of plasma membrane modulation in response to physiological shift in plant cells. Doctoral Thesis, Faculty of Science, Hokkaido University, Sappora, Japan

Uemura M, Yoshida S (1984) Involvement of plasma membrane alterations in cold acclimation of winter rye seedlings (*Secale cereale* L. cv. puma). Plant Physiol 75:818–826

Uemura M, Yoshida S (1986) Studies on freezing injury in plant cells. II. Protein and lipid changes in the plasma membranes of Jerusalem artichoke tubers during a lethal freezing in vivo. Plant Physiol 80:187–195
Vigh L, Horvath I, Horvath LI, Dudit D (1979) Protoplast plasmalemma fluidity of hardened wheats correlates with frost resistance. FEBS Lett 107:291–294
Willemot C (1975) Stimulation of phospholipid biosynthesis during frost hardening of winter wheat. Plant Physiol 55:356–359
Willemot C (1979) Chemical modification of lipids during hardening of herbaceous species. In: Lyons JM, Graham D, Raison JK (eds) Low temperature stress in crop plants. Academic Press, New York, pp 411–430
Willemot C (1979) Chemical modification of lipids during hardening of herbaceous species. In: Lyons JM, Graham D, Raison JK (eds) Low temperature stress in crop plants. Academic Press, New York, pp 411–430
Wilson RF, Rinne RW (1976) Effect of freezing and cold storage on phospholipids in developing soybean cotyledons. Plant Physiol 57:270–273
Wolfe J, Steponkus PL (1983) Tension in the plasma membrane during osmotic contraction. Cryo-Letters 4:315–322
Yoshida S (1979a) Freezing injury and phospholipid degradation in vivo in woody plant cells. I. Subcellular localization of phospholipase D in living bark tissues of the black locust tree (*Robinia pseudoacacia* L.). Plant Physiol 64:241–246
Yoshida S (1979b) Freezing injury and phospholipid degradation in vivo in woody plant cells. II. Regulator effects of divalent cations on activity of membrane-bound phospholipase D. Plant Physiol 64:247–251
Yoshida S (1979c) Freezing injury and phospholipid degradation in vivo in woody plant cells. III. Effect of freezing on activity of membrane-bound phospholipase D in microsome-enriched membranes. Plant Physiol 64:252–256
Yoshida S (1984a) Studies on freezing injury in plant cells. I. Relation between thermotropic properties of isolated plasma membrane vesicles and freezing injury. Plant Physiol 75:38–42
Yoshida S (1984b) Chemical and biophysical changes in the plasma membrane during cold acclimation of mulberry bark cells (*Morus bombycis* Koidz. cv. Goroji). Plant Physiol 76:257–265
Yoshida S, Sakai A (1974) Phospholipid degradation in frozen plant cells associated with freezing injury. Plant Physiol 53:509–511
Yoshida S, Uemura M (1984) Protein and lipid compositions of isolated plasma membranes from orchard grass (*Dactylis glomerata* L.) and changes during cold acclimation. Plant Physiol 75:31–37
Yoshida S, Uemura M (1986) Lipid composition of plasma membranes and tonoplasts isolated from etiolated seedlings of mung bean (*Vigna radiata* L.). Plant Physiol 82:807–812
Yoshida S, Washio K, Kenrick J, Orr G (1988) Thermotropic properties of lipids extracted from plasma membrane and tonoplast isolated from chilling-sensitive mung bean (*Vigna radiata* [L.] Wilczek). Plant Cell Physiol 29:1411–1416

Chapter 14 Role of the Plasma Membrane in Host-Pathogen Interactions

H. KAUSS[1]

[1]FB Biologie, Universität Kaiserslautern, Postfach 3049, D-6750 Kaiserslautern, FRG

Abbreviations: cAMP, 3′,5′-cyclic adenosine monophosphate; DP, degree of polymerization; HR, hypersensitive response or reaction.

C. Larsson, I.M. Møller (Eds):
The Plant Plasma Membrane

1 Introduction: The Meaning of *Specificity* in Host-Pathogen Interactions

When we attempt to rationalize a complex biological phenomenon we are always influenced by apparently similar phenomena where biochemical details are already established and the underlying molecular mechanisms are partly understood. In the field of plant pathology such influences clearly come from human and animal medicine, in which many fascinating observations have been made regarding the mechanisms causing resistance to pathogens.

The interaction between a member of an animal species and a certain pathogenic strain of a microorganism may result in a disease which is *specific* with respect to the two organisms involved and the resulting symptoms. The most striking discovery was that the antibodies formed against the respective microorganism are also *specific*. Subsequently, this specificity was shown to depend on the molecular structure of antigenic determinants on the microorganisms, and corresponding antigen binding sites on the antibodies. In this context the term *specific* is already used with different meanings: although antibody specificity may help the host to recognize a microorganism as an invader, the development of a disease depends on many additional parameters which also show *specificity,* but which act at different levels. In the immune system itself the generation of antibodies specific against a microbial determinant results from various molecular interactions which are specific in yet other respects, e.g., in the case of lymphocytes for the major histocompatibility complex at the cell surface or for soluble message factors like the interleukins. More recently, textbooks of immunology (e.g., Roitt et al. 1985) emphasize that in addition to antibody-mediated reactions other constitutive resistance factors, *not* related to antibodies, are also very important for the overall defense. Thus, there is some specificity of a different kind in the alternative complement pathway which is triggered by bacterial surface polymers, in C-reactive proteins which make microorganisms "sticky" for macrophage action, and in natural killer cells. In addition, development of a disease in animals also depends on other parameters which are specific in yet other respects, such as microbial enzymes required for invasion, or malnutrition of the host. Thus, the overall specificity of disease resistance in animals does not result from a single *specific* event, but from a combination of many steps each of which is specific but at quite different levels. It is becoming increasingly clear that this applies also to plants.

In many cases the development of a disease (nonresistance or susceptibility) in higher plants is also very "specific" with respect to the host and the particular pathogen involved. During breeding of crop plants certain varieties with a quite different resistance potential toward particular races of a pathogen species have been recognized and, vice versa, pathogenic races may evolve which can break the genetically determined resistance of a certain plant variety. To account for this phenomenon at a biochemical level, models with often quite speculative elements have been proposed. These will not be discussed here but have been dealt with recently in various reviews (Kuć and Rush 1985; Dixon 1986; Ebel 1986; Halverson and Stacey 1986; Callow et al. 1987; Doke et al. 1987; Ebel and

Grisebach 1988; Boller 1989; Hahn et al. 1989). There is general agreement that plants have preformed physical and biochemical barriers at their surface or within the cells which in most cases are highly effective. Specific enzymes are needed in the pathogen, e.g., to break the cuticle (Kolattukudy 1985) or detoxify secondary plant metabolites. These aspects of specificity have been neglected in most recent reviews. The authors have concentrated primarily on inducible defense systems, presumably because they involve sophisticated physiological regulation mechanisms.

Review authors whose background is traditional plant pathology consider the complex subject of inducible plant defense mainly to be an integrated system called the hypersensitive response (HR). Although this may reflect the biological situation, it may underestimate the extent to which the numerous physiological changes involved are triggered by one common event or individually, and also raises the crucial question of whether all the known component reactions are obligatory in any host-pathogen system. This appears to be rather important in order to explain the overall HR complex and to integrate the tremendous amount of work done with elicitors (see Sects. 2.4–2.6). Surprisingly, these substances are in most cases reported to trigger only one particular biochemical reaction sequence, or only one was monitored by each investigator. Characteristic for the HR is that it appears to be a general resistance system inducible by various types of biological attack and shows no selectivity with respect to most pathogens involved. This is illustrated by the fact that when one race of pathogen can avoid triggering HR in a certain genetically defined plant variety (resulting in susceptibility) another race of the same pathogen can often be found which evokes it (see Doke et al. 1987). Thus, the general ability to respond by HR (= resistance) appears to be inherited and susceptibility in a pathogen race-cultivar system appears to result from the suppression of the HR by different means acquired by the microorganism during its evolution. This view is also sustained by the now widely accepted notion that few if any of the elicitor substances that have been described are race-specific (see Sects. 2.4 and 2.5).

Plant physiologists and biochemists, in contrast to plant pathologists, prefer another experimental approach. Their research is based on elicitors capable of triggering defense reactions which can be monitored by biochemical methods. With this approach researchers tend to select simple experimental systems in the hope of studying only one particular induced reaction to be able to work out regulatory mechanisms. The respective reviews or discussions, therefore, often give the impression that resistance of plants simply results from switching on a set of "defense genes" by an elicitor typical for a host-pathogen pair. The danger with this approach is that one may fail to understand the mutual relationships between the various biochemical weapons used by a plant. These responses are well coordinated in time and space and are most likely also causally interrelated to a certain degree.

According to the discussion above, a potential plant pathogen appears to be successful when it either avoids triggering, or is able to suppress the initial steps of the plant's defense responses. In addition, specificity in the infection process is influenced both by the sequence of events and the speed with which host and pathogen respond during the early stages of infection. To avoid infection the host should have as sensitive and rapid defense reactions as possible, whereas a successful pathogen should rapidly produce substances to suppress the sensing system of the plant. The outcome of this interplay could be called *recognition*, mutual for microorganism and plant. Based, however, on the experience with animal resistance, as discussed above, it appears wise to avoid this term as it tends to provoke the idea that a single fitting of complementary molecules may explain plant disease resistance. A more detailed discussion of the complexity of the term recognition is given by Callow et al. (1987) and Mendgen et al. (1988), who also describe the many chemical and physical parameters which determine the recognition of a host plant by the pathogen.

The importance of timing during the initial host-pathogen interaction is best illustrated by the formation of papillae. These cell wall appositions are deposited locally from the plasma membrane below the sites of attempted penetration and are thought to arrest invading hyphae mainly by presenting a physical barrier, although antifungal substances such as the superoxide radical (see Sect. 2.2) may also contribute. When *Phytophthora infestans* infects tuber tissue of resistant potato cultivars, papillae form earlier than with susceptible cultivars and are more successful in preventing penetration (Hächler and Hohl 1984). More detailed information is available for the biotrophic powdery mildews. In response to an appressorium of *Erysiphe graminis* f. sp. *hordei*, epidermal cells of barley coleoptiles from a resistant line form papillae about 1 h *before* fungal penetration pegs become microscopically visible; whereas in the susceptible (= not resistant) near-isogenic line papillae form 0.7 h *after* the appearance of the penetration pegs (Gold et al. 1986). As a result, penetration frequency in the resistant line was only 5%, whereas in the susceptible line the papillae had obviously formed too late, were smaller, and penetration followed by the formation of a haustorium was observed in about 60% of the attempts. The ability to react rapidly enough can be enhanced by external mM concentrations of $Ca(NO_3)_2$, and inhibited by chlorotetracycline, an antibiotic with Ca^{2+}-ionophoric properties. Other observations suggest that a barley cell can also be preconditioned to respond early and successfully against barley mildew by a consistently unsuccessful penetration attempt on the same barley cell by *E. pisi*, which is a pathogen of pea and not of barley (Kunoh et al. 1985). Resistance or susceptibility are obviously not absolute events but can be subject to extensive modulation by external and internal physiological parameters. This is also well known from many other host-pathogen systems, as reflected in the influence of nutritional status and climatic factors on the expression of diseases.

2 The Role of the Plasma Membrane

Attempts to analyze the host-pathogen interaction in plants face several problems. In nature the critical stages of interaction occur very early and at the level of individual cells, and even there may only involve a small region of the cell surface. With pathogenic fungi, which cause most of the diseases of agricultural interest, the initial interaction does not involve fully developed hyphae but rare, specialized cells presumably involved in tip growth (either hyphal tips or penetration pegs) which probably have a cell wall surface of special chemical composition. For example, chitin or chitosan may be exposed at such growing sites but be covered in older parts of the hyphae (Mirelman et al. 1975, for literature see also Fink et al. 1988; Mendgen et al. 1988). It has been shown, e.g., with infection hyphae of rusts, that the appropriate techniques (binding by lectins, enzymes, or antibodies, and labeling with gold particles) are now available which enable the surface of encounter sites to be explored in more detail (Mendgen et al. 1988). Another problem results from the fact that the events occurring at different infection sites are not synchronized, and their exact location is not predictable (except perhaps in systems involving *Erysiphe* where penetration occurs below the fungal appressoria and at the surface of epidermal cells). These facts cause serious technical problems for research conducted at both the cytological and biochemical level, and only recently have methods been developed to partially circumvent these drawbacks.

For a long time information on the involvement of the plasma membrane in host-pathogen interactions remained descriptive and the membrane was considered to play a passive role, most likely as changes in membrane properties could only be observed at later stages of host-pathogen interaction which are characterized by gross structural damage. More recently, evidence has been accumulating to support a leading role of the plant plasma membrane in the regulation of cellular events in general (as outlined in other chapters of this book) and this also appears applicable to physiological changes involved in host-pathogen interactions. At present, the evidence for such a role remains, however, fragmentary. It comes from various systems and is in most cases only indirect. Some examples where the plant plasma membrane is obviously involved in host-pathogen interaction but details of the mechanism are still missing will be only briefly mentioned. For instance, spread of a virus in a plant appears to employ or change the plasma membrane in various ways and the features related to the HR are relatively well understood (Ponz and Bruening 1986; Doke and Ohashi 1988). However, the mechanism by which virus particles actually move from cell to cell remains obscure, although in tobacco mosaic virus a 30 kD protein has been sequenced and shown to be involved in viral movement using molecular biological techniques (Meshi et al. 1987). Similarly, in the course of pathogenesis plant plasmodesmata have to be closed in order to isolate the symplast of the uninfected tissue parts. Callose deposition may be one mechanism employed (see Sect. 2.5) but special plasma membrane proteins, which constitute a valve mechanism (Meiners et al. 1988) regulated via the inositolphosphate system (Tucker 1988), may also be involved.

2.1 Microscope Observations

Electron microscopy of fungal pathogen-host encounter sites is greatly hampered by timing and localization problems. Early studies, therefore, were often performed at late stages of infection. In addition, the use of chemical fixation methods, especially for cells with partly damaged membranes, may cause artifacts such as formation of membrane vesicles as well as an ondulation and detachment of the plasma membrane from the cell wall. Interpretation of such effects, therefore, was often controversial. In the meanwhile better electron microscope methods and new methods for fixation (e.g., high pressure freezing of whole leaves followed by freeze-substitution, Knauf et al. 1989) have appeared and cytochemical staining is increasingly applied (Mendgen et al. 1988; Chap. 4). Although the course of infection is characteristic for any host-pathogen combination, some common features with regard to the host plasma membrane are now evident, and two aspects appear to be particularly noteworthy.

2.1.1 Early Stages of Infection

During the early stages of infection, where the first signals are likely to be exchanged, the host plasma membrane does not appear to be greatly altered. When *Colletotrichum lindemuthianum*, for example, penetrates into epidermal cells of susceptible bean tissues, it forms infection vesicles from which primary infection hyphae grow and extend, usually between the cell wall and the plasma membrane of the host. This period lasts for 24–48 h and represents a transient phase of biotrophy of the fungus, during which the host plasma membrane appears to be unaltered as far as can be judged from electron micrographs and related cytochemistry (O'Connell et al. 1985; O'Connell 1987). Nevertheless, the physiological properties of the host plasma membrane may change to some extent in the course of this biotrophic phase as indicated by the observations that the bean cells lose their ability to plasmolyze and to exclude tannic acid (O'Connell et al. 1985). As typical for biotrophic situations (see below) the hyphal wall and the host plasma membrane are separated by a fine-grained amorphous material ("matrix") which is rich in carbohydrates (O'Connell 1987). Subsequently, the host plasma membrane increasingly shows signs of degeneration and the cells die, opening the necrotrophic phase of host-pathogen interaction. In resistant bean cultivars infected by *C. lindemuthianum* (O'Connell et al. 1985) the cells may have undergone the HR so rapidly that an appropriate electron microscope observation of possible early changes at the host plasma membrane was overlooked. Alternatively, some infection hyphae were completely encased by wall appositions; the respective cells appeared alive and no reference was made to an unusual appearance of their plasma membrane.

2.1.2 Membrane Changes on Interaction with Biotrophic Fungi

Striking changes in the ultrastructure of the host plasma membrane have been described for cases in which a successful infection by obligate biotrophic fungi (rusts, mildews) has been established. Under these conditions, the fungal haustoria feed for a long time on the living host cells without triggering dramatic defense reactions. Most interestingly, also in this case the fungal wall is always separated from the host plasma membrane by a matrix layer ("extrahaustorial matrix") which in many reports shows low electron contrast. This layer, therefore, has sometimes been regarded as an artifact but more advanced electron microscope techniques clearly prove its existence (Fig. 1). The matrix presumably acts as some kind of insulator to prevent direct contact between the fungal wall surface and the plasma membrane and thus may prevent a triggering

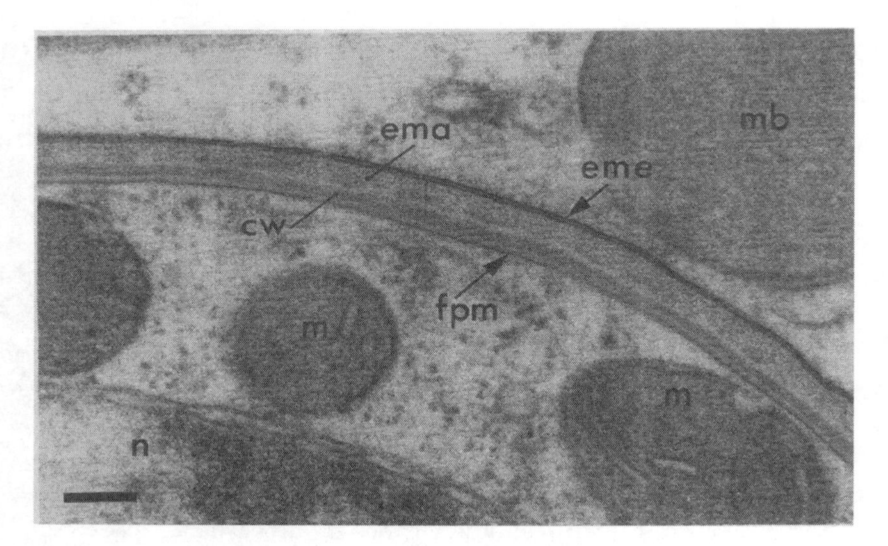

Fig. 1. The host-parasite interface between the rust fungus *Uromyces appendiculatus* in its dikaryotic phase and its host, *Phaseolus vulgaris*. The fungus has differentiated a haustorium (*lower part*) within the host cell. In the haustorium, a nucleus (*n*) and some mitochondria (*m*) can be recognized. The haustorium with the fungal plasma membrane (*fpm*) is surrounded by the fungal cell wall (*cw*), the extrahaustorial matrix (*ema*), and the invaginated plasmalemma of the host cell, called "extrahaustorial membrane" (*eme*). The host cytoplasm contains a microbody (*mb*); *bar* = 0.25 μm. For experimental details of the high-pressure freezing technique employed, see Knauf et al. (1989). The unpublished micrograph was kindly provided by K. Welter and K. Mendgen, Konstanz (FRG)

of defense reactions. The host plasma membrane surrounding the fungal haustorium has undergone several characteristic and stable changes and, therefore, is called the "extrahaustorial membrane" (Fig. 1). Detailed literature regarding these changes is cited by O'Connell (1987), Woods et al. (1988), and Knauf et al. (1989); the membrane alterations include changes in staining properties and the loss of intramembrane particles, sterols, and ATPase activity. In addition to the drastic changes in the extrahaustorial membrane, a certain decrease in the density of membrane particles has also been observed for the overall plasma membranes in cells of bean tissue infected by rust as compared to noninfected tissue (Knauf et al. 1989). These ultrastructural alterations of the plant plasma membrane strongly suggest that its physiological properties have changed and also that the biotrophic fungi influence the synthesis of the host plasma membrane.

2.1.3 Protoplasts as an Experimental Tool

When intact tissues are infected (as in the natural situation) the fungal hyphae always have to cope with the plant cell wall before contact is made with the protoplast. This may result in the production of endogenous soluble elicitors (Sect. 2.4) and, thereby, complicate experiments regarding the induction of defense reactions at the plasma membrane by contact with fungal walls. A new experimental system in which isolated soybean protoplasts are embedded in agarose and come in contact with growing hyphae of *Phytophthora* provides a means of studying contact in the absence of the cell wall (Odermatt et al. 1988). About 20% of such protoplasts reacted by deposition of wall-like polysaccharides in the contact area of the plasma membrane, others were surrounded by hyphae which made contact but induced no obvious reaction and some protoplasts were even penetrated right across. However, with this system there is an important difference compared with the natural situation: the race specificity is lost, although in nonhost combinations (*P. infestans* on soybean, *P. megasperma* on barley) production of extracellular plant polysaccharides occurred less frequently than in cultivar-specific systems. It would be of interest to determine whether or not the cells on removing their cell wall have lost their potential for race-specific reaction, or whether the comparatively larger diffusion space in the agarose gel resulted in a loss of fungal factors responsible under natural conditions for the specificity of the reaction.

2.2 The Hypersensitive Response

The hypersensitive response or reaction (HR) will be described briefly to give an impression of how the complex process of host-pathogen interaction might proceed over time and to illustrate how it represents a delicate interplay involving various biological properties of both host and pathogen. The HR is a multicomponent and dynamic sequence which involves the plasma membrane at several stages, and in different ways. Aspects of altered plant metabolism that occur in the course of the HR will be described in more detail below (Sects. 2.3–2.5).

2.2.1 Tissue Conditioning and Early Superoxide Production

Plant cells which have the genetically determined potential for HR may require a conditioning step before they gain the ability to really undergo HR on later contact. This has been documented in some detail for potato and *Phytophthora infestans* (Doke et al. 1987). In potato leaf tissue substances produced during spore germination may induce conditioning since it can also be achieved experimentally by means of a solution in which fungal spores were allowed to germinate. Alternatively, conditioning may result from wounding since potato tuber slices require an aging period before the HR can be triggered with fungal cell wall components. Conditioning is prevented by inhibitors of protein synthesis which suggests that signals of unknown nature have indeed been transmitted through or by the plasma membrane into the cytoplasm (Doke et al. 1987). An early manifestation of the conditioning process in potato appears to be the generation of superoxide anions ($O_2^{\cdot-}$) which is maximal in leaves about 2 h after the addition of germination fluid, and is presumed to occur on the outer surface of the plasma membrane (Doke et al. 1987). This oxygen radical might act as an antibiotic against the developing fungus and/or serve in the plant cells as a signal for further events, possibly due to its ability to cause lipid peroxidation via $O_2^{\cdot-}$ dismutation into H_2O_2 and final reduction to strong oxidants such as the OH radical ($OH^{\cdot}$). An $O_2^{\cdot-}$ regenerating NAD(P)H oxidase has been demonstrated in potato protoplasts but whether this enzyme is indeed responsible for the $O_2^{\cdot-}$ production and how it is activated during the above process remains unclear (Doke et al. 1987). The toxicity of $O_2^{\cdot-}$ suggests that it is generated outside the cell. Consistent with this idea is the location of an NAD(P)H oxidase at the outer surface of plant plasma membranes (Crane et al. 1985; see also Chap. 5). It remains, however, unclear how this enzyme in vivo is regulated and supplied with the necessary redox equivalents. Addition of external NAD(P)H and digitonin at rather high concentrations are necessary to produce $O_2^{\cdot-}$ in potato protoplasts (Doke et al. 1987). With cells, formation of NAD(P)H in the wall compartment by the wall-bound malate dehydrogenase represents one possibility (Elstner 1982); it appears, however, difficult to imagine how considerable concentrations of soluble nucleotides and organic acids are built up, for instance, in cell suspension cultures. Alternatively, the reduction of O_2 to $O_2^{\cdot-}$ might result from a transmembrane redox process (Chap. 5). Indirect evidence that such a pathway of $O_2^{\cdot-}$ production is possible comes from the observation that ferric reduction in bean roots grown under limited Fe supply is inhibited by the presence of superoxide dismutase (EC 1.15.1.1; Cakmak et al. 1987).

2.2.2 *Induction by Fungi and Characteristics of the Response*

The second step of HR induced in potato by *P. infestans* occurs during or shortly after the penetration and consists of a delicate and poorly analyzed complex of reactions. The interaction probably includes binding of the invading fungal hyphae to the host plasma membrane by means of potato lectin (Doke et al. 1987). This binding does not determine race specificity since it can also be observed with fungal cell wall components from *P. infestans* strains which do not trigger the HR in vivo. The most interesting but even less understood step is associated with, but not identical to, the lectin-like binding. It appears to initiate the HR, it can be triggered by certain unidentified fungal cell wall components, and it is suppressed on addition of water-soluble fungal 1,3/1,6-β-glucans. This latter effect is race-specific, although high concentrations of the suppressors are needed.

The actual HR, which is associated in vivo with penetration and which in potato experimentally is induced by fungal cell wall components, is accompanied by a second burst of $O_2^{\cdot-}$ production which could again act as an antibiotic agent and help to kill the invading pathogen. There are here obvious similarities to certain cells of the animal immune system such as macrophages in which the regulation of $O_2^{\cdot-}$ production is under hormonal control and mediated by the inositolphosphate system (Huang et al. 1988). In plants, superoxide production also appears to be very important for subsequent events associated with the HR. This reaction complex is accompanied by liberation of unsaturated free fatty acids which might be subject to further lipoxygenation to produce the endogenous Ca^{2+} ionophores, jasmonic and methyljasmonic acid, which in turn may speed up Ca^{2+} entry (Leshem 1987). Subsequent related events are ethylene production, membrane disintegration, electrolyte leakage, cell death, and tissue necrosis. As indicated above it is not clear yet to which extent parameters derived from these endogenous reactions act as elicitors for the synthesis of phytoalexins (Sect. 2.4), cell wall carbohydrates (Sect. 2.5), and proteins (Sect. 2.6) which occur simultaneously with HR. The list of events associated with the HR also includes the deposition of lignin or lignin-like polyphenolic substances. In wheat leaves such a lignification can be elicited by a single glycoprotein (the carbohydrate being the active part), which was isolated from wheat rust (Kogel et al. 1988). Alternatively, chitin oligomers or polymeric chitosan may induce this reaction in wounded wheat leaves (Barber et al. 1989). In leaves of various plants lignification was studied by fluorescence microscopy and it was always found to occur mainly during the later stages of the HR. In barley infected with mildew lignification continues after cell death and is accompanied by silicone accumulation from adjacent living cells into the walls lining the cell which was destroyed by the HR (Koga et al. 1988). Seen overall, the HR has many features in common with plant cell senescence (Leshem 1987), and HR has even been described as a greatly accelerated case of senescence (Ullrich-Eberius et al. 1988). Cells undergoing the HR do not simply die, but actually commit suicide in order to protect the entire tissue against further progression of the pathogen.

The profound metabolic changes and biological effects implicated in the HR are not restricted to the very few cells undergoing necrosis, but may spread over adjacent tissues and even the whole plant, a process involving the plasma membrane in several respects. This is exemplified by the phenomenon of systemic resistance in cucumber which can be induced by a local infection by tobacco necrosis virus or a fungal pathogen on a cotyledon or first leaf. These infections lead to HR-induced necrotic lesions and restriction of the pathogens to the area infected (Kuć and Rush 1985). Tobacco mosaic virus also undergoes a similar "localization" after infection as it is prevented from spreading (Cohen and Loebenstein 1975). In this case, however, the HR is not triggered, since no necrotic lesions appear. Only in the former case, where HR is associated with the first infection, do the new leaves which develop during the following weeks become partly or fully resistant to various pathogens (not only to those causing the primary infection). Thus it appears that the HR is required to induce systemic resistance. This mechanism implies the enhanced de novo formation, in the resistant leaves, of some enzymes presumably related to defense, such as chitinase (EC 3.2.1.14) or peroxidase, which are exported into the cell wall (Boller and Métraux 1988; Smith and Hammerschmidt 1988). The signal created during the HR in the infected first leaf must be of host nature and appears to be transported in the phloem (T. Boller, personal communication); some evidence is available that cytokinins play a role (Sziraki et al. 1980).

2.2.3 Elicitation by Bacteria

The HR induced by bacteria has also been studied in some detail. When plant tissues are invaded by certain strains of phytopathogenic bacteria (e.g., *Pseudomonas syringae*) these microorganisms slowly multiply in the extracellular space. They presumably trigger the HR by substances which are excreted and penetrate the plant cell wall. Induction of HR, presumably at the host plasma membrane, exhibits many features in common with those described above for the HR triggered in potato by a fungus. Early events of bacteria-induced HR are changes in membrane potential and ion transport and these will be discussed in more detail in Section 2.6. Further striking effects include membrane decay and lipid peroxidation which presumably are also initiated by $O_2^{\cdot-}$ (Keppler and Novacky 1987; Ullrich-Eberius et al. 1988; Keppler et al. 1989). The inducing bacterial substances may be quite varied and include toxins (Sect. 2.3) as well as exoenzymes which act on the plant cell wall, resulting in the release of oligosaccharides which are able to induce defense reactions ("endogenous elicitors"; Sect. 2.4). Of general interest to plant physiologists may be the observation that tobacco suspension cells undergoing HR due to the presence of pathogenic strains of *P. syringae* have a greatly diminished ability to take up fluorescein diacetate through the plasma membrane and, therefore, are less stainable (Keppler et al. 1988). This effect occurs before cell death and might indicate an early change in the lipid phase of the plasma membrane which results in reduced fluidity.

2.3 Toxins

To prove that a certain plant disease is caused by microorganisms the latter have to be isolated, cultured in vitro, and then shown to cause reinfection. In numerous cases, such cultured microorganisms have been the source of (nonenzymatic) substances which cause symptoms (wilting, chlorosis, necrosis) which indicate damage to the plant tissue. These compounds are called toxins and are normally grouped according to their host specificity: the majority, ranging in complexity from ammonia to very special organic molecules, are active both in hosts and nonhosts of the respective pathogen. There has been, and still is, a debate concerning the extent to which these are causally responsible for the respective disease. In some cases they appear to be at least necessary for a successful infection (Rudolph 1976; Scheffer and Briggs 1981). Only a few toxins, in contrast, are host-specific or -selective and cause symptoms only in those plants or varieties for which the respective microbes are pathogenic; in these cases their involvement in the disease is more obvious.

2.3.1 Nonselective Toxins

Nonselective toxins (Rudolph 1976; Stoessl 1981; Strobel 1982) often cause an increase in the loss of cellular ions and/or water, and it is difficult to ascertain whether this reflects a primary action on the plasma membrane or whether it is only a secondary effect due to cell decay or the HR. For instance, the toxin cercosporin, produced by *Cercospora beticola*, has been suggested to directly influence the H^+-ATPase of the plasma membrane (Blein et al. 1988). Its overall mechanism of action, however, must be more complex as it can also inhibit oxidative phosphorylation of isolated mitochondria (Macri et al. 1980) and can induce a light-dependent lipid peroxidation in leaf protoplasts which involves the formation of singlet oxygen (Youngman et al. 1983). In general, toxins with amphipathic, detergent-like structures (e.g., syringomycin, fusaric acid) might interact directly with the lipid phase of the plasma membrane, but a subsequent effect on endomembranes might also occur, depending on the concentrations used. This does not exclude the possibility that specific plasma membrane processes are preferentially affected, especially in the low concentration range (see also Sect. 2.5). Zinniol, for instance, is phytotoxic to a wide spectrum of plants and by itself promotes symptoms resembling those induced by fungi of the *Alternaria* group. This toxin has recently been found to bind to carrot protoplasts, and to stimulate at submicromolar concentrations the entry of Ca^{2+} (Thuleau et al. 1988). It is obvious that this effect might be linked to the regulation of physiological reactions (see Sect. 2.6). Toxins which are not host-selective, however, may also affect specific transport processes in the plasma membrane, as in the case of fusicoccin which is supposed to interfere with H^+ extrusion by the H^+-ATPase (Chaps. 6, 7, and 16).

2.3.2 *Host-Selective Toxins*

The less than 20 host-selective toxins that have been isolated to date have attracted considerable interest in the past as it was thought that their apparent role as primary determinants of disease would allow a more direct approach to studying the molecular mechanisms of pathogenesis. Early work, reviewed by Scheffer (1976) and Strobel (1982), have produced some insight into chemical structures. Helminthosporoside, for instance, is the toxin of *Helminthosporium sacchari*, a pathogen causing the eye-spot disease in sugarcane. It is a mixture of three isomers of a sesquiterpene bearing four β-galactosyl residues, all of which appear important for its binding to a plasma membrane protein which is suspected to be somehow involved in oligosaccharide transport. Although indirect evidence for effects on ATPases and ion transport were subsequently presented (see Strobel 1982), the causal link between these transport activities remained unclear. More recent electrophysiological studies have revealed that the *H. sacchari* toxin causes a loss of the H^+ gradient across the plasma membrane, but that the H^+-ATPase is unlikely to be the primary site of toxin action (Schröter et al. 1985). These observations may be explained by a reduced ATP supply due to an uncoupling of the mitochondria as recently reported for the toxin of *H. maydis*, race T, which is specific for male-sterile corn with T-cytoplasm (Holden and Sze 1987). The above results may fit into the current concept of H^+-driven solute transport; their implications for regulation of cell metabolism (Sect. 2.6) have to be further clarified. The host-selective toxins may indeed indirectly trigger metabolic reactions has also been shown for victorin, a mixture of related small peptides selective for certain oat cultivars. Susceptible protoplasts can be induced to produce extracellular polysaccharides at very low toxin concentrations (Walton and Earle 1985) and peeled leaf segments from sensitive cultivars only were induced to produce phytoalexins (Mayama et al. 1986).

2.4 Elicitors of Phytoalexin Production

During infection, plants often produce secondary metabolites which are absent, or only present in small amounts in healthy plants. Many of them have antibiotic properties and they are termed *phytoalexins*, since their production is regarded as a defense reaction. As seen below, evidence is accumulating that regulation of their synthesis occurs in part at the plasma membrane.

2.4.1 *Assay Systems*

To investigate the biochemical details of phytoalexin induction simple systems have been devised in which substances isolated from the pathogens (= *elicitors*) can be assayed on wounded surfaces of plant tissues, or in cell suspension cultures. This makes it possible to overcome the problems of timing and spacial orientation of the early events (see Sect. 2). Most importantly for biochemical studies, it also reduces the background of cells which are not involved and, therefore, show normal metabolism. Nevertheless, the assay systems used represent a compromise as they involve a somewhat artificial situation. Wounding provides additional signals which may condition the tissue (see Sect. 2.2). Suspension cultures represent cells artificially held at a certain stage of differentiation by using rather high concentrations and specific mixtures of phytohormones required for satisfactory growth. Any cell suspension culture used may contain an unpredictable set of potentially active genes which could be activated by treatment with elicitors. For instance, analysis of 76 cell clones derived from one leaf of *Catharanthus roseus* revealed a great variation in the alkaloid spectrum produced (Constabel et al. 1981). Of several established strains of *C. roseus* only the one which was able to produce traces of indole alkaloids could be induced by elicitor preparations from various fungi to greatly increase the production of these substances (Eilert et al. 1986). Similarly, four stable cell lines derived from a single seedling of *Eschscholtzia californica* were distinguished by their varying contents of benzophenanthridine alkaloids (Schumacher et al. 1987). One of these lines, characterized by a low alkaloid content could be elicited by a variety of substances (e.g., yeast maman, amphotericin B, poly-*L*-ornithine, Polymyxin B) to greatly increase its production of alkaloids.

2.4.2 Nature of Elicitors

The above approach turned out to be rather fruitful: many phytoalexins were discovered and details of their metabolism and molecular biology were unraveled (see reviews cited in Sect. 1). In summary, every plant species appears to produce a special set of phytoalexins, which often are chemically related compounds, and in various plants essentially any branch of secondary metabolism can be employed for this purpose. The elicitors used by the various investigators also show a vast diversity with respect to chemistry: polysaccharides, oligogalacturonides, glycoproteins, small peptides, fungal toxins or antibiotics, chitin, chitosan, digitonin, and other saponins, lipids, heavy metal ions, and simple organic compounds like reduced glutathione have proved to be active. It appears of special interest that enzymes secreted by pathogens, which can degrade the plant cell wall, or enzymes liberated from decaying plant cells, can produce oligosaccharides which themselves (*endogenous elicitors*) can induce phytoalexin synthesis (Boller 1989; Hahn et al. 1989). The best-studied examples are pectic enzymes producing oligogalacturonides (Collmer and Keen 1986). Some of the above-mentioned elicitors are classified by some authors as *abiotic* implying they are *nonphysiological*. In fact, however, all of these elicitors are simply experimental tools, and none have been proven to play a role in a normal physiological situation. It may transpire that only the simple ones finally provide insight into the black box of signal transmission at the plasma membrane (see below and Sect. 2.6).

Almost none of the reported elicitors are host-pathogen or race-cultivar specific and the few claims to the contrary (see Dixon 1986) might result from the fact that the elicitors could be isolated only from infected plants and not from in vitro cultured fungi. They, therefore, may have already contained part of the plant's answer. Alternatively, mixtures might have been used which contained specific suppressors of elicitor action, similar to those which affect induction of the HR in potato by wall components from *P. infestans* (see Sect. 2.2). It appears of interest in this context that the production of glyceollin elicited by unspecific β-glucans from *Phytophthora megasperma* in the soybean cotyledon assay can be inhibited in a race-specific manner by a mannan glycoprotein from the same fungus (Ziegler and Pontzen 1982).

Elicitors must be isolated from extracts recovered from the pathogen. In many cases only these crude mixtures have been used. Further purification is often accompanied by a considerable loss of activity and this might be related to the recent discovery that certain elicitors are active at low concentrations only if other substances are simultaneously present. This synergistic effect is clearly documented for oligogalacturonides and branched β-glucans in soybean (Davis et al. 1986) as well as for oligogalacturonides and a crude elicitor in parsley cell suspension cultures (Davis and Hahlbrock 1987). In the elicitor mixture used in the latter study the active component appears to be of protein nature (Parker et al. 1988). A similar synergistic effect has also been reported in potato using branched β-glucans from *Phytophthora infestans* and polyunsaturated free fatty acids (Preisig and Kuć 1985). The impact of these cooperative effects on the induction mechanism will be discussed in Section 2.6.

The above statement that phytoalexin elicitors are not race-cultivar specific does not mean that elicitors are not specific in other respects (see also Sect. 1). Elicitor preparations have often turned out to be more active in the system which was used for their assay during purification than in other systems, although this is not always true. There are also examples in which a certain crude elicitor preparation contains several active components. Parker et al. (1988) showed that extracts from *Phytophthora megasperma* f. sp. *glycinea* contain proteinase-sensitive components active in cell cultures from parsley but not from soybean. In contrast, the latter are induced by a glucan component presumably identical to the well-known soybean glucan elicitor for which a defined linkage arrangement of the glucose molecules has been established (Sharp et al. 1984b; Hahn et al. 1989). Such differences could be taken as an argument against the view that phytoalexin induction is part of the general defense system HR (see Sect. 2.2) but might also reflect differences in the specificity of the assay systems, which result from the fact that they employ tissues or suspension cells of a particular developmental stage.

2.4.3 *Induction of Protein Synthesis*

Evidence for de novo protein synthesis during pathogenesis comes mainly from three types of experimental approach. Firstly, the use of inhibitors, which has been mentioned in Section 2.2 in relation to the HR. Secondly, the analysis by electrophoretic methods or in vitro translation of proteins which are formed under stress conditions of various types, including infection by microorganisms (so-called pathogenesis-related proteins). A tremendous number of such proteins have been described (van Loon 1989). More recently, a few of these proteins have been identified, e.g., chitinases (EC 3.2.1.14) and β-1,3-glucanases (EC 3.2.1.59) which probably function in defense against fungi by degrading their cell walls (Kauffman et al. 1987; Legrand et al. 1987). In the context of this chapter it is of interest than in cucumber the systemically induced chitinase (Boller and Métraux 1988) and peroxidase (Smith and Hammerschmidt 1988) reside in the extracellular space. The same has been reported for constitutive chintinases and 1,3-β-glucanases in oat leaves (Fink et al. 1988). These enzymes, therefore, must be transported out through the plasma membrane. Hydroxyproline-rich glycoproteins, which accumulate in the cell walls of various infected or elicitor-treated dicotyledonous plants, must also be exported (Mazau and Esquerré-Tugayé 1986). How these latter proteins function in defense is less well established as even their role in the cell wall of the healthy plant is poorly understood (Cassab and Varner 1988) although they may make the cell wall more rigid, possibly in conjunction with lignification.

The third approach to study protein synthesis related to host-pathogen interactions is to search for suspected enzymes using biochemical methods and their respective mRNA using molecular biological techniques. In most of the experiments phytoalexin elicitors have been applied, and this revealed that induction of phytoalexin biosynthesis is, at least in part, regulated via de novo synthesis of enzymes (see Dixon 1986; Ebel 1986; Ebel and Grisebach 1988). How this process might be triggered at the plasma membrane will be discussed briefly in Section 2.6. This regulation process, however, clearly requires at least several hours to attain considerably increased enzyme and secondary metabolite levels. This again indicates that phytoalexins must function mainly during late stages of pathogenesis which explains why phytopathologists sometimes question the importance of phytoalexin synthesis for the overall defense. One problem is that phytoalexins normally have to be extracted from the infected tissue and the proportion of cells which actually produced them remains unclear. Improved methods for phytoalexin determination have recently allowed measurements in the vicinity of hyphae growing in soybean roots. Between 5–8 h after infection glyceollin reached concentrations sufficient to at least slow down the further progress of the disease and also sufficient to prevent secondary infection, e.g., by bacteria (see Ebel and Grisebach 1988).

2.4.4 Elicitor Receptors

Many authors speculate that the primary interaction of polymeric elicitors of phytoalexin synthesis occurs with *receptors* located on the plasma membrane. Use of the term *receptor* in this context requires some critical comments. As outlined in Chapter 9 of this book, it originates from animal hormone physiology, and in the present sense implies first a group-specific binding of an agonist known to arise upon physiological stimulation. That the binding causes the physiological response has to be further shown by inhibitors affecting both events. In addition, binding has to create a signal that can be transmitted to a target enzyme or physiological sequence. The ambivalence of the term *receptor* will be illustrated by two examples. Many of the animal receptors which are located in the plasma membrane bear glycoresidues of unknown function. If a physiological response which is normally induced by a hormone is triggered in vitro by a plant lectin, the term *receptor* is clearly used in a misleading sense: the lectin binds to the carbohydrate moiety of a *hormone* receptor and, at best, *mimics* the action of this hormone. There are surely no receptors in the proper sense for plant lectins on cells in the blood circulation system of animals, as these compounds are not members of a physiological signal chain. Similarly, a steadily increasing number of animal receptors which transmit signals via the G-protein family has been shown to have, as a common feature, seven hydrophobic disulfide-bridged transmembrane loops. Disulfide reduction with dithiothreitol or other thiol agents, and thus cleavage of these bridges, can "activate" the receptors, and cause the biological effect (Malbon et al. 1987). If this action had been discovered **before** the hormone effect was established, confusion would have been created by considering the existence of dithiothreitol receptors. These two examples show that the unreflected use of the term receptor in the field of plant physiology, where the chain of events is unclear, may provoke false assumptions on causal relationships.

For the elicitor-induced plant defense reactions no conclusive experimental evidence is yet available for the existence of receptors in the proper sense. To discuss the reasons for this and the enormous technical difficulties involved in such research is beyond the scope of this chapter. Clearly some progress is becoming visible now. Using labeled mycolaminarin from *Phytophthora* as ligand, apparently saturable binding sites on membrane fractions from soybean cotyledons have been reported (Yoshikawa et al. 1983). These studies have been extended using a more active β-glucan elicitor. Binding to soybean membranes (Schmidt and Ebel 1987) and protoplasts (Cosio et al. 1988) was effectively inhibited only by those polysaccharides that also inhibited phytoalexin induction in the soybean cotyledon bioassay, and the plasma membrane fractions of a linear sucrose gradient appeared to be enriched in the putative elicitor binding sites. Whether this type of study will finally result in the isolation of elicitor receptors in the proper sense remains to be seen. The few details known about the subsequent steps of induction, namely, the creation of a signal at the hypothetical receptors, and its transmission to target enzymes, will be discussed in Section 2.6.

2.5 Induction of Callose Synthesis

Alterations in the composition and structure of the plant cell wall due to host-pathogen interaction have often been reported. One striking example, linked to an obvious resistance mechanism, is the rather localized formation of papillae, wall appositions of various chemical composition but always rich in the 1,3-β-glucan callose (see Kauss 1987a). Callose deposition can occur in some systems within minutes and, therefore, appears unlikely to require de novo synthesis of enzymes but rather other more direct mechanisms for regulation.

2.5.1 *Properties of the 1.3-β-Glucan Synthase*

Work in my laboratory on the mechanism of callose formation began with the observation that the 1,3-β-glucan synthase (EC 2.4.1.34) located in the plasma membrane (Chaps. 2 and 3) is strictly dependent on Ca^{2+} in vitro. In the presence of polyamino compounds (e.g., spermine, spermidine, Ruthenium red, poly-*L*-ornithine) and/or Mg^{2+}, half of its activity is attained below 1 μM Ca^{2+}. This activation is readily reversible and, therefore, appears to be of allosteric nature (Hayashi et al. 1987; Kauss 1987a,b). The requirement for detergents to open right side-out membrane vesicles in the enzyme assay suggested that binding of the effectors Ca^{2+} and spermine as well as of the substrate UDP-glucose might occur at the cytoplasmic side of the membrane (Fink et al. 1987) and this has recently been confirmed (Fredrikson and Larsson 1989) using inside-out plasma membrane vesicles prepared by aqueous two-phase partitioning (Larsson et al. 1988).

The above regulatory properties suggested, as a working hypothesis, that callose deposition might be initiated by a change in permeability of the plasma membrane to Ca^{2+}, induced by the pathogen's attack, for instance, by surface-located chitosan or toxins. Alternatively, membrane-degrading processes occurring endogenously in the course of the HR might be the initial events. As a result, Ca^{2+} would flow into the cell following the concentration and electrochemical gradient. Thus, an increase in the concentration of free Ca^{2+} near the cytoplasmic side of the plasma membrane above the resting level of ~ 0.1 μM (Fig. 2; Kauss 1987a; Chap. 9) would occur, and activate the 1,3-β-glucan synthase directly in the area where the plasma membrane was perturbed.

2.5.2 *Induction of Callose Synthesis in Vivo*

To evaluate the above working hypothesis, methods were developed which allowed elicitation and determination of callose synthesis in suspension-cultured cells. Certain amphipathic compounds (e.g., digitonin, tomatin, filipin, amphotericin B, acylcarnitine, Polymyxin B, Echinocandine B) and polycations (e.g., chitosan, poly-*L*-ornithine, poly-*L*-lysine) are active elicitors, although their effectiveness varies greatly (Fig. 2). Digitonin and chitosan are examples of good elicitors, whereas Triton-X-100 and poly-*L*-lysine are poor elicitors (Kauss 1987a; Waldmann et al. 1988). It is, nevertheless, obvious from the difference in chemical composition of the above callose elicitors that their initial interaction with the cells was not likely to occur by interaction with sterically complementary sites of a receptor in the classical sense. The amphipaths are likely to act preferentially on the lipid phase of the membrane. Saponins

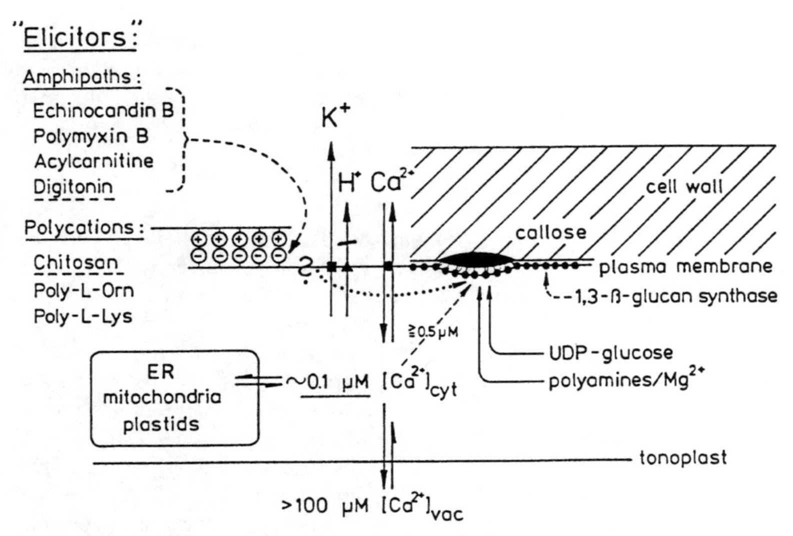

Fig. 2. Synoptic view of some reactions presumably involved in the induction of callose deposition. Polycations or certain amphipaths appear to perturb the plasma membrane, and trigger in an unknown manner (indicated by *?*) an increase in K^+ efflux and an external alkalinization which is possibly due to a decrease in H^+ export. The associated net Ca^{2+} uptake into the cell may increase the concentration of free Ca^{2+} in the cytoplasm from the resting level, normally held at about 0.1 μM by pumping Ca^{2+} into the vacuole, organelles, and the extracellular space, to a range above 0.5 μM, thus activating the 1,3-β-glucan synthase. This enzyme is in vitro strictly dependent on Ca^{2+}, and gains increased sensitivity toward Ca^{2+} by the presence of 200 μM spermine or 4 mM Mg^{2+}. Additional effectors may cooperate in vivo with Ca^{2+} (*dotted line*, for details see text). The events (depicted *above*, side by side) should be imagined to take place in the same cell surface region to explain the often very localized callose deposition. Note that conditions leading to callose synthesis can also induce phytoalexin synthesis, suggesting that Ca^{2+} is also one of the second messengers in this process (After Kauss 1987a)

(digitonin, tomatin) and polyene antibiotics (filipin, amphotericin B) are known to bind to sterols leading to changes in fluidity and other physiologically important membrane properties (Bolard 1986). Polymyxin B and acylcarnitine may intercalate in the phospholipid bilayer or deprive membrane proteins of essential lipids. In contrast, polycations must have a quite different mode of first interaction. Studies comparing chitosan fragments differing in degree of polymerization and N-acetylation showed that chitosan oligomers with a degree of polymerization (DP) up to 14 are almost inactive. The ability to induce callose formation in cells (Kauss et al. 1989) and protoplasts of *Catharanthus roseus* increases with chain length up to 2 μm (DP 4000) or more (Fig. 3). At a comparable DP the fragments exhibiting about 23% N-acetylation were less effective than fully deacetylated ones (Fig. 3; Kauss et al. 1989). If the row of amino groups in the polyglucosamine molecule (see Fig. 2) is interrupted at every fourth or fifth residue, that is, after 2–3 nm, the binding is obviously

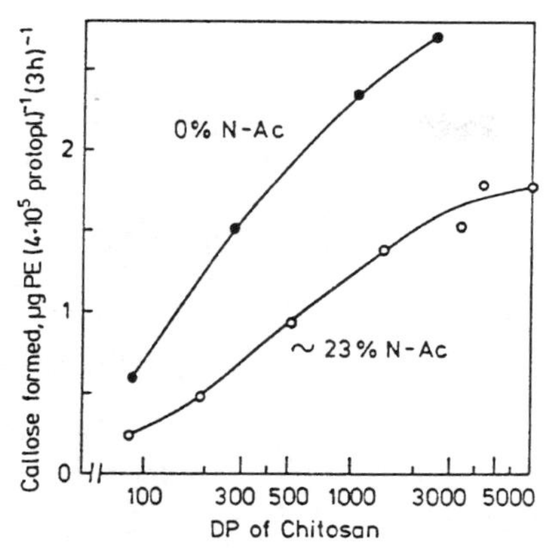

Fig. 3. Callose induction by fully deacetylated (●) and partially N-acetylated (○) chitosan fragments of different degrees of polymerization (*DP*). Protoplasts (4×10^5) from a suspension culture of *Catharanthus roseus* were suspended in 2 ml buffer containing 50 μM $CaCl_2$ and 0.4 M sucrose as an osmoticum. The samples were each supplied with 4 μg of the various chitosan fragments, the amount being standardized to the same polyglucosamine content. The callose formed within 3 h was determined fluorometrically and is expressed in μg pachyman equivalents (*PE*). A DP of 5000 corresponds to an average molecular weight of about 1000 kD and a chain length of about 2.5 μm. In similar experiments 4 μg of fully deacetylated chitosan oligomers of DP 8–14 induced only traces of callose. For experimental details, see Kauss et al. (1989)

disfavored. This suggests that the polyglucosamine binds to numerous points on the plasma membrane surface which must be within nm of each other and also occur regularly over long distances, conditions again fulfilled by phospholipids which could bind to the amino groups of chitosan by charge-charge or ternary interaction (Kauss et al. 1989). It is easy to imagine that such binding to a rigid polymer might change membrane fluidity to an extent dependent on the length of the molecule. Both types of callose elicitors appear, therefore, to exert their effects indirectly by interfering with general plasma membrane components rather than with group-specific receptor proteins. With the above working hypothesis on the role of Ca^{2+} in callose synthesis in mind, it is tempting to envisage parallels to stretch-activated Ca^{2+} channels shown by patch-clamp techniques to occur in plasma membranes of animals (for references, see Christensen 1987) and plants (Falke et al. 1988; Chap. 8). One could imagine that stretching might also change the density of boundary lipids near ion-transport proteins and thereby alter their activity. Such a mechanism is also compatible with the observation that callose synthesis, e.g., in epidermal or root hair cells, can be induced by physically bending the cell surface (see Kauss 1987a and references therein).

2.5.3 *Changes in Transport Rates of Ca^{2+} and Other Ions*

Early observations using complexing agents and cation-exchange beads indicated that at least several μM external Ca^{2+} are essential for the in vivo induction of callose synthesis (Köhle et al. 1985; Kauss 1987a). All the above callose elicitors also induce the leakage of ions, mainly K^+. Changes in the ion-transport properties as early events in callose induction were further substantiated by following the time course of K^+ release and the net Ca^{2+} uptake by digitonin-treated cells. K^+ release apparently preceded callose synthesis which occurred concomitantly with a net Ca^{2+} uptake (Waldmann et al. 1988). More recent experiments reveal that K^+ release is accompanied by an alkalinization of the external medium (T. Waldmann and H. Kauss, unpublished results). This might simply be due to the rules of electroneutrality if no counterion moves with K^+or, alternatively, might indicate a decrease in H^+ extrusion. On induction of callose synthesis with chitosan and many of the other elicitors mentioned in Section 2.5.2, a net uptake of Ca^{2+} can also be found (T. Waldmann and H. Kauss, unpublished results). The net Ca^{2+} uptake appears to be due to a change in the activity of Ca^{2+} channels as the process can be inhibited by nifedipine and flunarizine. Unfortunately, the methods used to study ion transport do not allow the determination of slight initial changes and, therefore, it is not possible to draw conclusions regarding causal relationships between the fluxes of K^+ and Ca^{2+}. The massive amounts of Ca^{2+} flowing into the cells appear to be sequestered mainly in the vacuoles (Fig. 2). Assuming that transport of Ca^{2+} into the vacuoles against a steep Ca^{2+} gradient is a limiting step (see Kauss 1987a), then the net Ca^{2+} uptake may indeed result in an increase in the concentration of Ca^{2+} in the cytoplasm. Although in some single plant cells the concentration of free Ca^{2+} in the cytoplasm can be determined (see Chap. 9), there is at present no reliable method available which would enable such an increase to be measured in cell suspension populations (Kauss 1987a) and this is required to make correlations with callose synthesis or other biochemical processes.

2.5.4 *Additional Possible Effectors*

The observed net uptake of Ca^{2+} and the presumed increase in cytoplasmic $[Ca^{2+}]$ are unlikely to be the only requirements for callose formation. When various callose inducers are compared, their ability to induce K^+ leakage does not correlate with the extent of callose synthesis (Waldmann et al. 1988). In addition, fusicoccin induces a moderate uptake of Ca^{2+} but does not cause callose synthesis, and the ionophore A-23187 causes a strong uptake of Ca^{2+} but only induces about 10% of the callose observed on addition of digitonin (Waldmann et al. 1988). Similarly, amphotericin B at low doses causes a striking Ca^{2+} uptake but low callose formation, whereas at higher doses both processes are induced (T. Waldmann, G. Euler, and H. Kauss, unpublished results). Therefore, it is obvious that in addition to Ca^{2+}, other still unknown effectors or regulatory parameters affect callose synthesis. Delmer (1987) (see also Chap. 11) has speculated that a decrease in the membrane potential would favor callose synthesis. However, preliminary results in my laboratory show that the opposite is true: the hyperpolarizing toxin fusicoccin, given in addition to chitosan or digitonin, stimulates callose synthesis, and membrane-depolarizing agents cause inhibiton (T. Waldmann and H. Kauss, unpublished results). Other possible regulators which in vivo might cooperate with Ca^{2+} to better activate the 1,3-β-glucan synthase are represented by certain polyunsaturated free fatty acids and amphipaths similar in structure to lysophosphatidylcholine. Such substances can, at low concentrations, activate the enzyme in vitro (Kauss and Jeblick 1986) and could be formed on membrane perturbation (see Sect. 2.2). However, it has not yet been established whether such compounds are produced during callose formation in the suspension-cell model system. In addition, there is some evidence from in vitro experiments that the polycationic elicitors as well as digitonin (for literature, see Kauss 1987a) might also directly influence the activity of the 1,3-β-glucan synthase. The degree to which this effect contributes to the in vivo callose formation in the suspension-cell model system, however, is difficult to evaluate, mainly due to the fact that the in vitro assay of the enzyme always requires some detergent to open the plasma membrane vesicles used.

2.6 Early Events Possibly Related to Signal Transduction

Although studies on signal transduction at the plant plasma membrane are at an early stage and generalizations are complicated by the variety of different experimental systems used, it appears worthwhile to highlight some common early events which may be involved, especially since some similarities to better characterized animal systems are just emerging.

2.6.1 K^+ Efflux and External Alkalinization

One relevant aspect is a rapid efflux of K^+, accompanied by an extracellular alkalinization. This pH effect can be induced by a glucan-type elicitor and measured directly in the diffusion droplet of the soybean cotyledon assay (Osswald et al. 1985). Using a similar elicitor and bean cell cultures, a corresponding decrease in intracellular pH has been monitored using ^{31}P-NMR (Ojalvo et al. 1987). Both a K^+ efflux and alkalinization were induced by a pectate lyase from *Erwinia* in tobacco suspension cells (Atkinson et al. 1986). Similar effects were also found when tobacco suspension cells underwent HR induced by pathogenic pseudomonads (M.M. Atkinson, personal communication). These processes appear to be correlated with a decrease in the size of the plasma membrane potential as monitored indirectly using fluorescent probes (Apostol et al. 1987), and directly using microelectrodes (Pavlovkin et al. 1986; Pelissièr et al. 1986; Ullrich-Eberius et al. 1988). In soybean cotyledonary tissue treated with a glucan-type elicitor, the plasma membrane potential depolarized within 2 min, followed after an additional 10 min by hyperpolarization (Mayer and Ziegler 1988). Dose-response curves and treatment with vanadate suggested that hyperpolarization contributes to signal transmission leading to phytoalexin synthesis.

There are different interpretations of the above data on K^+ leakage and pH-changes. A single antiporter may mediate the movement of both ions in opposite directions (Atkinson et al. 1986). Alternatively, the effects may be due to two indirectly coupled transport proteins, one of which might be the H^+-ATPase. It is also possible that the K^+ efflux is unspecific without participation of any transport protein, and that the observed concomitant alkalinization is a consequence of electroneutrality as well as, in some cases, an additional production of ammonia (Ullrich-Eberius et al. 1988).

2.6.2 Regulation of Ion Transport

Our own studies on the induction of callose synthesis (see Sect. 2.5) also showed an early K^+ efflux and external alkalinization and, in addition, a net Ca^{2+} uptake which indicates that Ca^{2+} is one of the second messengers involved in callose synthesis. The causal connections between these ion fluxes must now be clarified. Until a few years ago one would have considered only a coupling by the membrane potential (voltage-gated channels). The patch-clamp technique (Satter and Moran 1988; Hedrich and Schroeder 1989; Chap. 8) has recently added more possibilities, such as Ca^{2+} regulation of channels for other ions (Hedrich and Neher 1987), involvement of G-proteins in the direct regulation of K^+ (Yatani et al. 1987a) and Ca^{2+} channels (Yatani et al. 1987b), as well as a direct effect of membrane stretching on the activity of ion-transport proteins (Christensen 1987; Falke et al. 1988). In addition, signals derived from the inositol phosphate system (e.g., inositol-1,3,4,5-tetraphosphate, Houslay 1987) and endogenous Ca^{2+} ionophores resulting from lipid peroxidation (see Sect. 2.2) may also be involved in the regulation of ion fluxes. These exciting new results stress the importance of the plasma membrane in cellular regulation in general, and also, therefore, in defense reactions against plant pathogens. Clearly, however, it would be premature to speculate at this stage on the mechanisms which couple the abovementioned ion fluxes to each other, and to initial steps of plant-pathogen interaction.

2.6.3 Ca^{2+} and Other Possible Second Messengers

The observation that a net Ca^{2+} uptake occurs during the induction of callose synthesis (see Sect. 2.5) has also a bearing on the formation of phytoalexins (Köhle et al. 1985; Kauss 1987a,b). Under conditions at which chitosan induced callose synthesis in soybean suspension cells within hours, glyceollin accumulation occurred after 1–2 days (Köhle et al. 1984). The same effect could be observed using digitonin as an elicitor in soybean suspension cells (Kauss

1987a) and roots (Bonhoff and Grisebach 1988). Suspension cells from *Eschscholtzia californica* are another example that certain callose elicitors (e.g., Amphotericin B, Polymyxin B) can be used to induce secondary metabolites (Schumacher et al. 1987). Current research with parsley cells also indicates that furanocoumarin biosynthesis increases with the chain length of chitosan as described in Section 2.5 for callose formation in *Catharanthus* (H. Kauss, U. Conrath, A. Domard, unpublished results). This indicates that the conclusions drawn in Section 2.5 with respect to the unspecificity of the primary interaction of this polycation with the plasma membrane surface also apply to the induction of phytoalexin synthesis in these cell suspension systems. All the callose elicitors so far investigated in my laboratory have been found to cause a net Ca^{2+} influx, suggesting that an increase in the cytoplasmic concentration of free Ca^{2+} might be one of the signals triggering also phytoalexin synthesis, which has been shown to involve de novo synthesis of enzymes (Dixon 1986; Ebel 1986; see also Chap. 16). Similar conclusions were drawn from experiments with the Ca^{2+} ionophore A-23187 as an inducer, and La^{3+} as an inhibitor (Stäb and Ebel 1987). Ca^{2+} also has a marked effect on the elicitation by arachidonic acid and poly-*L*-lysine of rishitin and lubimin synthesis in potato (Zook et al. 1987). Protein kinases represent possible targets which might translate the Ca^{2+} signal into a cellular response involving enzyme synthesis (see Chap. 9). Interestingly, some of these enzymes are also activated by unsaturated fatty acids (Klucis and Polya 1987), substances which have been used as elicitors of phytoalexin synthesis in potato (Kuć and Rush 1985; Zook and Kuć 1987), and may arise endogenously in plants during the HR (Sect. 2.2) and/or be excreted from germinating fungal spores (Creamer and Bostock 1988).

If Ca^{2+} and possibly unsaturated fatty acids play a role as second messengers, the question arises whether other message systems, known from animal cells to reside in the plasma membrane and implicated in certain aspects of plant cell physiology (Chap. 9), may also be of relevance in host-pathogen interactions. A few scattered, recent results indeed suggest such a possibility. Production of the phytoalexin 6-methoxymellein in carrot suspension cells is enhanced not only by the Ca^{2+} ionophore A-23187 but also by dibutyryl cAMP and cholera toxin. This might indicate a role for cAMP and G-proteins, respectively (Kurosaki et al. 1987a). These authors also found that the addition of elicitors (a mixture of fragments prepared enzymatically from plant cell walls) to carrot cells transiently increased the cellular cAMP level; a maximum was reached after 30 min. These findings appear to represent evidence for a role of the cAMP system in plant defense reactions, although the current opinion is that cAMP does not play a role as second messenger in plant cells. In this connection it is of interest that the activity of adenylate cyclase (EC 4.6.1.1) is also rapidly increased in another type of plant-pathogen interaction, namely, in tobacco leaves undergoing HR due to infection by tobacco mosaic virus (Abad et al. 1986). In the aforementioned studies with carrot suspension cells, initiation of phytoalexin biosynthesis was also preceded by a rapid and transient activation of the inositol phosphate system. Furthermore, results which indirectly suggest the involvement of protein kinase C have been reported (Kurosaki et al. 1987b). It must be left to future research to show whether these various effects, which all indicate a regulatory role of the plasma membrane, can also be demonstrated in other systems involving the induction of phytoalexins, and in defense reactions of other types. There is some indication that regulatory mechanisms in different systems vary since cAMP has been reported not to be involved in soybean cells (Hahn and Grisebach 1983).

The importance of plasma membrane transport processes for the induction of defense reactions may help to explain certain observations made with phytoalexin assays. Cell suspension cultures, for instance, are sensitive to citrate (Apostol et al. 1987) and the standard soybean cotyledon assay to acetate and bicarbonate (Sharp et al. 1984a). These permeant anions affect the potential across the plasma membrane and may, thereby, interfere with ion transport. In addition, if the wound surface required for the soybean cotyledon assay is washed with water, the β-glucan elicitor becomes almost inactive but its activity can be restored by addition of some Ca^{2+} and a mixture of free amino acids which normally are present (or secreted) on the cut surface (Pontzen 1985). Similarly, increasing the Ca^{2+} concentration shifts the phytoalexin spectrum produced on treatment of potato tuber slices with poly-*L*-lysine from rishitin to lubimin (Zook et al. 1987). These and many similar observations indicate that the phytoalexin response triggered by elicitors does not represent an all-or-none reaction analogous to a switch turning on a light bulb, but instead is the result of a delicate balance between various physiological parameters.

3 Concluding Remarks

The above discussion of the various second messengers which might be involved in eliciting defense reactions again raises the question of the often supposed specificity of the experimental systems used, and the problems to be considered in using the term *specificity* (see Sect. 1). Elicitation by chitosan or digitonin of such different responses as callose synthesis (Sect. 2.5) and phytoalexin production (Sect. 2.4) can occur successively in the same suspension cell population. Participation of endogenous elicitors formed from decaying cells cannot be ruled out, but even so, this would not alter the impression that these experimental systems have some characteristics of the HR (Sect. 2.2): several defense reactions are triggered as a consequence of a comparatively unspecific perturbation of the plasma membrane. Similarly, induction of phytoalexin synthesis by interaction of thiol reagents with unknown components of the cell surface (Stössel 1984; Gustine 1987; Wingate et al. 1988) may be considered as another example of such unspecific interaction. In addition, as indicated in Section 2.4, the oligomeric or polymeric phytoalexin elicitors, derived from fungal or plant cell walls, are unspecific: they do not reflect phytopathologically defined host-pathogen relationships but appear to represent more general signals for "attack by fungi". These elicitors may or may not have a more group-specific initial binding to plasma membrane constituents (receptors?). Nevertheless, their primary interaction obviously triggers the subtle regulation of a general cellular reaction system. The fact that different elicitors can act synergistically in the same assay system (Preisig and Kuć 1985; Davis et al. 1986; Davis and Hahlbrock 1987; Sect. 2.4) may then reflect preferential stimulation of different messages. Their combined action might be required to result in optimal physiological effects. These messages have to be coordinated both at the level of short-term allosteric enzyme activation and long-term enzyme synthesis based on potentially active sets of genes. The complexity of the regulatory network required can be anticipated considering, for example, parsley suspension cells – or the respective protoplasts – which can be induced by fungal elicitors to produce furanocoumarins, and by UV light to synthesize flavonoids (Dangl et al. 1987). The two corresponding biochemical reaction sequences employ some common enzymes, whereas others are characteristic for each of them, suggesting differential regulation principles which are not yet understood. Research on host-pathogen interaction thus provides us with elegant experimental systems which will help us answer some open questions in general plant cell biology.

Acknowledgments. I would like to thank E.F. Elstner, H.R. Hohl, K. Mendgen, C.I. Ullrich-Eberius, T. Waldmann, and D.H. Young for critical comments and C. Jung for typing and patiently revising the manuscript several times. The experiments conducted in my laboratory were supported by the Deutsche Forschungsgemeinschaft and Fonds der Chemischen Industrie.

Note Added in Proof. The elicitor-induced decrease in fluorescence of presumed membrane probes (Sect. 2.6.1) appears not to indicate changes in pH and electric potential at the plasma membrane but results from a rapid production of H_2O_2 leading to a peroxidase-catalyzed destruction of the dyes (Apostol I, Heinstein PF, Low PS 1989, Plant Physiol 90: 109–116). Such a formation of H_2O_2 might play a direct role in pathogen defence and also provide a signal in phytoalexin induction as discussed for superoxide (Sect. 2.2.1).

The nonselective bacterial toxin syringomycin (Sect. 2.3.1), a linear acylated peptide, induces phosphorylation of proteins in isolated red beet plasma membrane vesicles (Bidwai AP, Takemoto JY 1989, Proc Natl Acad Sci USA 84:6755–6759). This toxin has recently been used in the authors laboratory to induce callose synthesis in suspension-cultured cells. The effect is saturated already at about 1 $\mu g\, ml^{-1}$, whereas K^+ leakage, external alkalinization, and Ca^{2+} uptake still increase at higher concentrations. These results contrast with those for amphotericin B (Sect. 2.5.4), but nevertheless again suggest that in addition to Ca^{2+} other signals might be involved in callose induction.

The net Ca^{2+} uptake induced by callose elicitors (Sect. 2.5.3) proceeds over several hours. In contrast, the phytoalexin elicitor effective in parsley suspension-cultured cells (Parker et al. 1988) induces a transient Ca^{2+} uptake which lasts 2–8 min and is subsequently even followed by some Ca^{2+} export (C. Colling, D. Scheel, K. Hahlbrock, T. Waldmann, H. Kauss, unpublished results).

References

Abad P, Guibbolini M, Poupet A, Lahlou B (1986) Occurrence and involvement of adenylate cyclase activity in the first step of tobacco mosaic virus infection of *Nicotiana tabacum* cv Xanthi nc leaves. Biochim Biophys Acta 882:44–50

Apostol I, Low PS, Heinstein P, Stipanovic RD, Altman DW (1987) Inhibition of elicitor-induced phytoalexin formation in cotton and soybean cells by citrate. Plant Physiol 84:1276–1280

Atkinson MM, Baker CJ, Collmer A (1986) Transient activation of plasmalemma K^+ efflux and H^+ influx in tobacco by a pectate lyase isozyme from *Erwinia chrysanthemi*. Plant Physiol 82:142–146

Barber MS, Bertram RE, Ride JP (1989) Chitin oligosaccharides elicit lignification in wounded wheat leaves. Physiol Mol Plant Pathol 34:3–12

Blein J-P, Bourdil I, Rossignol M, Scalla R (1988) *Cercospora beticola* toxin inhibits vanadate-sensitive H^+ transport in corn root membrane vesicles. Plant Physiol 88:429–434

Bolard J (1986) How do the polyene macrolide antibiotics affect the cellular membrane properties? Biochim Biophys Acta 864:257–304

Boller T (1989) Primary signals and second messengers in the reaction of plants to pathogens. In: Boss WF, Morré DJ (eds) Second messengers in plant growth and development. Alan R Liss, New York, pp 227–255

Boller T, Métraux JP (1988) Extracellular localization of chitinase in cucumber. Physiol Mol Plant Pathol 33:11–16

Bonhoff A, Grisebach H (1988) Elicitor-induced accumulation of glyceollin and callose in soybean roots and localized resistance against *Phytophthora megasperma* f. sp. *glycinea*. Plant Sci 54:203–209

Cakmak I, van de Wetering DAM, Marschner H, Bienfait HF (1987) Involvement of superoxide radical in extracellular ferric reduction by iron-deficient bean roots. Plant Physiol 85:310–314

Callow JA, Estrada-Garcia MT, Green JR (1987) Recognition of non-self: the causation and avoidance of disease. Ann Bot 60 (Suppl) 4:3–14

Cassab GI, Varner JE (1988) Cell wall proteins. Annu Rev Plant Physiol Mol Biol 39:321–353

Christensen O (1987) Mediation of cell volume regulation by Ca^{2+} influx through stretch-activated channels. Nature 330:66–68

Cohen J, Loebenstein G (1975) An electron microscope study of starch lesions in cucumber cotyledons infected with tobacco mosaic virus. Phytopathology 65:32–39

Collmer A, Keen NT (1986) The role of pectic enzymes in plant pathogenesis. Annu Rev Phytopathol 24:383–409

Constabel F, Rambold S, Chatson KB, Kurz WGM, Kutney JP (1981) Alkaloid production in *Catharanthus roseus* (L.) G. Don. VI. Variation in alkaloid spectra of cell lines derived from one single leaf. Plant Cell Rep 1:3–5

Cosio EG, Pöpperl H, Schmidt WE, Ebel J (1988) High-affinity binding of fungal β-glucan fragments to soybean (*Glycine max* L.) microsomal fractions and protoplasts. Eur J Biochem 175:309–315

Crane FL, Sun IL, Clark MG, Grebing C, Löw H (1985) Transplasma-membrane redox systems in growth and development. Biochim Biophys Acta 811:233–264

Creamer JR, Bostock RM (1988) Contribution of eicosapolyenoic fatty acids to the sesquiterpenoid phytoalexin elicitor activities of *Phytophthora infestans* spores. Physiol Mol Plant Pathol 32:49–59

Dangl JL, Hauffe KD, Lipphardt S, Hahlbrock K, Scheel D (1987) Parsley protoplasts retain different responsiveness to u.v. light and fungal elicitor. EMBO J 6:2551–2556

Davis KR, Hahlbrock K (1987) Induction of defense responses in cultured parsley cells by plant cell wall fragments. Plant Physiol 85:1286–1290

Davis KR, Darvill AG, Albersheim P (1986) Host-pathogen interactions XXXI. Several biotic and abiotic elicitors act synergistically in the induction of phytoalexin accumulation in soybean. Plant Mol Biol 6:23–32

Delmer DP (1987) Cellulose biosynthesis. Annu Rev Plant Physiol 38:259–290

Dixon RA (1986) The phytoalexin response: elicitation, signalling and control of host gene expression. Biol Rev 61:239–291
Doke N, Ohashi Y (1988) Involvement of an $O_2^{\cdot -}$ generating system in the induction of necrotic lesions on tobacco leaves infected with tobacco mosaic virus. Physiol Mol Plant Pathol 32:163–175
Doke N, Chai HB, Kawaguchi A (1987) Biochemical basis of triggering and suppression of hypersensitive cell response. In: Nishimura S, Vance CP, Doke N (eds) Molecular determinants of plant diseases. Japan Sci Soc Press, Tokyo/Springer, Berlin Heidelberg New York Tokyo, pp 235–251
Ebel J (1986) Phytoalexin synthesis: the biochemical analysis of the induction process. Annu Rev Phytopathol 24:235–264
Ebel J, Grisebach H (1988) Defense strategies of soybean against the fungus *Phytophthora megasperma* f. sp. *glycinea*: a molecular analysis. Trends Biochem Sci 13:23–27
Eilert U, Constabel F, Kurz WGW (1986) Elicitor-stimulation of monoterpene indole alkaloid formation in suspension cultures of *Catharanthus roseus*. J Plant Physiol 126:11–22
Elstner EF (1982) Oxygen activation and oxygen toxicity. Annu Rev Plant Physiol 33:73–96
Falke LC, Edwards KL, Pickard BG, Misler S (1988) A stretch-activated anion channel in tobacco protoplasts. FEBS Lett 237:141–144
Fink J, Jeblick W, Blaschek W, Kauss H (1987) Calcium ions and polyamines activate the plasma membrane-located 1,3-β-glucan synthase. Planta 171:130–135
Fink W, Liefland M, Mendgen K (1988) Chitinases and β-1,3-glucanases in the apoplastic compartment of oat leaves (*Avena sativa* L.). Plant Physiol 88:270–275
Fredrikson K, Larsson C (1989) Activation of 1,3-β-glucan synthase by Ca^{2+}, spermine and cellobiose. – Localization of activator sites using inside-out plasma membrane vesicles. Physiol Plant 77:196–201
Gold RE, Aist JR, Hazen BE, Stolzenburg MC, Marshall MR, Israel HW (1986) Effects of calcium nitrate and chlortetracycline on papilla formation, ml-o resistance and susceptibility of barley to powdery mildew. Physiol Plant Pathol 29:115–129
Gustine DL (1987) Induction of medicarpin biosynthesis in ladino clover callus by p-chloromercuribenzoic acid is reversed by dithiothreitol. Plant Physiol 84:3–6
Hächler H, Hohl HR (1984) Temporal and spatial distribution patterns of collar and papillae wall appositions in resistant and susceptible tuber tissue of *Solanum tuberosum* infected by *Phytophthora infestans*. Physiol Plant Pathol 24:107–118
Hahn MG, Grisebach H (1983) Cyclic AMP is not involved as a second messenger in the response of soybean to infection by *Phytophthora megasperma* f. sp. *glycinea*. Z Naturforsch 38c:578–582
Hahn MG, Bucheli P, Gervone F, Doares SH, O'Neill RA, Darvill A, Albersheim P (1989) The roles of cell wall constituents in plant-pathogen interactions. In: Nester E, Kosuge T (eds) Plant-microbe interactions, Vol 3. McGraw Hill, New York, in press
Halverson LJ, Stacey G (1986) Signal exchange in plant-microbe interactions. Microbiol Rev 50:193–225
Hayashi T, Read SM, Bussell J, Thelen M, Lin F-C, Brown Jr RM, Delmer DP (1987) UDP-Glucose:(1 → 3)β-glucan synthases from mung bean and cotton. Plant Physiol 83:1054–1062
Hedrich R, Neher E (1987) Cytoplasmic calcium regulates voltage-dependent ion channels in plant vacuoles. Nature 329:833–836
Hedrich R, Schroeder JI (1989) The physiology of ion channels and electrogenic pumps in higher plants. Annu Rev Plant Physiol Plant Mol Biol 40:539–569
Holden MJ, Sze H (1987) Dissipation of the membrane potential in susceptible corn mitochondria by the toxin *Helminthosporium maydis*, race T, and toxin analogs. Plant Physiol 84:670–676
Houslay MD (1987) Egg activation unscrambles a potential role for IP_4. Trends Biochem Sci 12:1–2
Huang SJ, Monk PN, Downes CP, Whetton AD (1988) Platelet-activating factor-induced hydrolysis of phosphatidylinositol 4,5-bisphosphate stimulates the production of reactive oxygen intermediates in macrophages. Biochem J 249:839–845
Kauffmann S, Legrand M, Geoffroy P, Fritig B (1987) Biological function of 'pathogenesis-related' proteins: four PR proteins of tobacco have 1,3-β-glucanase activity. EMBO J 6:3209–3212
Kauss H (1987a) Some aspects of calcium-dependent regulation in plant metabolism. Annu Rev Plant Physiol 38:47–72

Kauss H (1987b) Callose Synthese. Regulation durch induzierten Ca^{2+}-Einstrom in Pflanzenzellen. Naturwissenschaften 74:275–281

Kauss H, Jeblick W (1986) Influence of free fatty acids, lysophosphatidylcholine, platelet-activating factor, acylcarnitine, and Echinocandin B on 1,3-β-glucan synthase and callose synthesis. Plant Physiol 80:7–13

Kauss H, Jeblick W, Domard A (1989) Degree of polymerization and N-acetylation of chitosan determine its ability to elicit callose formation in suspension cells and protoplasts of *Cataranthus roseus*. Planta 178:385–392

Keppler LD, Novacky A (1987) The initiation of membrane lipid peroxidation during bacteria-induced hypersensitive reaction. Physiol Mol Plant Pathol 30:233–245

Keppler LD, Atkinson MM, Baker CJ (1988) Plasma membrane alteration during bacteria-induced hypersensitive reaction in tobacco suspension cells as monitored by intracellular accumulation of fluorescein. Physiol Mol Plant Pathol 32:209–219

Keppler LD, Atkinson MM, Baker CJ (1989) $O_2^{\cdot -}$-initiated lipid peroxidation in a bacteria-induced hypersensitive reaction in tobacco cell suspensions. Phytopathology 79:555–562

Klucis E, Polya GM (1987) Calcium-independent activation of two plant leaf calcium-regulated protein kinases by unsaturated fatty acids. Biochem Biophys Res Commun 147:1041–1047

Knauf GM, Welter K, Müller M, Mendgen K (1989) The haustorial plant parasite interface in rust-infected bean leaves after high pressure freezing. Physiol Mol Plant Pathol 34:519–530

Koga H, Zeyen RJ, Bushnell WR, Ahlstrand GG (1988) Hypersensitive cell death, autofluorescence, and insoluble silicon accumulation in barley leaf epidermal cells under attack by *Erysiphe graminis* f. sp. *hordei*. Physiol Mol Plant Pathol 32:395–409

Köhle H, Young DH, Kauss H (1984) Physiological changes in suspension-cultured soybean cells elicited by treatment with chitosan. Plant Sci Lett 33:221–230

Köhle H, Jeblick W, Poten F, Blaschek W, Kauss H (1985) Chitosan-elicited callose synthesis in soybean cells as a Ca^{2+}-dependent process. Plant Physiol 77:544–551

Kogel G, Beissmann B, Reisener HJ, Kogel RH (1988) A single glycoprotein from *Puccinia graminis* f. sp. *tritici* cell walls elicits the hypersensitive lignification response in wheat. Physiol Mol Plant Pathol 33:173–185

Kolattukudy PE (1985) Enzymatic penetration of the plant cuticle by fungal pathogens. Annu Rev Phytopathol 23:223–250

Kuć J, Rush JS (1985) Phytoalexins. Arch Biochem Biophys 236:455–472

Kunoh H, Hayashimoto A, Harui M, Ishizaki H (1985) Induced susceptibility and enhanced resistance at the cellular level in barley coleoptiles. I. The significance of timing of fungal invasion. Physiol Plant Pathol 27:43–54

Kurosaki F, Tsurusawa Y, Nishi A (1987a) The elicitation of phytoalexins by Ca^{2+} and cyclic AMP in carrot cells. Phytochemistry 26:1919–1923

Kurosaki F, Tsurusawa Y, Nishi A (1987b) Breakdown of phosphatidylinositol during the elicitation of phytoalexin production in cultured carrot cells. Plant Physiol 85:601–604

Larsson C, Widell S, Sommarin M (1988) Inside-out plant plasma membrane vesicles of high purity obtained by aqueous two-phase partitioning. FEBS Lett 229:289–292

Legrand M, Kauffmann S, Geoffroy P, Fritig B (1987) Biological function of pathogenesis-related proteins: four tobacco pathogenesis-related proteins are chitinases. Proc Natl Acad Sci USA 84:6750–6754

Leshem YY (1987) Membrane phospholipid catabolism and Ca^{2+} activity in control of senescence. Physiol Plant 69:551–559

Macri F, Vianello A, Cerana R, Rasi-Caldogno F (1980) Effects of *Cercospora beticola* toxin on ATP level of maize roots and on the phosphorylating activity of isolated pea mitochondria. Plant Sci Lett 18:207–214

Malbon CC, George ST, Moxham CP (1987) Intramolecular disulfide bridges: avenues to receptor activation? Trends Biochem Sci 12:172–175

Mayama S, Tani T, Ueno T, Midland SL, Sims JJ, Keen NT (1986) The purification of victorin and its phytoalexin elicitor activity in oat leaves. Physiol Mol Plant Pathol 29:1–18

Mayer MG, Ziegler E (1988) An elicitor from *Phytophthora megasperma* f. sp. *glycinea* influences the membrane potential of soybean cotyledonary cells. Physiol Mol Plant Pathol 33:397–407

Mazau D, Esquerré-Tugayé MT (1986) Hydroxyproline-rich glycoprotein accumulation in the cell walls of plants infected by various pathogens. Physiol Mol Plant Pathol 29:147–157

Meiners S, Baron-Epel O, Schindler M (1988) Intercellular communication – filling in the gaps. Plant Physiol 88:791–793

Mendgen K, Schneider A, Sterk M, Fink W (1988) The differentiation of infection structures as a result of recognition events between some biotrophic parasites and their hosts. J Phytopathol 123:259–279

Meshi T, Watanabe Y, Saito T, Sugimoto A, Maeda T, Okada Y (1987) Function of the 30 kd protein of tobacco mosaic virus: involvement in cell-to-cell movement and dispensability for replication. EMBO J 6:2557–2563

Mirelman D, Galun E, Sharon N, Lotan R (1975) Inhibition of fungal growth by wheat germ agglutinin. Nature 256:414–416

O'Connell RJ (1987) Absence of a specialized interface between intracellular hyphae of *Colletotrichum lindemuthianum* and cells of *Phaseolus vulgaris*. New Phytol 107:725–734

O'Connell RJ, Bailey JA, Richmond DV (1985) Cytology and physiology of infection of *Phaseolus vulgaris* by *Colletotrichum lindemuthianum*. Physiol Plant Pathol 27:75–98

Odermatt M, Röthlisberger A, Werner C, Hohl HR (1988) Interactions between agarose-embedded plant protoplasts and germ tubes of *Phytophthora*. Physiol Mol Plant Pathol 33:209–220

Ojalvo I, Rokem JS, Navon G, Goldberg I (1987) ^{31}P NMR study of elicitor treated *Phaseolus vulgaris* cell suspension cultures. Plant Physiol 85:716–719

Osswald WF, Zieboll S, Elstner EF (1985) Comparison of pH changes and elicitor induced production of glyceollin isomers in soybean cotyledons. Z Naturforsch 40c:477–481

Parker JE, Hahlbrock K, Scheel D (1988) Different cell wall components from *Phytophthora megasperma* f. sp. *glycinea* elicit phytoalexin production in soybean and parsley. Planta 176:75–82

Pavlovkin J, Novacky A, Ullrich-Eberius CI (1986) Membrane potential changes during bacteria-induced hypersensitive reaction. Physiol Mol Plant Pathol 28:125–135

Pelissier B, Thibaud JB, Grignon C, Esquerré-Tugayé MT (1986) Cell surfaces in plant-microorganism interactions. VII. Elicitor preparations from two fungal pathogens depolarize plant membranes. Plant Sci 46:103–109

Pontzen R (1985) Untersuchungen zum Einfluß von Glucanen und Mannanen aus *Phytophthora megasperma* F. Sp. *glycinea* auf die Akkumulation des Phytoalexins Glyceollin in Sojabohnen. PhD Thesis, Rheinisch-Westfälische Technische Hochschule Aachen, FRG

Ponz F, Bruening G (1986) Mechanisms of resistance to plant viruses. Annu Rev Phytopathol 24:355–381

Preisig CL, Kuć JA (1985) Arachidonic acid-related elicitors of the hypersensitive response in potato and enhancement of their activities by glucans from *Phytophthora infestans* (Mont) de Bary. Arch Biochem Biophys 236:379–389

Roitt IM, Brostoff J, Male DK (1985) Immunology. Gower Medical Publishing, London

Rudolph K (1976) Non-specific toxins. In: Heitefuss P, Williams PH (eds) Encyclopedia of plant physiology new series, vol 4. Springer, Berlin Heidelberg New York, pp 270–315

Satter RL, Moran N (1988) Ionic channels in plant cell membranes. Physiol Plant 72:816–820

Scheffer RP (1976) Host-specific toxins in relation to pathogenesis and disease resistance. In: Heitefuss P, Williams PH (eds) Encyclopedia of plant physiology new series, vol 4. Springer, Berlin, pp 247–269

Scheffer RP, Briggs SP (1981) A perspective of toxin studies in plant pathology. In: Durbin RD (ed) Toxins in plant disease. Academic Press, New York, pp 1–20

Schmidt WE, Ebel J (1987) Specific binding of a fungal glucan phytoalexin elicitor to membrane fractions from soybean *Glycine max*. Proc Natl Acad Sci USA 84:4117–4121

Schröter H, Novacky A, Macko V (1985) Effect of *Helminthosporium sacchari*-toxin on cell membrane potential of susceptible sugarcane. Physiol Plant Pathol 26:165–174

Schumacher H.-M, Gundlach H, Fiedler F, Zenk MH (1987) Elicitation of benzophenanthridine alkaloid synthesis in *Eschscholtzia* cell cultures. Plant Cell Reports 6:410–413

Sharp JK, Valent B, Albersheim P (1984a) Purification and partial characterization of a β-glucan fragment that elicits phytoalexin accumulation in soybean. J Biol Chem 259:11312–11320

Sharp JK, Mc Neil M, Albersheim P (1984b) The primary structures of one elicitor-active and seven elicitor-inactive hexa (β-D-glucopyranosyl)-D-glucitols isolated from the mycelial walls of *Phytophthora megasperma* f. sp. *glycinea*. J Biol Chem 259:11321–11336

Smith JA, Hammerschmidt R (1988) Comparative study of acidic peroxidases associated with induced resistance in cucumber, muskmelon and watermelon. Physiol Mol Plant Pathol 33:255–261

Stäb MR, Ebel J (1987) Effects of Ca^{2+} on phytoalexin induction by fungal elicitor in soybean cells. Arch Biochem Biophys 257:416–423

Stoessl A (1981) Structure and biogenetic relations: fungal nonhost-specific. In: Durbin RD (ed) Toxins in plant disease. Academic Press, New York, pp 109–219

Stössel P (1984) Regulation by sulfhydryl groups of glyceollin accumulation in soybean hypocotyls. Planta 160:314–319

Strobel GA (1982) Phytotoxins. Annu Rev Biochem 51:309–333

Sziraki J, Balàzs E, Kiraly Z (1980) Role of different stresses in inducing systemic acquired resistance to TMV and increasing cytokinin level in tobacco. Physiol Plant Pathol 16:277–284

Thuleau P, Graziana A, Rossignol M, Kauss H, Auriol P, Ranjeva R (1988) Binding of the phytotoxin zinniol stimulates the entry of calcium into plant protoplasts. Proc Natl Acad Sci USA 85:5932–5935

Tucker EB (1988) Inositol bisphosphate and inositol trisphosphate inhibit cell-to-cell passage of carboxyfluorescein in staminal hairs of *Setcreasea purpurea*. Planta 174:358–363

Ullrich-Eberius CI, Pavlovkin J, Schindel J, Fischer K, Novacky A (1988) Changes in plasmalemma functions induced by phytopathogenic bacteria. In: Crane FL, Morré DJ, Löw M (eds) Plasma membrane oxidoreductase in control of animal and plant growth. Plenum, New York, pp 323–332

van Loon LC (1989) Stress proteins in infected plants. In: Kosuge T, Nester EW (eds) Plant microbe interactions. Molecular and genetic perspectives, Vol 3. Mc Graw-Hill, New York, in press

Waldmann T, Jeblick W, Kauss H (1988) Induced net Ca^{2+} uptake and callose biosynthesis in suspension-cultured plant cells. Planta 173:88–95

Walton JD, Earle ED (1985) Stimulation of extracellular polysaccharide synthesis in oat protoplasts by the host-specific phytotoxin victorin. Planta 165:407–415

Wingate VPM, Lawton MA, Lamb CJ (1988) Glutathione causes a massive and selective induction of plant defense genes. Plant Physiol 87:206–210

Woods AM, Didehvar F, Gay JL, Mansfield JW (1988) Modification of the host plasmalemma in haustorial infections of *Lactuca sativa* by *Bremia lactucae*. Physiol Mol Plant Pathol 33:299–310

Yatani A, Codina J, Brown AM, Birnbaumer L (1987a) Direct activation of mammalian atrial muscarinic potassium channels by GTP regulatory protein G_k. Science 235:207–211

Yatani A, Codina J, Imoto Y, Reeves JP, Birnbaumer L, Brown AM (1987b) A G protein directly regulates mammalian cardiac calcium channels. Science 238:1288–1292

Yoshikawa M, Keen NT, Wang MC (1983) A receptor on soybean membranes for a fungal elicitor of phytoalexin accumulation. Plant Physiol 73:497–506

Youngman RJ, Schieberle P, Schnabl H, Grosch W, Elstner EF (1983) The photodynamic generation of singlet molecular oxygen by the fungal phytotoxin, cercosporin. Photobiochem Photobiophys 6:109–119

Ziegler E, Pontzen R (1982) Specific inhibition of glucan-elicited glyceollin accumulation in soybeans by an extracellular mannan-glycoprotein of *Phytophthora megasperma* f. sp. *glycinea*. Physiol Plant Pathol 20:321–331

Zook MN, Kuć J (1987) Arachidonic and eicosapentaenoic acids, glucans and calcium as regulators of resistance to a plant disease. In: Stumpf PK, Mudd JB, Nes WD (eds) The metabolism, structure and function of plant lipids. Plenum Publishing, New York, pp 75–82

Zook MN, Rush JS, Kuć JA (1987) A role for Ca^{2+} in the elicitation of rishitin and lubimin accumulation in potato tuber tissue. Plant Physiol 84:520–525

Chapter 15 The Role of the Plant Plasma Membrane in Symbiosis

N.J. BREWIN[1]

1 Introduction to Symbiosis

The symbiotic associations that will be considered in this chapter involve a stable physical relationship between two different organisms that is perceived to be mutually beneficial to both partners (Smith and Douglas 1987). In each case, intimate surface contact between a microorganism and a plant cell leads to invagination of the plant cell plasma membrane and sometimes to enclosure of

[1]John Innes Institute and AFRC Institute of Plant Science Research, Colney Lane, Norwich NR4 7UH, UK

Abbreviation: VA, vesicular arbuscular.

C. Larsson, I.M. Møller (Eds):
The Plant Plasma Membrane

the microsymbiont within the cytoplasm of the plant cell. However, although physically internalized, all stable endosymbionts remain, in a sense, topologically outside the plant cell cytoplasm because they are bounded by a host-derived membrane, the perisymbiotic membrane, which is a modified form of the plasma membrane. This review will focus on the plasma membrane as the agent for cellular morphogenesis during the establishment of the symbiosis, and as the mediator of signal and nutrient exchange between the two partners.

2 The *Rhizobium*-Legume Symbiosis

2.1 Overview

Nitrogen-fixing bacterial symbionts that induce legume nodules (Fig. 1) belong to three rhizobial genera, namely *Rhizobium, Bradyrhizobium* and *Azorhizobium*, but the infection mechanisms are essentially similar in all cases (Rolfe and Gresshoff 1988; Quispel 1988). In order to initiate nodule morphogenesis in response to rhizobial infection, two classes of plant response are needed: first the induction of cortical cell divisions that will ultimately give rise to the nodule meristem, and second a mechanism for invasion of the plant root surface which depends on the ability of rhizobia to modify and subvert the normal processes of plant cell wall growth, as directed by the plasma membrane. Incidentally, these two plant responses can be experimentally separated, because certain *Rhizobium* mutants that fail to synthesize extracellular polysaccharide still induce meristematic activity but fail to provoke intracellular penetration, and this results in small nodule-like structures which do not contain infecting bacteria (Finan et al. 1985).

A pre-condition for intracellular penetration by rhizobia appears to be continuing extension growth of the plant cell wall, and for this reason growing, expanding root hair cells are the most common point of rhizobial invasion (Bauer 1981), although alternative routes of rhizobial penetration are observed in some legumes (Chandler 1978; Chandler et al. 1982). The bacteria first induce curling and other growth deformations of the root hair cell. Subsequently, after a partial penetration of the plant cell wall, the bacteria then stimulate the plant to deposit new cell wall and plasma membrane as an inwardly growing tubular structure called an infection thread, which develops from the point of bacterial entry, growing towards the base of the root hair cell and ultimately into the underlying cortical cells. Inside the infection thread, the bacteria are arranged in roughly single file array, and their continued multiplication and growth provides a turgor pressure which could drive the extension growth of the infection thread structure. At a later stage in nodule development, bacteria are released from the thin-walled tips of infection thread branches, but remain surrounded by envelopes of plant plasma membrane. They then differentiate into bacteroids that synthesize nitrogenase (EC 1.18.6.1) and convert atmo-

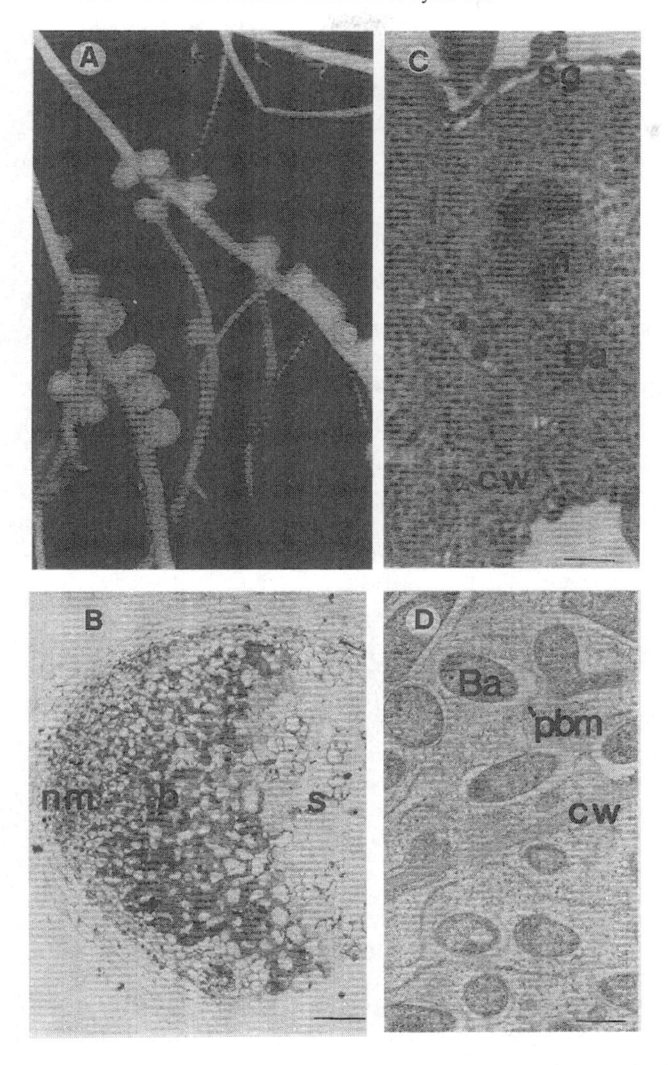

Fig. 1A-D. The *Rhizobium*-legume symbiosis, viewed at increasing levels of magnification. **A** Pea root nodules, ca. 4 mm long; **B** longitudinal section showing an apical nodule meristem (*nm*), a zone of infected cells containing bacteroids (*b*) and a zone of plant and bacterial cell senescence (*s*); (*bar* = 250 μm); **C** light micrograph showing many thousands of nitrogen-fixing bacteroids (*Ba*) filling the plant cytoplasmic space between nucleus (*n*) and cell wall (*cw*) (*sg* = starch grains; *bar* = 5 μm); **D** electron micrograph showing bacteroids (*Ba*) individually enclosed by a plant-derived peribacteroid membrane (*pbm*); *bar* = 1 μm)

spheric nitrogen into ammonia, and the peribacteroid membrane itself becomes specialized for the transport functions necessary to sustain the symbiosis (Verma et al. 1986).

2.2 The Root Hair Plasma Membrane

In most legumes, the first outward signs of rhizobial infection are distortions of cell wall growth leading to root hair curling and then the development of an inwardly growing infection thread. Root hair curling by *Rhizobium* is dependent on the activity of a small number of nodulation (*nod*) genes, whose transcription is triggered by flavone-like substances exuded from legume roots (Rolfe and Gresshoff 1988). In order to appreciate the mechanism by which rhizobia re-direct the process of cell wall deposition through the plasma membrane of a root hair cell, it is first necessary to consider the normal mechanism of apical growth in the root hair cylinder. Unfortunately, very little is known about this process, but a prominent feature of tip growth in root hair cells is that the nucleus often migrates in step with the extending tip. Immunofluorescence studies with anti-tubulin antibody have indicated that this relationship appears to be maintained by bundles of endoplasmic microtubules which progress from the nuclear region towards the apical dome, where they can be seen to fountain out into the cortex (Lloyd et al. 1987). Presumably this cytoskeletal system is responsible for targetting the Golgi vesicles that carry plasma membrane and cell wall precursors to the apical dome and for the oriented deposition of cellulose microfibrils in the cell wall. However, this relationship between nucleus and tip can be uncoupled experimentally by using anti-microtubule drugs such as cremart. After administration of this drug, root hair tip growth ceases and the nucleus migrates towards the base of the cell by a process that can itself be inhibited by the addition of cytochalasin D, a drug which is known to fragment the F-actin bundles. It is concluded from this interesting study that in normally growing root hairs microtubules connect the nucleus to the tip, but that F-actin is involved in basipetal migration of the nucleus.

Rhizobial attachment to the cell wall of the growing root hair (Kijne et al. 1988) perturbs subsequent growth in several ways. In the first place, there is marked curling of the root hair cell which Bauer (1981) has interpreted as a unilateral inhibition of further plant cell wall extension on the side of the root hair carrying attached rhizobia (Figs. 2 and 3). It is interesting to speculate that this localized growth inhibition might result from a localized disruption of the underlying cytoskeletal system that delivers Golgi vesicles and other components to the plasma membrane. This implies that the presence of rhizobia at the cell surface must be transduced through the plasma membrane into a secondary signal affecting further cell wall growth: perhaps, as in animal cells, a localized intracellular influx of Ca^{2+} could be sufficient to cause localized disruption of microtubules (Saunders and Jones 1988).

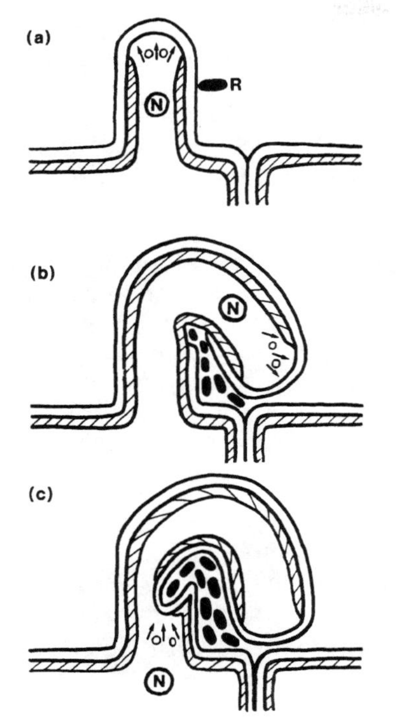

Fig. 2a-c. Model for root hair curling and the initiation of infection thread formation by *Rhizobium* (After Turgeon and Bauer 1985; Lloyd et al. 1987). **a** Root hair cell, showing Golgi vesicles being targetted to the apical growth point by the underlying microtubules (*small arrows*). The outer (α) layer of the cell wall is probably composed of pectic polysaccharides and hemicelluloses forming an amorphous flexible matrix containing randomly oriented cellulose microfibrils. The inner (β) layer develops underneath the newly synthesized α-layer and forms a rigid cellulosic cylinder extending almost to the hemispherical dome of the growing hair tip. Note the nucleus (*N*) associated with the growing tip through endoplasmic microtubules; rhizobia (*R*) attach to the surface of the root hair cell and embed in the mucigel. **b** Unilateral inhibition of cell wall growth in the root hair, leading to 'shepherd's crook' formation; bacteria grow and divide within the occluded infection site; partial penetration of the root hair cell wall by rhizobia, leads to formation of an infection sac. **c** Cell wall deposition ceases at the root hair tip; a new site of cell wall deposition develops at the tip of the inwardly growing infection thread; association of the nucleus with the growing point of infection thread leads to acropetal migration of the nucleus and growing infection thread tip towards the base of the root hair cell

The next consequence of deformed cell wall growth is often described as a hooked or 'shepherd's crook' structure in which the root hair is bent over through 180° and comes to press down against the epidermal cell surface or against a lower part of the root hair cell itself (Fig. 2). The entrapped rhizobia continue to grow and divide within this extracellular pocket and it is possible that the turgor pressure generated by bacterial expansion is an important force during the subsequent penetration of the plant cell wall (Turgeon and Bauer 1985).

It has been very difficult to achieve an ultrastructural analysis of the early infection events that lead to intracellular penetration at the root hair surface because these interactions occur at a single cell level and involve only a small percentage of host and bacterial cells. However, two careful studies of infection on clover (Callaham and Torrey 1981) and soybean roots (Turgeon and Bauer 1985) both reinforce the view that the origins of the infection thread depend partly on a process of invagination of the plasma membrane originating from the point of the infection site closure, and partly on a process of limited cell wall degradation and penetration.

In the soybean system, intracellular infections most commonly develop where rhizobia become entrapped at a three-way junction between the cell wall of a deformed root hair and the intercellular groove between adjacent epidermal cells (Fig. 2). Infections are initiated by bacteria which become embedded in an extracellular mucigel on the root epidermis. Subsequent infection thread formation appears to involve degradation of mucigel material and localized disruption of the outer layer of the cell wall of the folded root hair by one or more entrapped rhizobia. At the site of penetration, these rhizobia are still separated from the host cytoplasm by plasma membrane and a layer of wall material that appears similar to and continuous with the normal inner layer of the hair cell wall. However, it is not clear exactly how this limited amount of cell wall degradation is achieved as a prelude to rhizobial penetration. A role has been postulated for hypothetical enzymes of plant or bacterial origin, but it is hard to see how the activity of cell wall-degrading enzymes could be quite so localized and bring about the very abrupt cell wall discontinuities that are often observed. Moreover, if degradative enzymes are involved, it is not clear why it should always be the hair cell wall which is susceptible to degradation and not the wall of the contiguous epidermal cells. An obvious rationalization for these phenomena might be that the root hair cell wall is susceptible to disruption because, unlike the epidermal cell, it is still growing and expanding. Thus, the localized cell wall disruption is seen as a local perturbation of the normal cell wall growth mediated through the plasma membrane, and in some respects the initiation of an infection thread may resemble the creation of an intercellular space at a three-way junction between plant cells undergoing cell division (Jeffree et al. 1986).

2.3 The Infection Thread Membrane

Following this initial and partial penetration of the root hair cell wall, proliferation of the bacteria results in an irregular wall-bound sac near the site of penetration. A tubular infection thread (Fig. 3) emerges from this infection sac, apparently by involution of the plasma membrane. The wall material of the infection thread is continuous with that surrounding the infection thread itself and apparently of the same composition. The lumen of the infection thread contains rhizobia in roughly single file embedded in a matrix material (Fig. 4) composed, at least partly, of plant glycoprotein (Bradley et al. 1988; VandenBosch et al. 1989). Continued growth of the infection thread structure depends on the localized deposition of plasma membrane and cell wall precursors at the plasma membrane, and the underlying cytoplasm in this region is densely packed with the organelles associated with cell wall synthesis and deposition.

At this point in the development of the rhizobial infection, something very interesting must happen to the cytoskeleton of the root hair cell, which results in the uncoupling of the nucleus from the apical growing tip. Deposition of cell wall material ceases to take place at the tip and is instead targetted to the inwardly growing infection thread. Furthermore, the nucleus, which was

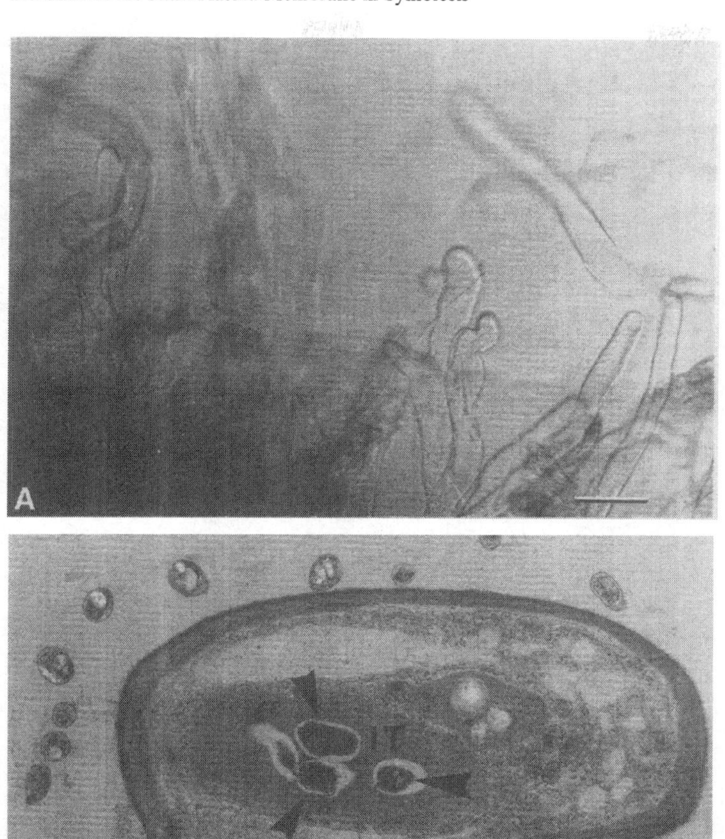

Fig. 3. Root hair curling and infection thread formation. **A** Curled and branched root hairs that have developed as a result of inoculation of peas with *Rhizobium* (*bar* = 100 μm); **B** electron micrograph showing a glancing section through a root hair cell containing an infection thread (*IT*) which contains bacteria (*arrowheads*) embedded in a matrix material (*bar* = 0.5 μm) (Photographs courtesy of M.F. LeGal)

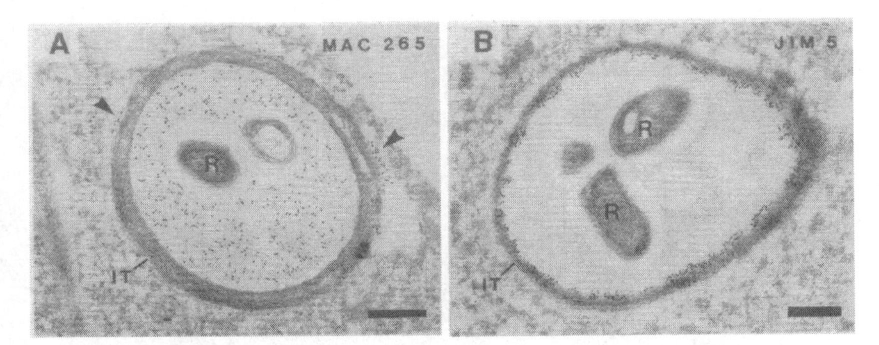

Fig. 4. Transverse section through infection threads (*IT*) containing rhizobia (*R*), and illustrating: **A** plant-derived matrix glycoprotein after immunogold staining with monoclonal antibody MAC 265; **B** pectic cell wall component after immunogold staining with monoclonal antibody JIM 5 (*bar* = 0.5 μm)

previously held about 5 μm behind the growing tip, now takes up an analogous position 5 μm ahead of the inwardly growing tip of the infection thread (C. W. Lloyd, personal communication; Fig. 3c). Although a cytoskeletal analysis of infected root hairs has not yet been undertaken, it seems reasonable to suggest that the nucleus is coupled to the inwardly growing plasma membrane of the infection thread by a system of microtubules similar to that described for the outwardly growing tip of an uninfected root hair (Lloyd et al. 1987).

Further inward growth of the infection thread follows a very irregular path towards the base of the hair cell and the infection thread is usually preceded in this migration by the cell nucleus. Hence, two different biophysical systems may be operating on the plasma membrane in the region of the infection thread tip. On the one hand, the continued growth and expansion of the entrapped rhizobia provides an inward turgor force causing involution of the plasma membrane and continued cell wall deposition at this point of expansion. On the other hand, the microfilament system, that was shown earlier (Lloyd et al. 1987) to be responsible for acropetal migration of the nucleus in root hair cells treated with cremart, could be responsible for migration of the infection thread towards the base of the cell, this migration being harnessed via the nucleus to the tip of the infection thread membrane through bundles of microtubules, as has been described for apical tip growth in a normal uninfected root hair.

In summary, therefore, it seems that the presence of extracellular rhizobia in the vicinity of a growing root hair cell causes several different perturbations to the plasma membrane and underlying cytoskeleton, and their cumulative effect is to generate the inwardly growing infection thread (Fig. 1). Firstly, the subapical microtubule system may be distorted and ultimately disrupted so as to generate a curled root hair that no longer incorporates new cell wall material at its tip. Secondly, as a result of plasma membrane involution from the infection sac, a new zone of membrane and cell wall growth arises at the point of inception of the infection thread, and this zone is underpinned by microtubules and other cytoskeletal structures previously observed at the apical tip. Thirdly, the oriented microfilament system is modified and harnessed to guide the growing infection thread acropetally towards the base of the root hair cell. Obviously this model is very speculative, but it illustrates several of the exciting opportunities that exist for investigating interactions between the cytoskeleton, the plasma membrane and the endomembrane system using the *Rhizobium*-legume symbiosis as a model system.

When the infection thread comes to the base of the root hair cell it is somehow able to fuse with the basal plasma membrane and plant cell wall so as to create a pore, thereby releasing the rhizobia from the intracellular tunnel into the intercellular space. Presumably this process is analogous to the way in which a nascent cell plate, formed after cell division grows into and fuses with the pre-existing plant cell wall (Jeffree et al. 1986). Subsequent penetration of the adjacent cortical cell wall by rhizobia released into the intercellular space appears to proceed in a manner similar to the initial infection event at the root hair surface, although the walls of these cortical cells are expanding radially rather than apically at the time of infection (Robertson and Farnden 1980). There are several interesting parallels between the growth and development of the cell plate and of the infection thread; perhaps the infection thread lumen should be thought of as being analogous to a fluid middle lamella which now harbours bacteria. In this connection, it is interesting to note that a plant glycoprotein identified within the matrix material of infection threads is also a component of the middle lamella and intercellular space (VandenBosch et al. 1989).

2.4 *The Infection Droplet Membrane*

As the infection thread penetrates into cells located deeper in the nascent nodule structure, a change in its pattern of wall growth is observed, which may be related to a change in function for the Golgi system such that the rate of membrane synthesis remains constant or increases, while the rate of synthesis or deposition of cell wall material decreases (Robertson et al. 1985; VandenBosch et al. 1989). This gives rise to an intracellular infection droplet structure in which the rhizobia, still embedded in plant matrix glycoprotein (Bradley et al. 1988), are now enclosed only by plasma membrane without any associated cell wall material. Throughout the whole process of plant cell infection by rhizobia this is the first time that the bacteria are exposed to naked plasma membrane, unmasked by wall material. The result is that some bacteria become very closely associated with the infection droplet membrane, although the nature of this close surface interaction is not yet understood.

2.5 The Peribacteroid Membrane

2.5.1 Morphogenesis

Phagocytosis is a very unusual phenomenon for plant cells because the plasma membrane is normally covered by a barrier of plant cell wall material. However, rhizobia from the infection droplet are unquestionably engulfed by the infection droplet membrane and internalized so that they come to occupy the cytoplasmic space, where they are still enclosed by a modified form of the plant plasma membrane, the peribacteroid membrane. For phagocytosis to occur, it has been suggested on the basis of ultrastructural examination that the outer surface of the rhizobia must adhere to the infection droplet membrane (Robertson and Lyttleton 1984). Perhaps this association then provides a focus for membrane-driven internalization, and it is interesting to consider whether a normal mechanism of membrane recycling might be involved, albeit in a modified form. However, the system of membrane recycling through clathrin-coated vesicles would not be suitable for engulfing large objects like rhizobia (Mellman et al. 1986), and a bulkier system of membrane internalization must be involved.

A common fate of endocytotic vesicles in animal cells is that they then fuse with lysosomal vesicles, but this does not appear to happen with membrane-enclosed rhizobia, at least not until very late in nodule development. Instead, the internalized bacteria continue to grow and divide until several thousand rhizobia come to occupy the cytoplasm of each infected plant cell. Every time the bacteria divide, there is an accompanying division of the peribacteroid membrane so that, in clover and pea nodules, every bacterial cell is individually enclosed (Robertson and Lyttleton 1984). However, in some legumes, e.g. soybean and *Phaseolus*, this synchrony breaks down in the final stages of bacterial growth, so that bacteroids come to be packaged by peribacteroid membranes in groups of a dozen rather than individually. In clover nodule cells it was shown that division of the peribacteroid membranes occurred by virtue of the fact that they tended to follow closely the contours of the dividing bacteria (Robertson and Lyttleton 1984). If this conclusion is correct, it implies that surface interactions between the bacteroid and peribacteroid membrane could be an important aspect of organization in nodule cells. Differences in the degree of adhesion between peribacteroid membranes and the bacteroid outer membrane could also explain why the number of bacteroids enclosed per peribacteroid membrane envelope can be different for the same rhizobial strain in nodules of different host legumes (Robertson et al. 1985). Moreover, it seems likely that the surface adhesion operating at this stage might be similar to that observed for the initial phagocytosis from the infection droplet, and could correspond to the physical association observed in vitro for isolated peribacteroid membrane fragments and lipopolysaccharide from the bacterial outer membrane (Bradley et al. 1986).

Once the rhizobia have been released into the plant cell cytoplasm, they begin to differentiate into nitrogen-fixing bacteroids, provided that other parameters such as the oxygen concentration are in the correct range. At the same time, there may also be some further specialization of the peribacteroid membrane relative to the plasma membrane in order to accomodate the special transport and surface interactions needed for the exchange of metabolites and morphogenetic signals between the plant and its microsymbiont. Several proteins (of unknown function) have been reported to be specifically targetted to the peribacteroid membrane, although the targetting signals that might be involved have not been elucidated (Verma et al. 1986, 1988; Werner et al. 1988).

It is the fate of all plant cells that have been infected by rhizobia that they will senesce more rapidly than uninfected cells of the same tissue. Breakdown of the peribacteroid membrane is an early event in this process, and this membrane may normally play an important part in protecting the rhizobia from plant defence reactions (Vance 1983). This is particularly apparent in the case of an abnormal *Bradyrhizobium japonicum* isolate (RH31, Marburg) which induces nodules on soybean where the peribacteroid membrane breaks down prematurely and host phytoalexins are induced as a defence response (Werner et al. 1984, 1985).

In all symbiotic associations involving nitrogenase, structural and physiological modifications are needed to protect the O_2-labile nitrogenase enzyme system from direct exposure to molecular oxygen. In herbaceous legume nodules, these modifications include differentiation of the nodule cortex, which controls O_2 diffusion through intercellular spaces within the nodule (Witty et al. 1986), bacteroid formation, leghaemoglobin production and function. In *Parasponium* and woody legume nodules, where rhizobia fix nitrogen within persistent infection threads, suberized thread walls may provide partial protection of the nitrogenase against O_2 within the bacteria (Smith et al. 1986; de Faria et al. 1987).

2.5.2 Cytochemistry

A considerable body of evidence supports the view that the mature peribacteroid membrane is a modified form of plasma membrane. In ultra-thin tissue sections only the peribacteroid and the plasma membrane (which both have a high lipid content) can be stained with phosphotungstic acid, whereas Golgi bodies and endoplasmic reticulum are stained by zinc iodide-osmium tetroxide (Robertson et al. 1978). Another feature shared by peribacteroid and plasma membranes is their association with coated pits and vesicles (Robertson and Lyttleton 1982) which presumably function to recycle membrane and to internalize the contents of the peribacteroid space.

The Golgi system has been implicated in the synthesis of peribacteroid membrane as well as plasma membrane (Robertson et al. 1978; Brewin et al. 1985). In an extensive immunochemical investigation (Bradley et al. 1988; Perotto et al. 1990) rat monoclonal antibodies were isolated that react with components of the peribacteroid membrane from pea nodules. Most of the antibodies recognized periodate-sensitive (carbohydrate) components of plant glycoproteins. These antibodies could be divided into seven separate groups on the basis of the distribution of antigens on SDS-polyacrylamide gels, as visualized by immunoblotting, and the distribution of antigens within the endomembrane system of pea root nodule cells, as visualized by immunogold electron microscopy. All the antibodies reacted, not only with the peribacteroid membrane, but also with the plasma membrane of infected nodule cells, and with the plasma membranes from uninfected cells. However, the different antibody groups showed differences in the relative abundance of their corresponding antigens on different membranes within the endomembrane system. These antibodies provide evidence of membrane trafficking into, or away from, the peribacteroid membrane compartment, and will permit the analysis of post-translational modifications for particular antigens within the plant endomembrane system.

2.5.3 Comparisons of Peribacteroid and Plasma Membranes

In some respects the peribacteroid membrane can be thought of as a simplified form of plasma membrane and in other respects, it should be thought of as enclosing a specialized compartment of the endomembrane system (Mellor and Werner 1987), with specialized functions concerned with transport to and from the endosymbiont (Udvardi et al. 1988), and with plant-microbe surface recognition phenomena (Bradley et al. 1986). Peribacteroid membranes can be released from membrane-enclosed bacteroids by osmotic shock and purified by flotation density gradient centrifugation (Domigan et al. 1988). The ATPase activity of the purified peribacteroid membrane material recovered from lupin nodules (EC 3.6.1.3) does not appear to be contaminated with the energy-transducing (F_oF_1)ATPases of rhizobia or plant mitochondria because it is insensitive to azide and oligomycin. Tonoplast ATPase can also be ruled out because the peribacteroid membrane ATPase is insensitive to nitrate ions. In contrast to these ATPases, the peribacteroid membrane ATPase has an acidic pH optimum (5.25), exhibits a high level of specificity for MgATP or MnATP, and is very sensitive to vanadate and diethylstilbestrol. In these and other respects, the peribacteroid membrane ATPase is similar to the root plasma membrane proton-pumping ATPase (Blumwald et al. 1985; Domigan et al. 1988).

A number of functions (or structures) that are present on plasma membranes are not found on peribacteroid membranes. For example, no plant cell wall material is associated with the peribacteroid membrane; neither is the plant-derived extracellular matrix glycoprotein which has been identified as a component of the infection thread (Bradley et al. 1988). In addition, Kinnbach et al. (1987) have demonstrated that an extracellular isoenzyme of α-mannosidase (isoenzyme III) is excluded from the peribacteroid space. Since the export of all of these components from the cell probably involves the targetting of Golgi vesicles to the plasma membrane, it implies that the peribacteroid membrane probably does not receive these particular classes of Golgi vesicles.

Moreover, in the immunochemical study of Perotto et al. (1990) several antigens were recognized which were much less abundant on the peribacteriod membrane than on the plasma membrane of infected cells.

Interestingly, the monoclonal antibody MAC 254 (Perotto et al. 1990) recognized a number of nodule-specific macromolecules, which could be identified after SDS-polyacrylamide gel electrophoresis (the molecular weights were 25, 28, 30, 32 and 35 kD). However, because this monoclonal antibody also recognized many other antigen bands that were common to both root and nodule material, a specific localization for these nodulin-like bands was not possible, although a strong localization of antigen in the Golgi compartment was noted (Fig. 5). Nevertheless, the monoclonal antibody will be useful as a recognition system with which to monitor the purification of these nodule-specific components as bands on SDS-polyacrylamide gels. Moreover, it would be interesting to investigate whether the particular epitope recognized by MAC 254 forms part of a targetting signal that determines the ultimate cell location of this particular group of glycoproteins, as for example mannose-6-phosphate has been shown to target glycoproteins to lysosomes in animals cells (Griffiths et al. 1988).

Several other workers have identified nodule-specific proteins or glycoproteins that are associated with the peribacteroid membrane (Verma et al. 1988; Werner et al. 1988). Nodulin 24 (Katinakis and Verma 1985) and nodulin 26 (Fortin et al. 1985) have been shown to be present in peribacteroid membranes, perhaps contributing to specialized transport processes (Sandal and Marcker 1988), although nodulin 26 is still expressed in nodules harbouring

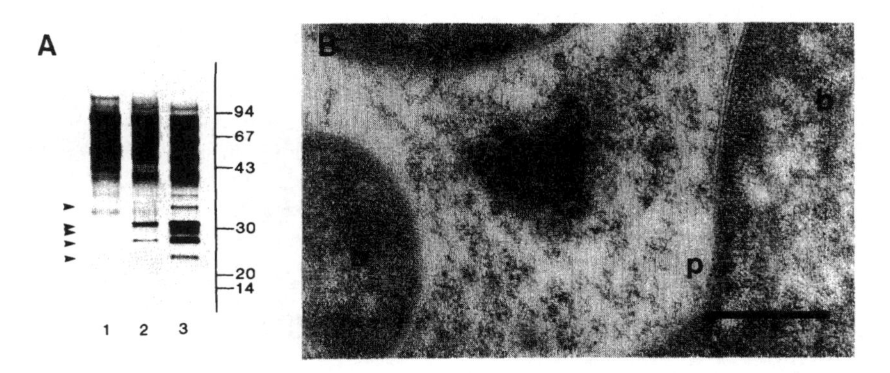

Fig. 5A,B. Derivation of peribacteroid material from the Golgi body, as illustrated by immunostaining with monoclonal antibody MAC 254. **A** Western blot after SDS-polyacrylamide gel electrophoresis, electroblotting to nitrocellulose and immunostaining with alkaline phosphatase conjugate: track *1*, pea root membranes; track *2*, nodule cytoplasmic fraction; track *3*, peribacteroid membrane fraction. The positions of nodule-specific components are indicated by *arrowheads*. **B** Localization of MAC 254 antigens by immunogold staining of pea nodule thin sections using 15 nm colloidal gold particles, showing intense labelling of Golgi bodies (*g*) and some labelling of peribacteroid membrane material (*p*) encircling bacteroids (*b*) (*bar* = 0.5 μm)

mutants of *Bradyrhizobium* that do not induce the formation of the peribacteroid compartment (Fortin et al. 1987; Morrison and Verma 1987). Nodulin 23 has also been shown to react with anti-serum specific for peribacteroid membrane (Mauro et al. 1985; Jacobs et al. 1987) and belongs to a small family of nodule-specific genes from soybean (Sandal et al. 1987). These genes share a conserved DNA sequence domain, coding for a putative signal peptide, and two other conserved domains, each containing four cysteine residues that are suggestive of a possible function involving metal binding or metal transport. It will be interesting to see whether any or all of these well-characterized nodulins share the Golgi-derived carbohydrate epitope recognized by MAC 254.

2.5.4 Possible Nature of the Peribacteroid Compartment

Although the peribacteroid membrane has many structural similarities to the plasma membrane, it is perhaps more helpful to consider it as enclosing a specialized compartment of the plant cell endomembrane system (Mellor and Wiemken 1988). This compartment is in many ways analogous to the animal cell phagosome (Mellman et al. 1986), or to the recently identified pre-lysosomal compartment (Griffiths et al. 1988). As with the phagosomal or pre-lysosomal compartments, the peribacteroid compartment can be thought of as being derived partly from a Golgi-derived pathway of membrane flow and partly from an endocytotic pathway stimulated by phagocytosis or receptor-mediated endocytosis. The convergence of these two pathways of membrane flow usually culminates in the development of acidic lytic compartments, such as the lysosome in animal cells (Anderson and Pathak 1985; Mellman et al. 1986), or the vacuole in plant cells (Boller and Wiemken 1986; Harris 1986).

Although the phagosome and pre-lysosomal compartments of animal cells are thought to be less acidic than the lysosome, their pH is still sufficiently low to activate many of the lytic enzymes. The pH of the peribacteroid space has not been measured, but it is perhaps significant that the contents of this space include a vacuolar isoenzyme of α-mannosidase with an isoelectric point of pH 5.4 (Kinnbach et al. 1987), a protease activity and a protease inhibitor (Mellor et al. 1984), and a transport system for dicarboxylic acids (Udvardi et al. 1988) similar to that found on the vacuolar membrane (Martinoia et al. 1985). Assuming that the peribacteroid space is equivalent to a phagosome or pre-lysosomal compartment, it is likely to be progressively acidified as a result of the action of membrane H^+-ATPases (Blumwald et al. 1985; Bassarab et al. 1986; Domigan et al. 1988) or by the inward transport of dicarboxylic acids (Udvardi et al. 1988).

The possible acidification of the peribacteroid compartment leads to an interesting model for pH control (Table 1), which could be very important in determining the survival, or otherwise, of the nitrogen-fixing endosymbiont. Decreasing the pH would in turn lead to the activation of acid hydrolases and other lytic systems in the peribacteroid compartment, and hence probably to the ultimate destruction of bacteria within the peribacteroid space. However, the

Table 1. Dynamic model for pH balance in the peribacteroid compartment

Plant-induced decrease of pH	Bacteroid-induced increase of pH
Succinate transport	Succinate metabolism
H^+-ATPase	NH_3 excretion

plant-induced acidification of the peribacteroid compartment could be counteracted by the metabolic activity of the bacteroids themselves. Firstly, succinate (or a similar dicarboxylic acid) is the major carbon source of bacteroids (Ronson et al. 1984; Finan et al. 1988) and the metabolism of this substrate would tend to increase the pH in the peribacteroid space as has been observed for free-living rhizobial cultures (Glenn and Dilworth 1981; Kannenberg et al. 1989). Furthermore, once they have differentiated into nitrogen-fixing forms, the bacteroids excrete ammonia (NH_3) as the product of nitrogen fixation (Bergersen and Turner 1967), and this would also tend to increase the pH of the peribacteroid space, counteracting the acidifying effects of plant-derived H^+-ATPases on the peribacteroid membrane. Hence, on this model, the continued survival of the bacteroids as endosymbionts may depend on their ability to maintain a dynamic equilibrium, balancing the progressive acidification of the peribacteroid space through the plant endomembrane system by the uptake and metabolism of dicarboxylic acids and the excretion of NH_3 by the nitrogen-fixing bacteroids. A prediction of this model is that mutants of rhizobia that, as bacteroids, fail to take up succinate or fail to excrete NH_3 through nitrogen fixation would be less able to compensate for the acidification of the peribacteroid space by the plant. These Fix^- mutant bacteroids would thus be expected to senesce prematurely, as has been observed in many cases (Werner et al. 1984; Fischer et al. 1986; Regensburger et al. 1986; Gardiol et al. 1987; Hirsch and Smith 1987; Finan et al. 1988).

2.5.5 The Peribacteroid Membrane as a Model Experimental System

From the above discussion, it is clear that the study of structure and function in the peribacteroid membrane is of central importance to our understanding of how the nitrogen-fixing symbiosis is established between *Rhizobium* and the host legume. But, insofar as the peribacteroid membrane can be viewed as a specialized component of the plant endomembrane system, which is common to all cells, it can be used as an experimental system to address many fundamental questions relating to the structure and function of plant membranes in general. Viewed as a simplified (more basic) form of plasma membrane, the peribacteroid membrane has the advantage that it can be obtained in pure form (although not in copious quantities) without prior digestion of plant cells with cell-wall degrading enzymes (Mellor and Werner 1986). Moreover, relative to the plasma membrane, it can be obtained 'inside out' as peribacteroid envelopes with the cytoplasmic face outermost (Udvardi et al. 1988). These properties may present some interesting opportunities for studies relating to the structure and function of the plasma membrane. Furthermore, insofar as the peribacteroid membrane is a specialized structure, it will be interesting to determine how proteins are specifically targetted into, or away from, this particular component of the endomembrane system. Glycosylation patterns of the constituent glycoproteins from these various membranes should help to chart the flow of material through the endomembrane system and provide some clues concerning the mechanisms of protein targetting in plants, and the use of monoclonal antibodies, in conjunction with appropriate carbohydrate analysis of their epitopes, should help to dissect the complexities of glycoprotein processing and targetting within the plant endomembrane system.

3 Actinorhizal Symbiosis

3.1 Mode of Infection

Frankia is a common soil bacterium that, unlike *Rhizobium*, is a filamentous non-motile actinomycete. It infects a broad and diverse range of host plants, forming nitrogen-fixing nodules (Benson 1988). The known hosts of *Frankia* include over 200 species in 24 genera, distributed among 8 families, all of them woody dicotyledonous plants (Moiroud and Gianinazzi-Pearson 1984). No species designations of *Frankia* strains have yet been assigned, although it is clear that different strains show different host specificities. This specificity may be analogous to the cross-inoculation groups that occur in the *Rhizobium*-legume symbiosis, but it is altogether a less well characterized phenomenon.

In *Alnus* and many other species (Berry et al. 1986), the mode of entry of *Frankia* is by root hair deformation which enhances the close cell-to-cell contact necessary to effect chemical dissolution of the root hair cell wall and thereby allows invasion to occur. Progression of the infection from cell to cell in *Alnus* involves invasive filaments of *Frankia,* which penetrate host cell walls and traverse the nodule meristem. Such cell-to-cell invasive filaments can be compared to the infection threads formed during *Rhizobium* infection of herbaceous legumes. In the same way, the invasive filaments of *Frankia* are sheated by plant cell wall and plant plasma membrane so that the bacterium, although inside the plant cell, is still topologically connected to the apoplastic space.

In *Elaeagnus*, infection of *Frankia* does not involve the root hairs, which may not even be deformed (Miller and Baker 1985). Instead, filaments of *Frankia* at the root surface penetrate intercellular material between epidermal cells, dissolving the existing middle lamella and thereby penetrating into the intercellular spaces of the root cortex. In some respects, such direct intercellular penetration is reminiscent of the process observed in some harbaceous legumes invaded by *Rhizobium*, as seen in *Arachis* or *Stylosanthes* (Chandler 1978; Chandler et al. 1982), and is also very similar to the early infection of the non-legume *Parasponia* by *Rhizobium* (Lancelle and Torrey 1984).

3.2 Physiological Adaptations for Nitrogen Fixation

In the case of actinorhizal root nodules, the problems of access to oxygen are somewhat different from those of legumes. *Frankia* cells grown in free-living culture in the absence of fixed nitrogen substrates develop terminal hyphal swellings termed vesicles, within which nitrogenase is formed (Tjepkema et. al. 1981). Under aerobic conditions, the vesicles possess a multi-laminate envelope which provides a physical barrier to oxygen diffusion and thereby protects the nitrogenase enzymes within the vesicles from oxygen damage. Thus, *Frankia* is unique among actinomycetes in being capable of fixing atmospheric nitrogen, even in culture.

Frankia also differentiates vesicles inside most actinorhizal root nodules, although their morphology depends on the particular type of host plant involved. In the Myricaceae, *Frankia* vesicles are elongate and club-shaped, whereas, in *Alnus* and *Elaeagnus*, spherical or pear-shaped structures are more common. In only one group of actinorhizal plants does *Frankia* fail to form vesicles at all. This unique group is the Casuarinaceae, where *Frankia* remains filamentous in the mature nodules. Interestingly, cells of *Casuarina* root nodules occupied by *Frankia* filaments are especially modified by cell wall thickening and the deposition of lignin-like materials not found in uninfected cells (Berg and MacDowell 1988). It is suggested that these host plant cell wall modifications present physical barriers to gaseous diffusion into the infected cells, thereby affording some measure of oxygen protection to nitrogenase within the actinorhizal filaments. In addition, oxygen-carrying haemoglobin proteins are synthesized in the cytoplasm of infected *Casuarina* root nodule cells (Tjepkema 1983); these plant haemoglobins presumably function to facilitate oxygen diffusion, as in *Rhizobium-legume and Rhizobium-Parasponia* symbioses.

4 Endomycorrhizal Symbiosis

4.1 Infection Strategies for Ericoid and Vesicular-Arbuscular Mycorrhizae

Endomycorrhizal fungi are easily distinguished from ectomycorrhizal forms because they not only penetrate the root but they also develop profusely within the parenchymatous host cell (Gianinazzi-Pearson 1986). Vesicular arbuscular (VA) and ericoid endomycorrhizae are symbiotic associations that are characterized by extensive bidirectional nutrient exchange between fungus and host root; mineral nutrients, especially phosphates, derived from the soil pass from fungus to plant, in exchange for carbon compounds derived from photosynthesis. The infected plant shows intense metabolic activity, with proliferation of its cytoplasm and organelles, and the host plasma membrane extends to surround the fungal haustorium so that an interfacial zone of close contact is established between fungus and plant, which is delimited by the plasma membranes of both organisms.

Ericoid endomycorrhizae are a fairly specific and self-contained group, but the VA mycorrhizae lack host specificity and several genera of fungi from one zygomycetous family can infect a wide range of plant species ranging from mosses to angiosperms. Endomycorrhizae only form in differentiated unsuberized tissues of host roots. The fungi do not infect meristematic or vascular regions, and haustorial invasion of the protoplast is limited to the parenchymatous cells surrounding the central vascular cylinder. With ericoid mycorrhizae, the haustoria within the protoplast develop as dense hyphal coils, whereas with VA mycorrhizae they develop into complex and multi-branched structures termed 'arbuscules'.

Initial penetration of the host root by VA mycorrhizal fungi involves a fairly well-defined haustorium. Hyphal penetration of the outer cell layers of the host root can be intercellular or intracellular. In the latter case, fungal hyphae that penetrate the protoplasts of the epidermal or outermost cortical cells often form a simple, unbranched hyphal coil. Subsequently, intercellular hyphae reemerge with increasing frequency as the fungus spreads into the inner cortex where the further invasions into plant protoplasts develop into characteristic arbuscule structures. It is assumed that these arbuscules represent the site of nutrient exchange between the endosymbiont and the enveloping plant cell membrane. After a few days, the arbuscule degenerates, the hyphae empty and the walls collapse, but the physiological triggers relating to this programmed senescence have not been elucidated. However, it is interesting to note that VA endomycorrhizal infections elicit phytoalexin production in roots (Morandi and Gianinazzi-Pearson 1986).

4.2 Apoplastic and Symplastic Specializations of the Host Plasma Membrane

When the fungal symbiont of VA mycorrhizae is present in the intercellular spaces of the root tissue, the host plant cells show very little response to the presence of hyphae. However, as soon as the hyphae break through the cell wall and come into contact with the host plasma membrane, this membrane elongates and deposits new wall material around the invading fungus (Gianinazzi-Pearson et al. 1981). This wall-building activity appears to result from an unspecialized response of the host plant cell, designed to accommodate the increasing surface area of its protoplast. However, it results in rather different intracellular structures for infected cells in the outer and inner root cortex.

In the case of the intracellular coils developing in the outer cell layers of the root, the fungal wall is still relatively thick and chitinous, and is continuously separated from the host plasma membrane by wall material of host origin. The latter forms a barrier separating the fungus from the host protoplast, and in this respect the host-fungus interface resembles the apoplastic relationship established with intercellular hyphae and ectomycorrhizal hyphae. [Incidentally, there are also close morphological parallels with the *Rhizobium*-induced infection thread structures in legume root nodules (Sect. 2.3)]. At this stage the plasma membrane at the host-fungus interface does not appear to be specialized for nutrient exchange, and the host cell cytoplasm similarly shows no signs of physiological specializations.

By contrast, in the inner root cortex, where endomycorrhizal fungi form intracellular arbuscules, there appear to be profound changes in the morphology and physiology of the host plant cell. As the hyphae ramify deep into the host protoplast, the fungal wall becomes extremely thin and simplified in the arbuscule and the amount of plant cell wall material deposited against the invading fungus is greatly reduced around the fine arbuscule branches. Thus, the space between fungal and host plasma membranes is drastically reduced and a close relationship is established between the cell surface of the two symbionts, which presumably facilitates nutrient exchange. It is interesting to note that the host membrane which surrounds the arbuscule still shows the same morphology and cytochemical properties as the peripheral membrane (Gianinazzi-Pearson 1986) even though it is not associated with the synthesis of cell wall material. However, neutral phosphatases (which are characteristic of sites of polysaccharide synthesis) are specifically localized on the host plasma membrane at the arbuscule interface, and furthermore, plasmalemmasomes, which are often associated with regions of active cell wall synthesis, can also be seen in the interfacial matrix surrounding the fine arbuscule branches. It therefore seems reasonable to conclude that the host membrane surrounding arbuscular hyphae may still have at least some components of the wall-building machinery, but that the processes of wall deposition and organization are somehow affected by the actively growing fungal hyphae. Diversion of wall precursors or localized hydrolysis of wall material could provide an important supplementary source of carbohydrate for the endomycorrhizal fungus.

In the inner cortex of the root, the plasma membrane of uninfected parenchymatous cells shows little or no Mg^{2+}-ATPase activity. However, when an endomycorrhizal fungus infects these cells, the peripheral plasma membrane remains inactive, but the newly synthesized host plasma membrane surrounding the fine arbuscule branches becomes the site of intense Mg^{2+}-ATPase activity (Smith and Smith 1986). A high level of ATPase activity on a membrane normally indicates a high transport capacity, and therefore it is highly probable that the plasma membrane surrounding the fungal haustorium is specialized for the transfer of phosphate from fungal to host cell. The release of phosphate by the internal mycelium of VA mycorrhizae may involve hydrolysis of polyphosphate from the fungal vacuoles being triggered by a suitable signal (Smith and Smith 1986).

5 Ectomycorrhizal Symbiosis

Ectomycorrhizae are symbiotic associations between fungi and the roots of at least 5000 higher plant species (Malloch et al. 1980), including most of the important temperate gymnosperm and angiosperm trees and a few forest species in tropical ecosytems. The interaction of ectomycorrhizal fungi with plant roots is characterized structurally by the formation of a mantle of fungal hyphae encasing the root surface and a network of intercellular hyphae, the Hartig net, which forms most of the interface between fungus and root tissue. Hyphae emanating from the outer mantle form the extraradical mycelium, an absorbing interface with the soil, which is capable of transferring nutrients from the soil into the adjoining root cells.

A compatible mycorrhizal association must involve a certain degree of mutual recognition and specificity between plant and fungus, although in general a very low level of host specificity seems to be involved. A striking feature of ectomycorrhizal establishment in, for example, the *Pinus strobus-Pisolithus tinctorius* association (Piche et al. 1983) is the extensive colonization of short lateral roots when fungal hyphae come into close contact with the root surface, suggesting that these roots may be releasing exudates that stimulate hyphal proliferation.

Following the growth of a hyphal mass on the root surface, the inner mantle hyphae begin to penetrate the root, initiating an intercellular Hartig net in the outer root cortex (Piche et al. 1983). This is achieved by penetrating between root cells that are at a particular stage of development. It is not known whether cells in this zone have an altered metabolism which favours fungal growth or whether the middle lamella is particularly susceptible to fungal penetration in this region. *Pinus* root cells do not show pronounced structural changes when interacting with the Hartig net hyphae. However, in the symbiosis between *Alnus crispa* and *Alpova diplophloeus*, Hartig net hyphae change dramatically by synthesizing large amounts of rough endoplasmic reticulum, and the adjacent root cells respond by synthesizing additional wall material in the form of ingrowths along the cell walls that interface with Hartig net hyphae (Massicotte et al. 1986). Presumably the resulting labyrinthine system of branched hyphae is a structural adaptation that facilitates the exchange of nutrients between symbionts. An important feature of this symbiosis is that senescence occurs very quickly, as determined by changes in cellular organelles in both symbionts and the deposition of secondary cell wall material over the wall ingrowths. This perhaps implies that only a restricted portion of the mycorrhizal organ is involved in transport of nutrients at any particular time.

6 General Conclusions

6.1 Host Specificity

In each of the plant-microbe symbioses that have been discussed, some form of host-specific interaction is observed. In some cases, host range is narrow and restricted, as for example in the interaction of *Rhizobium* with temperate legumes; in other cases, e.g. with VA endomycorrhizae, the same microorganism can infect an extraordinarily wide range of host plants. In no case has the molecular basis for host-endophyte recognition been fully established, although the effects of legume-derived flavones on the induction of early nodulation genes in *Rhizobium* is now well documented. However, very little is known about the role of the plant cell plasma membrane in mediating the specific surface interactions which give rise to any of the symbiosis-specific structures.

6.2 Initial Penetration Through the Apoplastic Space

The first intimate surface contacts between host and endophyte involve invasion of the apoplastic space. Sometimes this involves the dissolution of the middle lamella in the region of three-way junctions between the walls of adjacent cells, followed by penetration of the endophyte between these cell walls. Alternatively, specialized ingrowths can arise which conduct the endophyte across the plant cell through an intracellular apoplastic tunnel which is still sheathed by plant plasma membrane and plant cell wall. The *Rhizobium*-induced infection thread is a good example of such an apoplastic tunnel; this structure arises initially from a three-way junction between a curled root-hair cell and two epidermal cell walls, and subsequently traverses from cell to cell within the root cortex. The luminal matrix of the infection thread has been shown to contain the same plant glycoprotein that is localized at the three-way junction between dividing plant cells (Jeffree et al. 1986; VandenBosch et al. 1989) and may be associated with localized cell wall dissolution. Apoplastic intracellular tunnels are also seen in the early stages of root invasion by the actinomycete *Frankia* and by VA mycorrhizal haustoria localized in the outer cortex of the root.

6.3 Symplastic Differentiation of Internalized Plasma Membrane

For most of the symbiotic interactions that have been considered, invasion of the endophyte culminates in a specialized intracellular structure which appears to optimize biotrophic interactions between the symbiotic partners. The *Rhizobium* bacteroid, the actinorhizal vesicle cluster and the arbuscule of VA mycorrhizae all involve a major invasion of the cytoplasmic space of the host plant, and the host plant cells also show a high degree of specialization for symbiotic interactions. Each of these endosymbiotic structures is still bounded by a host-derived membrane that is structurally related to the plasma membrane; in most cases, these two membrane systems are physically continuous, although the endosymbiotic *Rhizobium* bacteroids are individually packaged by plant-derived peribacteroid membrane, which is therefore not continuous with the plasma membrane. In all cases however, this 'perisymbiotic' membrane is greater in total surface area than the plasma membrane of the same cell (about 50 times greater for host legume cells containing *Rhizobium* bacteroids). In the absence of intervening plant cell wall material, the membranes of host and endosymbiont come into very close contact. The host membrane appears to have acquired specialized functions concerned with surface recognition of the endosymbiont, and with transport into and out of the compartment which encloses the endosymbiont. These specialized functions suggest the involvement of a specialized targetting system directed towards the perisymbiotic membrane rather than to the plasma membrane (and vice versa). At this point, it becomes a matter of semantics whether an internalized plasma membrane should cease to be considered as plasma membrane or whether it should instead be thought of as a specialized component of the endomembrane system.

Acknowledgements. I thank K.A. VandenBosch and M.F. LeGal for providing electron micrographs and K.A. VandenBosch and C.W. Lloyd for discussions.

References

Anderson RGW, Pathak RK (1985) Vesicles and cisternae in the trans-Golgi apparatus of human fibroblasts are acidic compartments. Cell 40:635–643

Bassarab S, Mellor RB, Werner D (1986) Evidence for two types of Mg^{++} ATPase in the peribacteroid membrane from *Glycine max* root nodules. Endocyt Cell Res 3:189–196

Bauer WD (1981) Infection of legumes by rhizobia. Annu Rev Plant Physiol 32:407–449

Benson DR (1988) The genus *Frankia*: actinomycete symbiont of plants. Microbiol Sci 5:9–13

Berg RH, McDowell L (1988) Cytochemistry of the wall of infected cells in *Casuarina* actinorhizae. Can J Bot 62:2038–2047

Bergensen FJ, Turner GL (1967) Nitrogen fixation by the bacteroid fraction of breis of soybean nodules. Biochim Biophys Acta 141:507–515

Berry AM, McIntyre L, McCully ME (1986) Fine structure of root hair infection leading to nodulation in the *Frankia-Alnus* symbiosis. Can J Bot 64:292–305

Blumwald E, Fortin MG, Rea PA, Verma DPS (1985) Presence of host plasma membrane type H^+-ATPase in the membrane envelope enclosing the bacteroids in soybean root nodules. Plant Physiol 78:665–672

Boller T, Wiemken A (1986) Dynamics of vacuolar compartmentation. Annu Rev Plant Physiol 37:137–164

Bradley DJ, Butcher GW, Galfre G, Wood E, Brewin NJ (1986) Physical association between the peribacteroid membrane and lipopolysaccharide from the bacteroid outer membrane in *Rhizobium*-infected pea root nodule cells. J Cell Sci 85:47–61

Bradley DJ, Wood EA, Larkins AP, Galfre G, Butcher GW, Brewin NJ (1988) Isolation of monoclonal antibodies reacting with peribacteroid membranes and other components of pea root nodules containing *Rhizobium leguminosarum*. Planta 173:149–160

Brewin NJ, Robertson JG, Wood EA, Wells B, Larkins AP, Galfre G, Butcher GW (1985) Monoclonal antibodies to antigens in the peribacteroid membrane from *Rhizobium*-induced root nodules of pea cross-react with plasma membranes and Golgi bodies. EMBO J 4:605–611

Callaham DA, Torrey JG (1981) The structural basis for infection of root hairs of *Trifolium repens* by *Rhizobium*. Can J Bot 62:2375–2384

Chandler MR (1978) Some observations of infection of *Arachis hypogaea* by *Rhizobium*. J Exp Bot 29:749–755

Chandler MR, Date RA, Roughley RJ (1982) Infection and root-nodule development in *Stylosanthes* species by *Rhizobium*. J Exp Bot 33:47–57

Domigan NM, Farnden KJF, Robertson JG, Monk BC (1988) Characterization of the peribacteroid membrane ATPase of lupin root nodules. Arch Biochem Biophys 264:564–573

deFaria SM, McInroy SG, Sprent JI (1987) The occurrence of infected cells with persistent infection threads in legume root nodules. Can J Bot 65:553–558

Finan TM, Hirsch AM, Leigh JA, Johansen E, Kuldau GA, Deegan S, Walker GC, Signer ER (1985) Symbiotic mutants of *R. meliloti* that uncouple plant from bacterial differentiation. Cell 49:869–877

Finan TM, Oresnik I, Bottacin A (1988) Mutants of *R. meliloti* defective in succinate metabolism. J Bacteriol 170:3396–3403

Fischer HM, Alvarez-Morales A, Hennecke H (1986) Pleiotropic nature of symbiotic regulatory mutants. *B. japonicum nifA* gene is involved in control of *nif* gene expression and formation of determinate symbiosis. EMBO J 5:1165–1173

Fortin MG, Zelechowska M, Verma DPS (1985) Specific targetting of membrane nodulins to the bacteroid-enclosing compartment in soybean nodules. EMBO J 4:3041–3046

Fortin MG, Morrison N, Verma DPS (1987) Nodulin 26, a peribacteroid membrane nodulin is expressed independently of the development of the peribacteroid compartment. Nucleic Acids Res 15:813–824

Gardiol AE, Truchet GL, Dazzo FB (1987) Requirement for succinate dehydrogenase activity for symbiotic bacteroid differentiation of *Rhizobium meliloti* root nodules. Appl Environ Microbiol 53:1947–1950

Gianinazzi-Pearson V (1986) Cellular modifications during host-fungus interactions on endomycorrhizae. In: Bailey JA (ed) Biology and molecular biology of plant pathogen interactions. NATO ASI Series Vol H1. Springer, Berlin Heidelberg New York Tokyo, pp 29–37

Gianinazzi-Pearson V, Morandi D, Dexheimer J, Gianinazzi S (1981) Ultrastructural and ultracytochemical features of a *Glomus tenuis* mycorrhiza. New Phytol 88:633–639

Glenn AR, Dilworth MJ (1981) Oxidation of substrates by isolated bacteroids and free-living cells of *Rhizobium leguminosarum* 3841. J Gen Microbiol 126:243–247

Griffiths G, Hoflack B, Simons K, Mellman I, Kornfeld S (1988) The mannose-6-phosphate receptor and the biogenesis of lysosomes. Cell 52:329–341

Harris N (1986) Organization of the plant endomembrane system. Annu Rev Plant Physiol 37:73–92

Hirsch AM, Smith CA (1987) Effects of *Rhizobium meliloti nif* and *fix* mutants on alfalfa root nodule development. J Bacteriol 169:1137–1146

Jacobs FA, Zhang M, Fortin MG, Verma DPS (1987) Several nodulins of soybean share structural domains but differ in their subcellular locations. Nucleic Acids Res 15:1271–1280

Jeffree CE, Dale JE, Fry SC (1986) The genesis of intercellular spaces in developing leaves of *Phaseolus vulgaris*. Protoplasma 132:90–98

Kannenberg EL, Brewin NJ (1989) Expression of a cell surface antigen is regulated by oxygen and pH in *Rhizobium leguminosarum* strain 3841. J Bacteriol 171:4543–4548

Katinakis P, Verma DPS (1985) Nodulin 24 gene of soybean codes for a peptide of the peribacteroid membrane. Proc Natl Acad Sci USA 82:4157–4161

Kijne JW, Smit G, Diaz CL, Lugtenberg EJJ (1988) Lectin-enhanced accumulation of manganese limited *Rhizobium leguminsarum* cells of pea root hair tips. J Bacteriol 170: 2994–3000

Kinnbach A, Mellor RB, Werner D (1987) Alpha-mannosidase II isoenzyme in the peribacteroid space of *Glycine max* root nodules. J Exp Bot 38:1373–1377

Lancelle SA, Torrey JG (1984) Early development of *Rhizobium*-induced root nodules of *Parasponia rigida*. Infection and early nodule initiation. Protoplasma 123:26–37

Lloyd CW, Pearce KJ, Rawlings DJ, Ridge RW, Shaw PJ (1987) Endoplasmic microtubules connect the advancing nucleus to the tip of legume root hairs, but F-actin is involved in basipetal migration. Cell Mot Cytoskel 8:27–36

Malloch DW, Pirozynski KA, Raven PH (1980) Ecological and evolutionary significance of mycorrhizal symbioses in vascular plants (a review). Proc Natl Acad Sci USA 77:2113–2118

Martinoia E, Flügge UI, Kaiser G, Heber U, Heldt HW (1985) Energy-dependent uptake of malate into vacuoles isolated from barley mesophyll protoplasts. Biochim Biophys Acta 806:311–319

Massicotte HB, Peterson RL, Ackerly CA, Piche Y (1986) Structure and ontogeny of *Alnus crispa-Alpova diplophloeus* ectomycorrhizae. Can J Bot 64:177–192

Mauro VP, Nguyen T, Katinakis P, Verma DPS (1985) Primary structure of the soybean nodulin-23. Nucleic Acids Res 13:239–249

Mellman I, Fuchs R, Helenius A (1986) Acidification of the endocytic and exocytic pathways. Annu Rev Biochem 55:663–671

Mellor RB, Werner D (1986) The fractionation of *Glycine max* root nodule cells: a methodological overview. Endocyt Cell Res 3:317–336

Mellor RB, Werner D (1987) Peribacteroid membrane biogenesis in mature legume root nodules. Symbiosis 3:75–100

Mellor RB, Wiemken A (1988) Peribacteroid organelles as organ-specific forms of lysosomes. In: Bothe H (ed) Nitrogen fixation: Hundred years after. Gustav Fischer, Stuttgart, pp 528

Mellor RB, Morschel E, Werner D (1984) Proteases and protease inhibitors present in the peribacteroid space. Z Naturforsch 39c:123–125

Miller IM, Baker DD (1985) The initiation, development and structure of root nodules in *Elaeagnus angustifolia*. Protoplasma 128:107–119

Moiroud A, Gianinazzi-Pearson V (1984) Symbiotic relationships in *Antinorhizae*. In: Verma DPS, Hohn T (eds) Genes involved in microbe-plant interactions. Springer, Berlin Heidelberg New York Tokyo, pp 205–223

Morandi D, Gianinazzi-Pearson V (1986) Influence of mycorrhizal infections and phosphate nutrition on secondary metabolite content of soybean roots. In: Gianinazzi-Pearson V, Gianinazzi S (eds) Physiological and genetical aspects of mycorrhizae. INRA, Paris, pp 787–791

Morrison N, Verma DPS (1987) A block in the endocytosis of *Rhizobium* allows cellular differentiation in nodules but affects the expression of some peribacteroid membrane nodulins. Plant Mol Biol 9:185–196

Perotto S, VandenBosch KA, Butcher GW, Brewin NJ (1989) Immunological analysis of the peribacteroid membrane from pea root nodules and its relationship to other components of the plant endomembrane system. *submitted*

Piche Y, Peterson RL, Howarth MJ, Fortin JA (1983) A structural study of the interaction between the ectomycorrhizal fungus, *Pisolithus tinctorius* and *Pinus strobus* roots. Can J bot 61:1185–1193

Quispel A (1988) Bacteria-plant interactions in symbiotic nitrogen fixation. Physiol Plant 74:783–790

Regensburger B, Meyer L, Filser M, Weber J, Studer D, Lamb JW, Fischer HM, Hahn M, Hennecke H (1986) *Bradyrhizobium japonicum* mutants defective in root nodule bacteroid development and nitrogen fixation. Arch Microbiol 144:355–366

Robertson JG, Farnden KJF (1980) Ultrastructure and metabolism of the developing legume root nodule. In: Stumpf PK, Conn EE (eds) The biochemistry of plants Vol. 5, Academic Press, New York, pp 65–113

Robertson JG, Lyttleton P (1982) Coated and smooth vesicles in the biogenesis of cell walls, plasma membranes, infection threads and peribacteroid membranes in root hairs and nodules of white clover. J Cell Sci 58:63–78

Robertson JG, Lyttleton P (1984) Division of peribacteroid membrane in root nodules of white clover. J Cell Sci 69:147–157

Robertson JG, Lyttleton P, Bullivant S, Grayston GF (1978) Membranes in lupin root nodules. 1. The role of Golgi bodies in the biogenesis of infection threads and peribacteroid membranes. J Cell Sci 30:129–150

Robertson JG, Wells Brewin NJ, Wood EA, Knight CD, Downie JA (1985) The legume-*Rhizobium* symbiosis: a cell surface interaction. J Cell Sci Suppl 2:317–331

Rolfe BG, Gresshoff PM (1988) Genetic analysis of legume nodule initiation. Annu Rev Plant Physiol 39:297–319

Ronson CW, Astwood PM, Downie JA (1984) Molecular cloning and genetic organization of C_4-dicarboxylate transport genes from *Rhizobium leguminosarum*. J Bacteriol 160:903–909

Sandal NN, Marcker KA (1988) Soybean nodulin 26 is homologous to the major intrinsic protein on the bovine lens fiber membrane. Nucleic Acids Res 16:9347–9348

Sandal NN, Bojsen K, Marcker KA (1987) A small family of nodule-specific genes from soybean. Nucleic Acids Res 15:1507–1519

Saunders MJ, Jones KJ (1988) Distortion of cell plate formation by the intracellular calcium antagonist TMB-8. Protoplasma 144:92–100

Smith CA, Skvirsky RC, Hirsch AM (1986) Histochemical evidence for the presence of a suberin-like compound in *Rhizobium*-induced nodules of the non-legume *Parasponia rigida*. Can J Bot 64:1474–1483

Smith DC, Douglas AE (1987) The biology of symbiosis. Edward Arnold, London

Smith FA, Smith SE (1986) Movement across membranes: physiology and biochemistry. In: Gianinazzi-Pearson V, Gianinazzi S (eds) Physiological and genetical aspects of mycorrhizae. INRA, Paris, pp 75–84

Tjepkema JD (1983) Hemoglobins in the nitrogen fixing root nodules of actinorhizal plants. Can J Bot 61:2924–2929

Tjepkema JD, Ormerod W, Torrey JG (1981) Factors affecting vesicle formation and acetylene reduction (nitrogenase activity) in *Frankia*. Can J Microbiol 27:815–823

Turgeon BG, Bauer WD (1985) Ultrastructure of infection thread development during the infection of soybean by *Rhizobium japonicum*. Planta 163:328–349

Udvardi MK, Price DG, Gresshoff PM, Day DA (1988) A dicarboxylate transporter on the peribacteroid membrane of soybean nodules. FEBS Lett 231:36–40

Vance CP (1983) *Rhizobium* infection: a beneficial plant disease. Annu Rev Microbiol 37:399–424

VandenBosch KA, Bradley DJ, Knox JP, Perotto S, Butcher GW, Brewin NJ (1989) Common components of the infection thread matrix and the intercellular space identified by immunocytochemical analysis of pea root nodules and uninfected roots. EMBO J 8:335–342

Verma DPS, Fortin MG, Stanley J, Mauro VP, Purokit S, Morrison N (1986) Nodulins and nodulin genes of *Glycine max*: a perspective. Plant Mol Biol 7:51–61

Verma DPS, Delauney A, Kuhse J, Hirel B, Schafer R, Raju K (1988) Metabolites and protein factors controlling nodulin gene expression. In: Bothe H (ed) Nitrogen fixation — 100 years after. Gustav Fischer, Stuttgart, pp 599–604

Werner D, Mörschel E, Kort R, Mellor RB, Bassarab S (1984) Lysis of bacteroids in the vicinity of the host cell nucleus in an ineffective (Fix^-) root nodule of soybean. Planta 162:8–16

Werner D, Mellor RB, Hahn MG, Grisebach H (1985) Glyceollin I accumulation in an ineffective type of soybean nodule with an early loss of peribacteroid membrane. Z Naturforsch 40c:179–181

Werner D, Mörschel E, Garbers C, Bassarab S, Mellor RB (1988) Particle density and protein composition of the peribacteroid membrane from soybean root nodules is affected by mutation in the microsymbiont *B. japonicum*. Planta 174:263–270

Witty JF, Minchin FR, Skøt L, Sheehy JE (1986) Nitrogen fixation and oxygen in legume root nodules. Oxford Survey Plant Mol Cell Biol 3:275–314

Chapter 16 Molecular Biology of the Plasma Membrane – Perspectives

P. Kjellbom, J. Chory, and C.J. Lamb[1]

[1]Plant Biology Laboratory, The Salk Institute for Biological Studies, P.O. Box 85800, San Diego, CA 92138, USA

Abbreviations: ABA, abscisic acid; cDNA, complementary DNA; DAG, diacylglycerol; ER, endoplasmic reticulum; G-protein, GTP-binding protein complex; HR, hypersensitive reaction; HSV, herpes simplex virus; IP_2, inositol bisphosphate; IP_3, inositol trisphosphate; kb, kilobases; λ gt11, bacteriophage lambda gt11; NEM, N-ethylmaleimide; PAGE, polyacrylamide gel electrophoresis; PCR, polymerase chain reaction; PI, phosphatidylinositol; PIP, phosphatidylinositol 4-phosphate; PIP_2, phosphatidylinositol 4,5-bisphosphate; polyA RNA, polyadenylated RNA; rbcS, ribulose-1,5-bisphosphate carboxylate/oxygenase, small subunit; RFLP, restriction fragment length polymorphism; SDS, sodium dodecyl sulfate

C. Larsson, I.M. Møller (Eds):
The Plant Plasma Membrane

1 Introduction

Our understanding of the structure and function of the plant plasma membrane is far from complete, although it is well recognized that the membrane is involved in many important functions relating to interactions between plant cells and their environment. Implicated biological functions of the membrane have so far not led to any extensive identification and purification of plasma membrane proteins involved in these kinds of interactions. Recent advances that permit the isolation of highly pure plasma membranes have enabled the confirmation of the presence of several enzyme activities previously thought to be localized to the membrane, as well as the discovery of new plasma membrane-localized activities. The plasma membrane purification techniques now at hand will enable us to isolate plasma membrane proteins of interest, and several examples of the successful purification of plant plasma membrane proteins have recently been reported. This together with molecular techniques and genetic approaches described in this chapter will facilitate the cloning of genes coding for plasma membrane proteins, and will allow us to study the expression and regulation of these genes.

The current view of the general structure of the plant plasma membrane is covered in Chapter 1. We wish to point out that in view of the membrane flow model (Franke et al. 1971), i.e., the flow of membrane vesicles between different cell compartments, such as the endoplasmic reticulum (ER), the Golgi apparatus, and the plasma membrane, it is not surprising that certain enzymes are common to the ER, the Golgi apparatus, and the plasma membrane, although with very different specific activities (Kjellbom and Larsson 1984). This would also be the result if one favors the "shuttle" theory (Rothman 1981) as a model for intracellular transport of proteins. In this case membrane vesicles return to the organelle from whence they came and thus merely act as a vehicle of transportation. Whether one or the other viewpoint is favored, the result is that the plasma membrane most probably shares some properties with other cellular organelles. This should be kept in mind while elucidating the molecular biology of the plasma membrane.

2 Known and Putative Constituents of the Plasma Membrane

2.1 Purified Plasma Membrane Proteins

The electrogenic proton pumping ATPase is one of a few plant plasma membrane proteins isolated and characterized to any extent. It is K^+-stimulated, Mg^{2+}-dependent, and can be inhibited by vanadate (Chap. 6). It is a ca 100 kD transmembrane polypeptide with the ATP-binding site on the cytoplasmic side of the membrane, and it is transiently phosphorylated during ATP hydrolysis. It shows homologies, both functionally and on the amino acid level, with several other eukaryotic ATPases (Chap. 6). The enzyme is also phosphorylated by an endogenous, plasma membrane-bound, Ca^{2+}-stimulated protein kinase (Schaller and Sussman 1988). Kinase-mediated phosphorylation of the enzyme could be a means by which the activity is regulated, e.g., in response to blue light or fusicoccin. Plant H^+-ATPases are known to respond rapidly to these stimuli. Recently, a calmodulin-stimulated ($Ca^{2+} + Mg^{2+}$)-ATPase has been identified and localized to the plasma membrane of maize leaves (Robinson et al. 1988), and it has also been purified from a microsomal fraction (Briars et al. 1988). A polypeptide of 140 kD was isolated which is recognized by polyclonal antibodies to a mammalian calmodulin-stimulated Ca^{2+}-pumping ATPase. The molecular weight is similar to that of the Ca^{2+}-pumping ATPase of the mammalian erythrocyte membrane (138 kD) to which the antibodies were raised. An auxin-stimulated NADH oxidase has been partially purified from plasma membranes of soybean hypocotyls. Three prominent polypeptides, of 36, 52, and 72 kD, comigrate with the enzyme activity (Brightman et al. 1988). A fusicoccin-binding protein complex has also recently been purified from oat roots (de Boer et al. 1989), and consists of one 30 kD and two 31 kD polypeptides.

2.2 Activities Known to be Associated with the Plasma Membrane

2.2.1 Polyglucan Synthesis

The presence of different polyglucans (e.g., callose and cellulose) at the surface of plant cells implicate the plasma membrane as a potential site of synthesis (Chaps. 11 and 14). Very strong indications point to the plasma membrane as the site of the cellulose synthase and this activity correlates with the rosettes visualized by freeze-fracture techniques (Delmer 1987). These protein complexes seem to be localized at the ends of the growing microfibrils. The substrate of the cellulose synthase is UDP-glucose, the same as for the callose synthase of the plasma membrane. Rates of cellulose synthesis are generally high in living, undisturbed plant cells. In disturbed cells, or membrane preparations, cellulose synthesis is not detectable, but instead a high rate of callose (β-1,3-polyglucan) synthesis is induced. The enzyme responsible for this activity is the plasma membrane-localized glucan synthase II. The high in vitro synthesis of callose by pure plasma membrane fractions (Kjellbom and Larsson 1984), using the same substrate as the cellulose synthase, and appearing when cellulose synthase activity disappears, favors the hypothesis that the two activities reside in the same enzyme complex, and that there might be a switch from β-1,4- to β-1,3-polyglucan synthesis upon disturbing the cells (Delmer 1987). This view has been strengthened by some recent reports suggesting that the change in specificity is due to the action of an endogenous protease (Delmer 1987; Girard and Maclachlan 1987).

Regardless of the nature of these similarities, the callose synthase activity is a true plasma membrane property since the activity is strongly enriched in pure plasma membrane fractions compared to microsomal fractions or crude homogenates of plant tissue (Larsson et al. 1987). The recovery of the enzyme activity in the plasma membrane fraction is quantitative. Solubilization of plasma membrane preparations using digitonin show protein bands on native polyacrylamide gels capable of incorporating glucose from UDP-glucose into glucan polymers (Thelen and Delmer 1986). The proteins present in this band can be eluted and further resolved by SDS-PAGE, to reveal several major polypeptides as putative constituents of the enzyme complex responsible for callose synthesis (P. Kjellbom and C.J. Lamb, unpublished results). This should provide a handle for further characterization of the callose synthase by allowing the production of antisera or oligonucleotides based on partial amino acid sequence data. Antibodies might also allow the investigation of the relationship between the callose synthase and the cellulose synthase activities. Immunogold labeling should allow localization of these enzyme complexes and confirmation of their occurrence in plasma membrane rosettes.

Another area of investigation would be race-cultivar-specific differences in callose deposition following infection of plants with fungi. The soybean-*Phytophthora megasperma* f.sp. *glycinea* couple is one example (Bonhoff et al. 1987), and barley-powdery mildew another (Skou et al. 1984), where an early and extensive callose deposition in the incompatible plant-fungus reaction could be a resistance-determining factor. An induced callose deposition can also be seen after treatment of plantlets or suspension cultured cells with elicitors or digitonin (Kauss 1987; Bonhoff and Grisebach 1988; Waldmann et al. 1988). Antibodies to some of the components of the callose synthase complex should allow the isolation of corresponding cDNA clones, and this would enable the determination of whether the induction of callose deposition is due to transcriptional or translational activation, or rather is a result of a stimulation of preexisting callose synthase or due to the suggested switch from β-1,4- to β-1,3-polyglucan synthesis upon pathogen/elicitor treatment.

2.2.2 *Hormone Receptors/Translocators*

There are five major classes of low-molecular-weight compounds in plants that regulate plant growth. These are the auxins, the gibberellins, the cytokinins, abscisic acid, and ethylene. The concept of plant hormone action is that each hormone has its own specific receptor. After the hormone binds to the receptor a biochemical response is triggered which ultimately evokes a physiological response. Since all of the plant hormones are more or less polar it is possible that their initial sites of interaction are on the plasma membrane (see also Chap. 9).

In the case of mammalian hormones, except for the steroid hormones, there is a receptor localized in the plasma membrane, and these receptors are usually integral glycoproteins that span the membrane. After hormone binding to the outside of the plasma membrane a second messenger response is triggered at the cytoplasmic side of the membrane, e.g., activation of adenylate cyclase (EC 4.6.1.1) which then synthesizes cAMP from ATP; cAMP in turn activates specific intracellular cAMP-dependent protein kinases, which thereafter phosphorylate defined cell components, leading to the physiological response. With steroid hormones there might also be an interaction at the plasma membrane, but in this case the function would be to transport the hormone across the membrane. The mammalian system provides a model for plant hormone action, even if the role of cAMP as a second messenger has so far not been directly established. However, cAMP-dependent protein kinase activities have been reported in plants (Janistyn 1988).

Membrane-localized binding sites for auxin have been found (Hertel et al. 1972; Löbler and Klämbt 1985a,b; Shimomura et al. 1986; Venis 1987), and one influx carrier and one efflux carrier have putatively been identified at the plasma membrane (Hertel et al. 1983; Benning 1986; Sabater and Rubery 1987). Auxin-binding proteins from maize coleoptile and shoot membranes have been characterized (Löbler and Klämbt 1985a,b; Shimomura et al. 1986), although these are probably not plasma membrane proteins (Shimomura et al. 1988). Auxin stimulates H^+ extrusion by the plasma membrane H^+-ATPase (Senn and Goldsmith 1988). The fungal toxin fusicoccin also stimulates H^+ extrusion. The plasma membrane binding site for this compound has been partly characterized (Feyerabend and Weiler 1988; Senn and Goldsmith 1988), and recently purified (de Boer et al. 1989), and found to be distinct from the auxin binding site. The native ligand for the fusicoccin binding site is still unknown, but the receptor itself seems to be a heterotrimer with a total molecular weight of ca 92 kD (de Boer et al. 1989). While auxin binding sites can also be found in locations other than the plasma membrane, the synthetic auxin transport inhibitor 1-N-naphthylphthalamic acid only binds to the plasma membrane (Chap. 2). The endogenous ligand is unknown, but the binding site seems to be closely connected or identical to the auxin efflux transporter (Heyn et al. 1987; Voet et al. 1987). The effects of the synthetic auxin transport inhibitor can be mimicked by endogenous flavonoids which thus might be candidates for endogenous ligands (Jacobs and Rubery 1988).

Indications of cytokinin and gibberellin binding sites have also been found, but the situation here is less clear since it is not known where these binding sites are located. Gibberellic acid promotes stem elongation and the hormone-insensitive mutants of pea and maize could be receptor mutants or at least allow the components of the signal transduction pathway to be delineated (Lenton et al. 1987; MacMillan 1987).

Abscisic acid (ABA) is needed for normal seed development and stomatal function. During water stress, stomata close in response to ABA. The K^+ concentration in the guard cells decreases and since K^+ is the main osmoticum in the cells they close (Schauf and Wilson 1987). High-affinity binding sites for ABA have been found in the plasma membrane of guard cells (Hornberg and Weiler 1984), and these sites seem to be closely connected to receptor function. Three different polypeptides could be identified as being involved in the binding using photoactivated crosslinking of ABA to the binding site. By using transposon tagging of a ABA insensitive vivipary mutant (Vp 1) of maize, it was possible to isolate the Vp 1 allele (McCarty et al. 1989). The mutant phenotype suggests that the wild-type allele codes for an ABA receptor protein, though it may also code for another component in the signal transduction pathway.

There is no clue as yet to cytokinin receptor functions. The only mutants connected to cytokinin effects are overproduction mutants, and these are not likely to have anything to do with a putative cytokinin receptor.

No ethylene mutants which point to the existence of an ethylene receptor at the plasma membrane are available, although the presence of such a receptor has been suggested (Evans et al. 1982). An ethylene receptor might be of such an importance that a mutation in the gene is lethal. Ripening, abscission, and senescence are all phenomena in which ethylene is thought to be involved. Mutations corresponding to these developmental stages have been identified and it should be possible to probe whether ethylene actually is involved, and if so, whether some of the mutants represent receptor mutations, or blocks in the biosynthetic pathway (McGlasson 1985). Recently, an ethylene-insensitive mutant of *Arabidopsis thaliana* has been isolated that promises insight into the mechanism of ethylene responses (Bleecker et al. 1988). It is a dominant mutation, and all of the ethylene responses tested were affected by this mutation regardless of the plant organ involved. This might suggest that the mutation is in a gene for a single ethylene receptor common for the different tissues. It was also shown that this mutant only had 20% of the ethylene-binding capacity of the wild-type plant. Considering this and the dominant nature of the mutation, one suggestion is that the mutated gene codes for a monomer of a multimeric ethylene receptor complex, although the presence of more than one ethylene-binding protein cannot be ruled out since 20% of the binding capacity still remained. Alternatively, a mutation in a suppressor of ethylene responses, possibly a DNA-binding protein, would also fit the observations. In the wild-type plant ethylene would abolish the inhibitory effect of the suppressor, while the mutated gene product would be unable to respond to ethylene (Bleecker et al. 1988).

2.2.3 Components of Plasma Membrane Signal Transduction Systems

Mammalian cells interact with a large variety of signal molecules such as hormones, light and growth factors. These molecules control cellular functions such as cell proliferation, excretion, contraction, phototransduction, metabolism, and cell growth. The activation of specific plasma membrane receptors by a specific signal molecule leads to signal transduction across the plasma membrane and stimulation of a second intracellular signal, which ultimately causes the physiological response. The best-studied mammalian signal transduction system involves cAMP as the intracellular second messenger. One example where this cascade system is used is the β-adrenergic receptor that binds catecholamine hormones. This type of receptor belongs to a group of plasma membrane receptors that includes the muscarinic-acetylcholine receptor and rhodopsin. These are all integral glycoproteins with transmembrane α-helical domains. The amino terminal ends of the polypeptide chains are located at the extracellular side of the plasma membrane and glycosylated at two asparagine residues. The carboxyl-terminal ends are located at the cytoplasmic side and contain clusters of serine and threonine residues as possible target sites for protein kinases. This class of receptors works via heterotrimeric GTP-binding protein complexes (G-proteins) located at the cytoplasmic side of the plasma membrane (Neer and Clapham 1988). The G-protein is activated by the ligand-receptor complex, involving the substitution of a bound GDP for a GTP, whereafter the activated G-protein stimulates adenylate cyclase to synthesize cAMP.

Another class of G-protein-linked receptors acts through the so-called phosphatidylinositol response (PI response), e.g., the α-adrenergic receptor and the muscarinic cholinergic receptor (Abdel-Lafit 1986; Berridge 1986, 1987). In this case, a rapid response to hormone-receptor interaction results in an increased Ca^{2+} concentration in the cytoplasm. The intracellular Ca^{2+} concentration is normally less than 1 μM, but it can increase very rapidly due to different stimuli acting on plasma membrane-localized receptors. The release of Ca^{2+}, stored in the ER, is mediated through the hydrolysis of plasma membrane lipid components. Phosphatidylinositol (PI) is a minor lipid constituent of the plasma membrane, comprising 1–10% of the total phospholipids (Chap. 1). Two separate plasma membrane lipid kinases catalyze the consecutive synthesis of phosphatidylinositol 4-phosphate (PIP) and phosphatidylinositol 4,5-bisphosphate (PIP_2) from PI. Upon binding of ligand (e.g., a hormone) to the plasma membrane receptor, and the subsequent activation of a G-protein, a plasma membrane-localized polyphosphoinositide phosphodiesterase (phospholipase C) is activated, resulting in the hydrolysis of PIP_2 to form inositol trisphosphate (IP_3) and diacylglycerol (DAG). IP_3 is the component responsible for the release of Ca^{2+} from internal stores, i.e., the ER, whereas DAG operates within the membrane activating a plasma membrane-localized protein kinase C. Released Ca^{2+} activates calmodulin which in turn activates a variety of enzymes including Ca^{2+}-dependent protein kinases (Chap. 9).

Protein kinase C is able to regulate a variety of events such as ion conductance, gene regulation, and cell proliferation depending upon its target (Sibley et al. 1987). It can phosphorylate both soluble and membrane-bound proteins and there seems to be two pools of the enzyme, one associated with the plasma membrane, and one cytoplasmic. Protein kinase C is also involved in down-regulation, i.e., desensitizing receptors or other components of the signal transduction pathway, by phosphorylation. This is often mediated through other protein kinases, themselves activated by protein kinase C. This negative feedback control is believed to be used for long-term responses such as cell proliferation. Down-regulation by receptor phosphorylation can also be seen in the cAMP receptor of *Dictyostelium* (Klein et al. 1988), the T-cell receptor of T-lymphocytes and the β-adrenergic receptor (Sibley et al. 1987). Protein kinase C also influences the hydrolysis of PIP_2 to IP_3 and DAG, and thus regulates intracellular Ca^{2+} levels. It might also activate an IP_3 phosphatase and a plasma membrane-localized Ca^{2+}-ATPase, thus stimulating the removal of Ca^{2+} from the cytoplasm (Berridge and Irvine 1984; Briars et al. 1989).

Many of the components of the outlined signal transduction pathway have recently been found in plants and some of them have also been localized to the plant plasma membrane. This is the case with the plasma membrane localized lipid kinases catalyzing the synthesis of PIP and PIP_2, as well as the plasma membrane polyphosphoinositide phosphodiesterase (phospholipase C), hydrolyzing PIP_2 to IP_3 and DAG (Sandelius and Sommarin 1986; Melin et al. 1987; Pfaffman et al. 1987; Sommarin and Sandelius 1988).

The presence of microsomal and plasma membrane-localized GTP-binding proteins has recently been reported (Hasunuma and Funadera 1987; Hasunuma et al. 1987; Blum et al. 1988; Drøbak et al. 1988; Jacobs et al. 1988). They share antigenic homologies with mammalian G-proteins (Blum et al. 1988; Jacobs et al. 1988). Enzymes similar to the Ca^{2+} and phospholipid-stimulated protein kinase C of mammalian cells have also been found in microsomal and plasma membrane-enriched fractions from plants (Schäfer et al. 1985; Elliott and Skinner 1986; Lucantoni and Polya 1987; Klucis and Polya 1988).

The second messenger function of IP_3 in influencing the release of Ca^{2+} from internal stores has been verified in plants (Ranjeva et al. 1988) as well as IP_3 influencing the transport of Ca^{2+} across the plasma membrane (Drøbak and Ferguson 1985). Possible targets for the regulation of Ca^{2+} transport over the plasma membrane could be a Ca^{2+}-ATPase (Rasi-Caldogno et al. 1987; Briars et al. 1989). It is also interesting to note that gradients of free Ca^{2+} have been implicated in developmental processes in plants, such as pollen tube growth (Hepler and Wayne 1985; Nobiling and Reiss 1987).

In addition to the presence in plants of several components of well-characterized signal transduction systems of mammalian cells, many examples can be found in the recent literature pointing to the type of receptors and ligands responsible for the initiation of signal transduction pathways. Auxin, for instance, has been reported to cause very rapid changes in the concentration of IP_3 and IP_2 (Ettlinger and Lehle 1988), as well as causing changes in the Ca^{2+} concentration in the cytoplasm (Felle 1988), thus pointing to the existence of an auxin receptor in the plant plasma membrane initiating the PI response. Likewise, Ca^{2+} has been shown to be involved in the signal transduction pathway by which phytochrome acts (Das and Sopory 1985). Phytohormones other than auxin, such as gibberellins and cytokinins, also induce changes in phosphorylation patterns in plants (Ranjeva and Boudet 1987).

2.2.4 Plasma Membrane-Bound Protein Kinases

Besides protein kinase C, there are other protein kinase activities associated with the plant plasma membrane (Blowers and Trewavas 1987; Ranjeva and Boudet 1987; Battey and Venis 1988; Klucis and Polya 1988; Chap. 9). Such proteins in the plant plasma membrane may correspond to the mammalian receptors for a variety of growth factors. These mammalian receptors are themselves tyrosine kinases (Hanks et al. 1988), and they can be regulated both by autophosphorylation and by phosphorylation by other protein kinases. In the case of the insulin receptor, tyrosine phosphorylation enhances its ability to phosphorylate substrates while phosphorylation on serine and threonine residues decreases insulin binding and reduces tyrosine kinase activity (Sibley et al. 1987). It is presently unknown whether plants have tyrosine kinases, although a plant phospholipid similar to the mammalian platelet-activating factor, an ether phospholipid, was recently found in zucchini (Scherer et al. 1988). One effect of this plant lipid was the stimulation of a soluble protein kinase activity. This activity could also be stimulated by Ca^{2+}, and increased protein kinase activity coincided with a stimulation of H^+ transport in microsomal membrane vesicles.

2.2.5 Coated Pits, Coated Vesicles, and Receptor Mediated Endocytosis

Coated pits, coated vesicles, and receptor-mediated endocytosis are well documented phenomena in animal cells (Pearse 1987), and recent years have seen an increasing documentation of coated pits and coated vesicles in plant tissue (Chap. 10). The occurrence of receptor-mediated endocytosis is still not firmly established for plant cells even if an increasing number of reports point in that direction (Hübner et al. 1985; Hillmer et al. 1986). Possible plasma membrane-localized receptors, which could initiate such a response upon ligand binding include the putative auxin receptor (Löbler and Klämbt 1985a,b), the fusicoccin-binding protein (de Boer et al. 1989), and a putative ABA receptor (Hornberg and Weiler 1984; see also Chap. 9). In mammalian systems receptor phosphorylation precedes receptor internalization. This might in some cases result in a down-regulation of the receptor, although in other cases it is linked to the receptor action, i.e., carries the ligand-receptor complex itself into the cell, eventually releasing the ligand before the receptor is returned to the plasma membrane (Sibley et al. 1987). Redistribution of receptors back to the plasma membrane is coupled to dephosphorylation.

In plants there is some evidence for endocytosis as a means to internalize extracellular material (Herman et al. 1990), and this endocytotic pathway may be connected to a class of plasma membrane-associated glycoproteins (Herman et al. 1990; Norman et al. 1990). These glycoproteins run between ca. 135 and 180 kD on SDS-PAGE gels, but when deglycosylated give a single polypeptide of ca. 50 kD. The heterogeneous carbohydrate portion is rich in arabinose and galactose, and the protein is localized at the extracellular side of the plasma membrane. Using electron microscopy and immunogold labeling, internalized vesicles can be found containing this protein. Partly disrupted vesicles, still staining for the antigen, can be seen in the vacuole, thus pointing to a possible internalization pathway for cell wall or plasma membrane material (Herman et al. 1990).

One class of receptors that might be subject to endocytosis are those involved in the binding of pathogen-derived or endogenous elicitors. Such an internalization might induce the plant defense response, e.g., the hypersensitive reaction (HR), by presenting the ligand to an internal target, thus inducing the genes coding for enzymes in the phenylpropanoid pathway responsible for the synthesis of phytoalexins and lignin (Lamb et al. 1989), as well as other defense genes. Soybean plasma membranes harbor proteins with high-affinity binding sites for fungal β-glucan fragments (Cosio et al. 1988). In addition to inducing the synthesis of phytoalexins (e.g., glyceollin), elicitors are able to induce callose synthesis, possibly through the mediation of increased Ca^{2+} levels in the cytoplasm. In fact, Ca^{2+} might be a major signal transmitter for gene activation during defense responses as well as more directly activating the callose synthase (Chap. 14).

2.2.6 Sugar Translocators

Transport systems for sugars are important for higher plants. Sugars derived from photosynthesizing cells are loaded into the phloem and transported to other plant cells. Sucrose is the main sugar transported by the phloem and in order to reach other cells the sucrose has to be transported into the phloem and later also transported into the target cells. There is considerable evidence for transport of sucrose across the plasma membrane in the cases where symplastic transport is not available (M'Batchi et al. 1987; Schmalstig and Hitz 1987; Lemoine et al. 1988; M'Batchi and Delrot 1988). There is some evidence for hydrolysis of sucrose to fructose and glucose by a cell wall invertase (EC 3.2.1.26) before sugar uptake (Stanzel et al. 1988a,b). This hexose carrier-mediated uptake is evidently not the only way for sucrose to cross the plant plasma membrane since sucrose analogs, not susceptible to hydrolysis by invertase, can cross the membrane with the same uptake rate as sucrose (Schmalstig and Hitz 1987; Damon et al. 1988).

There have been several reports of sucrose-binding proteins in the plasma membrane. One of the approaches made use of photolyzable derivatives of sucrose to identify the putative carrier in microsomal fractions (Ripp et al. 1988). A 62 kD polypeptide present in microsomal membrane fractions was identified and antisera generated to this protein allowed immunocytological localization at the plasma membrane. Another approach made use of the differential labeling of highly pure plasma membrane fractions from sugarbeet leaves, using ^{3}H-NEM and ^{14}C-NEM in the presence of sucrose or the nontransportable sucrose analog palatinose. Using this technique a 42 kD polypeptide was differentially labeled (Gallet et al. 1989). Polyclonal antibodies raised against this polypeptide were effective in inhibiting sucrose transport in a protoplast system, while transport of hexoses or amino acids was not disturbed (Lemoine et al. 1989). This antiserum was used to obtain cDNA clones coding for the putative sucrose carrier at the plant plasma membrane using a λ gt11 library (Sect. 3.1.1) made from a rice suspension culture (P. Kjellbom, unpublished results). These cDNAs are now being characterized.

As mentioned, carriers for glucose and fructose are also present in the plant plasma membrane (Damon et al. 1988). It will be important to understand the complex variation in plants and tissues which have acquired one or more of these transport systems. Presently it is not very clear what is the driving force for these sugar translocators, even if electrochemical gradients often are implicated (Getz et al. 1987). A convenient system for studying plasma membrane transport involves single-cell organisms such as the green alga *Chlorella*. Several plasma membrane proteins are induced in *Chlorella* if glucose is present in the growth medium (Sauer and Tanner 1987). By differential screening of a cDNA library generated from induced *Chlorella kessleri*, a cDNA clone was isolated coding for a putative glucose transporter at the plasma membrane. This cDNA clone was used to isolate homologous sequences in *Arabidopsis thaliana* (N. Sauer, personal communication).

2.2.7 Amino Acid Translocators

There is good evidence for the presence of an H^+/amino acid symport system in the plasma membrane (Wyse and Komor 1984; Bush and Langston-Unkefer 1988). The energy required for transport is probably derived from the electrochemical gradient generated by the plasma membrane H^+-ATPase (Chap. 6). A 65 kD plasma membrane polypeptide which is involved in alanine transport across the plasma membrane has been identified in *Chlorella* (Sauer and Tanner 1987). One interesting aspect of amino acid transporters is that some plant pathogen-produced toxins seem to be transported into plant cells by these systems (Bush and Langston-Unkefer 1988).

2.2.8 Components of Inorganic Carbon Assimilation Systems

Cyanobacteria like *Synechococcus* seem to have separate transport systems for CO_2 and HCO_3^- (Espie et al. 1988). A 42 kD membrane-associated polypeptide accumulates during adaptation to low CO_2 conditions (Schwarz et al. 1988). Antibodies raised against this polypeptide cross-react with a 45 kD plasma membrane polypeptide from *Synechocystis* PCC 6803 (Ogawa et al. 1987; Omata et al. 1987). A 42 kD plasma membrane polypeptide is accumulated when the cyanobacterium *Anacystis nidulans* adapts to low CO_2 concentrations (Omata et al. 1987). *Chlamydomonas reinhardii* accumulate 44 and 46 kD polypeptides during a similar adaptation (Manuel and Moroney 1988). These polypeptides are distinct from the simultaneously induced enzyme carbonic anhydrase (EC 4.2.1.1). The location of the polypeptides is unknown. It remains to be proven whether the polypeptides from cyanobacteria and the green alga are related, and indeed involved in inorganic carbon assimilation, although many indications point in this direction (Kaplan 1985). A homologous system has not yet been identified in higher plants.

2.2.9 *Ion Translocators/Channels*

2.2.9.1 *Nitrate Translocator*

The transport of nitrate across the plasma membrane appears to be carrier-mediated, and it is believed to be driven by the electrochemical gradient generated by the plasma membrane H^+-ATPase. Several hydrophobic polypeptides, putatively of plasma membrane origin, are induced when nitrate-starved maize seedlings are transferred to a medium containing 5 mM nitrate (Dhugga et al. 1988). A polypeptide of ca. 30 kD, and a number of polypeptides of around 40 kD appear to be specifically induced. Whether these polypeptides are localized to the plasma membrane and represent components of a nitrate translocator remains to be shown.

2.2.9.2 *Na^+, K^+, and Ca^{2+} Translocators*

Several other ion transport activities are operating in the plasma membrane of plants, although none of the transporters involved have so far been isolated or characterized. By analogy with voltage-gated channels for Na^+, K^+, and Ca^{2+} in mammalian tissues (Maelicke 1988), guard cell plasma membranes from *Vicia faba* have K^+-selective, voltage-dependent channels controlling the outward and inward flow of K^+ in a potential-dependent way (Schroeder et al. 1984, 1987; Chap. 8). In *Neurospora crassa* a K^+/H^+-symport system has been reported. This carries both ions inward while H^+ is extruded by the plasma membrane H^+-ATPase to maintain the charge balance (Blatt and Slayman 1987). In analogy with Na^+/K^+-ATPase-driven exchange of Na^+ for K^+ in animal cells, an Na^+/H^+-antiporter has been reported for plant membranes, possibly located at the plasma membrane (Braun et al. 1988).

A calmodulin-stimulated (Ca^{2+} + Mg^{2+})-ATPase associated with the plasma membrane of maize leaves has been reported (Robinson et al. 1988), and recently a similar (or identical) ATPase was purified from maize coleoptiles (Briars et al. 1988). A Ca^{2+}-ATPase with similar properties has also been reported in plasma membranes of *Commelina communis* L. (Gräf and Weiler 1989) and was suggested to be responsible for maintaining a low cytoplasmic Ca^{2+} level since inside-out plasma membrane vesicles showed an ATP-driven Ca^{2+} uptake. This may represent a system similar to the mammalian Ca^{2+}-ATPase, which is known to be transiently phosphorylated during ATP hydrolysis, and which is regulated by the Ca^{2+}/calmodulin regulatory pathway.

The latest addition to the list of putative ion channel proteins in the plant plasma membrane, by analogy with mammalian counterparts, is the stretch-activated anion, and possibly, cation channel of the tobacco plasma membrane (Falke et al. 1988). Such channels might monitor hydrostatic pressure, gravity, and changes in the osmotic pressure of the apoplast.

It is interesting to note that plant pathogens have been shown to interfere with ATPase-driven uptake systems. By destroying the electrochemical gradient created by the plasma membrane H^+-ATPase the pathogens reduce the driving force for uptake, leading to the accumulation of sugars, amino acids, and inorganic ions in intercellular spaces, and the pathogens can subsequently use these metabolites and nutrients themselves (Atkinson and Baker 1987).

2.2.10 Redox Activities

Plant plasma membranes contain a number of redox activities (Chap. 5). On addition of NAD(P)H and an electron acceptor to plasma membrane vesicles, electron transport will occur, ultimately reducing the electron acceptor (e.g., ferricyanide or cytochrome c). This may result in a net transport of protons, thus generating an electrochemical gradient. This gradient can be used to drive energy-requiring processes, and represents an alternative (or complement) to the electrochemical gradient created by the plasma membrane H^+-ATPase. The NAD(P)H binding site is located at the cytoplasmic surface of the membrane and a recent report shows that the acceptor sites also are located at the cytoplasmic surface (Askerlund et al. 1988; Chap. 2). It is concluded that if transmembrane e^--transport occurs, it only constitutes a small proportion of the total plasma membrane redox activity.

One possible photoreceptor for blue-light photomorphogenesis has tentatively been identified as a flavoprotein-b-type-cytochrome complex (Briggs and Iino 1983). Components of the plasma membrane redox system might thus be involved in transducing the blue-light signal into a physiological response, but the nature of the photoreceptor is so far not established. A plant equivalent to rhodopsin is another possible candidate for a blue-light receptor, especially since Ca^{2+}, a known second messenger of rhodopsin-like receptor systems, also seems to be needed for the blue-light response (Shinkle and Jones 1988).

Recent results show that a protein from pea, putatively associated with the plasma membrane, is phosphorylated when vesicles are supplied with ATP. Upon illumination with blue light the phosphorylation declines (Gallagher et al. 1988). The protein is most probably not phytochrome, and thus represents another possible component of the suggested blue-light receptor of the plasma membrane.

Despite recent progress in the characterization and partial purification of components of the plasma membrane redox systems, the components involved in blue-light signal transduction still await to be established. However, blue-light responses and redox processes may not only be similar but have common elements, since the addition of ferricyanide enhances guard cell swelling (which is a result of K^+ uptake) and this is a well-known blue-light response (Zeiger 1983; Roth-Bejerano et al. 1988).

3 Molecular Genetic Approaches to Characterize Components of the Plant Plasma Membrane

In order to study the regulation of components of the plant plasma membrane, it will be necessary to isolate and characterize genes for proteins located in the plant plasma membrane. A large array of techniques is now available for the cloning of such genes. For example, the cloned gene may be isolated on the basis of its pattern of expression (Rebagliati et al. 1985), by its ability to code for a particular antigen (Young and Davis 1983), or by hybridization to heterologous probes to genes from nonplant systems (Huynh et al. 1985). Using "reverse genetic" approaches, one can clone a gene based only on the phenotype of a mutant using either transposons or chromosomal walking techniques. Given the recent improvements in the purification of the plant plasma membrane (Chap. 3), the identification of monoclonal antibodies to plasma membrane components (Norman et al. 1986), and the identification of enzyme activities associated with the plasma membrane (Chap. 2), several genes for plant plasma membrane components will likely be cloned in the next few years. It is the intent of this portion of the review to discuss three general strategies for the cloning of genes coding for plant plasma membrane proteins. The first takes advantage of knowledge of the purified protein or an antibody against it. In the second approach, screening is done based on some knowledge of the DNA sequence, using either degenerate oligonucleotides based on sequences published for similar genes from other eukaryotic systems or heterologous gene probes. The third approach allows the investigator to clone a gene based only on the existence of a mutant in plasma membrane function.

3.1 From Protein to Gene

3.1.1 Expression Libraries

If one has an antibody to a component of the plasma membrane, the cloning of its gene is quite straightforward using expression vectors such as λ gt11 which is capable of producing a polypeptide specified by the DNA insert fragment (Young and Davis 1983; Huynh et al. 1985). The antibody can be obtained by several methods:

1. If the protein of interest has an enzyme activity that can be assayed in vitro, then the protein can be purified from plasma membrane preparations. For instance, the plasma membrane ATPase was purified in this manner. The purified protein can then be injected into rabbits or mice for raising antibodies.
2. Whole plant plasma membrane preparations can be injected into rabbits to raise antibodies against membrane components. This should yield an array of antibodies to various components of the plasma membrane which can be used to screen cDNA expression libraries.

3. Preparative SDS gels can be used as a source of a particular antigen for the raising of antibodies. Separated protein bands can be cut from gels, the proteins eluted from the acrylamide, and the purified proteins used to make antibodies.

4. Given recent improvements in microsequencing technology (e.g., Aebersold et al. 1987), very tiny amounts of protein can be used to obtain a peptide sequence. The techniques are now so sensitive that protein spots can be simply cut out of two-dimensional gels and subjected to microsequencing. Synthetic peptides can then be made and used as a source of antigen to make antibodies to plasma membrane components.

5. Protoplasts can be used as a source of antigen to make monoclonal antibodies to plant plasma membrane surface components. These might be very useful for identifying genes for receptors.

Once an antibody or antibodies are available, the procedures for cloning genes are quite straightforward given the available techniques and vectors for cloning genes by expression of protein antigens (Huynh et al. 1985). The successful strategy involves isolating a DNA copy of the mRNA encoded by that gene (a cDNA clone) (Gubler and Hoffman 1983). The general strategy is as follows: A cDNA library representing the mRNA population is constructed using polyA RNA made from the appropriate tissue or cell type. The cDNA clone of interest is identified from the total population of cDNA clones by screening the library with an antibody probe. Since mRNAs for plasma membrane proteins may be rare, it will probably be necessary to construct very large cDNA libraries, and for that purpose the λ gt11 vector is probably the most appropriate. Taking advantage of the highly reproducible and efficient in vitro packaging of λ DNA, cDNA libraries containing 10^5–10^7 recombinants per μg of starting mRNA can be routinely obtained. Thus, the successful isolation of a cDNA clone of interest will not be limited by either the low abundance of the mRNA or by the availability of tissue from which to prepare the polyA RNA.

3.1.2 A General Strategy for the Cloning of Plant Plasma Membrane Receptors

Very recently, a series of elegant papers (Aruffo and Seed 1987a,b; Seed and Aruffo 1987) has described an approach for the high-efficiency cloning of cell surface antigens from animal cells. This technique is based on the transient expression in animal fibroblast cells of genes coding for membrane receptors and adherence of the cells expressing the surface antigen to antibody-coated dishes. This approach has allowed a large number of surface antigen cDNAs to be cloned in a short period of time (Aruffo and Seed 1987a,b; Seed 1987; Seed and Aruffo 1987; Stengelin et al. 1988).

The large success of the approach described above is due to the construction of a new expression vector for the construction of libraries (Aruffo and Seed 1987a, b). Three features of the vector make it particularly suitable for use in cloning: (1) The eukaryotic transcription unit (a chimeric cytomegalovirus/human immunodeficiency virus promoter) allows extremely high expression in COS cells (a monkey cell line) of coding sequences placed under its control; (2) the small size of the plasmid (3.9 kbp) and particular arrangement of sequences in the plasmid permit high-level replication in COS cells from an SV40 origin of replication; and (3) an efficient oligonucleotide-based strategy was developed in order to promote high-efficiency cloning into the vector. Indeed, the library construction efficiencies these authors observed were between $0.5-2 \times 10^6$ recombinants per μg of mRNA, which is comparable with those described for phage vectors.

Surface antigen cDNAs can be isolated from these libraries using an antibody-enrichment method (Seed and Aruffo 1987). The method of spheroplast fusion is used to introduce the library into COS cells where it replicates and expresses its inserts. The cells are harvested, treated with monoclonal antibodies specific for the surface antigens desired, and distributed in dishes coated with affinity-purified antibody to mouse immunoglobulins (Seed and Aruffo 1987). Cells expressing surface antigen adhere, and the remaining cells can be washed away [a technique called "panning" in the immunological literature (Wysocki and Sato 1978)]. From the cells that adhere, a Hirt fraction of episomal DNA is prepared (Hirt 1967), and this DNA is used to transform *E. coli* for further rounds of fusion and selection. (The vector also contains an origin of replication that works in *E. coli* and a supF-selectable marker). The authors reported that after three rounds of fusion and selection, approximately 50% of the colonies screened contained the clone of the gene of interest. The entire screening procedure takes about 2 weeks.

We would like to propose that the system described above can also be used for the cloning of surface components of the plant plasma membrane. The use of COS cells is necessary because there will be no endogenous expression of the plant membrane components in animal cells. The one implicit assumption is that plant plasma membrane proteins will be cell-surface localized in COS cells. This is not unlikely, however, because numerous biochemical studies support the notion that plant and animal plasma membranes contain many similar functional components (Sect. 2.2).

3.2 Homology Probing

If genes for plasma membrane proteins (e.g., protein kinases, G-proteins) have been isolated from other eukaryotic organisms, then isolation of genes for plant plasma membrane proteins may be straightforward. One could use heterologous cDNA probes to screen a plant cDNA library in order to identify the corresponding plant gene. For instance, the *Arabidopsis thaliana* gene for acetohydroxy acid synthase was cloned using heterologous yeast probes (Mazur et al. 1988). The H^+-ATPase has been cloned from *Neurospora*, and use of the *Neurospora* clone under low stringency hybridization conditions may lead to the identification of the corresponding plant gene. The only drawbacks to this kind of experiment are (1) the shortage of heterologous probes for plasma membrane protein genes; and (2) the difficulty in estimating the amount of homology between the plant sequence and the heterologous cDNA. If there is not much homology between the genes, it may be difficult, if not impossible, to clone genes using heterologous cDNA probes.

If a gene of interest has been cloned from several different organisms, then examination of the sequences may indicate that only small discrete regions of the gene are conserved. These regions most probably will be localized at or near the active site or will be involved in substrate or cofactor binding (Hanks et al. 1988). A set of degenerate oligonucleotides that have homology to these conserved regions could then be used to identify the gene of interest (Wood et al. 1985). In general, one screens with oligonucleotides to one conserved region to identify a group of homologous cDNAs from a library. These clones are then subjected to a second round of screening using oligonucleotides to a second conserved region. Clones that show homology to two discrete conserved regions will most likely be the corresponding plant gene. This approach has been recently applied to clone plant genes for protein kinases (Lawton et al. 1989).

A powerful extension to homology probing using degenerate oligonucleotides is provided by use of the polymerase chain reaction (PCR) (Scharf et al. 1986; Mullis and Faloona 1987; Saiki et al. 1988). The PCR technique is capable of producing a selective enrichment of a specific DNA sequence by a factor of 10^6, which greatly facilitates high-efficiency cloning of genomic sequences without the use of libraries. PCR amplification involves two oligonucleotide primers that flank the DNA segment to be amplified (see Mullis and Faloona 1987, for details). If one had knowledge of two highly conserved regions of DNA within a gene of interest, cloning of the gene would be straightforward. This approach is more powerful than homology probing using degenerate oligonucleotides as described above, because oligonucleotide probes which are as much as 10^6-fold degenerate can be used as the source of primer DNA for the PCR. Kinase genes from many different systems have been cloned using the PCR; there is no reason to believe that this technique will not also work for the cloning of plant genes.

3.3 Assignment of a Function to Cloned Genes

If one clones a gene using heterologous probes or antibodies to a plasma membrane protein of unknown function, the problem of identifying the function of the gene still remains. Even if one clones a gene for a protein kinase by homology probing, for instance, the function of the particular kinase may still be unknown. The classical approach for the identification of gene function has been to inactive the gene and examine the effect this has on the organism. In bacteria and yeast, this can be done by homologous exchange of a mutated form of the gene for the wild-type copy (e.g., Ruvkun and Ausubel 1981); however, this experiment is still technically impossible in plants.

One approach that has worked for at least some plant genes involves the disruption of gene function at the level of RNA using "antisense" RNA (Izant and Weintraub 1984; Melton 1985). Antisense RNA blocks the expression of a gene by preventing the translation of its sense transcripts (Izant and Weintraub 1984). This method has been used with some notable successes in plants, for example, flower color "mutants" of *Petunia* were made by expression in transgenic plants of the antisense RNA for the chalcone synthase (EC 2.3.1.74) gene (van der Krol et al. 1988); a tomato-ripening "mutant" was created by the antisense expression of the polygalacturonase (EC 3.2.1.15) gene (Smith et al. 1988); and a "mutant" in photosynthetic function was made by producing antisense transcripts to the tobacco rbcS RNA (Rodermel et al. 1988). In general, the severity of the "mutant" phenotype is inversely correlated with the levels of RNA for the gene of interest. As gene transfer technology is improved and the availability of highly expressed and inducible plant promoters is increased, further successes should result from the use of this method.

A new strategy in which the function of a gene is blocked at the protein level was proposed by Herskowitz in 1987. In this approach, the cloned gene is manipulated and overexpressed such that it disrupts the function of the wild-type gene product. The new "mutations" are dominant negative mutations because not only will the phenotype be manifested in the presence of the wild-type gene, but the activity of the wild-type gene function will be inhibited. This approach should be feasible for the functional analysis of DNA-binding proteins, multimeric enzymes, enzymes with well-defined domains, e.g., protein kinases, or structural proteins (Herskowitz 1987). The approach has led to the functional analysis of a transcription factor, VP16, from herpes simplex virus (HSV; Friedman et al. 1988). Overexpression of a truncated form of VP16, lacking the activating domain, led to the selective impairment of the lytic infectious cycle of HSV, thus showing that the VP16 gene product was necessary for initiating events associated with HSV infection (Friedman et al. 1988).

3.4 Reverse Genetic Approach

An alternative method for the identification of genes for plasma membrane components is by the identification of mutants that are lacking a particular plasma membrane protein or possess an altered function for one of the proteins. If a map position for the mutant was known, then chromosomal walking could be used to isolate the gene. Alternatively, if the mutant was made by insertion of a transposon into a gene of interest, one could clone out the transposon and isolate the surrounding sequences in the gene of interest. The approach taken would depend on the plant species chosen. The power of genetic approaches is that the phenotype of the mutant would be known, thus knowledge of the function of a plasma membrane component would be immediately available.

Two types of screens to identify mutants can be employed. First, one can employ a biochemical genetic strategy to identify mutants that are lacking a particular enzymatic activity. This strategy would employ a mass screening approach on extracts from individual seedlings from a population of mutagenized seeds. This strategy presupposes that an easy enzymatic or chromatographic assay be available to the researcher or a clever selection strategy be designed. [For instance, selection for chlorate resistance has resulted in plants with reduced nitrate reductase activity (Oostindiër-Braaksma and Feenstra 1973); allyl alcohol resistance has resulted in alcohol dehydrogenase (EC 1.1.1.1) mutants (Freeling and Birchler 1981)]. The plant of choice here is *Arabidopsis thaliana*, a small cruciferous plant (Meyerowitz 1987). The small size, short generation time, high seed set, ease of growth, and self- or cross-fertilization at will are attributes which make *Arabidopsis* a convenient subject for studies in classical and biochemical genetics. If one uses a relatively heavy mutagen dose, the frequency of desired mutants is about 1 per 1000 M_2 seed (Somerville and Ogren 1982). Thus, even if one does not have a selection, screening 1000 mutagenized seedlings is not too tedious. Several lipid mutants and starchless mutants of *Arabidopsis* have been obtained by mass screening protocols (Browse et al. 1985; Lin et al. 1988).

Secondly, one could postulate that certain types of signal perception and transduction elements will be present in plant plasma membranes and then design screens or selections to identify mutations in such functions. For instance, hormone receptors might be located in plasma membranes, and mutants could be selected which had the phenotype of a receptor mutant. Indeed, such mutants already exist in several plant species. For instance, in *Arabidopsis*, potential receptor mutants for auxin, ethylene, gibberellin, and abscisic acid have already been identified (Koornneef et al. 1984, 1985; Estelle and Somerville 1987; Bleecker et al. 1988). These mutants have all been mapped, and presumably chromosomal walking strategies are or will be employed to clone genes defined by the mutant loci. A potential gibberellin-receptor mutant also exists in maize (Phinney 1961), and procedures to transposon-tag this locus are currently being employed. Certain photomorphogenic mutants might also be deficient in plasma membrane elements; for instance, phototropic-minus mutants of *Arabidopsis* might be deficient in the blue-light receptor or an element in the blue-light transduction pathway (Chory and Ausubel 1987; Poff et al. 1987). A recent report has identified a phosphorylated protein putatively localized in the plasma membrane as a possible component of the blue-light signal transduction pathway (Gallagher et al. 1988).

Arabidopsis thaliana is uniquely suited for chromosomal walking experiments (see Meyerowitz 1987, for review). It has the smallest known genome size among the higher plants, the haploid genome size being 70 Mbp. This is only 5 times the size of the yeast, *Saccharomyces cerevisiae*, and about 12 times the size of the *E. coli* genome; thus only 16 000 random λ clones of 20 kb average insert size need be screened for a 99% probability of obtaining any *Arabidopsis* DNA fragment. In addition to its low content of DNA, *Arabidopsis* also has a very low content of repeated sequences, with an average of 1 kb repeat element every 125 kb (Meyerowitz 1987). A restriction fragment length polymorphism (RFLP) map has been constructed and there are already over 100 markers mapped on it. Taking into account the size of the *Arabidopsis* genome, this means that one could typically expect to find a λ clone from which to begin a chromosomal walk within 500 kbp of any mapped locus. Once one has a probe from which to begin a chromosomal walk, then one must isolate overlapping probes that cover the contiguous region between the starting probe and the gene of interest. Several genes in *Drosophila* and mammals have been isolated by such strategies (Steinmetz et al. 1982; Bender et al. 1983). Finally, the clone that contains the gene of interest must be identified from the group of overlapping clones. This can be accomplished by transformation of the homozygous recessive mutant with the wild-type copy of the gene to complement the mutant phenotype. *Arabidopsis* is also well suited for this experiment because high-efficiency root transformation protocols for *Agrobacterium tumefaciens* transformation of *Arabidopsis* have been published (Valvekens et al. 1988).

The organism of choice for transposon tagging to isolate a gene of interest is maize. The maize genome contains several well-characterized promiscuous transposon families. Several interesting regulatory genes have been cloned using transposon tagging protocols (Cone et al. 1986; Schmidt et al. 1987). Since at least one of the maize transposons, *Ac*, can be introduced into other plant genomes where it becomes mobile, use of maize transposons to tag genes of interest in tobacco, tomato, and *Arabidopsis* might soon be feasible (Baker et al. 1987; van Sluys et al. 1987).

Once a clone for the presumed receptor or signal transduction mutant has been obtained, sequence data may provide information on the role of the gene product. In situ hybridization to thin sections of plant material to localize mRNA to specific cell types could give insight into the mechanism of action of the gene product. Finally, use of the cloned gene to overexpress the protein, and use of the protein as an antigen would allow the subcellular localization of the gene product, and may lead to hypotheses for its function within the cell.

4 Conclusions

Recent technical advances that permit the isolation of highly pure plasma membrane preparations now provide the basis for unambiguous identification and purification of the proteins and activities associated with this cell structure. Thus, by using a combination of conventional protein purification methods together with the molecular techniques and genetic approaches described above it should be possible to clone the genes coding for these plasma membrane proteins, and thus enable us to study the regulation of these genes in response to different environmental stimuli such as light and pathogens. The availability of molecular clones encoding plasma membrane proteins, coupled with the now routine methods for introduction of genes into plant cells, will allow the function of these proteins to be addressed either by altering the pattern of expression of the encoding genes, or by the expression of in vitro mutated genes. By homology probing it should also be possible to isolate the plant versions of genes known to encode plasma membrane "house-keeping" proteins in other eukaryotes. These cloned sequences can be expressed in *E. coli* to produce immunogen for raising antisera, which will be of considerable value for studying the function and localization of the corresponding plant proteins.

Acknowledgments. Research was supported by the Swedish Council for Forestry and Agricultural Research, the Prince Charitable Trusts, and the Samuel Roberts Noble Foundation.

Note Added in Proof. The existence of receptor-mediated endocytosis in plants (Sect. 2.2.5) was recently supported by two articles. One shows the internalization of high-affinity plasma membrane receptor-elicitor complexes (Horn MA, Heinstein PF, Low PS, 1989, Plant Cell 1:1003–1009). The other article, using a monoclonal antibody to a plasma membrane epitope at the apoplastic surface (Pennell RI, Knox JP, Scofield GN, Selvendran RR, Roberts K, 1989, J Cell Biol 108:1967–1977), presents results very similar to those of Norman et al. (1990) and Herman et al. (1990). Sucrose transport (Sect. 2.2.6) into plasma membrane vesicles was recently shown to be driven by the proton motive force (Bush DR, 1989, Plant Physiol 89:1318–1323; Lemoine R, Delrot S, 1989, FEBS Lett 249:129–133).

References

Abdel-Lafit AA (1986) Calcium-mobilizing receptors, polyphosphoinositides, and the generation of second messengers. Pharmacol Rev 38:227–272

Aebersold RH, Leavitt J, Saavedra RA, Hood LE, Kent SBH (1987) Internal amino acid sequence analysis of proteins separated by one- or two-dimensional gel electrophoresis after in situ protease digestion on nitrocellulose. Proc Natl Acad Sci USA 84:6970–6974

Aruffo A, Seed B (1987a) Molecular cloning of a CD28 cDNA by a high-efficiency COS cell expression system. Proc Natl Acad Sci USA 84:8573–8577

Aruffo A, Seed B (1987b) Molecular cloning of two CD7 (T-cell leukemia antigen) cDNAs by a COS cell expression system. EMBO J 6:3313–3316

Askerlund P, Larsson C, Widell S (1988) Localization of donor and acceptor sites of NADH dehydrogenase activities using inside-out and right-side-out plasma membrane vesicles from plants. FEBS Lett 239:23–28

Atkinson MM, Baker CJ (1987) Alteration of plasmalemma sucrose transport in *Phaseolus vulgaris* by *Pseudomonas syringae* pv. *syringae* and its association with K^+/H^+ exchange. Phytopathology 77:1573–1578

Baker B, Coupland G, Fedoroff N, Starlinger P, Schell J (1987) Phenotypic assay for excision of the maize controlling element *Ac* in tobacco. EMBO J 6:1547–1554

Battey NH, Venis MA (1988) Calcium-dependent protein kinase from apple fruit membranes is calmodulin-independent but has calmodulin-like properties. Planta 176:91–97

Bender W, Spierer P, Hogness DS (1983) Chromosome walking and jumping to isolate DNA from the *Ace* and *rosy* loci and the bithorax complex in *Drosophila melanogaster*. J Mol Biol 168:17–33

Benning C (1986) Evidence supporting a model of voltage-dependent uptake of auxin into *Curcubita* vesicles. Planta 169:228–237

Berridge MJ (1986) Cell signalling through phospholipid metabolism. J Cell Sci Suppl 4:137–153

Berridge MJ (1987) Inositol triphosphate and diacylglycerol: two interacting second messengers. Annu Rev Biochem 56:159–193

Berridge MJ, Irvine RF (1984) Inositol triphosphate, a novel second messenger in cellular signal transduction. Nature 312:315–321

Blatt MR, Slayman CL (1987) Role of "active" potassium transport in the regulation of cytoplasmic pH by nonanimal cells. Proc Natl Acad Sci USA 84:2737–2741

Bleecker AB, Estelle MA, Somerville C, Kende H (1988) Insensitivity to ethylene conferred by a dominant mutation in *Arabidopsis thaliana*. Science 241:1086–1089

Blowers DP, Trewavas AJ (1987) Autophosphorylation of plasma membrane bound calcium calmodulin dependent protein kinase from pea seedlings and modification of catalytic activity by autophosphorylation. Biochem Biophys Res Commun 143:691–696

Blum W, Hinsch K-D, Schultz G, Weiler EW (1988) Identification of GTP-binding proteins in the plasma membrane of higher plants. Biochem Biophys Res Commun 156:954–959

Bonhoff A, Grisebach H (1988) Elicitor-induced accumulation of glyceollin and callose in soybean roots and localized resistance against *Phytophthora megasperma* f.sp. *glycinea*. Plant Sci 54:203–209

Bonhoff A, Rieth B, Golecki J, Grisebach H (1987) Race cultivar-specific differences in callose deposition in soybean roots following infection with *Phytophthora megasperma* f.sp. *glycinea*. Planta 172:101–105

Braun Y, Hassidim M, Lerner HR, Reinhold L (1988) Evidence for Na^+/H^+ antiporter in membrane vesicles isolated from roots of the halophyte *Atriplex nummularia*. Plant Physiol 87:104–108

Briars SA, Kessler F, Evans DE (1988) The calmodulin-stimulated ATPase of maize coleoptiles is a 140,000 M_r polypeptide. Planta 176:283–285

Briggs WR, Iino M (1983) Blue-light-absorbing photoreceptors in plants. Philos Trans R Soc Lond B 303:347–359

Brightman AO, Barr R, Crane FL, Morré DJ (1988) Auxin-stimulated NADH oxidase purified from plasma membrane of soybean. Plant Physiol 86:1264–1269

Browse J, McCourt P, Somerville CR (1985) A mutant of *Arabidopsis* lacking a chloroplast-specific lipid. Science 227:763–765

Bush DR, Langston-Unkefer PJ (1988) Amino acid transport into membrane vesicles isolated from zucchini. Plant Physiol 88:487–490

Chory J, Ausubel FM (1987) Genetic analysis of photoreceptor action pathways. In: Book of abstracts, 3rd International Meeting on *Arabidopsis,* p 56

Cone KC, Burr FA, Burr B (1986) Molecular analysis of the maize anthocyanin regulatory locus C1. Proc Natl Acad Sci USA 83:9631–9635

Cosio EG, Pöpperl H, Schmidt WE, Ebel J (1988) High-affinity binding of fungal β-glucan fragments to soybean (*Glycine max* L.) microsomal fractions and protoplasts. Eur J Biochem 175:309–315

Damon S, Hewitt J, Neider M, Bennett AB (1988) Sink metabolism in tomato fruit. II. Phloem unloading and sugar uptake. Plant Physiol 87:731–736

Das R, Sopory SK (1985) Evidence of regulation of calcium uptake by phytochrome in maize protoplasts. Biochem Biophys Res Commun 128:1455–1460

de Boer AH, Watson BA, Cleland RE (1989) Purification and identification of the fusicoccin binding protein from oat root plasma membrane. Plant Physiol 89:250–259

Delmer DP (1987) Cellulose biosynthesis. Annu Rev Plant Physiol 38:259–290

Dhugga KS, Waines JG, Leonard RT (1988) Correlated induction of nitrate uptake and membrane polypeptides in corn roots. Plant Physiol 87:120–125

Drøbak BK, Ferguson IB (1985) Release of Ca^{2+} from plant hypocotyl microsomes by inositol-1,4,5-triphosphate. Biochem Biophys Res Commun 130:1241–1246

Drøbak BK, Allan EF, Comerford JG, Roberts K, Dawson AP (1988) Presence of guanine nucleotide-binding proteins in a plant hypocotyl microsomal fraction. Biochem Biophys Res Commun 150:899–903

Elliott DC, Skinner JD (1986) Calcium-dependent, phospholipid-activated protein kinase in plants. Phytochemistry 25:39–44

Espie GS, Miller AG, Birch DG, Canvin DT (1988) Simultaneous transport of CO_2 and HCO_3^- by the cyanobacterium *Synechococcus* UTEX 625. Plant Physiol 87:551–554

Estelle MA, Somerville CR (1987) Auxin-resistant mutants of *Arabidopsis thaliana* with an altered morphology. Mol Gen Genet 206:200–206

Ettlinger C, Lehle L (1988) Auxin induces rapid changes in phosphatidylinositol metabolites. Nature 331:176–178

Evans DE, Bengochea T, Cairns AJ, Dodds JH, Hall MA (1982) Studies on ethylene binding by cell-free preparations from cotyledons of *Phaseolus vulgaris* L.: subcellular localization. Plant Cell Environ 5:101–107

Falke LC, Edwards KL, Pickard BG, Misler S (1988) A stretch-activated anion channel in tobacco protoplasts. FEBS Lett 237:141–144

Felle H (1988) Auxin causes oscillations of cytosolic free calcium and pH in *Zea mays* coleoptiles. Planta 174:495–499

Feyerabend M, Weiler EW (1988) Characterization and localization of fusicoccin-binding sites in leaf tissues of *Vicia faba* L. probed with a novel radioligand. Planta 174:115–122

Franke WW, Morré DJ, Deumling B, Cheetham RD, Kartenbeck J, Jarasch ED, Zentgraf H-W (1971) Synthesis and turnover of membrane proteins in rat liver: an examination of the membrane flow hypothesis. Z Naturforsch 26b:1031–1039

Freeling M, Birchler J (1981) Mutants and variants of the alcohol dehydrogenase-1 gene in maize. In: Setlow J, Hollaender A (eds) Genetic engineering principles and methods, vol 3. Plenum, New York, pp 223–264

Friedman AD, Triezenberg SJ, McKnight SL (1988) Expression of a truncated viral *trans*-activator selectively impedes lytic infection of its cognate virus. Nature 335:452–454

Gallagher S, Short TW, Ray PM, Pratt LH, Briggs WR (1988) Light-mediated changes in two proteins found associated with plasma membrane fractions from pea stem sections. Proc Natl Acad Sci USA 85:8003–8007

Gallet O, Lemoine R, Larsson C, Delrot S (1989) The sucrose carrier of the plant plasma membrane. I. Differential affinity labeling. Biochim Biophys Acta 978:56–64

Getz H-P, Schulte-Altedorneburg M, Willenbrink J (1987) Effects of fusicoccin and abscisic acid on glucose uptake into isolated beet root protoplasts. Planta 171:235–240

Girard V, Maclachlan G (1987) Modulation of pea membrane β-glucan synthase activity by calcium, polycation, endogenous protease, and protease inhibitor. Plant Physiol 85:131–136

Gräf P, Weiler EW (1989) ATP-driven Ca^{2+}-transport in sealed plasma membrane vesicles prepared by aqueous two-phase partitioning from leaves of *Commelina communis* L. Physiol Plant 75:469–478

Gubler U, Hoffman BJ (1983) A simple and very efficient method for generating cDNA libraries. Gene 25:263–269

Hanks SK, Quinn AM, Hunter T (1988) The protein kinase family: conserved features and deduced phylogeny of the catalytic domains. Science 241:42–52

Hasunuma K, Funadera K (1987) GTP-binding protein(s) in green plant, *Lemna paucicostata*. Biochem Biophys Res Commun 143:908–912

Hasunuma K, Furukawa K, Tomita K, Mukai C, Nakamura T (1987) GTP-binding proteins in etiolated epicotyls of *Pisum sativum* (Alaska) seedlings. Biochem Biophys Res Commun 148:133–139

Hepler PK, Wayne RO (1985) Calcium and plant development. Annu Rev Plant Physiol 36:397–439

Herman EM, Fitter MS, Norman PM, Lamb CJ (1990) Plant cell internalization pathway defined by a monoclonal antibody to periplasmic vesicles and plasmalemmasomes. (submitted)

Herskowitz I (1987) Functional inactivation of genes by dominant negative mutations. Nature 329:219–222

Hertel R, Thomson K-St, Russo VEA (1972) In-vitro auxin binding to particulate cell fractions from corn coleoptiles. Planta 107:325–340

Hertel R, Lomax TL, Briggs WR (1983) Auxin transport in membrane vesicles from *Curcubita pepo* L. Planta 157:193–201

Heyn A, Hoffman S, Hertel R (1987) In-vitro auxin transport in membrane vesicles from maize coleoptiles. Planta 172:285–287

Hillmer S, Depta H, Robinson DG (1986) Confirmation of endocytosis in higher plant protoplasts using lectin-gold conjugates. Eur J Cell Biol 41:142–149

Hirt B (1967) Selective extraction of polyoma DNA from infected mouse cell cultures. J Mol Biol 26:365–369

Hornberg C, Weiler EW (1984) High-affinity binding sites for abscisic acid on the plasmalemma of *Vicia faba* guard cells. Nature 310:321–324

Hübner R, Depta H, Robinson DG (1985) Endocytosis in maize root cap cells. Evidence obtained using heavy metal salt solutions. Protoplasma 129:214–222

Huynh TV, Young RA, Davis RW (1985) Constructing and screening cDNA libraries in λ gt10 and λ gt11. In: Glover DM (ed) DNA cloning. Vol I. A practical approach. IRL, Oxford, pp 49–78

Izant JG, Weintraub H (1984) Inhibition of thymidine kinase gene expression by anti-sense RNA: a molecular approach to genetic analysis. Cell 36:1007–1015

Jacobs M, Rubery PH (1988) Naturally occurring auxin transport regulators. Science 241:346–349

Jacobs M, Thelen MP, Farndale RW, Astle MC, Rubery PH (1988) Specific guanine nucleotide binding by membranes from *Curcubita pepo* seedlings. Biochem Biophys Res Commun 155:1478–1484

Janistyn B (1988) Stimulation by manganese(II)sulphate of a cAMP-dependent protein kinase from *Zea mays* seedlings. Phytochemistry 27:2735–2736

Kaplan A (1985) Adaptation to CO_2 levels: Induction and the mechanism for inorganic carbon uptake. In: Lucas WJ, Berry JA (eds) Inorganic carbon uptake by aquatic photosynthetic organisms. The American Society of Plant Physiologists, Rockville, MD, pp 325–338

Kauss H (1987) Some aspects of calcium-dependent regulation in plant metabolism. Annu Rev Plant Physiol 38:47–72

Kjellbom P, Larsson C (1984) Preparation and polypeptide composition of chlorophyll-free plasma membranes from leaves of light-grown spinach and barley. Physiol Plant 62:501–509

Klein PS, Sun TJ, Saxe III CL, Kimmel AR, Johnson RL, Devreotes PN (1988) A chemoattractant receptor controls development in *Dictyostelium discoideum*. Science 241:1467–1472

Klucis E, Polya GM (1988) Localization, solubilization and characterization of plant membrane-associated calcium-dependent protein kinases. Plant Physiol 88:164–171

Koornneef M, Reuling G, Karssen CM (1984) The isolation and characterization of abscisic acid-sensitive mutants of *Arabidopsis thaliana*. Physiol Plant 61:377–384

Koornneef M, Elgersma A, Hangart CJ, Van Loeneu-Martinet van Rijn L, Zeevaart JAD (1985) A gibberellin insensitive mutant of *Arabidopsis thaliana*. Physiol Plant 65:33–41

Lamb CJ, Lawton MA, Dron M, Dixon RA (1989) Signals and transduction mechanisms for activation of plant defenses against microbial attack. Cell 56: 215–224

Larsson C, Widell S, Kjellbom P (1987) Preparation of high-purity plasma membranes. Methods Enzymol 148:558–568

Lawton MA, Yamamoto RT, Hanks SK, Lamb CJ (1989) Molecular cloning of plant transcripts encoding protein kinase homologs. Proc Natl Acad Sci USA 86:3140–3144

Lemoine R, Daie J, Wyse R (1988) Evidence for the presence of a sucrose carrier in immature sugar beet roots. Plant Physiol 86:575–580

Lemoine R, Delrot S, Gallet O, Larsson C (1989) The sucrose carrier of the plant plasma membrane. II. Immunological characterization. Biochim Biophys Acta 978:65–71

Lenton JR, Hedden P, Gale MD (1987) Gibberellin insensitivity and depletion in wheat – consequences for development. In: Hoad GV, Lenton JR, Jackson MB, Atkin RK (eds) Hormone action in plant development. A critical appraisal. Butterworths, London, pp 45–160

Lin TP, Caspar T, Somerville C, Preiss J (1988) Isolation and characterization of a starchless mutant of *Arabidopsis thaliana* (L) Heynh lacking ADP glucose pyrophosphorylase activity. Plant Physiol 86:1131–1135

Löbler M, Klämbt D (1985a) Auxin-binding protein from coleoptile membranes of corn (*Zea mays* L.). I. Purification by immunological methods and characterization. J Biol Chem 260:9848–9853

Löbler M, Klämbt D (1985b) Auxin-binding protein from coleoptile membranes of corn (*Zea mays* L.). II. Localization of a putative auxin receptor. J Biol Chem 260:9854–9859

Lucantoni A, Polya GM (1987) Activation of wheat embryo calcium-regulated protein kinase by unsaturated fatty acids in the presence and absence of calcium. FEBS Lett 221:33–36

MacMillan J (1987) Gibberellin-deficient mutants of maize and pea and the molecular action of gibberellins. In: Hoad GV, Lenton JR, Jackson MB, Atkin RK (eds) Hormone action in plant development. A critical appraisal. Butterworths, London, pp 73–87

Maelicke A (1988) Structural similarities between ion channel proteins. Trends Biochem Sci 13:199–202

Manuel LJ, Moroney JV (1988) Inorganic carbon accumulation by *Chlamydomonas reinhardii*. New proteins are made during adaption to low CO_2. Plant Physiol 88:491–496

Mazur BJ, Chris CF, Harnett ME, Mauvais JE, McDevitt RE, Knowlton S, Smith J, Falco S (1988) Second Int Congress of Plant Mol Biol, Jerusalem Nov. 13–18, abstract 112

M'Batchi B, Delrot S (1988) Stimulation of sugar exit from leaf tissues of *Vicia faba* L. Planta 174:340–348

M'Batchi B, Pichelin D, Delrot S (1987) Selective solubilization of membrane proteins differentially labeled by p-chloromercuribenzenesulfonic acid in the presence of sucrose. Plant Physiol 83:541–545

McCarty DR, Carson CB, Stinard PS, Robertson DS (1989) Molecular analysis of *viviparous-1*: an abscisic acid-insensitive mutant of maize. Plant Cell 1:523–532

McGlasson WB (1985) Ethylene and fruit ripening. Hortscience 20:51–54

Melin P-M, Sommarin M, Sandelius AS, Jergil B (1987) Identification of Ca^{2+}-stimulated polyphosphoinositide phospholipase C in isolated plant plasma membranes. FEBS Lett 223:87–91

Melton DA (1985) Injected anti-sense RNAs specifically block messenger RNA translation in vivo. Proc Natl Acad Sci USA 82:144–148

Meyerowitz EM (1987) *Arabidopsis thaliana*. Annu Rev Genet 21:93–111

Mullis KB, Faloona FA (1987) Specific synthesis of DNA in vitro via a polymerase-catalyzed chain reaction. Methods Enzymol 155:335–350

Neer EJ, Clapham DE (1988) Roles of G protein subunits in transmembrane signalling. Nature 333:129–134

Nobiling R, Reiss H-D (1987) Quantitative analysis of calcium gradients and activity in growing pollen tubes of *Lilium longiflorum*. Protoplasma 139:20–24

Norman PM, Wingate VPM, Fitter MS, Lamb CJ (1986) Monoclonal antibodies to plant plasma membrane antigens. Planta 167:452–459

Norman PM, Kjellbom P, Bradley DJ, Hahn MG, Lamb CJ (1990) Immunoaffinity purification and biochemical characterization of a plasma membrane associated arabinogalactan-rich protein (submitted)

Ogawa T, Kaneda T, Omata T (1987) A mutant of *Synechococcus* PCC7942 incapable of adapting to low CO_2 concentration. Plant Physiol 84:711–715

Omata T, Ogawa T, Marcus Y, Friedberg D, Kaplan A (1987) Adaptation to low CO_2 level in a mutant of *Anacvstis nidulans* R2 which requires high CO_2 for growth. Plant Physiol 83: 892–894

Oostindiër-Braaksma FJ, Feenstra WJ (1973) Isolation and characterization of chlorate-resistant mutants of *Arabidopsis thaliana*. Mutat Res 19:175–184

Pearse BMF (1987) Clathrin and coated vesicles. EMBO J 6:2507–2512

Pfaffman H, Hartmann E, Brightman AO, Morré DJ (1987) Phosphatidylinositol specific phospholipase C of plant stems. Membrane associated activity concentrated in plasma membranes. Plant Physiol 85:1151–1155

Phinney BO (1961) Dwarfing genes in *Zea mays* and their relation to the gibberellins. In: Plant Growth Regulation. Iowa State Univ Press, Ames, pp 489–501

Poff KL, Best T, Gregg M, Ren Z (1987) Mutants of *Arabidopsis thaliana* with altered phototropism and/or altered geotropism. In: Book of abstracts, 3rd International Meeting of *Arabidopsis*, pp 79

Ranjeva R, Boudet AM (1987) Phosphorylation of proteins in plants: regulatory effects and potential involvement in stimulus/response coupling. Annu Rev Plant Physiol 38:73–93

Ranjeva R, Carrasco A, Boudet AM (1988) Inositol triphosphate stimulates the release of calcium from intact vacuoles isolated from *Acer* cells. FEBS Lett 230:137–141

Rasi-Caldogno F, Pugliarello MC, De Michelis MI (1987) The Ca^{2+}-transport ATPase of plant plasma membrane catalyzes a nH^+/Ca^{2+} exchange. Plant Physiol 83:994–1000

Rebagliati MR, Weeks DL, Harvey RP, Melton D (1985) Identification and cloning of localized maternal RNAs from *Xenopus* eggs. Cell 42:769–777

Ripp KG, Viitanen PV, Hitz WD, Franceschi VR (1988) Identification of a membrane protein associated with sucrose transport into cells of developing soybean cotyledons. Plant Physiol 88:1435–1445

Robinson C, Larsson C, Buckhout TJ (1988) Identification of a calmodulin-stimulated (Ca^{2+} + Mg^{2+})-ATPase in a plasma membrane fraction isolated from maize (*Zea mays*) leaves. Physiol Plant 72:177–184
Rodermel SR, Abbott MS, Bogorad L (1988) Nuclear-organelle interactions: nuclear antisense gene inhibits ribulose bisphosphate carboxylase enzyme levels in transformed tobacco plants. Cell 55:673–681
Roth-Bejerano N, Nejidat A, Rubinstein B, Itai C (1988) Effect of ferricyanide on potassium uptake by intact epidermal tissue and guard cell protoplast. Plant Cell Physiol 29:677–682
Rothman JE (1981) The Golgi apparatus: two organelles in tandem. Science 213:1212–1219
Ruvkun GB, Ausubel FM (1981) A general method for site-directed mutagenesis in prokaryotes. Nature 289:85–88
Sabater M, Rubery PH (1987) Auxin carriers in *Curcubita* vesicles. II. Evidence that carrier-mediated routes of both indole-3-acetic acid influx and efflux are electroimpelled. Planta 171:507–513
Saiki RK, Gelfand DH, Stoffel S, Scharf SJ, Higuchi R, Horn GT, Mullis KB, Erlich HA (1988) Primer-directed enzymatic amplification of DNA with a thermostable DNA polymerase. Science 239:487–491
Sandelius AS, Sommarin M (1986) Phosphorylation of phosphatidylinositols in isolated plant membranes. FEBS Lett 201:282–286
Sauer N, Tanner W (1987) Inducible sugar and amino acid transport proteins in *Chlorella vulgaris*. In: Gohiin SE (ed) Membrane proteins. Proceedings of the membrane protein symposium, San Diego, pp 483–490
Schäfer A, Bygrave F, Matzenauer S, Marmé D (1985) Identification of a calcium- and phospholipid-dependent protein kinase in plant tissue. FEBS Lett 187:25–28
Schaller GE, Sussman MR (1988) Phosphorylation of the plasma-membrane H^+-ATPase of oat roots by a calcium-stimulated protein kinase. Planta 173:509–518
Scharf SJ, Horn GT, Erlich HA (1986) Direct cloning and sequence analysis of enzymatically amplified genomic sequences. Science 233:1076–1078
Schauf CL, Wilson KJ (1987) Effects of abscisic acid on K^+ channels in *Vicia faba* guard cell protoplasts. Biochem Biophys Res Commun 145:284–290
Scherer GFE, Martiny-Baron G, Stoffel B (1988) A new set of regulatory molecules in plants: a plant phospholipid similar to platelet-activating factor stimulates protein kinase and proton-translocating ATPase in membrane vesicles. Planta 175:241–253
Schmalstig JG, Hitz WD (1987) Transport and metabolism of a sucrose analog (1′-fluorosucrose) into *Zea mays* L. endosperm without invertase hydrolysis. Plant Physiol 85:902–905
Schmidt RJ, Burr FA, Burr B (1987) Transposon tagging and molecular analysis of the maize regulatory locus *opaque*-2. Science 238:960–963
Schroeder JI, Hedrich R, Fernandez JM (1984) Potassium-selective single channels in guard cell protoplasts of *Vicia faba*. Nature 312:361–362
Schroeder JI, Raschke K, Neher E (1987) Voltage dependence of K^+ channels in guard-cell protoplasts. Proc Natl Acad Sci USA 84:4108–4112
Schwarz R, Friedberg D, Kaplan A (1988) Is there a role for the 42 kilodalton polypeptide in inorganic carbon uptake by cyanobacteria? Plant Physiol 88:284–288
Seed B (1987) An LFA-3 cDNA encodes a phospholipid-linked membrane protein homologous to its receptor CD2. Nature 329:840–842
Seed B, Aruffo A (1987) Molecular cloning of the CD2 antigen, the T-cell erythrocyte receptor, by a rapid immunoselection procedure. Proc Natl Acad Sci USA 84:3365–3369
Senn AP, Goldsmith MHM (1988) Regulation of electrogenic proton pumping by auxin and fusicoccin as related to the growth of *Avena* coleoptiles. Plant Physiol 88:131–138
Shimomura S, Sotobayashi T, Futai M, Fukui T (1986) Purification and properties of an auxin-binding protein from maize shoot membranes. J Biochem 99:1513–1524
Shimomura S, Inohara N, Fukui T, Futai M (1988) Different properties of two types of auxin-binding sites in membranes from maize coleoptiles. Planta 175:558–566
Shinkle JR, Jones RL (1988) Inhibition of stem elongation in *Cucumis* seedlings by blue light requires calcium. Plant Physiol 86:960–966

Sibley DR, Benovic JL, Caron MG, Lefkowitz RJ (1987) Regulation of transmembrane signaling by receptor phosphorylation. Cell 48:913–922

Skou JP, Jørgensen JH, Lilholt U (1984) Comparative studies on callose formation in powdery mildew compatible and incompatible barley. Phytopathol Z 109:147–168

Smith CJS, Watson CF, Ray J, Berd CR, Morris PC, Schuch W, Grierson D (1988) Antisense RNA inhibition of polygalacturonase gene expression in transgenic tomatoes. Nature 334:724–726

Somerville CR, Ogren WL (1982) Isolation of photorespiration mutants in *Arabidopsis thaliana*. In: Edelman J (ed) Methods in chloroplast molecular biology. Elsevier Biomedical, Amsterdam, pp 129–138

Sommarin M, Sandelius AS (1988) Phosphatidylinositol and phosphatidylinositolphosphate kinases in plant plasma membranes. Biochim Biophys Acta 958:268–278

Stanzel M, Sjolund RD, Komor E (1988a) Transport of glucose, fructose and sucrose by *Streptanthus tortuosus* suspension cells. I. Uptake at lower sugar concentration. Planta 174:201–209

Stanzel M, Sjolund RD, Komor E (1988b) Transport of glucose, fructose and sucrose by *Streptanthus tortosus* suspension cells. II. Uptake at high sugar concentration. Planta 174:210–216

Steinmetz M, Minard K, Horvath S, McNicholas J, Srelinger J, Wake C, Long E, Mach B, Hood LE (1982) A molecular map of the immune response region from the major histocompatibility complex of the mouse. Nature 300:35–42

Stengelin S, Stamenkovic I, Seed B (1988) Isolation of cDNAs for two distinct human Fc receptors by ligand affinity cloning. EMBO J 7:1053–1059

Thelen MP, Delmer DP (1986) Gel-electrophoretic separation, detection, and characterization of plant and bacterial UDP-glucose glucosyltransferases. Plant Physiol 81:913–918

Valvekens D, Van Montagu M, Van Lijsebettens M (1988) *Agrobacterium tumefaciens*-mediated transformation of *Arabidopsis thaliana* root explants by using kanamycin selection. Proc Natl Acad Sci USA 85:5536–5540

van der Krol AR, Lenting PE, Veenstra J, van der Meer IM, Koes RE, Gerats AGM, Mol JNM, Stuitje AR (1988) An anti-sense chalcone synthase gene in transgenic plants inhibits flower pigmentation. Nature 333:866–869

Van Sluys MA, Tempe J, Fedoroff N (1987) Studies on the introduction and mobility of the maize *Activator* element in *Arabidopsis thaliana* and *Daucus carota*. EMBO J 6:3881–3889

Venis MA (1987) Hormone receptor sites and the study of plant development. In: Hoad GV, Lenton JR, Jackson MB, Atkin RK (eds) Hormone action in plant development. A critical appraisal. Butterworths, London, pp 53–61

Voet JG, Howley KS, Shumsky JS (1987) 5′-Azido-N-1-naphthylphthalamic acid, a photolabile analog of N-1-naphthylphthalamic acid. Synthesis and binding properties in *Curcubita pepo* L.. Plant Physiol 85:22–25

Waldmann T, Jeblick W, Kauss H (1988) Induced net Ca^{2+} uptake and callose biosynthesis in suspension-cultured plant cells. Planta 173:88–95

Wood WI, Gitschier J, Lasky LA, Lawn RM (1985) Base composition-independent hybridization in tetramethylammonium chloride: a method for oligonucleotide screening of highly complex gene libraries. Proc Natl Acad Sci USA 82:1585–1588

Wyse RE, Komor E (1984) Mechanism of amino acid uptake by sugarcane suspension cells. Plant Physiol 76:865–870

Wysocki LJ, Sato VL (1978) "Panning" for lymphocytes: a method for cell selection. Proc Natl Acad Sci USA 75:2844–2848

Young RA, Davis RW (1983) Efficient isolation of genes by using antibody probes. Proc Natl Acad Sci USA 80:1194–1198

Zeiger E (1983) The biology of stomatal guard cells. Annu Rev Plant Physiol 34:441–475

Subject Index

Key terms refer to plasma membrane (PM) where nothing else is specified. Page numbers in **bold** type indicate main entries.

Printed by Publishers' Graphics LLC